Novel approaches to improving
high temperature corrosion resistance

European Federation of Corrosion Publications
NUMBER 47

Novel approaches to improving high temperature corrosion resistance

Edited by
M. Schütze and W. J. Quadakkers

Published for the European Federation of Corrosion
by Woodhead Publishing and Maney Publishing
on behalf of
The Institute of Materials, Minerals & Mining

CRC Press
Boca Raton Boston New York Washington, DC

WOODHEAD PUBLISHING LIMITED
Cambridge England

Woodhead Publishing Limited and Maney Publishing Limited on behalf of
The Institute of Materials, Minerals & Mining

Published by Woodhead Publishing Limited, Abington Hall, Granta Park,
Great Abington, Cambridge CB21 6AH, England
www.woodheadpublishing.com

Published in North America by CRC Press LLC, 6000 Broken Sound Parkway, NW,
Suite 300, Boca Raton, FL 33487, USA

First published 2008 by Woodhead Publishing Limited and CRC Press LLC
© 2008, Institute of Materials, Minerals & Mining
The authors have asserted their moral rights.

British Library Cataloguing in Publication Data
A catalogue record for this book is available from the British Library.

Library of Congress Cataloging in Publication Data
A catalog record for this book is available from the Library of Congress.

Woodhead Publishing ISBN 978-1-84569-238-4 (book)
Woodhead Publishing ISBN 978-1-84569-447-0 (e-book)
CRC Press ISBN 978-1-4200-7959-3
CRC Press order number: WP7959
ISSN 1354-5116

The publishers' policy is to use permanent paper from mills that operate a
sustainable forestry policy, and which has been manufactured from pulp
which is processed using acid-free and elementary chlorine-free practices.
Furthermore, the publishers ensure that the text paper and cover board used
have met acceptable environmental accreditation standards.

Project managed by Macfarlane Book Production Services, Dunstable, Bedfordshire,
England (e-mail: macfarl@aol.com)
Typeset by Replika Press Pvt Ltd, India
Printed by T J International Limited, Padstow, Cornwall, England

Contents

Contributor contact details   *xvii*

Series introduction   *xxvii*

Volumes in the EFC series   *xxix*

Part I   Alloy modification

1   Design strategies for oxidation-resistant intermetallic
         and advanced metallic alloys   3

         MICHAEL P BRADY, PETER F TORTORELLI, KARREN L MORE, E ANDREW
         PAYZANT, BETH L ARMSTRONG, HUA-TAY LIN and MICHAEL J LANCE,
         Oak Ridge National Laboratory, USA and FENG HUANG and
         MARK L WEAVER, University of Alabama, USA

1.1   Introduction   3
1.2   Self-graded metallic precursor coating concepts   4
1.3   Surface modification approaches   8
1.4   Functional surface formation by gas reactions   13
1.5   Concluding remarks   15
1.6   Acknowledgements   16
1.7   References   16

2   Alloying with copper to reduce metal dusting rates   19

         J ZHANG, DMI COLE and DJ YOUNG, University of
         New South Wales, Australia

2.1   Introduction   19
2.2   Experimental methods   20
2.3   Results   20
2.4   Discussion   29
2.5   Conclusions   34
2.6   Acknowledgement   35
2.7   References   35

3   Performance of candidate gas turbine abradeable
        seal materials in high temperature combustion
        atmospheres   36

        NJ Simms, Cranfield University, UK; JF Norton, Consultant,
        UK and G McColvin, Siemens Industrial Turbomachinery Ltd, UK

3.1   Introduction   36
3.2   Experimental methods   39
3.3   Results and discussion   47
3.4   Conclusions   62
3.5   Acknowledgements   63
3.6   References   63

4   Two calculations concerning the critical aluminium
        content in Fe-20Cr-5Al alloys   65

        G Strehl and G Borchardt, Institut für Metallurgie, Germany;
        P Beaven, GKSS-Forschungszentrum, Germany and B Lesage,
        Université Paris XI, France

4.1   Introduction   65
4.2   The critical aluminium content   67
4.3   Three-dimensional Al depletion profile   70
4.4   Conclusions   77
4.5   Acknowledgments   77
4.6   References   77

5   The different role of alloy grain boundaries on the
        oxidation mechanisms of Cr-containing steels and
        Ni-base alloys at high temperatures   80

        VB Trindade and U Krupp, University of Applied Science,
        Germany; Ph E-G Wagenhuber, S Yang and H-J Christ,
        Universität Siegen, Germany

5.1   Introduction   80
5.2   Materials and experimental methods   81
5.3   Results and discussion   82
5.4   Conclusions   90
5.5   Acknowledgments   91
5.6   References   91

6   Improving metal dusting resistance of Ni-X alloys
        by an addition of Cu   93

        Y Nishiyama, K Moriguchi, N Otsuka and T Kudo, Sumitomo
        Metal Industries Ltd, Japan

6.1   Introduction   93

6.2   Experimental methods   94
6.3   Theoretical analysis   96
6.4   Results   97
6.5   Discussion   102
6.6   Conclusions   108
6.7   References   108

Part II Surface treatment

7   Effects of minor additions and impurities on oxidation
       behaviour of FeCrAl alloys in the development of
       novel surface coatings compositions (SMILER)   113

V KOCHUBEY, Forschungszentrum Jülich GmbH, Germany;
H AL-BADAIRY and G TATLOCK, University of Liverpool, UK;
J LE-COZE, Ecole Nationale Supérieure des Mines de Saint-Etienne,
France and D NAUMENKO and WJ QUADAKKERS, Forschungszentrum
Jülich GmbH, Germany

7.1   Introduction   113
7.2   Experimental methods   114
7.3   Results and discussion   115
7.4   Alloys with adhesive mode of scale spallation: 'MRef', '+C'   118
7.5   Alloys with cohesive mode of scale spallation: '+Zr', '+Hf'   121
7.6   Implications for alloy/coating development   124
7.7   Conclusions   127
7.8   Acknowledgment   127
7.9   References   127

8   Lifetime extension of FeCrAlRE-alloys in air.
       Potential role of enhanced Al reservoir and surface
       pre-treatments (SMILER)   129

JR NICHOLLS, MJ BENNETT and NJ SIMMS, Cranfield University, UK;
H HATTENDORF, Thyssen Krupp VDM GmbH, Germany; D BRITTON
and AJ SMITH, Diffusion Alloys Ltd, UK; WJ QUADAKKERS,
D NAUMENKO and V KOCHUBEY, Forschungszentrum Jülich GmbH
Germany; R FORDHAM and R BACHORCZYK, JRC Petten,
The Netherlands and D GOOSSENS, Bekaert Technology Centre,
Belgium

8.1   Introduction   129
8.2   Experimental methods   132
8.3   Results   136
8.4   Air oxidation of foil and sheet coupons   136
8.5   Air oxidation of near real-service components   147
8.6   Discussion   155
8.7   Conclusions   158

8.8   Acknowledgements   159
8.9   References   159

9   Effect of gas composition and contaminants on the
       lifetime of surface-treated FeCrAlRE alloys (SMILER)   161

GJ Tatlock and H Al-Badairy, University of Liverpool, UK and
R Bachorczyk-nagy and R Fordham, JRC, Petten, The Netherlands

9.1   Introduction   161
9.2   Experimental methods   162
9.3   Results   164
9.4   Discussion   171
9.5   Conclusions   174
9.6   Acknowledgements   174
9.7   References   175

10   Development of novel diffusion coatings for 9–12%
       Cr ferritic-martensitic steels (SUNASPO)   176

V Rohr and M Schütze, Karl-Winnacker-Institut der DECHEMA
e.V., Germany; E Fortuna and DN Tsipas, Aristotle University of
Thessaloniki, Greece and A Milewska and FJ Pérez, Universidad
Complutense Madrid, Spain

10.1   Introduction   176
10.2   Experimental methods   177
10.3   Experimental results   179
10.4   Discussion   188
10.5   Conclusion   191
10.6   Acknowledgements   191
10.7   References   191

11   Steam oxidation and its potential effects on creep
       strength of power station materials (SUNASPO)   193

M Schütze and V Rohr, Karl-Winnacker-Institut der DECHEMA
e.V., Germany and L Nieto Hierro, PJ Ennis and WJ Quadakkers,
Forschungszentrum Jülich GmbH Germany

11.1   Introduction   193
11.2   Steam oxidation tests   193
11.3   Model alloys   200
11.4   Effect of oxidation on creep behaviour   203
11.5   Creep tests on oxidised Alloy 800 and P92   204
11.6   Acknowledgement   207
11.7   References   208

12   The erosion-corrosion resistance of uncoated and
      aluminized 12% chromium ferritic steels under
      fluidized-bed conditions at elevated temperature
      (SUNASPO)   210

      E Huttunen-Saarivirta, Institute of Materials Science, Finland;
      S Kalidakis and FH Stott, Corrosion and Protection Centre,
      UMIST, UK and V Rohr and M Schütze, Karl-Winnacker-Institut
      der DECHEMA e.V., Germany

12.1 Introduction   210
12.2 Experimental methods   211
12.3 Results   214
12.4 Discussion   230
12.5 Conclusions   233
12.6 Acknowledgements   234
12.7 References   234

13   Silicon surface treatment via CVD of the inner
      surface of steel pipes   236

      K Berreth, K Maile and A Lyutovich, Universität Stuttgart,
      Germany

13.1 Introduction   236
13.2 Experimental methods   237
13.3 Results and discussion   240
13.4 Conclusion   247
13.5 Acknowledgment   248
13.6 References   248

14   Performance of thermal barrier coatings on γ-TiAl   249

      R Braun, German Aerospace Centre (DLR), Germany; C Leyens,
      Technical University of Brandenburg at Cottbus, Germany and
      M Fröhlich, German Aerospace Centre (DLR), Germany

14.1 Introduction   249
14.2 Experimental methods   250
14.3 Results and discussion   252
14.4 Conclusions   262
14.5 Acknowledgments   262
14.6 References   263

15   Looking for surface treatments with potential
      industrial use in high temperature oxidation conditions   264

      G Bonnet, JM Brossard and J Balmain, University of La Rochelle,
      France

15.1 Introduction   264

15.2   Experimental methods   264
15.3   Results   266
15.4   Conclusions   274
15.5   Acknowledgments   275
15.6   References   275

Part III Test methods and service conditions

16   Reliable assessment of high temperature oxidation
      resistance by the development of a comprehensive
      code of practice for thermocycling oxidation testing
      (COTEST)   279

M SCHÜTZE and M MALESSA, Karl-Winnacker-Institut der
DECHEMA e.V., Germany

16.1   Introduction   279
16.2   The concept of COTEST   284
16.3   Work package 1: Evaluation of current test procedures and
        experimental facilities for cyclic oxidation testing   285
16.4   Work package 2: Evaluation of existing cyclic oxidation
        data available from the literature and those supplied by the
        various project contractors   285
16.5   Work package 3: Development of a set of test procedures
        suitable for standardisation   286
16.6   Work package 4: Reference materials in the test
        programme   287
16.7   Work package 5: Experimental investigation of selected
        materials under cyclic oxidation conditions and
        evaluation of oxidation behaviour   290
16.8   Work package 6: Development of a draft code of practice 293
16.9   Work package 7: Experimental validation of the draft
        code of practice   293
16.10   Work package 8: Formulation and fine-tuning of the final
        version of the code of practice   294
16.11   Concluding remarks   295
16.12   Acknowledgement   295
16.13   References   296

17   Variation in cyclic oxidation testing practice and data:
      The European situation before COTEST   297

S OSGERBY, Alston Power, UK; Formerly National Physical
Laboratory, UK and R PETTERSSON, Swedish Institute for Metals
Research, Sweden

17.1   Introduction   297

17.2	Methodology	297
17.3	Differences in testing practice	299
17.4	Cyclic oxidation data	305
17.5	Conclusions	310
17.6	Acknowledgements	310
17.7	References	311

18   Designing experiments for maximum information from cyclic oxidation tests and their statistical analysis using half normal plots (COTEST)   312

SY COLEMAN, University of Newcastle upon Tyne, UK and JR NICHOLLS, Cranfield University, UK

18.1	Introduction	312
18.2	Experimental method and response summaries	313
18.3	Influential parameters	315
18.4	Design of experiments, results and operational observations	315
18.5	Analysis of results for Alloy 800	317
18.6	Orthogonal contrasts	320
18.7	Synergies	321
18.8	Method of half normal plotting	322
18.9	Prediction of oxidation trends	324
18.10	Summary	327
18.11	Appendices	328
18.12	References	333

19   Influence of cycling parameter variation on thermal cyclic oxidation testing of high temperature materials (COTEST)   334

M SCHÜTZE and M MALESSA, Karl-Winnacker-Institut der DECHEMA e.V., Germany; SY COLEMAN, University of Newcastle upon Tyne, UK and L NIEWOLAK and WJ QUADAKKERS, Forschungszentrum Jülich GmbH, Germany

19.1	Introduction	334
19.2	Experimental methods	335
19.3	Results	336
19.4	Discussion	359
19.5	Conclusions	361
19.6	Acknowledgement	361
19.7	References	362

20   Rapid cyclic oxidation tests, using joule heating of
        wire and foil materials (COTEST)   363

        JR Nicholls and T Rose, Cranfield University, UK and R Hojda,
        Thyssen Krupp VDM GmbH, Germany

20.1   Introduction   363
20.2   Background   364
20.3   Design of a joule heated, ultra-short dwell cycle test
        facility capable of materials testing in controlled
        environments   367
20.4   Statistical design of a test matrix to investigate critical
        parameters controlling ultra-short dwell cyclic oxidation
        tests   371
20.5   Statistical analysis of the rapid thermal cycle tests on
        Kanthal A1 wire/foil samples tested in laboratory air   375
20.6   Conclusions   382
20.7   Acknowledgements   382
20.8   References   383

21   Thermo-mechanical fatigue – the route to
        standardisation (TMF-STANDARD)   384

        T Beck, Forschungszentrum Jülich GmbH, Germany; P Hähner,
        Joint Research Centre, The Netherlands; H-J Kühn, BAM, Labor
        V 21, Germany; C Rae, University of Cambridge, UK; E E Affeldt,
        MTU AeroEngines GmbH & Co. KG, Germany; H Andersson,
        SIMR, Sweden; A Köster, ENMSP/ARMINES, France and
        M Marchionni, CNR-IENI – TEMPE, Italy

21.1   Introduction   384
21.2   Test material   386
21.3   Reference TMF cycle   386
21.4   Pre-normative research work   386
21.5   Conclusions and outlook   397
21.6   References   398

22   Oxidation behaviour of Fe-Cr-Al alloys during
        resistance and furnace heating   400

        H Echsler, Forschungszentrum Jülich GmbH, Germany;
        H Hattendorf, Thyssen Krupp VDM GmbH, Germany and
        L Singheiser and WJ Quadakkers, Forschungszentrum Jülich
        GmbH, Germany

22.1   Introduction   400
22.2   Experimental methods   401
22.3   Results   402
22.4   Discussion   407

22.5	Conclusions	413
22.6	References	413

23	Service conditions and their influence on the oxide scale formation on metallic high temperature alloys for the application in innovative combustion processes	415

G TENEVA-KOSSEVA, H KÖHNE, and H ACKERMANN,
Oel-Wärme-Institut gGmbH, Germany and M SPÄHN, S RICHTER
and J MAYER, Central Facility for Electron Microscopy, Germany

23.1	Introduction	415
23.2	Experimental methods	416
23.3	Results	419
23.4	Discussion	424
23.5	Conclusion	426
23.6	References	427

24	Reducing superheater corrosion in wood-fired boilers	428

PJ HENDERSON C ANDERSSON, H KASSMAN and J HÖGBERG, Vattenfall
Research and Development, Sweden and P SZAKÁLOS and
R PETTERSSON, Corrosion and Metals Research Institute, Sweden

24.1	Introduction	428
24.2	Experimental methods	429
24.3	Results	432
24.4	Discussion	442
24.5	Conclusions	443
24.6	Acknowledgements	444
24.7	References	444

25	Hot erosion wear and carburization in petrochemical furnaces	445

RL DEUIS, AM BROWN and S PETRONE, Quantiam Technologies
Inc., Canada

25.1	Introduction	445
25.2	Experimental methods	446
25.3	Results	453
25.4	Discussion	466
25.5	Conclusions	472
25.6	Acknowledgements	473
25.7	References	473

26   High temperature corrosion of structural materials
      under gas-cooled reactor helium   474

C CABET, A TERLAIN, P LETT, L GUÉTAZ and J-M GENTZBITTEL,
CEA, France

26.1   Introduction   474
26.2   Overview of Ni-Cr-Mo alloy and corrosion in GCR helium   475
26.3   Experimental methods   480
26.4   Results   483
26.5   Summary and conclusion   489
26.6   Acknowledgement   489
26.7   References   489

27   Geometry effects on the oxide scale integrity during
      oxidation of the Ni-base superalloy CMSX-4 under
      isothermal and thermal cycling conditions   491

R OROSZ and H-J CHRIST, Universität Siegen, Germany and
U KRUPP, University of Applied Science, Germany

27.1   Introduction   491
27.2   Experimental methods   492
27.3   Results   493
27.4   Discussion   497
27.5   Conclusions   499
27.6   Acknowledgements   499
27.7   References   499

28   What are the right test conditions for the simulation
      of high temperature alkali corrosion in biomass
      combustion?   501

T BLOMBERG, ASM Microchemistry Ltd, Finland

28.1   Introduction   501
28.2   Biomass vs fossil fuels   502
28.3   Discussion   510
28.4   Conclusions   512
28.5   References   513

Part IV Modelling

29   Optimisation of in-service performance of boiler
      steels by modelling high-temperature corrosion
      (OPTICORR)   517

L HEIKINHEIMO, VTT Industrial Systems, Finland; D BAXTER, JRC,
Petten, The Netherlands; K HACK, GTT-Technologies, Germany;
M SPIEGEL, Max-Planck-Institut für Eisenforschung, Germany;

M HÄMÄLÄINEN, Helsinki University of Technology, Finland;
U KRUPP, University of Applied Science Osnabrück, Germany and
M ARPONEN, Rautaruukki Steel, Finland

29.1   Introduction   517
29.2   Thermodynamic data collection   518
29.3   Corrosion and oxidation modelling   519
29.4   Summary   530
29.5   Acknowledgements   531
29.6   References   531

30   Influence of gas phase composition on the kinetics
        of chloride melt induced corrosion of pure iron
        (OPTICORR)   533

A RUH and M SPIEGEL, Max Planck-Institut für Eisenforschung
GmbH, Germany

30.1   Introduction   533
30.2   Experimental methods   535
30.3   Results   535
30.4   Discussion   541
30.5   Conclusions   547
30.6   References   548

31   Development of toolboxes for the modelling of hot
        corrosion of heat exchanger components
        (OPTICORR)   550

K HACK and T JANTZEN, GTT-Technologies, Germany

31.1   Introduction   550
31.2   Available software tools at the onset of the project 552
31.3   Database work   552
31.4   Calculational results   558
31.5   Summary   566
31.6   Acknowledgement   566
31.7   References   567

32   Computer-based simulation of inward oxide scale
        growth on Cr-containing steels at high temperatures
        (OPTICORR)   568

U KRUPP, University of Applied Sciences Osnabrück, VB TRINDADE,
P SCHMIDT, S YANG, H-J CHRIST, U BUSCHMANN and W WIECHERT,
Universität Siegen, Germany

32.1   Introduction   568
32.2   Numerical modeling of diffusion-controlled corrosion
        processes   570

32.3 Simulation results and discussion 575
32.4 Summary 578
32.5 Acknowledgements 580
32.6 References 580

33 High temperature oxidation of γ-NiCrAl modelling
and experiments 582

TJ NIIDAM, NM VAN DER PERS and WG SLOOF, Delft University of
Technology, The Netherlands

33.1 Introduction 582
33.2 Experimental methods 583
33.3 Data evaluation 584
33.4 Results and discussion 590
33.5 Conclusions 596
33.6 Acknowledgement 597
33.7 References 597

Index 599

Contributor Contact Details

(*=main contact)

Editors

M. Schütze
Karl-Winnacker-Institut der
DECHEMA e.V.
Theodor-Heuss-Allee-25
60486 Frankfurt am Main
Germany

E-mail: schuetze@dechema.de

W. J. Quadakkers
Forschungszentrum Jülich GmbH
Institute for Energy Research
Microstructure and Properties of
Materials (IEF-2) D-52425 Jülich
Germany

E-mail: j.quadakkers@fz-juelich.de

Chapter 1

Michael P. Brady*, Peter F. Tortorelli,
Karren L. More, E. Andrew Payzant,
Beth L. Armstrong, Hua-Tay Lin
and
Michael J. Lance
Oak Ridge National Laboratory
Oak Ridge TN 37831-6115
USA

E-mail: bradymp@ornl.gov
        tortorellipf@ornl.gov

Feng Huang and Mark L. Weaver
University of Alabama
Tuscaloosa AL 35487-0202
USA

Chapter 2

Jianqiang Zhang*, Daniel M. I.
Cole and David J. Young
School of Materials Science and
    Engineering
University of New South Wales
Sydney 2052
Australia

E-mail j.q.zhang@unsw.edu.au
       d.young@unsw.edu.au

Chapter 3

N. J. Simms*
Cranfield University
Power Generation Technology
    Centre
Cranfield MK43 0AL
UK

E-mail: n.j.simms@cranfield.ac.uk

J. F. Norton
Consultant in Corrosion Science
    and Technology
Hemel Hempstead HP1 1SR
UK

E-mail: j.f.norton@cranfield.ac.uk

G. McColvin
Siemens Industrial Turbomachinery
    Ltd
Lincoln LN5 7FD
UK

E-mail:
gordon.mccolvin@industrial-
turbines.siemens.com

Chapter 4

Dr Gernot Strehl*
S+C MÄRKER GmbH
Kaiserau 2
51789 Lindlar
Germany

E-mail: g.strehl@schmidt-clemens.de

G. Borchardt
Institut für Metallurgie
TU Clausthal
38678 Clausthal-Zellerfeld
Germany

P. Beaven
GKSS-Forschungszentrum
Geesthacht
21502 Geesthacht
Germany

B. Lesage
LEMHE
CNRS UMR 8647
Université Paris XI
91405 Orsay
France

Chapter 5

V. B. Trindade and U. Krupp*
University of Applied Science
Osnabrück
Faculty of Engineering and
    Computer Science
Albrechtstr. 30
49009 Osnabrück
Germany

E-mail: u.krupp@fh-osnabrueck.de
    krupp@ifwt.mb.umi-
    siegen.de

Ph. E.-G. Wagenhuber, S. Yang and
H.-J. Christ
Universität Siegen
Institut für Werkstofftechnik
Paul-Bonatz Str. 9-11
57068-Siegen
Germany

Chapter 6

Y. Nishiyama*, K. Moriguchi,
N. Otsuka and T. Kudo
Sumitomo Metal Industries Ltd
Corporate Res & Dev Labs
1-8 Fusocho
Amagasaki 660-0891
Japan

E-mail:
nishiyam-yst@sumitomometals.co.jp
mori@kiso.amaken.sumitomometals.co.jp
ootsuka-nbo@sumitomometals.co.jp
kudou-tko@sumitomometals.co.jp

Chapter 7

V. Kochubey, D. Naumenko* and
W. J. Quadakkers
Forschungszentrum Jülich GmbH
Institute for Energy Research
Microstructure and Properties of
Material (IEF-2)
D-52425 Jülich
Germany

E-mail: d.naumenko@fz-juelich.de
            j.quadakkers@fz-juelich.de

H. Al-Badairy and G. Tatlock
Department of Materials Science
    and Engineering
University of Liverpool
Brownlow Hill
Liverpool L69 3GH
UK

E-mail: g.j.tatlock@liverpool.ac.uk

J. Le-Coze
Ecole des Mines de Saint-Etienne
UMR CNRS 5146
158 Cours Fauriel
42023 Saint-Etienne
Cedex 2
France

Chapter 8

J. R. Nicholls*, M. J. Bennett and
N. J. Simms
Cranfield University
Cranfield
Bedford MK43 0AL
UK

E-mail: j.r.nicholls@cranfield.ac.uk
            n.j.simms@cranfield.ac.uk

H. Hattendorf
Thyssen Krupp VDM GmbH
D-5877 Werdohl
Germany

E-mail: u.krupp@fh-osnabrueck.de
krupp@ifwt.mb.umi-siegen.de

D. Britton and A. J. Smith
Diffusion Alloys Ltd
Hatfield AL9 5JW
UK

R. Fordham and R. Bachorczyk
Institute of Energy
JRC Petten
1755 ZG Petten
The Netherlands

D. Goossens
Bekaert Technology Centre 6030
B-8550
Belgium

Chapter 9

G. J. Tatlock* and H. Al-Badairy
Department of Materials Science
    and Engineering
University of Liverpool
Brownlow Hill
Liverpool L69 3GH
UK

E-mail: g.j.tatlock@liverpool.ac.uk

R. Bachorczyk-Nagy and R. Fordham
European Commission
Joint Research Centre
Institute for Energy
Postbus 2
1755ZG Petten
The Netherlands

Chapter 10

V. Rohr and M. Schütze*
Karl-Winnacker-Institut der
DECHEMA e.V.
Theodor-Heuss-Allee-25
60486 Frankfurt am Main
Germany

E-mail: schuetze@dechema.de

E. Fortuna and D. N. Tsipas
Aristotles University of
  Thessaloniki
Thessaloniki 540 06
Greece

A. Milewska and F. J. Pérez
Universidad Complutense Madrid
28040 Madrid
Spain

Chapter 11

M. Schütze* and V. Rohr
Karl-Winnacker-Institut der
DECHEMA e.V.
Theodor-Heuss-Allee-25
60486 Frankfurt am Main
Germany

E-mail: schuetze@dechema.de

L. Nieto Hierro and P. J. Ennis and
W. J. Quadakkers
Forschungszentrum Jülich GmbH
Institute for Energy Research
Microstructure and Properties of
  Material (IEF-2)
D-52425 Jülich
Germany

E-mail: j.quadakkers@fz-juelich.de
    p.j.ennis@fz-juelich.de

Chapter 12

E. Huttunen-Saarivirta*
Institute of Materials Science
Tampere University of Technology
P.O. Box 589
33101 Tampere
Finland

E-mail: elina.huttunen-saarivirta@tut.fi

S. Kalidakis and F. H. Stott
Corrosion and Protection Centre
School of Materials
University of Manchester
P.O. Box 88
Sackville Street
Manchester M60 1QD
UK

V. Rohr and M. Schütze
Karl-Winnacker-Institut der
  DECHEMA e.V.
Theodor-Heuss-Allee 25
60486 Frankfurt am Main
Germany

E-mail: schuetze@dechema.de

Chapter 13

K. Berreth*, K. Maile and A.
  Lyutovich
Materials Testing Institute (MPA)
University of Stuttgart
Pfaffenwaldring 32
70569 Stuttgart
Germany

E-mail:
Karl.Berreth@mpa.uni-stuttgart.de
Karl.Maile@mpa.uni-stuttgart.de
Abraham.Lyutovich@mpa.uni-
stuttgart.de

Chapter 14

R. Braun* and M. Fröhlich
German Aerospace Center (DLR)
Institute of Materials Research
D-51170 Köln
Germany

E-mail: reinhold.braun@dlr.de
maik.froehlich@dlr.de

C. Leyens
Technical University of
Brandenburg at Cottbus
Chair of Physical Metallurgy and
   Materials Technology
Konrad-Wachsmann-Allee 17
D-03046 Cottbus
Germany

E-mail: leyens@tu-cottbus.de

Chapter 15

G. Bonnet*, J. M. Brossard and
   J. Balmain
Laboratoire d'Etude des Matériaux
en Milieux Agressifs
EA3167, Université de La Rochelle
Bâtiment Marie Curie
Avenue Michel Crépeau
17042 La Rochelle Cedex 01
France

E-mail: gbonnet@univ-lr.fr

Chapter 16

M. Schütze and M. Malessa
Karl-Winnacker-Institut der
DECHEMA e.V.
Theodor-Heuss-Allee 25
D-60486 Frankfurt/Main,
Germany

E-mail: schuetze@dechema.de

Chapter 17

S. Osgerby*
Ferrous Alloy Development Group
ALSTOM Power
Newbold Rd
Rugby CV21 2NH
UK

E-mail:
steve.osgerby@power.alstom.com

R. Pettersson
Swedish Institute for Metals
Research AB
Drottning Kristinasväg 48
SE 114 28 Stockholm
Sweden

Chapter 18

S. Y. Coleman*
Industrial Statistics Research Unit
King's Walk
University of Newcastle upon Tyne
Newcastle upon Tyne NE1 7RU
UK

E-mail:
shirley.coleman@newcastle.ac.uk

J. R. Nicholls
School of Applied Sciences
Cranfield University
Cranfield MK43 0AL
UK

E-mail: j.r.nicholls@cranfield.ac.uk

Chapter 19

M. Schütze* and M. Malessa
Karl-Winnacker-Institut der
    Dechema e.V.
Theodor-Heuss-Allee 25
D-60486 Frankfurt/Main
Germany

E-mail: schuetze@dechema.de

L. Niewolak and W. J. Quadakkers
Forschungszentrum Jülich GmbH
Institute for Energy Research
Microstructure and Properties of
    Material (IEF-2)
D-52425 Jülich
Germany

E-mail: j.quadakkers@fz-juelich.de

S. Y. Coleman
King's Walk Industrial Statistics
Research Unit
University of Newcastle upon Tyne
Newcastle upon Tyne NE1 7RU
UK

E-mail:
shirley.coleman@newcastle.ac.uk

Chapter 20

J. R. Nicholls* and T. Rose
School of Applied Sciences
Cranfield University
Cranfield
Bedford MK43 0AL
UK

E-mail: j.r.nicholls@cranfield.ac.uk

R. Hojda
Thyssen Krupp VDM GmbH
D-5877 Werdohl
Germany

Chapter 21

T. Beck*
Forschungszentrum Jülich GmbH
IEF-2
52425 Jülich
Germany

E-mail: t.beck@fz-juelich.de

Peter Hähner
European Commission
Joint Research Centre
Institute for Energy
NL-1755 ZG Petten
The Netherlands

E-mail: peter.haehner@jrc.nl

H.-J. Kühn
BAM, Labor V 21
Unter den Eichen 87
D-12205 Berlin
Germany

E-mail: hans-joachim-kuehn@bam.de

C. Rae
University of Cambridge
Department of Materials Science
Metallurgy
Pembroke Street
Cambridge CB2 3QZ
UK

E-mail: cr18@cam.ac.uk

E. E. Affeldt
MTU AeroEngines GmbH & Co.
KG
Turbine Materials (TEWT)
Dachauer Str. 665
D-80995 München
Germany

E-mail: ernst.affeldt@muc.mtu.com

H. Andersson
SIMR
Drottning Kristinasväg 48
S-11428 Stockholm
Sweden

E-mail: henrik.andersson@simr.se

A. Köster
ENMSP/ARMINES
B. P. 87
F-91003 Evry Cedex
France

E-mail: alain-koster@enmsp.fr

M. Marchionni
CNR-IENI – TEMPE
Via Cozzi 53
I-20125 Milano
Italy

E-mail: marchionni@ieni.cnr.it

Chapter 22

H. Echsler*
Forschungszentrum Jülich GmbH
Institute of Energy Research
Fuel Cells (IEF-3)
D-52425 Jülich
Germany

E-mail: h.echsler@fz-juelich.de

H. Hattendorf
Thyssen Krupp VDM GmbH
Kleffstr. 23
D-58762 Altena
Germany

E-mail:
heike.hattendorf@thyssenkrupp.com

L. Singheiser and W. J. Quadakkers
Forschungszentrum Jülich GmbH
Institute of Energy Research
Microstructure and Properties of
    Material (IEF-2)
D-52425 Jülich
Germany

E-mail: j.quadakkers@fz-juelich.de
l.singheiser@fz-juelich.de

Chapter 23

G. Teneva-Kosseva*, H. Köhne and
H. Ackermann
Oel-Wärme-Institut GmbH
Kaiserstr. 100
52134 Herzogenrath
Germany

E-mail: g.teneva@owi-aachen.de

M. Spähn, S. Richter and J. Mayer
Central Facility for Electron
    Microscopy
GFE, RWTH Aachen
Ahornstr. 55
52074 Aachen
Germany

Chapter 24

P. J. Henderson*, C. Andersson
and, H. Kassman J. Högberg
Vattenfall Research and
  Development
162 87 Stockholm
Sweden

E-mail:
pamela.henderson@vattenfall.com

P. Szakálos and R. Pettersson
Corrosion and Metals Research
  Institute
KIMAB
Drottning Kristinasväg 48
114 28 Stockholm
Sweden

Chapter 25

R. L. Deuis, A. M. Brown and S.
Petrone*
Quantiam Technologies Inc.
8207 Roper Road
Edmonton
Alberta, T6E 6S4
Canada

E-mail: spetrone@quantiam.com

Chapter 26

C. Cabet*, A. Terlain and P. Lett
Laboratoire d'Etude de la
Corrosion Non Aqueuse
Bât. 458
DEN/DANS/DPC/SCCME
CEA Saclay
91191 Gif-sur-Yvette
France

E-mail: celine.cabet@cea.fr

L. Guétaz and J.-M. Gentzbittel
DRT/LITEN/DTEN
17 rue des martyrs
CEA Grenoble
F-38054 Grenoble
France

Chapter 27

R. Orosz and H.-J. Christ
Universität Siegen
Institut für Werkstofftechnik
Paul-Bonatz-Str. 9-11
57068 Siegen
Germany

U. Krupp*
University of Applied Science
Osnabrück
Faculty of Engineering and
  Computer Science
Albrechtstr. 30
49009 Osnabrück
Germany

E-mail: u.krupp@fh-osnabrueck.de

Chapter 28

T. Blomberg
Process Engineer
ASM Microchemistry Ltd
Kiertomäentie 11 As. 1
01260 Vantaa
Finland

E-mail: tom.blomberg@asm.com

Chapter 29

L. Heikinheimo*
VTT Industrial Systems
Espoo
Finland
P.O. Box 1000
FI-02044 VTT
Finland

E-mail: liisa.heikinheimo@vtt.fi

D. Baxter
JRC Petten
Institute for Energy
1755 ZG Petten
The Netherlands

E-mail:david.baxter@jrc.nl

K. Hack
Managing Director
GTT-Technologies
Kaiserstrasse 100
D-52134 Herzogenrath
Germany

E-mail: kh@gtt-technologies.de

M. Spiegel
Max-Planck-Institute für
Eisenforschung GmbH
Abt. Physikalische Chemie
Max-Planck-Str.1
40237 Düsseldorf

E-mail: m.spiegel@du-szmf.de

M. Hämäläinen
Helsinki University of Technology
Espoo
P.O. Box 1100
FI-02015 TKK
Finland

U. Krupp
University of Applied Science
Osnabrück
Faculty of Engineering and
    Computer Science
Albrechtstr. 30
49009 Osnabrück/Germany

E-mail: u.krupp@fh-osnabrueck.de

M. Arponen
Rautaruukki Steel
PL 138 (Suolakivenkatu 1)
00811 Helsinki
Finland

Chapter 30

A. Ruh and M. Spiegel*
Max-Planck-Institut für
Eisenforschung GmbH
Max-Planck-Str. 1
D-40237 Düsseldorf
Germany

E-mail: ruh@mpie.de; spiegel@mpie.de
m.spiegel@du-szmf.de

Chapter 31

K. Hack* and T. Jantzen
GTT-Technologies
Kaiserstrasse 100
D-52134 Herzogenrath
Germany

E-mail: KH@GTT-Technologies.de

Chapter 32

U. Krupp*
University of Applied Sciences
Osnabrück
Faculty of Engineering and
    Computer Science
Albrechtsr.
30, 49009 Osnabrück
Germany

E-mail: krupp@fh-osnabrück.de

V. B. Trindade, P. Schmidt, H.-J.
    Christ
Institut für Werkstofftechnik
Universität Siegen
Germany

U. Buschmann and W. Wiechert
Institut für Systemtechnik
Universität Siegen
Germany

Chapter 33

T. J. Nijdam, N. M. van der Pers
    and W. G. Sloof*
Department of Materials Science
    and Engineering
Delft University of Technology
2628 CD Delft
The Netherlands

E-mail: W.G.Sloof@tudelft.nl

European Federation of Corrosion (EFC) publications: Series introduction

The European Federation of Corrosion (EFC), incorporated in Belgium, was founded in 1955 with the purpose of promoting European co-operation in the fields of research into corrosion and corrosion prevention.

Membership of the EFC is based upon participation by corrosion societies and committees in technical Working Parties. Member societies appoint delegates to Working Parties, whose membership is expanded by personal corresponding membership.

The activities of the Working Parties cover corrosion topics associated with inhibition, education, reinforcement in concrete, microbial effects, hot gases and combustion products, environment sensitive fracture, marine environments, refineries, surface science, physico-chemical methods of measurement, the nuclear industry, the automotive industry, computer-based information systems, coatings, tribo-corrosion and the oil and gas industry. Working Parties and Task Forces on other topics are established as required.

The Working Parties function in various ways, e.g. by preparing reports, organising symposia, conducting intensive courses and producing instructional material, including films. The activities of Working Parties are co-ordinated, through a Science and Technology Advisory Committee, by the Scientific Secretary. The administration of the EFC is handled by three Secretariats: DECHEMA e.V. in Germany, the Société de Chimie Industrielle in France, and The Institute of Materials, Minerals and Mining in the United Kingdom. These three Secretariats meet at the Board of Administrators of the EFC. There is an annual General Assembly at which delegates from all member societies meet to determine and approve EFC policy. News of EFC activities, forthcoming conferences, courses, etc., is published in a range of accredited corrosion and other specialist journals throughout Europe. More detailed descriptions of activities are given in a newsletter prepared by the Scientific Secretary.

The output of the EFC takes various forms. Papers on particular topics, for example reviews or results of experimental work, may be published in scientific and technical journals in one or more countries in Europe. Conference proceedings are often published by the organisation responsible for the conference. In 1987, the then Institute of Metals was appointed as the official EFC publisher. Although the arrangement is non-exclusive and other routes for publication are still available, it is expected that the Working Parties of the EFC will use The Institute of Materials, Minerals and Mining for publication of reports, proceedings, etc., wherever possible.

The name of The Institute of Metals was changed to The Institute of Materials on 1 January 1992 and to The Institute of Materials, Minerals and Mining with effect from 26 June 2002. The series is now published by Woodhead Publishing and Maney Publishing on behalf of The Institute of Materials, Minerals and Mining.

P. McIntyre
EFC Series Editor, The Institute of Materials, Minerals and Mining, London, SW1Y 5DB UK

EFC Secretariats are located at:
Dr B A Rickinson
European Federation of Corrosion, The Institute of Materials, Minerals and Mining, 1 Carlton House Terrace, London SW1Y 5DB, UK

Dr J P Berge
Fédération Européene de la Corrosion, Société de Chimie Industrielle, 28 rue Saint-Dominique, F-75007 Paris, France

Professor Dr G Kreysa
Europäische Föderation Korrosion, DECHEMA e.V., Theodor-Heuss-Allee 25, D-60486 Frankfurt, GERMANY

Volumes in the EFC series

1 **Corrosion in the nuclear industry**
Prepared by Working Party 4 on Nuclear Corrosion

2 **Practical corrosion principles**
Prepared by Working Party 7 on Corrosion Education (out of print)

3 **General guidelines for corrosion testing of materials for marine applications**
Prepared by Working Party 9 on Marine Corrosion

4 **Guidelines on electrochemical corrosion measurements**
Prepared by Working Party 8 on Physico-chemical Methods of Corrosion Testing

5 **Illustrated case histories of marine corrosion**
Prepared by Working Party 9 on Marine Corrosion

6 **Corrosion education manual**
Prepared by Working Party 7 on Corrosion Education

7 **Corrosion problems related to nuclear waste disposal**
Prepared by Working Party 4 on Nuclear Corrosion

8 **Microbial corrosion**
Prepared by Working Party 10 on Microbial Corrosion

9 **Microbiological degradation of materials and methods of protection**
Prepared by Working Party 10 on Microbial Corrosion

10 **Marine corrosion of stainless steels: chlorination and microbial effects**
Prepared by Working Party 9 on Marine Corrosion

11 **Corrosion inhibitors**
Prepared by the Working Party on Inhibitors (out of print)

12 **Modifications of passive films**
    Prepared by Working Party 6 on Surface Science and Mechanisms of Corrosion and Protection

13 **Predicting CO$_2$ corrosion in the oil and gas industry**
    Prepared by Working Party 13 on Corrosion in Oil and Gas Production (out of Print)

14 **Guidelines for methods of testing and research in high temperature corrosion**
    Prepared by Working Party 3 on Corrosion by Hot Gases and Combustion Products

15 **Microbial corrosion (Proceedings of the 3rd International EFC Workshop)**
    Prepared by Working Party 10 on Microbial Corrosion

16 **Guidelines on materials requirements for carbon and low alloy steels for H$_2$S-containing environments in oil and gas production**
    Prepared by Working Party 13 on Corrosion in Oil and Gas Production

17 **Corrosion resistant alloys for oil and gas production: guidance on general requirements and test methods for H$_2$S service**
    Prepared by Working Party 13 on Corrosion in Oil and Gas Production

18 **Stainless steel in concrete: state of the art report**
    Prepared by Working Party 11 on Corrosion of Steel in Concrete

19 **Sea water corrosion of stainless steels – mechanisms and experiences**
    Prepared by Working Party 9 on Marine Corrosion and Working Party 10 on Microbial Corrosion

20 **Organic and inorganic coatings for corrosion prevention – research and experiences**
    Papers from EUROCORR '96

21 **Corrosion – deformation interactions**
    CDI '96 in conjunction with EUROCORR '96

22 **Aspects of microbially-induced corrosion**
    Papers from EUROCORR '96 and EFC Working Party 10 on Microbial Corrosion

23 **CO$_2$ corrosion control in oil and gas production – design considerations**
    Prepared by Working Party 13 on Corrosion in Oil and Gas Production

24 **Electrochemical rehabilitation methods for reinforced concrete structures – a state of the art report**
Prepared by Working Party 11 on Corrosion of Steel in Concrete

25 **Corrosion of reinforcement in concrete – monitoring, prevention and rehabilitation**
Papers from EUROCORR '97

26 **Advances in corrosion control and materials in oil and gas production**
Papers from EUROCORR '97 and EUROCORR '98

27 **Cyclic oxidation of high temperature materials**
Proceedings of an EFC Workshop, Frankfurt/Main, 1999
Edited by M. Schütze and W.J. Quadakkers

28 **Electrochemical approach to selected corrosion and corrosion control**
Papers from 50th ISE Meeting, Pavia, 1999

29 **Microbial Corrosion (Proceedings of the 4th International EFC Workshop)**
Prepared by the Working Party on Microbial Corrosion

30 **Survey of literature on crevice corrosion (1979–1998): mechanisms, test methods and results, practical experience, protective measures and monitoring**
Prepared by F. P. Ijsseling and Working Party 9 on Marine Corrosion

31 **Corrosion of reinforcement in concrete: corrosion mechanisms and corrosion protection**
Papers from EUROCORR '99 and Working Party 11 on Corrosion of Steel in Concrete

32 **Guidelines for the compilation of corrosion cost data and for the calculation of the life cycle cost of corrosion – a working party report**
Prepared by Working Party 13 on Corrosion in Oil and Gas Production

33 **Marine corrosion of stainless steels: testing, selection, experience, protection and monitoring**
Edited by D Féron on behalf of Working Party 9 on Marine Corrosion

34 **Lifetime modelling of high temperature corrosion processes**
Proceedings of an EFC Workshop 2001. Edited by M. Schütze, W. J. Quadakkers and J. R. Nicholls

35 **Corrosion inhibitors for steel in concrete**
Prepared by B. Elsener with support from a Task Group of Working Party 11 on Corrosion of Steel in Concrete

36 **Prediction of long term corrosion behaviour in nuclear waste systems**
Edited by D. Féron and Digby D. Macdonald on behalf of Working Party 4 on Nuclear Corrosion

37 **Test methods for assessing the susceptibility of prestressing steels to hydrogen-induced stress corrosion cracking**
Prepared by B. Isecke on behalf of Working Party 11 on Corrosion of Steel in Concrete

38 **Corrosion of reinforcement in concrete: mechanisms, monitoring, inhibitors and rehabilitation techniques**
Edited by M. Raupach, B. Elsener, R. Polder and J. Mietz on behalf of Working Party 11 on Corrosion of Steel in Concrete

39 **The use of corrosion inhibitors in oil and gas production**
Edited by J. W. Palmer, W. Hedges and J. L. Dawson on behalf of Working Party 13 on Corrosion in Oil and Gas Production

40 **Control of corrosion in cooling waters**
Edited by J. D. Harston and F. Ropital on behalf of Working Party 15 on Corrosion in the Refinery Industry

41 **Corrosion by carbon and nitrogen: metal dusting, carburisation and nitridation**
Edited by H. Grabke and M. Schütze on behalf of Working Party 3 on Corrosion by Hot Gases and Combustion Products

42 **Corrosion in refineries**
Edited by J. D. Harston and F. Ropital on behalf of Working Party 15 on Corrosion in the Refinery Industry

43 **The electrochemistry and characteristics of embeddable reference electrodes for concrete**
Prepared by R. Myrdal on behalf of Working Party 11 on Corrosion of Steel in Concrete

44 **The use of electrochemical scanning tunnelling microscopy (EC-STM) in corrosion analysis: reference material and procedural guidelines**
Prepared by R. Lindström, V. Maurice, L. Klein and P. Marcus on behalf of Working Party 6 on Surface Science

45 **Local probe techniques for corrosion research**
Edited by R. Oltra, V. Maurice, R. Akid and P. Marcus on behalf of Working Party 8 on Physico-Chemical Methods of Corrosion Testing

46 **Amine unit corrosion in refineries**
Edited by J. D. Harston and F. Ropital on behalf of Working Party 15 on Corrosion in the Refinery Industry

47 **Novel approaches to improving high temperature corrosion resistance**
Edited by M. Schütze and W. Quadakkers on behalf of Working Party 3 on Corrosion by Hot Gases and Combustion Products

48 **Corrosion of metallic heritage artefacts: investigation, conservation and prediction of long term behaviour**
Edited by P. Dillmann, G. Béranger, P. Piccardo and H. Matthiessen on behalf of Working Party 4 on Nuclear Corrosion

49 **Electrochemistry in light water reactors: reference electrodes, measurement, corrosion and tribocorrosion issues**
Edited by R.-W. Bosch, D. Féron and J.-P. Celis on behalf of Working Party 4 on Nuclear Corrosion

50 **Corrosion behaviour and protection of copper and aluminium alloys in seawater**
Edited by D. Féron on behalf of Working Party 9 on Marine Corrosion

51 **Corrosion issues in light water reactors: stress corrosion cracking**
Edited by D. Féron and J.-M. Olive on behalf of Working Party 4 on Nuclear Corrosion

52 **(Not yet published)**

53 **Standardisation of thermal cycling exposure testing**
Edited by M. Schütze and M. Malessa on behalf of Working Party 3 on Corrosion by Hot Gases and Combustion Products

54 **Innovative pre-treatment techniques to prevent corrosion of metallic surfaces**
Edited by L. Fedrizzi, H. Terryn and A. Simões on behalf of Working Party 14 on Coatings

55 **Corrosion-under-insulation (CUI) guidelines**
Edited by S. Winnik on behalf of Working Party 13 on Corrosion in Oil and Gas Production and Working Party 15 on Corrosion in the Refinery Industry

Part I

Alloy modification

1

Design strategies for oxidation-resistant intermetallic and advanced metallic alloys

MICHAEL P BRADY, PETER F TORTORELLI, KARREN L MORE, E ANDREW PAYZANT, BETH L ARMSTRONG, HUA-TAY LIN and MICHAEL J LANCE, Oak Ridge National Laboratory, USA and FENG HUANG, and MARK L WEAVER, University of Alabama, USA

1.1   Introduction

Protective scale formation for low and high temperature environments remains a key materials challenge to the successful implementation and advancement of energy production and related technologies. In keeping with the theme of the 2004 European Federation of Corrosion conference on 'Novel Approaches to the Improvement of High Temperature Corrosion Resistance', this chapter will put forward some thoughts on how coatings or surface modification can alter near-surface composition and microstructure in potentially beneficial ways to achieve corrosion resistance using principles of gas–metal reactions. Extrapolation of these concepts to functional applications will also be discussed. This is the third in a series of related works focusing on design strategies for oxidation/corrosion resistance and the control of surface chemistry for functional applications [1,2]. These works are not meant to be an exhaustive compilation of established approaches or a rigorous discussion of various oxidation mechanisms; rather, they speculate on some concepts and highlight some preliminary experimental phenomenological observations that show potential for possible future development.

In a simplistic sense, control of surface composition can generally be accomplished in one of two ways:

1.   by deposition of coatings to modify surface composition relative to the substrate, or
2.   by application of internal and/or selective gas reactions (oxidation, nitridation, carburization, etc.) to precipitate new phases, modify existing phases, or preferentially segregate one or more elements from the substrate alloy to the surface.

The latter phenomena encompass the basis for protective oxide scale formation, including the subsequent formation of scales on deposited coating alloys. However, it can also be extended to the formation of protective

3

and/or functional nitride, carbide, etc., surface layers by elevated temperature gas reactions. These aspects will be discussed in Section 1.3.

1.2   Self-graded metallic precursor coating concepts

A key advantage to coatings is that the substrate alloy can be optimized for properties other than corrosion resistance (i.e. ductility, formability, creep strength, etc.). The key limitation is that the physical/chemical/mechanical differences between the substrate and the coating can lead to detrimental interactions, which can limit lifetime. Chief among these is interdiffusion between coating and substrate, which can result in the formation of brittle phases or the loss of protective scale-forming elements. For example, in thermal barrier coating (TBC)/superalloy systems, a major problem is the loss of Al from the bond coat to the substrate superalloy over time at temperature. This ultimately contributes to an inability to maintain Al_2O_3 growth and subsequent TBC failure [3,4]. Design of coatings in consideration of the substrate, as well as substrate alloy design in consideration of the fact that it will be coated, has received increased attention in recent years [e.g. 5, 6]. For example, recent work by Gleeson et al. [4] has identified Ni–Al–Pt bond coat compositions for which Al diffuses from the superalloy substrate into the bond coat, helping to maintain a reservoir of Al for Al_2O_3 formation by the bond coat, rather than loss of Al from the bond coat into the substrate. Consideration of coating/substrate compatibility issues has also been extended to include a number of design and processing approaches, many of which utilize functionally graded structures to manage chemical and thermomechanical incompatibilities [7,8].

For applications utilizing oxide coatings, a potentially useful alternative to direct deposition of oxide compositions may be the deposition and subsequent complete oxidation of a thin metallic precursor alloy layer. The advantage is that the deposited metallic layer may self grade as it is converted to oxide, with the possibility of obtaining improved adherence and/or chemical compatibility with the substrate. For proof of principle exploration, a series of Al_2O_3-forming alloys based on Cr_2Al-Y (79.7Cr-20Al-0.3Y weight percent, wt.%), NiAl-Hf (69.6Ni-30Al-0.4Hf wt.%), NiCrAlY (79.7Ni-10Cr-10Al-0.3Y wt.%), and FeCrAlY (74.7Fe-20Cr-5Al-0.3Y wt.%) were sputtered deposited as thin layers (1–5 micron thick) onto Kyocera's SN282 and Honeywell Ceramic Components NT154 Si_3N_4 and carborundum hot-pressed SiC (hexoloy) substrates. The technological driving force for such a coating was to explore the possibility of using Al_2O_3 as a volatility barrier for SiO_2-forming ceramics in high-temperature oxidizing environments containing water vapor. SiO_2 scales can be severely compromised due to volatilization and accelerated oxidation in the presence of water vapor, especially under the high pressure/high gas flow conditions encountered in gas turbines applications [9,10]. Al_2O_3 scales are more resistant to water vapor [11], but also have a much higher coefficient of thermal expansion than do the Si_3N_4 and SiC substrates (~9 vs ~3 × 10^{-6} °C), which leads to cracking, spallation, and coating failure. This work was focused on trying to form a graded duplex scale, consisting of an outer Al_2O_3 layer overlying an inner graded oxide layer of the base metal(s) of the alloy (e.g. Cr, Ni, Fe), Si, and Al. Such grading may lead to enhanced chemical stability and, conceivably, reduce stress generation in the surface layers(s) during cooling because changes in CTE may occur more gradually throughout the volume of the modified material.

The coated substrates were pre-oxidized at 1150–1250°C for up to 1 h in pure O_2 to completely convert the deposited alloy layers to oxides. Representative cross-sections of the deposited coatings after the oxidation pretreatment are shown in Fig. 1.1. The microstructures of the oxidized coatings were qualitatively similar, and consisted of a 0.5 µm outer layer of Al_2O_3, overlying an inner Al-Si-O base layer 2–3 µm thick. The Al-Si-O layers also contained the base alloy components (Cr, Fe, Ni), and the levels of Si and Al varied across the layer, from more Si-rich near the substrate interface to more Al-rich at the interface with the outer Al_2O_3 layer. Little Al migrated into the ceramic. All the coatings were adherent after the oxidation pretreatment, although some surface cracking was observed, especially in the NiCrAlY and Cr_2AlY precursor samples. Photo-stimulated luminescence spectroscopy (PSLS) stress measurements [12,13] of the Al_2O_3 scales (Fig. 1.2) indicated significant tensile stresses in the outer Al_2O_3 layers, suggesting

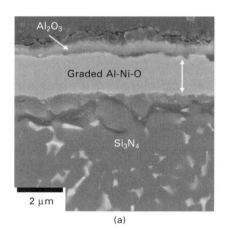
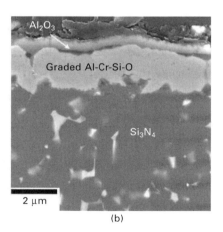

(a)   (b)

1.1 Scanning electron cross-section images of sputtered deposited metallic precursor coatings on SN282 Si_3N_4 after a 30 minute/1150°C/ O_2 exposure: (a) NiAl-Hf, (b) Cr_2Al-Y. Qualitatively similar microstructures were formed on all precursor metallic coating/ substrate combinations studied.

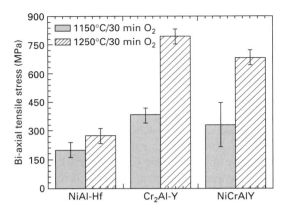

1.2 Typical room-temperature PSLS stress measurements of the Al_2O_3 scales formed on oxidized metallic precursor coatings on SN282 Si_3N_4. Tensile stresses were observed for all precursor metallic coating/substrate combinations studied, with the exception of one NiAl-Hf/SiC sample, which showed a biaxial compressive stress of ~200 MPa.

that whatever grading occurred was insufficient to greatly mitigate coefficient of thermal expansion (CTE) mismatch issues (on the other hand, such high stresses do indicate good coating adhesion). A short term 1000°C, 72 h oxidation exposure resulted in demixing of the oxide phases in the graded layer and/or significant spallation in the NiCrAlY and Cr_2Al-Y precursor coatings; however, the NiAl-Hf and FeCrAlY precursor coatings remained adherent. Figure 1.3 shows a cross-section of a NiAl-Hf precursor coating after a 1200°C, 500 h exposure in air + 15% water vapor at 10 atm. (details of typical exposure conditions provided in reference 14). The coating did not act as a water vapor barrier, with significant accelerated/non-protective SiO_2 growth evident below. Although these results indicate this approach is not promising for environmental barries coating EBC applications, self-graded oxide coatings were successfully formed from the deposited metallic precursor alloys, and there may be other applications where such an approach may be of interest.

For the EBC application, a variation on conventional metallic coating strategy was subsequently investigated [13]. Alloy coatings are typically sufficiently thick such that they remain primarily metallic and can reform their protective oxide scale in the event of cracking or damage. Alumina-forming compositions based on Fe or Ni were not considered attractive options for Si-based ceramics because they may react with Si to form low melting point (~ ≤ 1200°C) eutectics. In the case of Si_3N_4 substrates, Al/AlN is more stable than Si/Si_3N_4 (at unit activities) and the loss of Al from the coating to the substrate would limit coating lifetime, as well as potentially

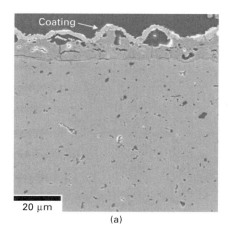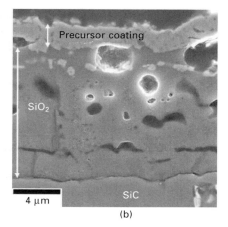

1.3 Scanning electron cross-section image of a sputter deposited/ oxidized NiAl-Hf precursor coating on SiC after 500 h at 1200°C in air + 15% water vapor at 10 atm. Penetration of water vapor through the coatings resulted in rapid, non protective SiO_2 formation on the SiC substrate. (a) Low magnification overview, (b) remaining coating on oxidized substrate (SiC)

degrade the mechanical properties of the Si_3N_4. In an attempt to manage this issue, an Al_2O_3-forming Ti-based alloy, Ti-51Al-12Cr atomic percent (at.%) was selected for evaluation [13,15]. The nitrides of Ti are more stable than those of Al and Si and the lowest melting point eutectic in the Ti-Si system is around 1330°C. The goal was for the TiAlCr coating to self grade, such that a Ti-N rich barrier layer is formed at the coating/substrate interface, leaving the Al to act as a reservoir for Al_2O_3 scale formation.

Figure 1.4 shows an oxidized cross-section of an initially 8–10 μm thick TiAlCr (Ti-51Al-12Cr at.%) [13] sputtered deposited metallic bond coat on SN282 Si_3N_4. The coating self graded in the desired manner. At the onset of oxidation, Al was selectively oxidized from the TiAlCr coating to form Al_2O_3. Concurrently, Ti reacted with the substrate Si_3N_4 to form a Ti-rich nitride. The movement of the Ti to the Si_3N_4 was sufficiently more rapid than the consumption of Al to form Al_2O_3 such that the unusual and beneficial situation of Al enrichment (relative to the initial coating composition) underneath the Al_2O_3 scale resulted. The Cr rejected from the oxidation front and Si moving out from the Si_3N_4 reacted with Ti to form a Ti-Si-Cr-rich phase as an intermediate layer between the Al-rich outer layer and the Ti-rich inner layer. Effectively, the coating alloy formed a Ti-N-rich barrier *in situ* at the substrate/coating interface, which prevented loss of Al, at least under the short-term exposure conducted. Significantly more work for the EBC application will be necessary to determine if this approach is feasible for long lifetimes. The growth rate of Al_2O_3 would likely also limit this type of

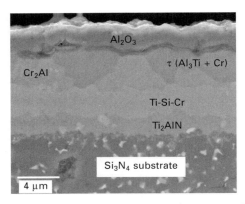

1.4 Scanning electron image of a sputter deposited Ti-51Al-12Cr at.% coating on SN 282 Si$_3$N$_4$ after three 100 h cycles at 1000°C in air [13]. The τ region underneath the scale contained ~ 65 at.% Al, and the Ti$_2$AlN region at the interface with the Si$_3$N$_4$ contained ~ 60 at.% Ti (phase identification based on composition).

coating to no higher than 1100–1150°C for the 40,000 h lifetimes needed. (Details are provided in reference 13.) However, the observed pattern of behavior does demonstrate the potential of this form of self grading to manage coating/substrate interdiffusion.

1.3   Surface modification approaches

1.3.1   Continuous protective nitride and carbide surface layer formation

Protective oxide scale forming alloy design principles can readily be extended to the formation of continuous layers of nitrides, carbides, borides, sulfides, etc. Under high-temperature conditions in the presence of oxygen, these phases will subsequently form oxides. However, there are low- and high-temperature situations where such initial dense (nitride, carbide, etc.) layer formation may be desirable.

One such example is the formation of protective layers for metallic bipolar plates in proton exchange membrane fuel cells (PEMFCs). Bipolar plates serve to electrically connect the anode of one cell to the cathode of the next in a fuel cell stack to achieve a useful voltage. Metallic alloys would be ideal as bipolar plates because they are amenable to low-cost/high-volume manufacturing, offer high thermal and electrical conductivities, and can be made into thin sheet or foil form (0.1–1 mm thick) to achieve high power densities [16-18]. However, most metals exhibit inadequate corrosion behavior in PEMFC environments (aqueous/acidic in the 60–80°C temperature range) due to formation of passive oxide layer(s), which increase cell resistance, and contamination of the polymer membrane by metallic ion dissolution (as little as 10 ppm can degrade behavior) [19-21]. Metal nitrides are of interest as protective coatings because they offer an attractive combination of high electrical conductivity and corrosion resistance [21]. However, conventional coating methods tend to leave pin-hole defects [22], which result in local corrosion and unacceptable behavior. A potential solution is the use of thermal nitridation reactions to form the protective nitride surface layer [23, 24]. Pinhole defects are not expected because at elevated temperatures thermodynamic and kinetic factors favor reaction of all exposed metal surfaces. Rather, the key issues are nitride layer composition, morphology, and adherence, which can potentially be controlled through proper selection of alloy composition/microstructure and nitridation conditions. Thermal nitridation is also a relatively inexpensive, industrially viable technique (usually used to impart wear resistance) and is viable for covering complex surface geometries such as bipolar plate flow field features.

Figure 1.5 shows a cross-section of a model Ni-50Cr alloy that was nitrided at 1100°C for 2 h in N_2 [24]. A dense, continuous CrN/Cr_2N surface layer was formed, overlying an extensive internally nitrided zone. This material exhibited excellent behavior during a 4000 h corrosion exposure under simulated anodic and cathodic PEMFC conditions and 1000 h of single-cell fuel cell testing, with no evidence of significant pin-hole defects, no increase in surface contact resistance, and virtually no metal ion dissolution [24]. This result provides strong proof of principle evidence that thermal nitridation can be used to form dense, pin-hole free nitride layers to provide protection under aqueous corrosion conditions. Current work for this application is

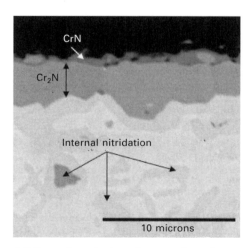

1.5 Scanning electron image of continuous, external Cr-nitride formed on Ni-50Cr after nitridation at 1100°C for 2 h in N_2 [23–25].

focused on forming such nitride layers on lower cost Fe-Cr base alloys for
PEMFC bipolar plates [25, 26].

Other continuous nitride layers can also readily be formed by this approach.
Figure 1.6 shows typical cross-sections for Fe-(5-15)Ti, Ni-(5-15)Ti, and
Ni-10Nb-5V alloys which formed continuous, external layers of binary TiN,
and a Ni-Nb-V nitride by thermal nitridation. Note that a wide range of
nitride microstructures, from fine, equiaxed TiN grains to coarse, columnar
Ni-Nb-V nitride base phase grains are possible. Again, the potential advantage
of this approach is the growth of pin-hole free layers on components with
complex surface geometries to provide corrosion protection in aqueous
environments.

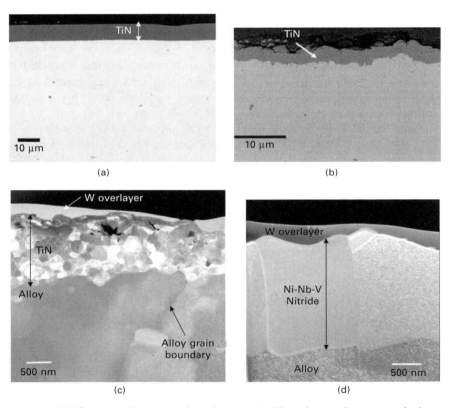

1.6 Cross-section scanning electron (a, b) and scanning transmission
electron (c, d) images illustrating a range of microstructures in
continuous nitride layers. (a) Continuous TiN layer on Ni-10Ti-2.5W-
0.15Zr wt.%, 1100°C/24 h/N_2; (b) continuous TiN layer on Fe-15Ti
wt.%, 1100°C/36 h/N_2; (c) nanoscale equiaxed TiN on Ni-10Ti-0.15Zr
wt.%, 1100°C, 24 h, N_2 (d) coarse, columnar structure of a ternary Ni-
Nb-V nitride on Ni-10Nb-5V wt.%, 1100°C/24 h/N_2. The W overlayer in
(c, d) was for sample preparation purposes.

As discussed by Rubly and Douglass [27], much higher levels of Cr are needed in Ni-Cr alloys to form an external Cr-nitride layer on nitridation, as compared to the level of Cr needed to form an external Cr-oxide layer on oxidation. Other than a possible improvement in wear resistance due to the high hardness of the Cr-nitride surface layer, there would be no advantage to pre-nitriding Ni-Cr alloys to improve high temperature oxidation resistance. However, some situations may exist where it may be possible to preferentially segregate a protective oxide scale forming element to the surface by a nitridation, carburization, etc., reaction, at a lower alloying level than is needed for oxidation. It is interesting to note that the TiN and related surface layers shown in Fig. 1.6 were formed on Ni and Fe base alloys at relatively low levels of additions, much lower than that needed with Cr/Cr-nitrides.

Following the above reasoning, the level of Al needed to form a continuous AlN scale on nitridation may (hypothetically) be lower for a particular alloy than the level needed to form an Al_2O_3 scale in oxygen/air. On exposure to a high temperature oxidizing environment, the AlN layer could then subsequently form a protective Al_2O_3 scale. The advantage of this approach over a conventional pre-oxidation treatment to form Al_2O_3 directly is that the nitride layer could heal the Al_2O_3 scale if it became cracked or otherwise damaged, whereas a scale formed on pre-oxidation may not be able to reform the Al_2O_3. This speculated protection scheme clearly is not viable for all applications, and there would certainly be limitations based on the thickness of the initial AlN layer that could be formed, and CTE and related issues with AlN.

1.3.2 Ternary nitride and carbide protective surface layer formation

Nitridation, carburization, etc., surface treatments to improve high temperature oxidation resistance are an area of increasing interest, especially for intermetallic phases such as TiAl which are borderline for protective oxide scale formation [28–31]. Results to date for nitridation and related surface treatments have been mixed, with true protective Al_2O_3 scale formation by such surface-treated TiAl alloys generally not reported. (Some success has been achieved by the introduction of Cl to the surface of TiAl alloys; see reference 32 for further details.) At the same time, there has been great recent interest in layered ternary nitride and carbide phases, due to a unique combination of properties, including their ability to be easily machined [33, 34]. This class of materials is referred to as MAX phases, and follow a general formula of $M_{n+1}AX_n$, where $n = 1, 2,$ or 3, M is an early transition metal, A is an A group element, and X is carbon or nitrogen [33, 34]. Some of these phases, notably Ti_3AlC_2 and Ti_2AlC, also form continuous protective Al_2O_3 scales on exposure in air, despite their low Al content (comparable to or less than that of Ti_3Al, which does not form Al_2O_3) [35–37]. Reduced oxygen permeability in these structures compared with the corresponding binary Ti_3Al phases of equivalent Al content is likely a key contributor to their ability to form protective Al_2O_3. A number of these ternary phases with transition/refractory metals exist and are potentially capable of protective scale formation.

These MAX phase observations suggest the possibility of a carburization pretreatment to improve the oxidation resistance of related intermetallic phases [2]. For example, one could envision surface treatment of a Ti_3Al alloy under carburization conditions to form a case layer of Ti_3AlC_2, which would then be capable of forming protective Al_2O_3 [2]. An interesting question is whether such ternary phases can be formed directly as continuous surface layers by gas reactions. For direct ternary interstitial analogs of intermetallic phases, such ternary phase formation may occur readily. However, phases such as Ti_3AlC_2 are not simply carbon interstitial analogs of Ti_3Al and would require atomic rearrangement during the gas reaction to form. The formation of anti-perovskite Cr_3PtN by nitridation of intermetallic Cr_3Pt was recently demonstrated and it was speculated that structural and bonding aspects of intermetallic compounds may favor formation of related ternary carbide and nitride phases on gas reaction [2, 38]. The formation of Cr_3PtN also depended greatly on the stoichiometry of the Cr_3Pt, such that a nearly single phase continuous external Cr_3PtN layer formed on Cr-25Pt at.%, but a complex internal lamellar structure formed on Cr-17Pt at.% (Fig. 1.7) [38]. Inoue *et al.* recently reported the plasma nitridation of TiAl to form Ti_2AlN, and by quantum chemical simulation found evidence that formation of the ternary nitride was influenced primarily by chemical interaction of the N with the Ti sub-lattice [39, 40]. The formation of Ti_2AlC was also reported by plasma carburization of TiAl [41], although oxidation resistance of the carburized surface was apparently not studied.

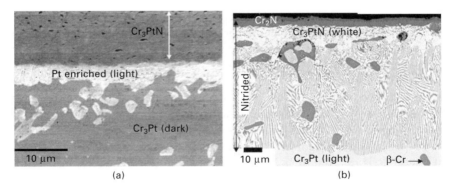

1.7 Scanning electron cross-section images of nitrided Cr_3Pt alloys showing effect of Cr:Pt stoichiometry [38]. (a) 75Cr-25Pt at.%, 24 h/ 950°C/N_2 (b) 83Cr-17Pt, 24 h/1000°C/N_2.

Figure 1.8 shows the cross-section of an initial attempt to carburize a Ti$_3$Al alloy to form a ternary carbide surface to improve oxidation resistance. An outer layer of TiC was formed, overlying an inner region consisting of Ti, Al, and C. X-ray diffraction data suggested that the inner layer may have contained Ti$_3$AlC$_2$ and/or Ti$_2$AlC. However, the data was not definitive and cross-section transmission electron microscopy (TEM) will be needed for positive phase identification. A simple air oxidation screening at 900°C showed no evidence of improved oxidation resistance via the carburization treatment. A close examination of the as-carburized structure indicated that fingers of TiC penetrated well into the inner layer, i.e. the inner Ti-Al-C containing layer was not continuous and unlikely to be able to form a protective Al$_2$O$_3$ layer. Further work will clearly be needed, however, it is believed that the potential at least exists to improve oxidation resistance by formation of surface layers of MAX and related phases via carburization pretreatments.

1.4   Functional surface formation by gas reactions

A possible extension of the aforementioned phenomena is the use of gas reactions as a synthesis approach to functional near-surface structures [2, 38, 42]. Two reaction schemes are of particular interest:

1. the use of multi-phase alloys as templates to form composite surfaces, and
2. exploiting the atomic level mixing and structure of intermetallics as precursors to form complex ceramic phases.

The potential technological utility of these phenomena as a synthesis route is yet to be determined. However, early results appear to show some promise.

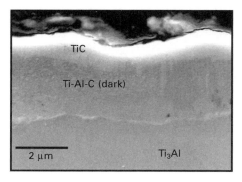

1.8 Scanning electron cross-section image of Ti-25Al at.% after carburization at 950°C for 24 h in ~0.1 atm. CH$_4$.

Figure 1.9 shows the surface of a vapor deposited Ag-Si thin film on a single crystal MgO substrate after oxidation. The Ag-Si system was selected as a model system because these elements are immiscible, Si readily forms an oxide in air while Ag is relatively noble, and Ag-SiO$_2$ composite structures are of general interest for functional applications such as nonlinear optical devices [43]. The film was adherent after the oxidation treatment, and went from mirror-like as deposited to tinted/transparent after oxidation. Preliminary energy dispersive spectroscopy (EDS) analysis in the scanning electron microscope (SEM) suggested the formation of regions containing 75–100 nm size range 'spheres' of Ag dispersed in SiO$_2$ (note that TEM analysis to confirm this structure has not yet been performed). The original intention was to form this type of structure by in-place internal oxidation. However, Ag has a tendency for surface migration and the structure may have resulted from surface diffusion of the Ag rather than an internal oxidation process.

Figure 1.10 shows surface micrographs of CoMo intermetallic alloys after carburization in an attempt to synthesize a bimetallic (ternary) carbide. Such phases are of increasing interest as catalysts for a variety of industrial processes, including hydro-treating to remove S, N, and O impurities from fossil fuels, ammonia synthesis, water gas shift reaction to produce hydrogen, and fuel cell catalysts [44–47]. They are typically formed by gas reactions of complex oxide precursors or by molecular precursor or chemical synthesis [48–50] routes. The use of intermetallic precursors offers the potential to leverage a different set of precursor stoichiometries and structures to form new complex

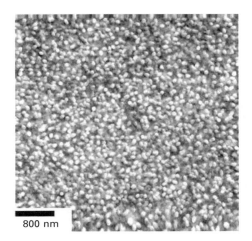

800 nm

1.9 Scanning electron surface image of a ~250 nm Si-Ag layer vapor deposited on single-crystal MgO after oxidation at 725°C for 10 min. in air. Preliminary analysis suggested that the bright phase is Ag-base and the dark phase is SiO$_2$. Note that regions of coarse, micron sized Ag-base 'spheres' were also observed.

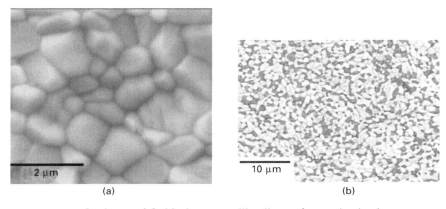

(a)   (b)

1.10 Surfaces of CoMo intermetallic alloys after carburization illustrating range of surface features possible [51]. (a) Secondary electron image of Co-45Mo consistent with single phase $Co_6Mo_6C_2$ formed on CoMo; (b) back scatter electron image of a two-phase Co(Mo) + CoMo alloy Co-25Mo showing $Co_6Mo_6C_2$-light and Co(Mo)-dark phase formation (preliminary identification).

carbide and nitride phases not attainable by currently used synthesis routes [2, 38, 42]. It also offers the opportunity to leverage the initial metallic phase equilbria of the precursor alloy to control the morphology and structure of the phases that are formed; for example, to synthesize a near-surface composite structure based on an initial two-phase alloy precursor microstructure [2, 38, 42]. Preliminary analysis of the surface of several carburized CoMo alloys in an attempt to form the $Co_6Mo_6C_2$ phase [44] suggests that both single-phase $Co_6Mo_6C_2$ and composite mixtures of $Co_6Mo_6C_2$ with Co(Mo) could be formed, depending on the initial precursor alloy composition (Fig. 1.10) [51]. Detailed microstructural characterization of the surfaces and evaluation of catalytic properties to test this approach are currently in progress.

1.5   Concluding remarks

This chapter combined speculation regarding new approaches to mitigating oxidation/corrosion with preliminary results in order to highlight oxidation phenomena of potential interest for future development. Many of these approaches appear to show promise for controlling surface chemistry. However, the key to their ultimate technological utility for applications as protective scales lies in the ability to form these phases as dense, continuous surface layers. This appears to be possible, particularly for the growth of protective nitride layers on Ni and Fe base alloys. Of scientific interest will be to study the similarities and differences that arise from the growth of surface nitride, carbide, etc., layers on multi-component/multi-phase alloys, compared to the better studied oxide phase formation. For functional applications, gas reactions of complex multi-component/multi-phase alloys to form complex near-surface structures appear to be of both scientific and technological interest. Such efforts are an extension of the surface modification strategies currently employed in semiconductor and related industries, as well as for catalysts, sensors, etc., generally for the formation of simple (binary) oxide phases by gas reactions of pure metals [52, 53]. The driving force is the potential gain possible through improved control of the resultant surface via the manipulation of the precursor alloy composition and microstructure (i.e. leveraging the metallurgical variables of the precursor alloy), as well as the possibility to form new phases based on heretofore untapped precursor metallic and intermetallic structures.

1.6   Acknowledgements

The authors gratefully acknowledge ongoing discussions with T. Besmann, B. Gleeson, J. A. Haynes, B. A. Pint, R. A. Rapp, J. L. Smialek, and I. G. Wright in the area of high temperature oxidation reactions, and T. Brummett, C. A. Carmichael, L. D. Chitwood, K. Cooley, G. W. Garner, M. Howell, P. Sachenko, and L. R. Walker for assistance with much of the experimental work presented in this chapter. The authors also thank J. H. Schneibel and M. Radovic for reviewing this manuscript. This research was sponsored by the US Department of Energy, Fossil Energy Advanced Research Materials (ARM) Program, the Assistant Secretary for Energy Efficiency and Renewable Energy for Distributed Energy Programs, and the Hydrogen, Fuel Cells, and Infrastructure program. Oak Ridge National Laboratory is managed by UT-Battelle, LLC for the US Department of Energy under contract DE-AC05-00OR22725.

1.7   References

1. M.P. Brady, B. Gleeson and I.G. Wright, *JOM* **2000**, 52 (1), 16.
2. M.P. Brady and P.F. Tortorelli, *Intermetallics* **2004**, 12 (7–9): 779.
3. J.L. Smialek and C.E. Lowell, *J. Electrochem. Soc.* **1974**, 121(6) 800.
4. B. Gleeson, H. Wang, S. Hayashi and D. Sordelet, *Mat. Sci. Forum* **2004**, 461–464, 213.
5. J. Nicholls, *MRS Bulletin* **2003**, 28 (9), 659.
6. J.E. Morral and M.S. Thompson, *Surf. Coat. Tech.* **1990**, 43–4(1–3), 371.
7. F. Erdogan, *Composites Eng.* **1995**, 5 (7), 753.
8. M. Niino and S. Maeda, *ISIJ Int.* **1990**, 30 (9), 699.
9. E.J. Opila, *J. Am. Ceram. Soc.* **2003**, 86 (8), 1238.
10. P.F. Tortorelli and K.L. More, *J. Am. Ceram. Soc.* **2003**, 86 (8), 1249.
11. E.J. Opila and D.L. Myers, *J. Am. Ceram. Soc.* **2004**, 87(9), 1701.
12. D.M. Lipkin and D.R. Clarke, *Oxid. Met.* **1996**, 45(3–4), 267.

13. M.P. Brady and B.L. Armstrong, H.T. Lin, M.J. Lance, K.L. More and L.R. Walker, *Scripta Mater.* **2005**, 52(5), 393.

14. K.L. More, P.F. Tortorelli, L.R. Walker, paper 2003-GT-38923 in *Proc. 48th ASME Turbo Expo,* ASME International, 2003.

15. M.P. Brady, W.J. Brindley, J.L. Smialek and I.E. Locci, *JOM* **1996**; 8(11), 50.

16. R.C. Makkus, A.H.H. Janssen, F.A., de Bruijn, R.K. Mallant, *Fuel Cells Bulletin* **2000**, 3,5.

17. D.P. Davies, P.L. Adcock, M. Turpin and S.J. Rowen, *J. Appl. Electrochem.* **2000**, 30, 101.

18. H. Wang, M. Sweikart and J.A. Turner, *J. Power Sources* **2003**, 115, 243.

19. C.L. Ma, S. Warthesen and D.A. Shores, *J. New Mater. Electrochem. Sys.* **2000**, 3, 221.

20. J. Scholta, B. Rohland and J. Garche, in *New Materials for Fuel Cell and Modern Battery Systems II*, O. Savadogo, P.R. Roberge (eds), Editions de l'Ecole Polytechnique de Montreal, Quebec, Canada **1997**, 330.

21. R. Borup and N.E. Vanderborgh, in *Mat Res. Soc. Symp. Proc. 393*, D.H. Doughty, B. Vyas, T. Takamura, J.R. Huff (eds), Material Research Society, Pittsburgh, PA, USA **1995**, 151.

22. W. Brandl and C. Gendig, *Thin Solid Films* **1996**, 290–291, 343.

23. M. P. Brady, K. Weisbrod, C. Zawodzinski, I. Paulauskas, R. A. Buchanan and L. R. Walker, *Electrochemical and Solid-State Letters* **2002**, 5, A245.

24. M.P. Brady, K. Weisbrod, I. Paulauskas, R.A. Buchanan, K.L. More, H. Wang, M. Wilson, F. Garzon, L.R. Walker, *Scripta Mater* **2004**, 50 (7), 1017.

25. H. Wang, M P. Brady, K.L. More, H.M. Meyer and J. A. Turner, *J. Power Sources* **2004**, 138 (1–2), 75.

26. M.P. Brady, H. Wang, I. Paulauskas, B. Yang, P. Sachenko, P.F. Tortorelli, J.A. Turner, R.A. Buchanan, *Proceedings of The 2nd International Conference on Fuel Cell Science, Engineering and Technology*, Rochester, NY (June 14–16, 2004).

27. R.P. Rubly and D.L. Douglass, *Oxid. Met.* **1991**, 35 (3–4), 259.

28. B. Zhao, J. Wu, J. Sun, B. Tu, F. Wang, *Intermetallics* **2001**, 9, 697.

29. T Narita, T. Izumi, M. Yatagai and T. Yoshioka, *Intermetallics* **2000**, 8(4), 371.

30. P. Perez, P. Advea, *Oxid Met* **2001**, 56 (3–4), 271.

31. S Kim, Y. Yoon, H. Kim, K. Park, *Mat Sci Tech* **1998**, 14 (5), 435.

32. G. Schumacher, F. Dettenwanger, M. Schutze, U. Hornauer, E. Richter, E. Wieser and W. Moller, *Intermetallics* **1999**, 7 (10), 1113.

33. M.W. Barsoum, M. Radovic, 'Mechanical Properties of Mn+1 AXn', in *Encyclopedia of Materials Science and Technology*, K.H.J. Buschow, R.W. Cahn, M.C. Flemings, E.J. Kramer, J. Mahajan and P. Veyssiere (eds), Elsevier Science, Amsterdam **2004**.

34. M.W. Barsoum, *Prog. Sol. State Chem.* **2000**, 28, 201.

35. X.H. Wang and Y.C. Zhou, *Corr. Sci.* **2003**, 45, 891.

36. X.H. Wang and Y.C. Zhou, *Oxid. Met.* **2003**, 59, 303.

37. M. Sunberg, G. Malmqyvist, A. Magnusson, T. El-Raghby, *Ceram. Internat.* **2004**, 30, 1899.

38. M.P. Brady, S.K. Wrobel, T.A. Lograsso, E.A. Payzant, D.T. Hoelzer, J.A. Horton, and L.R. Walker, *Chem. Mater.* **2004**, 16 (10), 1984.

39. M. Inoue, M. Nunogaki and T. Yamamoto, *Mater Manufact. Processes* **2002**, 17(4), 553.

40. M. Inoue, M. Nunogaki and K. Suganuma, *J Solid State Chem* **2001**, 157, 339.

41. T. Noda, M. Okabe, S. Isobe, *Mat Sci Eng A* **1996**, 213, 157.

42. M.P. Brady, D.T. Hoelzer, E.A. Payzant, P.F. Tortorelli, J.A. Horton, I.M. Anderson, L.R. Walker and S.K. Wrobel, *J. Mater. Res.* **2001**, 16(10), 2784.
43. T. Li, J. Moon, A.A. Morrone, J.J. Mecholosky, D.R. Talham and J.H. Adair, *Langmuir* **1999**, 15 (13), 4328.
44. S. Korlann, B. Diaz and M.E. Bussell, *Chem Mater* **2002**, 14, 4049.
45. V. Schwartz V, S.T. Oyama, J.G.G. Chen, *J. Phys. Chem. B.* **2000**, 104 (37), 8800.
46. J. Patt, D.J. Moon, C. Phillips and L. Thompson, *Catalysis Letters* **2000**, 65 (4), 193.
47. R. Kojima, K. Aika, *Applied Catalysis A–General* **2001**, 219 (1–2), 141.
48. S.H. Elder, L.H. Doerrer, F.J. DiSalvo, J.B. Parise, D. Giyomard and J.M. Tarascon, *Chem. Mater.* **1992**, 4, 928.
49. R. Marchand, Y. Laurent, J. Guyader, P. L'Haridon and P. Verdier, *J. Europ. Ceram. Soc.* **1991**, 8, 197.
50. K.S. Weil and P.N. Kumta, *Mat. Sci. Eng. B* **1996**, 38 (1–2), 109.
51. M. P. Brady, E. A. Payzant, P. F. Tortorelli, C. T. Liu, and L. R. Walker, *Proceedings of the 18th Annual Conference on Fossil Energy Materials*, June 2–4, 2004 Knoxville, TN
52. K. Thürmer, E. Williams and J. Reutt-Robey, *Science* **2002**, 297, 2033.
53. H. Over and A.P. Seitsonen, *Science* **2002**, 297, 2003.

2

Alloying with copper to reduce metal dusting rates

J ZHANG, D M I COLE and D J YOUNG,
University of New South Wales, Australia

2.1 Introduction

Metal dusting is a catastrophic corrosion process leading to the disintegration of iron-, nickel- and cobalt-based alloys in strongly carburising gas atmospheres (carbon activity $a_C > 1$) at moderately high temperatures (400–800°C). For iron-based alloys, the proposed mechanism [1–4] involves the super-saturation of iron with carbon and subsequent formation of cementite at the surface. This metastable cementite decomposes as the carbon activity in the cementite/graphite interface is lowered to unity when graphite deposits on the cementite surface. The decomposition of cementite produces fine metal particles which then strongly catalyse further carbon deposition. In some cases [5–7] the fine catalytic particles have been identified as cementite. A separate mechanism that does not involve carbide formation has been formulated for the metal dusting of nickel-based alloys [4, 8, 9]. As for iron-based alloys, the process begins with carbon saturation of the alloy. The subsequent graphite precipitation, however, is thought to result in the disintegration of the saturated metal matrix into small metal particles, which catalyse further carbon deposition.

Current methods of protection against metal dusting are either directed to the process conditions – temperature, gas composition and sulphur content – or the development of a dense adherent oxide layer on the surface of the alloy by selective oxidation [5, 10–12]. This is done by the addition of chromium, and to a lesser extent aluminium and silicon. However, carbon can still dissolve in the base metal via defects in the oxide scale, meaning that metal dusting is not stopped, but slowed and delayed.

Copper is thought to be noncatalytic to carbon deposition in all gas atmospheres, and owing to the extremely low solubility of carbon in copper, inert to the metal dusting reaction. Copper-based alloys have recently been reported in the patent literature [13, 14] to be resistant or immune to carburisation, metal dusting and coking. Thus, the addition of copper to nickel, which forms a near perfect solid solution, may be able to suppress or greatly retard the metal dusting of the alloy, without the need for a protective oxide scale on the surface.

Binary nickel-copper alloys have already been investigated as substitutes for nickel catalysts in order to reduce the amount of coking because of the noncatalytic nature of copper [15–18]. Nickel-copper alloys, however, have never been tested for their resistance to metal dusting. The aim of this work was to study the effects of adding copper on the behaviour of nickel metal dusting and associated coking.

2.2   Experimental methods

Test alloys were prepared from high purity constituent metals, 99.99% Ni and 99.96% Cu, by argon arc melting. Pure nickel and five alloys were prepared in this manner: Ni, Ni-2.5Cu, Ni-5Cu, Ni-10Cu, Ni-20Cu, Ni-50Cu (all in weight percentage). The alloys were then annealed for 24 h at 1000°C in flowing varigon gas (5% H_2 in Ar), and cut into a rectangular sample shape, $14 \times 7 \times 1.5$ mm. Pure copper was available in bar form, from which samples of dimension $15 \times 12 \times 1.5$ mm were cut. All samples were ground on SiC paper to 1200 grit and were ultrasonically cleaned before reaction.

Carburising reactions were carried out in a vertical tube furnace at 680°C. All seven materials were reacted at the same time. After purging the reactor with argon, the reaction gas, 68%CO-31%H_2-1%H_2O with a calculated carbon activity of 19.0 [19], was introduced. The H_2 and CO flow rates were set using mass flow controllers, and the H_2 plus CO mixture passed through saturated LiCl water solution at 40°C to obtain 1% water vapour [20]. The total gas flow rate was fixed at 500 mL/min to establish a linear gas flow rate of 0.3 m/min in the reaction zone.

After reaction, the samples were carefully removed and their weight changes were measured. The samples were subsequently examined using X-ray diffraction (XRD), scanning electron microscopy (SEM), optical metallography (OM), and transmission electron microscopy (TEM).

2.3   Results

Weight changes after carburisation are shown in Fig. 2.1. In general, increasing the alloy Cu content decreased the rate of carbon uptake. This decrease was very significant at Cu levels of less than 10%. At Cu levels above 20%, the weight gain was negligible, even after 150 h reaction. Surprisingly, a slight weight loss was found for all pure Cu samples after the carburisation reaction.

The appearance of the reacted samples is shown in Fig. 2.2. Clearly, the amount of surface carbon deposit decreased with increasing Cu levels, in agreement with the results shown in Fig. 2.1.

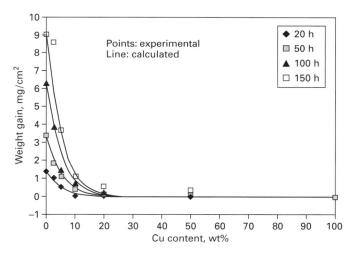

2.1 Weight gains of Ni-Cu alloys carburised in the gas mixture of 68%CO-31%H_2-1%H_2O at 680°C. Continuous lines calculated from Eq. 2.1 with $N = 20$.

Figure 2.3 shows metallographic cross-sections of Ni, Ni-10Cu and Ni-50Cu after 150 h carburisation. A thick coke layer formed at the surface of the Ni sample. The carbon–metal interface was irregular, and relatively large metal inclusions were occluded into the coke. On Ni-10Cu, a thin coke layer developed a serrated interface with the underlying alloy (Fig. 2.3b). In the case of Ni-50Cu, however, the sample surface remained smooth, without any visible coke layer formation (Fig. 2.3c).

The surface morphologies of the carbon deposits were further analysed by SEM. Figure 2.4 shows results for pure Ni after 20 h carburisation. A low magnification image (Fig. 2.4a) shows some bright carbon deposits on the grey surface. At high magnifications (Figs 2.4b, d) this bright carbon deposit is seen to be made up of graphite particle clusters, protruding from the surface. The size of the particles in this cluster was in the range of 0.1–0.4 µm. The grey surface between the cluster deposits was not bare (Fig. 2.4c); instead, carbon filaments developed on the surface. EDX analysis showed that this grey surface contained very high carbon, representing a general graphite deposition on the surface. A thin graphite layer at the surface of Ni in the metallographic cross-section (Fig. 2.5) further testified to this general graphite deposition.

Increasing the reaction time to 50 h led to most of the sample surface being covered by graphite particle clusters, as shown in Fig. 2.6a. Filamentous carbon was found in some carbon particle clusters (Fig. 2.6b). After a further increase in reaction time to 150 h, fine filaments became the main visible graphite on the surface. Filament diameters were about 20 nm. Graphite

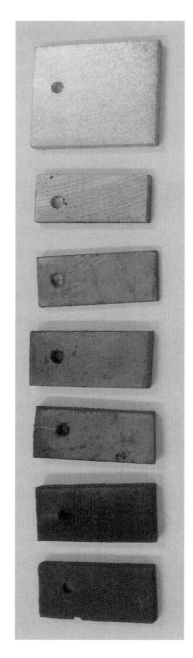

2.2 Ni-Cu samples after 50 h carburisation. From left to right: Ni, Ni-2.5Cu, Ni-5Cu, Ni-10Cu, Ni-20Cu, Ni-50Cu, and Cu.

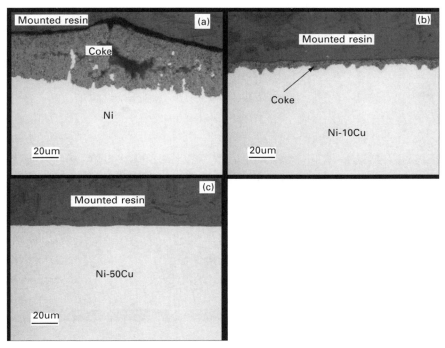

2.3 Metallographic cross-sections of (a) Ni, (b) Ni-10Cu and (c) Ni-50Cu after 150 h carburisation (samples without etching).

particle clusters could still be observed in some areas together with filaments, as seen in Fig. 2.6d.

The morphological development of carbon on Ni-2.5Cu was similar to that on the pure metal. After 20 h reaction, graphite particle clusters had already formed, together with some graphite filaments distributed on the surface (Fig. 2.7a). After 150 h reaction, more graphite particle clusters were found, and the filaments were longer and more numerous (Fig. 2.8a and 2.8b). The deposition pattern on Ni-5Cu was different. Even after 50 h reaction, graphite particle clusters had not developed; only carbon filaments were formed on the surface (Fig. 2.7b). After 150 h reaction, a small amount of graphite particle clusters could be found (Fig. 2.8c) but most of the surface was covered with a filamentary network (Fig. 2.8d). The fraction of the surface decorated with graphite particle clusters was very limited, and much less than on Ni or Ni-2.5Cu.

For Ni-10Cu, no graphite particle clusters were found on the surface even after 150 h reaction. After 20 h reaction, some short filaments and small, bright particles formed on the surface (Fig. 2.9a). Increasing reaction time to 50 h resulted in the formation of longer filaments (Fig. 2.9b), and after 150 h reaction, the surface was covered by long filaments. Ni-20Cu and Ni-50Cu

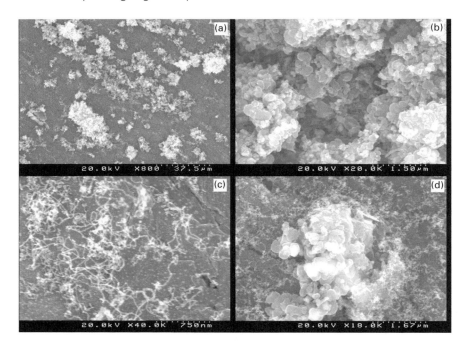

2.4 Surface of Ni after 20 h carburisation: (a) low magnification image showing the distribution of carbon deposits, (b) and (c) high magnification images showing bright carbon deposit area and grey surface region in (a), respectively, and (d) a typical protruding graphite particle cluster.

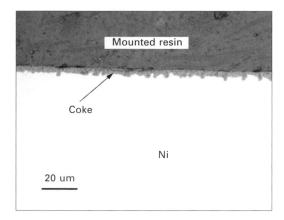

2.5 Metallographic cross-section of Ni after 20 h carburisation (samples without etching).

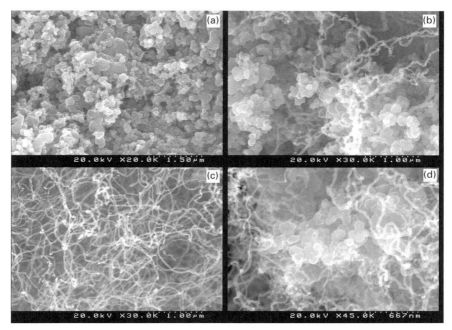

2.6 Surface observation of Ni samples after carburisation for (a–b) 50 h: (a) graphite particle cluster, (b) graphite particle cluster mixed with carbon filaments; and (c–d) 150 h: (c) carbon filament network, (d) graphite particle cluster together with filaments.

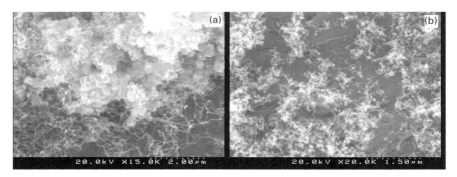

2.7 Surface observation of (a) Ni-2.5Cu sample after 20 h and (b) Ni-5Cu sample after 50 h carburisation.

did not develop surface carbon particle clusters either. Only filamentous carbon was detected, as shown in Fig. 2.10. The amount of filamentary coke was found to decrease with increasing copper concentration.

Figure 2.11a shows the surface of pure copper after 20 h carburisation. It was found that some parts of surface had experienced grain boundary grooving.

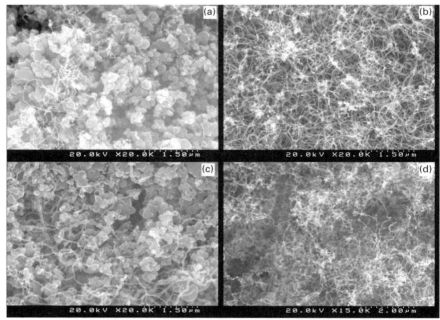

2.8 Typical coke morphologies (graphite particle clusters and filaments) observed on the surfaces of (a-b) Ni-2.5Cu and (c-d) Ni-5Cu after 150 h carburisation.

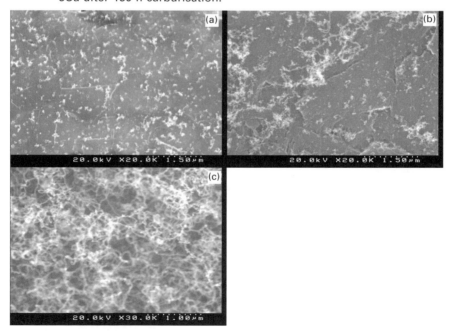

2.9 Surface observation of Ni-10Cu carburised for (a) 20 h, (b) 50 h and (c) 150 h.

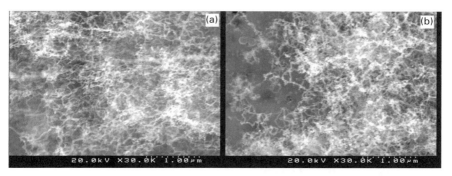

2.10 Surface observation of the samples (a) Ni-20Cu and (b) Ni-50Cu after 150 h carburisation.

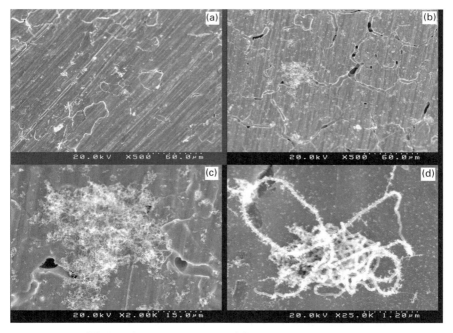

2.11 SEM surface observation of Cu samples carburised for (a) 20 h and (b–d) 50 h; (c) and (d) highlight the filamentous carbon-rich areas in (b).

The scratches from sample grinding can still be seen. After 50 h reaction, the sample showed heavily grooved grain boundaries (Fig. 2.11b), but also, at higher magnifications showed some fine filaments randomly deposited on the surface (Fig. 2.11c and 2.11d). After 150 h reaction, more filaments were found on the surface (Fig. 2.12a). The morphologies and distributions of these filaments can be seen in Figs 2.12b and 2.12c.

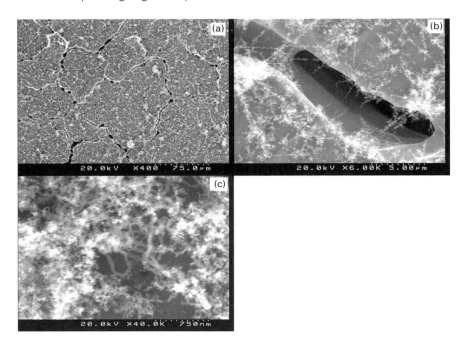

2.12 SEM surface observation of Cu samples carburised for 150 h,
(a) low magnification image showing the whole surface and (b–c)
high magnification images showing the distribution and morphology
of filamentous carbon.

TEM analysis of coke samples showed that the graphite particle clusters
formed on the surface of Ni and Ni-5Cu contained embedded metal particles,
as shown in Figs 2.13a and 2.13b, respectively. Graphite encapsulated the
metal particles to form a shell structure. The diameters of these metal particles
varied from 20–200 nm. Figure 2.13c shows the morphology of filaments
formed on the surface of Ni-50Cu. All filaments were bent, often twisted and
appeared to be solid. Metal particles were found to be at filament tips or
encapsulated along their lengths. The diameters of these particles were in the
range 20–50 nm. Figure 2.13d shows TEM observation of coke from a pure
copper sample after carburisation, revealing almost transparent carbon filament
tubes. Inside these nanotubes, there were no clear metal particles.

The alloys were analysed by XRD after 150 h carburisation. The results
are presented in Fig. 2.14. The austenite fcc structure was found in all cases.
Graphite was detected in all cases except pure Cu. Increasing Cu content
resulted in a shift of all fcc peaks to lower angles, representing a slight
increase of crystal lattice parameters due to Cu alloying.

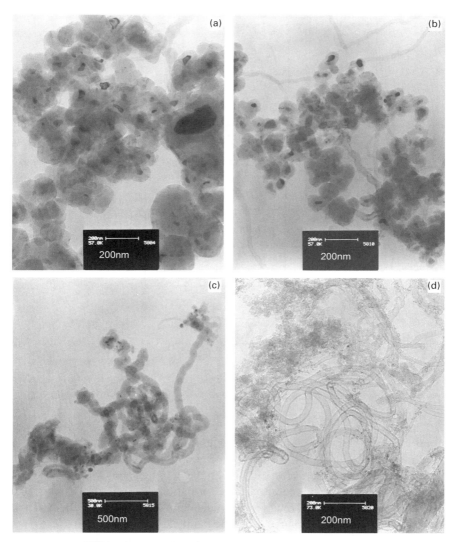

2.13 TEM micrographs of carbon deposits in the cokes of the samples of (a) Ni, (b) Ni-2.5%Cu, (c) Ni-50%Cu and (d) Cu carburised for 150 h.

2.4   Discussion

Metal dusting of Ni-based alloys starts from carbon absorption, dissolution and further supersaturation on the metal surface. Unlike Fe-based alloys, there is no formation of metastable carbide as an intermediate. The subsequent graphite deposition is generally accepted as the process which produces nanoscale particles for further carbon deposition – forming metal dusting.

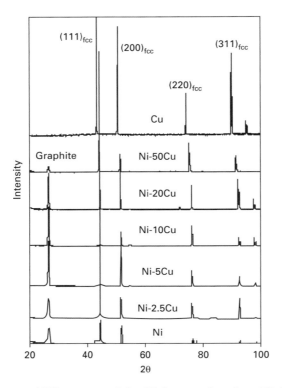

2.14 XRD spectra of the Ni-Cu samples after 150 h carburisation.

However, there is disagreement as to the way in which fine metal particles are formed. Pippel *et al.* [4] proposed that nickel atoms migrate between the graphite basal planes to the surface and agglomerate into small particles. This suggestion was also accepted by Grabke *et al.* [8] and Chun *et al.* [9] as a main microprocess of metal particle formation. Zeng and Natesan [21], however, thought an internal graphite deposition actually resulted in the disintegration of nickel matrix to form fine nickel particles.

The present investigation reveals two typical morphologies of carbon – graphite particle clusters and graphite filaments – on the surface of Ni and dilute Ni-Cu alloys. The metal particle size in graphite particle clusters is not as uniform as that in filaments, and is sometimes very large as shown in Fig. 2.4b and Fig. 2.13a. This observation suggests that metal particles in graphite particle clusters and those in filaments could be the products of different mechanisms. A large protrusion of graphite particle clusters on the surface (Fig. 2.4d) suggests an eruption of metal particles encapsulated by graphite shells from a localised severe carburising area. The formation of this type of deposit could be due to the release of localised stresses caused by graphite precipitation within the bulk metal. This kind of internal graphite precipitation has been reported to be found in the transition metals Ni, Co and Fe [21, 22]. The appearance of these graphite particle clusters supports the hypothesis of Zeng and Natesan [21] that internal graphite precipitation plays an important role in the dusting of nickel.

In contrast, the metal particles found in filaments are rather uniform (Figs 2.6c, 2.8b, 2.8d, 2.13c), and less likely to have come from mechanical disintegration due to internal graphite precipitation. Filaments started to form at a very early stage of reaction, and developed in parallel with the growth of graphite particle clusters. The competitive growth of filaments and graphite particle clusters resulted in different local occurrence of the clusters (Figs 2.4b, 2.6a), filaments (Figs 2.4c, 2.6c, 2.7b, 2.8b, 2.8d) and sometimes mixtures of both (Figs 2.6b, 2.6d, 2.8a, 2.8c) at different reaction stages. At an early stage, filaments were uniformly but only sparsely distributed, together with some graphite particle clusters scattered on the surface (Fig. 2.4). After 50 h reaction of pure nickel, graphite particle clusters had developed over the whole surface (Fig. 2.6a). After long reaction duration, e.g. 150 h, filaments covered most of the surface as in Fig. 2.6c, indicating an outside growth of filaments on top of the carbon particle clusters. This observation cannot easily be explained by the hypothesis that the metal atoms migrate between the graphite basal planes and agglomerate to form particles on the surface of the coke layer [4].

When copper levels were more than 10%, no graphite particle clusters, only filaments were observed. The number of filaments was limited, especially in samples with more than 20% Cu. This corresponded to the very low weight increases observed kinetically (Fig. 2.1). This observation indicates that when copper levels are above 10%, internal graphite precipitation does not cause disintegration of the metal to form graphite particle clusters containing metal fragments. The lack of graphite particle clusters in the coke on Ni-Cu alloys is attributed to the diluting effect of copper, which reduces the rate of carburisation. The mechanism of this effect is now examined.

Copper alloying could result in the change of carbon solubility in nickel, which may affect carburisation and metal dusting. According to the German Copper Institute [23], the solubility of carbon in nickel (max. 0.18%) is severely reduced as copper content increases – it is about 0.01% with copper content of 90%. However, Mclellan and Chraska's early 1970s work [24] showed that carbon solubility was unaffected by the presence of up to 40% copper. From these contrusting reports, it is not possible to draw a firm conclusion. But more likely, adding copper will decrease carbon solubility in nickel because of extremely low solubility of carbon in copper, and therefore reduce the rate of carburisation. The measurement of carbon solubilities in nickel-copper alloys were not done in this work.

The effect of copper alloying on the carburisation of nickel is significant, as shown in Fig. 2.1. The rate of carburisation decreases markedly with increasing copper concentration up to 10%. At copper levels of 20% or more, the rate of carburisation was very low, and not sensitive to alloy composition. Similar observations have been reported by others [15, 16, 18] in their research on Ni-Cu catalysts in carburising gas atmospheres. Based on this observation, it could be concluded that the effect of copper alloying on carburisation, carbon deposition and metal dusting is mainly achieved by reducing the catalytic effect of nickel. If the effect of copper is just to dilute the catalytic nickel sites, the rate of carbon deposition could be described by the following equation [15, 17, 18]:

$$r = r_{Ni} (1-x)^N \hspace{3cm} 2.1$$

where r_{Ni} is the reaction rate when the catalyst is pure nickel; x is the copper fraction in the metallic phase; and N is the number of unoccupied neighbouring nickel atoms needed for graphite nucleation. The group of N atoms forms an 'active ensemble'. The larger the value of N in the ensemble, the higher the sensitivity of the reaction to surface dilution by copper. The carbon deposition rate reduced near two orders of magnitude for copper concentrations higher than 20% in this work, suggesting a relatively large value of N.

Equation 2.1 was used to describe the present kinetic data, using different values of N. The resulting standard deviations are shown in Fig. 2.15, where it is seen that the best fit was obtained for N in the range 16–20. Predictions based on $N = 20$ are shown as continuous lines in Fig. 2.1, where agreement with measurement is seen to be good except at copper concentrations above 20%. Figure 2.16 shows the epitaxial relationship between a graphite (0002) monolayer and nickel (111) surface. A stable graphite nucleus requires several graphite hexagonal units to reach a critical size, which results in a large value of N as shown in Fig. 2.16. Also because the surface is not completely

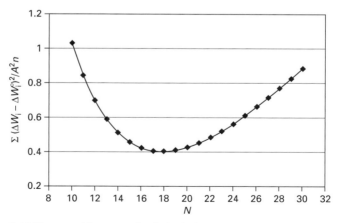

2.15 The resulting standard deviations obtained from Eq. 2.1 using different values of N.

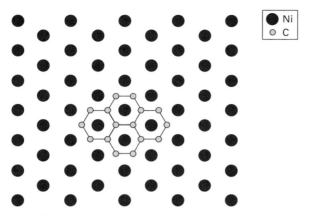

2.16 The epitaxial relationship between a graphite (0002) monolayer and a Ni (111) surface.

flat, nickel atoms in two or three layers near the surface will contribute to graphite nucleation, which in turn enlarge the N value.

The majority of the weight uptake measured on nickel and low copper content alloys corresponded to carbon in the form of graphite particle clusters. It is concluded that this process is successfully described using Eq. 2.1. However, this model underpredicts the amount of deposition occurring on high copper alloys. This is a consequence of the fact that the carbon deposits on the latter alloys are made up solely of filaments and these grow by a mechanism which differs from that of the clusters. It is therefore suggested that graphite particle precipitation is a co-operative phenomenon, involving a large number of nearest neighbour nickel atoms, whereas filament growth involves a small number of nickel atoms and is therefore less sensitive to copper dilution.

Tavares *et al.* [25] found that small copper additions, say 1 atm.%, caused the rate of carbon deposition to slightly increase. This enhancement of the catalytic activity of nickel was attributed to an increase in lattice defects which provided a closer match crystallographically between graphite and the alloy [26]. This phenomenon was not observed in this work, probably because the lowest copper content of 2.5 wt.%, was large enough to avoid this effect.

Copper is thought to be non-catalytic to carbon deposition in most environments. Based on this idea, two patents [13, 14] on copper-based alloys to resist metal dusting have recently been published. However, the present SEM observations showed the presence of filamentous carbon, suggesting that copper may not be completely non-catalytic to carbon deposition in aggressive metal dusting environments. TEM observation of the coke formed on copper, showed almost transparent tubular nanotube morphology, unlike the thick-walled tubes grown on the Ni-Cu alloys. Although metal particles were not clearly visible inside the nanotubes, the nanotubes grown on copper appear to be somewhat similar to those catalysed on Ni-based alloys. The difference was the very limited number of carbon nanotubes produced on copper. Similar observations were made by Bernardo *et al.* [15] during their experiments with Cu in CH_4-H_2 mixture. These authors found a carbon tube structure that was long, thin and contained elongated copper particles within the tubes. The observation of carbon filaments on pure copper suggests that metal particles for filament formation are formed directly on the metal surface, because no graphite layer was formed.

Although there was obvious filament formation on the surface, the carburising kinetics of pure copper are seen in Fig. 2.1 to indicate a slight weight loss. Of course, the amount of filamentous carbon was extremely small, even after 150 h reaction, and could not be detected by XRD (Fig. 2.14). The observed weight loss could be attributed to the removal of oxides or other impurities at high temperature in strong reducing gas atmospheres, as indicated by cavity formation at grain boundaries as shown in Fig. 2.12b.

2.5   Conclusions

The effects of copper alloying on the metal dusting of nickel were investigated by exposing alloys, along with pure metals, to 68%CO-31%H_2-1%H_2O gas mixture corresponding to $a_C = 19$ at 680°C. Two types of carbon deposit, graphite particle clusters and filaments, were observed on nickel and Ni-Cu alloys with low Cu contents ($\leq$ 5wt.%). The high copper concentration alloys ($\geq$ 10 wt.%) showed very little coking, but did develop a small deposit in the form of filaments. TEM analyses showed that graphite particle clusters consisted of metal particles encapsulated by graphite shells. In the carbon filaments, the metal particles were found at the tip or along the filament lengths. Although copper is generally thought not to be catalytic to carbon formation, carbon nanotubes were detected on the pure metal surface, along with grooved grain boundaries.

A kinetic investigation showed that the rate of metal dusting decreased significantly with increasing copper contents up to 10 wt.%. When the copper content was more than 20 wt.%, the rate of metal dusting was very low and was not sensitive to alloy composition. This very strong retarding effect of copper on metal dusting of nickel was attributed to a dilution effect by copper. A model based on reaction catalysed by large ensembles of near neighbour nickel atoms accounted for the effect of copper additions on the rate of graphite cluster growth. It did not account for the small amount of filamentary coke, which is thought to be produced by a different mechanism.

2.6   Acknowledgement

Support of this study by the Australian Research Council is gratefully acknowledged.

2.7   References

1. R. F. Hochman, *Proceedings of the Materials Engineering and Sciences Division Biennial Conference*, **1970**, 401.
2. J. C. Nava Paz and H. J. Grabke, *Oxid. Met.*, **1993**, *39*, 437.
3. H. J. Grabke, R. Krajak and J. C. Nava Paz, *Corrosion Sci.*, **1993**, *35*, 1141.
4. E. Pippel, J. Woltersdorf and R. Schneider, *Mater. Corros.*, **1998**, *49*, 309.
5. C. H. Toh, P. R. Munroe and D. J. Young, *Oxid. Met.*, **2002**, *58*, 1.
6. Z. Zeng, K. Natesan and V. A. Maroni, *Oxid. Met.*, **2002**, *58*, 147.
7. J. Zhang, A. Schneider and G. Inden, *Corros. Sci.*, **2003**, *45*, 1329.
8. H. J. Grabke, R. Krajak, E. M. Müller-Lorenz and S. Strauβ, *Mater. Corros.*, **1996**, *47*, 495.
9. C. M. Chun, J. D. Mumford and T. A. Ramanarayanan, *J. Electrochemical Soc.*, **2000**, *147*, 3680.
10. A. Schneider, G. Inden, H. J. Grabke, Q. Wie, E. Pippel and J. Woltersdorf, *Steel Res.*, **2000**, *71*, 179.
11. C. Rosado and M. Schütze, *Mater. Corros.*, **2003**, *54*, 831.
12. A. Schneider and J. Zhang, *Mater. Corros.*, **2003**, *54*, 778.
13. T. A. Ramanarayanan, C. M. Chun and J. D. Mumford, US Patent 20030029528A1.
14. P. Szakalos, M. Lundberg and J. Hernblom, US Patent 20040005239A1.
15. C. A. Bernardo, I. Alstrup and J. R. Rostrup-Nielsen, *J. Catal.*, **1985**, *96*, 517.
16. I. Alstrup, M. T. Tavares, C. A. Bernardo, O. Sørensen and J. R. Rostrup-Nielsen, *Mater. Corros.*, **1998**, *49*, 367.
17. J. A. Dalmon and G. A. Martin, *J. Catal.*, **1980**, *66*, 214.
18. M. T. Tavares, I. Alstrup and C. A. Bernard, *Mater. Corros.*, **1999**, *50*, 681.
19. J. Zhang and A. Schneider and G. Inden, *Corros. Sci.*, **2003**, *45*, 281.
20. N. A. Gokcen, *J. Am. Chem. Soc.*, **1951**, *73*, 3789.
21. Z. Zeng and K. Natesan, *Chem. Mater.*, **2003**, *15*, 872.
22. H. March and A. P. Warburton, *J. Appl. Chem.*, **1970**, *20*, 133.
23. DKI German Copper Institute Booklet: Copper Nickel Alloys: Properties, Processing, Application. http://www.copper.org/applications/cuni/txt_DKI.html.
24. R. B. Mclellan and P. Chraska, *Mater. Sci. Eng.*, **1970**, *6*, 176.
25. M. T. Tavares, I. Alstrup, C. A. Bernard and J. R. Rostrup-Nielsen, *J. Catal.*, **1996**, *158* 402.
26. Y. Nishiyama and Y. Tamai, *J. Catal.*, **1974**, *33*, 98.

3

Performance of candidate gas turbine abradeable seal materials in high temperature combustion atmospheres

N J S I M M S, Cranfield University, UK, J F N O R T O N, Consultant, UK and G M cC O L V I N, Siemens Industrial Turbomachinery Ltd, UK

3.1   Introduction

There is on-going world-wide interest in improving the efficiency of gas turbines, in order to minimise fuel consumption and environmental emissions, as well as improving their economic use. Approaches to improved efficiency include increased firing temperatures and reduced use of cooling air for some hot gas path components. The seals within the hot gas path of gas turbines are critical components for gas turbine operations [1, 2]. Many of these seals are at least partially shielded from the hot gas path environment and operate at relatively low metal temperatures. However, abradeable seals, which are used to reduce inter-stage leakage of the main combustion gas flow, are in direct contact with the hot environment as well as the rotating components within the hot gas path (e.g. blade tips).

The development of abradeable gas turbine seals for higher temperature duties has been the target of an EU-funded R&D project, ADSEALS, with the aim of moving towards seals that can withstand surface temperatures as high as ~1100°C for periods of at least 24,000 h [2]. The ADSEALS project has investigated the manufacturing of traditional honeycomb abradeable seal designs (Fig. 3.1) using a range of alternative materials and also potential novel alternative designs (Fig. 3.2) for abradeable seals [2]. The performance of candidate materials for higher temperature seals and of seals manufactured from selected materials combinations has been investigated by participants in the ADSEALS project in terms of oxidation, thermal cycling, abrasion and mechanical properties, as well as in larger test rigs and engine trials.

This chapter reports selected results of the performance of materials within honeycomb and seal structures from two of the series of oxidation and thermal cycling exposure tests carried out as part of the evaluation of seal materials within the ADSEALS project:

- A series of three screening tests carried out in laboratory furnaces, using simulated combustion gases and weekly thermal cycles. This involved

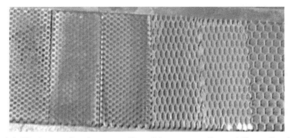

(a) Photograph of conventional honeycomb seal (top view)

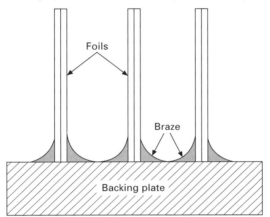

(b) Diagram of cross-section through conventional
honeycomb seal

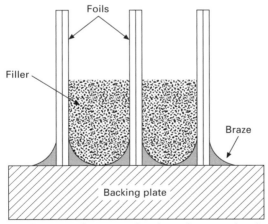

(c) Diagram of cross-section through conventional
honeycomb seal filler

3.1 Conventional honeycomb seals: (a) photograph of conventional
honeycomb seal (top view), (b) diagram of cross-section through
conventional honeycomb seal, (c) diagram of cross-section through
conventional honeycomb seal with filler.

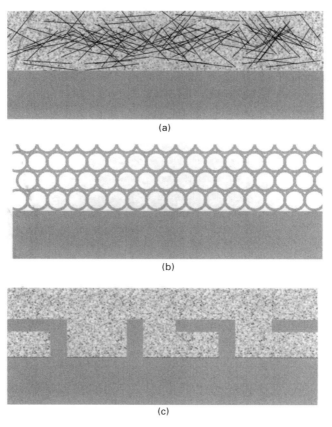

3.2 Novel seal configurations: (a) fibre mat seals, (b) hollow sphere seals, (c) thermal barrier coated metal grid seals.

testing a range of different fabricated honeycomb sections, with and without filler materials, brazed onto a thicker section backing plate material.

• A series of three tests in a natural gas fired 'ribbon furnace', with three-hourly thermal cycles. Samples were cooled on their rear faces to produce a temperature gradient through the seals. This test rig allowed the simultaneous exposure of up to 60 samples of candidate seal structures (including honeycombs, novel hollow sphere structures and porous ceramics).

For the first of these series of tests, materials performance was monitored by mass change measurements. However, the main method of determining materials performance in both tests was by destructive examination of samples discontinued after a predetermined number of cycles (or after an intermediate number of cycles, if this was considered more appropriate, e.g. due to sample failure). Metal damage measurements on these cross-sections enabled meaningful comparisons to be made of materials performance and allowed assessments to be made of the potential of the various alloys to resist breakaway oxidation and the performance of the joining techniques (e.g. brazing) used in the seal manufacture. Breakaway oxidation was a particular concern for the materials used for the thin honeycomb parts of many of the seals, as this type of rapid, often catastrophic materials degradation has been frequently reported for the candidate materials being considered for these parts of the seals [3–11]. This type of damage is triggered by a depletion of element(s) (for the materials in this study, aluminium) which have the potential to form stable protective oxides, and is usually the result of either a gradual loss during the formation of a thickening surface scale, or during repeated scale spallation and reformation during thermal cycling.

3.2   Experimental methods

3.2.1   Laboratory furnace screening tests

Materials and sample preparation

The materials and product forms selected for these screening tests (following on from earlier foil-only screening tests) were:

- Simple honeycomb structures manufactured from alloys PM2000 and Haynes 214; with foil wall thicknesses of 127 μm and cell sizes of 1.6 mm.
- Seals manufactured using either PM2000 or Haynes 214 honeycombs (with a 1.6 mm cell size and 127 μm wall thickness) brazed onto Haynes 214 backing plates, using a braze alloy to BS EN1044 (Ni-19Cr-10Si-0.1C-0.03B). During the assembly of the seals, care should have been taken to exclude any brazing material from an outer 1.5 mm thick zone of the honeycomb, i.e. from the face that would be in direct contact with the combustion gas during the high temperature exposure. Seals were tested with a variety of surface treatments/fillers:
  - untreated
  - pre-oxidised (1 h at 1050°C)
  - aluminised
  - MCrAlY powder (applied by HVOF, electroplating, powder filled/ sintered)
  - Ni-Al powder
  - Bentonite powder (pre-oxidised for 1 h at 1050°C).

The nominal compositions of the honeycomb and backing materials are shown in Table 3.1 and the combinations of these with the surface treatments and fillers are given Table 3.2.

Table 3.1 Nominal alloy compositions

Material	Cr	Fe	Ni	Al	Mo	W	Y	C	Mn	Si	Others
PM2000	20	Bal		5.5	<0.02		0.5 Y_2O_3		0.15	0.09	0.5 Ti
PM2Hf	20	Bal		5.5	<0.02		0.5 Y_2O_3		0.15	0.09	0.5 Ti
Aluchrom YHf	20.2	Bal	0.16	5.8	<0.01		0.05		0.21	0.28	0.05 Zr, 0.05 Hf, <0.002 B
Haynes 214	16	3	Bal	4.5			0.01	0.04			<0.1 Zr, <0.01 B
Haynes 230	22	<3	Bal	0.3	2	14		0.05	<0.5	<0.2	0.02 La, <5 Co, <0.015 B
IN738LC	16		Bal	3.4	1.7	2.6		0.10	0.5	0.4	3.4 Ti, 0.05 Zr, 8.5 Co, 0.01 B
MarM247	8.3		Bal	5.5	0.7	10		0.14			1.0 Ti, 1.5 Hf, 0.05 Zr, 0.010 B, 10 Co

Table 3.2 Laboratory screening tests – material combinations and treatments

Honeycomb material	Treatment/fill		Backing material
	Treatment	Free cell depth (nominal)	Comments	Backing material
---	---	---	---	---
Haynes 214	–	–		Haynes 214
Haynes 214	Pre-oxidised	–	Heat to 1050°C for 1 h	Haynes 214
Haynes 214	Westaim 2312	1.7 mm	Ni-Al thermal spray coating	Haynes 214
Haynes 214	Westaim 2501	1.7 mm	Bentonite powder fill and sinter	Haynes 214
Haynes 214	Aluminised coating	Not applicable	Nominal thickness 25 μm	Haynes 214
Haynes 214	MCrAlY HVOF	1.7 mm		Haynes 214
Haynes 214	MCrAlY electroplated	Not applicable	Nominal thickness 25 μm	Haynes 214
Haynes 214	MCrAlY powder fill	1.7 mm	Powder fill then sinter	Haynes 214
PM2000	Pre-oxidised	–	Heat to 1050°C for 1 h	Haynes 214

Test facilities

Three high temperature horizontal furnaces were used for this study. These were of a design utilising an externally heated central pure Al_2O_3 reaction tube in which the samples were placed. Figure 3.3 is a schematic illustration of such a furnace showing the major features. By careful control of the furnace and surrounding insulation, it was possible to establish a hot zone of better than $T \pm 10°C$ at the highest exposure temperature (even better at the lower temperatures) over the entire length of the specimen assembly, in line with TESTCORR recommended practice [12]. The combustion gas was passed over the samples on a 'once-through' basis, at a velocity of ~4 mm/s.

A ceramic T-section carrier was designed and manufactured for these honeycomb and seal samples. Minor ends of each sample were located in longitudinal grooves along the base-plate of the carrier and the samples gently held in position using restraining wires before insertion into the reaction tube. In this way, triplicate samples of each alloy and material combination could be exposed to the combustion gas during the test period at each temperature, enabling direct comparisons of material performance to be made and allowing their reproducibility to be assessed. However, for these samples weight changes measured did not include any spalled product (and hence only net weight change could be determined).

Test gas composition

The gas composition selected for these investigations simulates the natural gas combustion environment considered relevant to a commercial gas turbine application. Basically it is a N_2-based gas mixture containing nominally

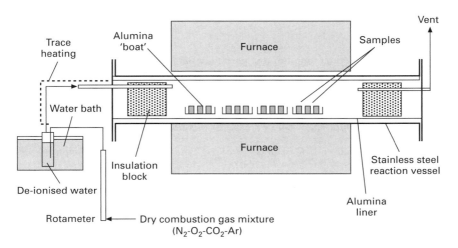

3.3 Horizontal furnace used for oxidation screening tests [10].

15%O_2, 3.4%CO_2 and 1%Ar with 5–7.8% water subsequently added by bubbling the gas through a humidification chamber held at constant temperature.

Pre-mixed dry gas was purchased from a commercial supplier, BOC Gases. The range of compositions from the bottles of gas used (taken from analysis certificates) and the furnace gas compositions (after humidification) are shown in Table 3.3.

Test programme

Three honeycomb and seal samples from each material combination were exposed at three temperatures (1050, 1100 and 1150°C) for up to 15 cycles of a weekly (168 h) duration, giving a maximum total exposure period of 2520 h. In each series of tests, samples were discontinued after, nominally, 5, 10 and 15 cycles, although this was reviewed during the experiments and, in some cases, varied, depending upon the oxidation behaviour of the material.

Procedure

Pre-exposure specimen preparation

Honeycomb and seal samples were not measured due to their complex and variable geometry and therefore the weight changes recorded on these samples remain uncorrected for area differences. Each honeycomb or seal specimen was given a two-stage degreasing treatment at room temperature in an ultrasonic bath, i.e., complete immersion in Volasil 344 followed by IPA (10–15 min. for each stage). All samples were subsequently allowed to dry naturally in laboratory air for 15 min. and then stored in a desiccator. Handling of these clean samples was done using tweezers or gloved hands.

Table 3.3 Compositional range of test gases (as-purchased and after humidification) [13]

	Nominal gas bottle composition (vol.%)	Actual gas bottle composition (vol.%)*	Nominal furnace gas composition (sitions vol.%)	Furnace gas composition (vol.%)
Nitrogen	Balance	Balance	Balance	Balance
Oxygen	15	14.74–15.07	13.8	13.6–14.3
Carbon dioxide	3.4	3.33–3.48	3.1	3.1–3.3
Argon	1	0.95–1.02	0.9	0.9–1.0
Water vapour	–	–	7.8	5–7.8

*Taken from analysis certificates provided by the supplier.

Specimen weighing

Specimens were weighed and re-weighed after exposure using an internally calibrated Sartorius microbalance. In addition, a reference weight (of approximately the same weight as the samples) was weighed in order to check balance reproducibility between intermittent weighing of the specimen batches. Specimens were weighed to an accuracy of ±0.00001 g on a clean aluminium foil to prevent any spalled products remaining on the balance pan after weighing.

Furnace operation and test procedure

The honeycomb and seal samples were assembled and secured on the T-section carrier, as described earlier. Prior to the introduction of samples, furnaces were heated to the appropriate test temperature and the whole system left to equilibrate. Before entering the reaction tube, the test gas was first bubbled through a humidification chamber controlled at 40°C. Heated supply lines connected the humidifier to the reaction tube inlet in order to avoid water condensation. The correct flow rate was established and the specimen assembly slid into the reaction chamber; normally, the samples took about 30 min. to reach the test temperature. In a similar manner, after the samples had been exposed for the required period, they were slowly removed from the furnace (again taking about 30 min.) and cooled to room temperature in laboratory air before desiccation and re-weighing.

3.2.2   Thermal cycling (ribbon furnace) tests

Materials

Materials for this series of tests were selected after considering the results of the screening tests described above, and also other data that became available during the course of the ADSEALS project, including the novel seal designs that had been successfully developed through the manufacturing processes to produce realistically sized parts. The nominal compositions of the materials used in this series of tests are given in Table 3.1 and the combinations of materials used in these tests are given in Table 3.4.

Variations on conventional designs of seal consisted of:

• Honeycombs (manufactured from HA214, Aluchrom YHf or PM2Hf) attached to a substrate using a low vacuum braze (Nicrobraz 210 (Co-19Cr-17Ni-8Si-4W-0.8B-0.4C) or Ni/Pd). A pre-oxidation treatment was given to the majority of these seals prior to their exposure to the combusted natural gas environment (as detailed in Table 3.4).
• A plasma spray overlay coating (LCO22, Co-32Ni-21Cr-8Al-0.5Y) on substrates.

Table 3.4 Thermal cycling tests – selected materials combinations and treatments

Honeycomb/grid	Braze	Substrate	Filler/coating	Pre-oxidation treatment
HA214	Nicrobraz 210	HA214	–	Y
PM2Hf	Nicrobraz 210	HA214	–	Y
Aluchrom YHf	Nicrobraz 210	HA214	–	Y
Aluchrom YHf	Pd/Ni	HA214	–	Y
Aluchrom YHf	Nicrobraz 210	HA230	–	Y
Aluchrom YHf (stretched)	Nicrobraz 210 (HV)	HA230	–	Y
Aluchrom YHf (thicker wall)	Nicrobraz 210 (HV)	HA230	–	Y
HA214	Co101 (HV)	HA214	–	Y
PM2Hf	Co101 (HV)	HA214	–	Y
Aluchrom YHf	Co101 (HV)	HA214	–	Y
–	–	HA230	LCO22 MCrAlY	N
HA214 (large cell)	Nicrobraz 210	HA214	–	N
HA214 (large cell)	Nicrobraz 210	HA214	–	Y
Aluchrom YHf fibres	Nicrobraz 210	HA230	–	Y
Aluchrom YHf hollow spheres	Nicrobraz 210	HA230	–	Y
Aluchrom	Nicrobraz 210	HA230	TBC + bond coat	N
IN738 cast grid		IN738LC	TBC + bond coat	N
MarM247 cast grid		IN738LC	TBC + bond coat	N

Novel designs of seals included:

- Samples manufactured from fibre mats [2, 14] and metallic spheres [2, 15] joined to backing plates.
- 25% porous thermal barrier coatings (TBCs, 8 wt.% yttria stabilised zirconia) sprayed onto cast and etched grids brazed onto substrates.

Test facilities

The horizontal ribbon furnace used allows up to a total of 60 samples to be exposed simultaneously (including instrumented samples). A schematic cross-sectional view and an illustration of part of the facility are shown in Fig. 3.4. Up to six samples were installed in each of the ten sample holders before insertion in the furnace. The backing plates of all samples were air-cooled and the performance of the test rig was continuously monitored during the test using two instrumented samples, each with three thermocouples attached.

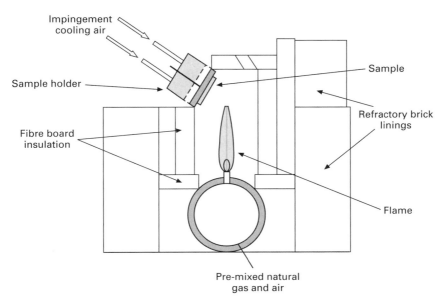

Impingement
cooling air

Sample holder

Fibre board
insulation

Sample

Refractory brick
linings

Flame

Pre-mixed natural
gas and air

3.4 Ribbon furnace used for thermal cycle oxidation tests: (a) cross-section of novel experimental set-up for ADSEALS tests,
(b) photograph of novel experimental set-up before start of an ADSEAL test.

Test programme

Three tests were carried out in the ribbon furnace with seal face temperatures of ~1000°C (test 1), ~1100°C (test 2) and ~1180°C (test 3). Cycles were of 3 h duration (gas firing) followed by 15 min. 'combustion air' cooling via the flame nozzles (i.e. the combustion air fan was kept running). It had been intended to expose the seals for nominally 100, 250 and 350 thermal cycles; however, severe oxidation of many of the samples resulted in shorter exposures. The rear faces of all samples were cooled by compressed air at all times.

Procedure

Pre-exposure specimen preparation

Honeycomb and seal samples were not measured due to their complex and variable geometry and therefore the weight changes recorded on these samples remain uncorrected for area differences. Each honeycomb or seal specimen was given a two-stage degreasing treatment at room temperature in an ultrasonic bath, i.e., complete immersion in Volasil 344 followed by IPA (10–15 min. for each stage). All samples were subsequently allowed to dry naturally in laboratory air for 15 min. and then stored in a desiccator. Handling of these clean samples was done using tweezers or gloved hands.

Furnace operation and test procedure

In the case of tests carried out in the ribbon furnace, the furnace was heated to a temperature of ~1400°C by the combustion of a natural gas/air mixture. For each series of tests, the seal face temperature was achieved by adjusting the air flow rate used to cool the rear face of the samples. During the test run, the operation of the rig was interrupted occasionally to enable visual inspection of the seal samples to be carried out. This was necessary since, from a visual perspective, the appearance of the seals ranged from total failure to simple surface discoloration.

3.2.3   Post exposure examinations

Honeycomb and seal assemblies exposed in the conventional furnaces were examined visually at each thermal cycle, in addition to being weighed. Specimens from each alloy/material combination were also discontinued after 5 and 10 cycles, as well as at the termination of the test, for more detailed surface examinations and cross-sectional studies. In some situations, however, where oxidation was proceeding rapidly, samples were discontinued after an intermediate numbers of cycles so that a particular event could be monitored more closely. Similarly, samples which had been exposed in the ribbon furnace facility were examined and weighed at predetermined cycles and samples selected for cross-sectional examinations and metal loss measurements.

Samples selected for cross-sectional examinations were vacuum-mounted in epoxy resin and polished to a 1 μm diamond finish for examination using a conventional optical microscope (and in some cases a scanning electron microscope, SEM, fitted with energy dispersive x-ray analysis, EDX). Sections were taken at approximately one third the distance from one end and after carefully aligning the specimen normal to the section thickness, polished back to approximately mid-section. Oxide thickness and metal loss measurements were carried out and the more important features recorded photographically.

3.3   Results and discussion

3.3.1   Laboratory furnace screening tests

Weight change data

Samples of two honeycombs (Haynes 214 and PM2000) were tested alongside the more complex seal configurations at all test temperatures [13]. These allowed the performances of the basic honeycomb structures to be compared without the complications of brazing, backing plates, surface treatments and/or fillers of the seals.

At the highest test temperature, 1150°C, the Haynes 214 honeycomb had gained less weight than the PM2000 after a total exposure of 2520 h (15 × 168 h cycles) (Fig. 3.5). The PM2000 honeycombs showed a large weight change between 840 and 1008 h; this is characteristic of the breakaway oxidation of this material (the subsequent weight loss of one of the PM2000 honeycombs was due to the almost fully oxidised sample partially breaking up). The Haynes 214 honeycomb samples showed no sign of breakaway oxidation. The multiple samples of each alloy produced similar weight change data and, in the case of PM2000, similar times to breakaway.

Differences in the weight change results were less marked at the lower test temperatures. At 1100°C, the 15 cycle PM2000 sample started gaining weight more rapidly after ~1848 h exposure (indicating the start of breakaway corrosion on this sample). At 1050°C, PM2000 samples steadily gained more weight than Haynes 214 samples, but neither material showed any sign of breakaway damage [13].

For the seal samples that were just honeycomb brazed to a backing plate:

- Haynes 214 honeycomb samples (with and without pre-oxidation) had low weight changes after exposure at all temperatures (e.g. Fig. 3.6).
- PM2000 honeycomb samples had to be removed from the two higher

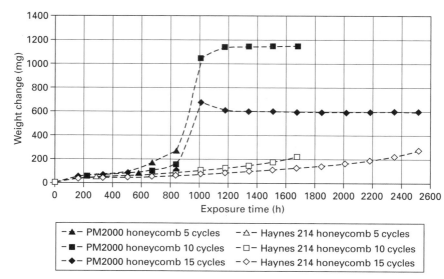

3.5 Weight change data from honeycomb exposure at 1150°C [13].

temperature tests after the honeycombs became detached from their backing plates (e.g. Fig. 3.6, from 1150°C exposure where samples had to be removed after two cycles). However, at the lowest temperature test there were only low weight gains.

For the filled and/or coated samples:

- Aluminised samples and those filled with HVOF MCrAlY showed low weight changes during all three tests (e.g. Fig. 3.6).
- Samples filled with Ni-Al spray coating, bentonite and MCrAlY powder all showed large weight changes, often with an initial jump in weight followed by progressive increase (e.g. Fig. 3.6), though the amount of progressive increase was decreased at the lower temperatures. In some cases there was a large drop in weight when large parts of the coating/ fill became detached from the surface (see below and Fig. 3.7).
- Samples with electroplated MCrAlY also showed significant weight changes (e.g. weight loss in Fig. 3.6), but these were more erratic than the other fills and may indicate coating/fill losses during the exposures.

Visual examinations

The honeycombs and seal configurations were given a visual examination after each five-cycle period. It was noted that some of the seals became bent during their exposures: Fig. 3.7(a)–(c) gives typical examples of these different

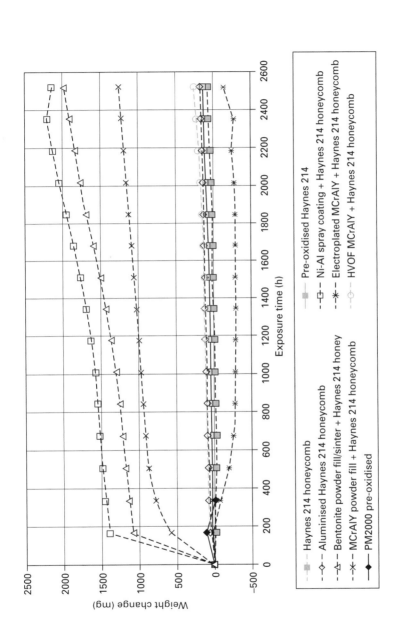

3.6 Weight change data from seals exposed at 1150°C.

Legend:
- Haynes 214 honeycomb
- Aluminised Haynes 214 honeycomb
- Bentonite powder fill/sinter + Haynes 214 honey
- MCrAlY powder fill + Haynes 214 honeycomb
- PM2000 pre-oxidised
- Pre-oxidised Haynes 214
- Ni-Al spray coating + Haynes 214 honeycomb
- Electroplated MCrAlY + Haynes 214 honeycomb
- HVOF MCrAlY + Haynes 214 honeycomb

Exposure time (h)

Weight change (mg)

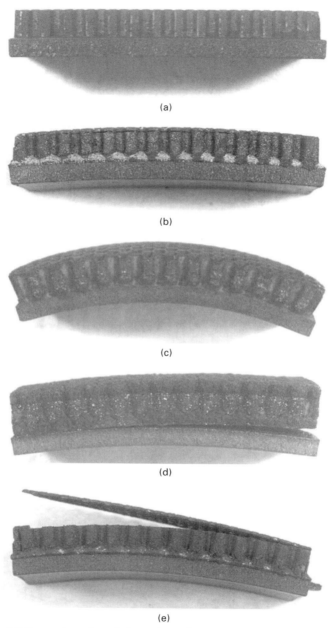

3.7 Examples of seal distortion after exposure in isothermal laboratory tests: (a) little or no distortion (1050°C, 5 cycles); (b) small amount of bending (1100°C, 5 cycles); (c) large amount of bending (1150°C, 15 cycles); (d) honeycomb separated from backing plate (1150°C, 15 cycles); (e) small amount of bending and 'filler' separating from honeycomb (1100°C, 15 cycles).

degrees of bending observed, from which it can be seen that bending always appeared concave with respect to the backing plate/seal configuration. Figure 3.7 also shows two of the types of seal failures that were observed: (d) honeycomb separation and (e) outer coating/filler layer separation from the top of the honeycomb. The latter was observed on samples where the filler/coating had built up on the top edges of the honeycombs, as well as partially filling the insides of the honeycombs.

At the lowest exposure temperature, little or no bending of the seal assemblies was noted. However, as the exposure temperature of the seal assemblies was raised, the incidence of bending increased for several of the seal combinations: in particular samples filled with Ni-Al spray coating, bentonite and MCrAlY powder. The pre-oxidised PM2000 seal configuration appeared highly susceptible to separation of the honeycomb from the substrate after only a few cycles at 1100 and 1150°C. The Haynes 214 seal assemblies in both untreated and pre-oxidised conditions and those with an aluminised coating or MCrAlY electroplated showed very little or no sign of bending at any of the test temperatures. However, the aluminised honeycombs showed at tendency to distort, with the originally hexagonal shaped cells becoming more rectangular as the single wall membrane parts of the cells bent.

Cross-sectional examinations

Cross-sectional examinations were carried out on selected samples from which the most important, industrially relevant information could be derived. Therefore, detailed microscopic studies were focused on seals which had been exposed for the full 15 cycles. In addition, the behaviour of several seals was evaluated after only a few cycles if failures were observed after shorter exposures.

Seals of PM2000 honeycombs/Haynes 214 substrates failed (by parting of the honeycombs from the substrates) after 11, 7 and 2 cycles at 1050, 1100 and 1150°C, respectively. Figure 3.8 shows the appearance of an unexposed seal and those from the two highest temperature exposures. There has clearly been severe oxidation/nitridation damage in the honeycomb/braze region that has resulted in the separation of the honeycombs from their backing plates. In addition, severe oxidation along the foil and at the foil tip had occurred for samples exposed at the highest temperature.

The effect of a pre-oxidation treatment on the performance of Haynes 214 honeycomb/Haynes 214 substrate seal combination is illustrated in Fig. 3.9. Comparing the performance in Fig. 3.9(a), no pre-oxidation, with Fig. 3.9(b), pre-oxidised, both after exposure for 15 cycles (i.e. 2520 h) at 1100°C shows the beneficial effect of this treatment. Significant internal precipitation both in the honeycomb foil section and in the substrate had taken place in the untreated seal in contrast to a lack of attack on the pre-oxidised seal assembly.

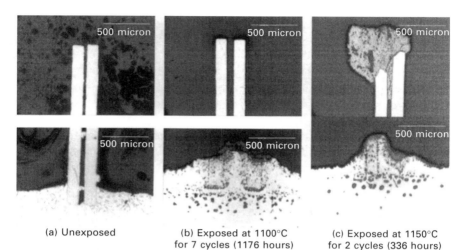

(a) Unexposed

(b) Exposed at 1100°C
for 7 cycles (1176 hours)

(c) Exposed at 1150°C
for 2 cycles (336 hours)

3.8 Effect of oxidation in laboratory screening tests on pre-oxidised
PM2000 honeycombs with Haynes 214 substrates.

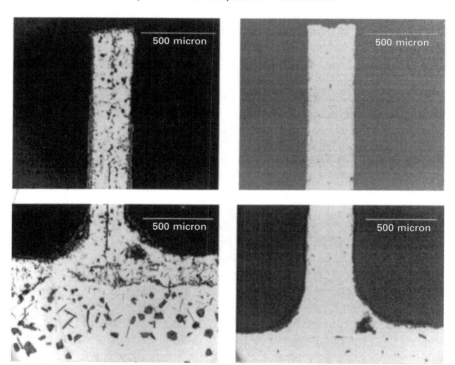

(a) Haynes 214 honeycomb/Haynes 214
substrate oxidised at 1100°C for 15 cycles
(2520 hours)

(b) Pre-oxidised Haynes 214 honeycomb/
Haynes 214 substrate oxidised at 1100°C
for 15 cycles (2520 hours)

3.9 Effect of pre-oxidation on Haynes 214 honeycombs exposed at
1100°C in laboratory oxidation screening tests.

The performance of both of these types of seals is also notably superior to the PM2000 honeycomb based seals shown in Fig. 3.8(b).

The behaviour of Ni-Al powder-fill in a Haynes 214 honeycomb is shown in Fig. 3.10. The filler has been completely oxidised after exposure at 1100°C for 15 cycles (i.e. 2520 h) and there is significant internal precipitation both in the honeycomb foil section and in the substrate.

The effect of electroplating the surfaces of the Haynes 214 honeycomb with MCrAlY is shown in Fig. 3.11. from which it can be seen that the oxidation attack is significantly enhanced as the exposure temperature is

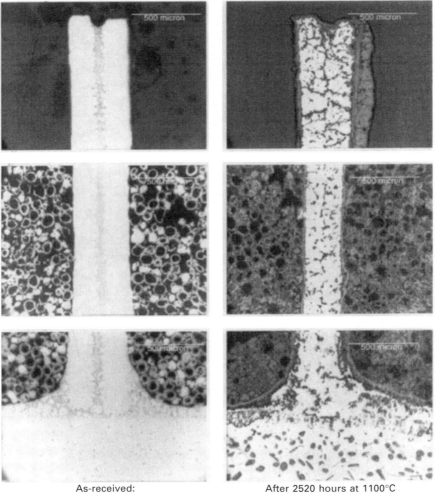

As-received: After 2520 hours at 1100°C
Haynes 214 honeycomb with NiAl fill

3.10 Performance of Ni-Al filler in honeycomb structure.

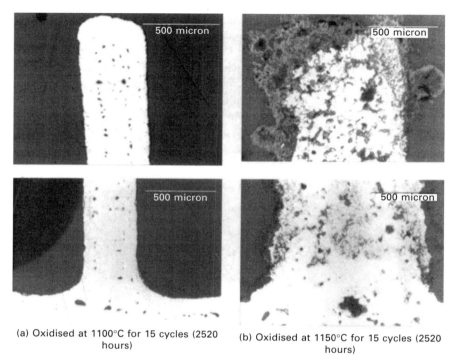

(a) Oxidised at 1100°C for 15 cycles (2520 hours)

(b) Oxidised at 1150°C for 15 cycles (2520 hours)

3.11 Performance of MCrAlY electroplate on Haynes 214 honeycomb and substrate.

raised from 1100 to 1150°C. Clearly, the application of an electroplated MCrAlY coating has not conferred lasting protection at the highest temperature with severe oxidation apparent, particularly at the edges and corners of the thinner honeycomb foil section.

3.3.2   Thermal cycling (ribbon furnace) tests

These tests were carried out in a much more realistic environment than the screening tests outlined above (i.e. using combusted natural gas and a heat flux through the samples, as well as regular thermal cycling). However, the effect of using increased realism in such testing can be accompanied by a lower degree of control over the exposure conditions. Figure 3.12 illustrates an extract from the automated logging carried out during the first of the thermal cycling tests: this shows that the gas temperature heats up to ~1400°C, while two sets of metal temperatures from instrumented samples vary by ±10°C between cycles after the initial heating period.

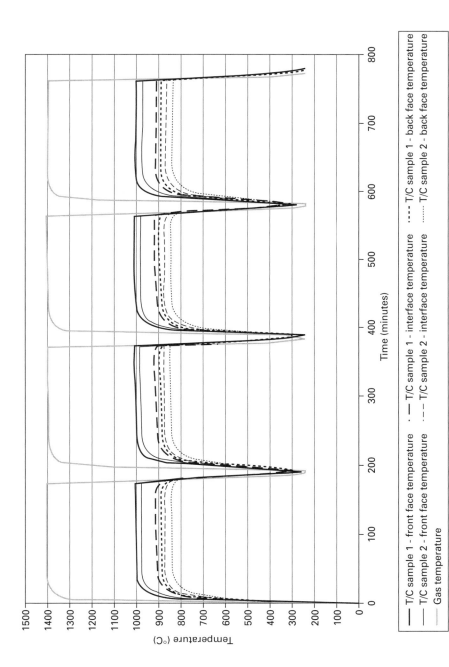

3.12 Example of temperature logging from first ribbon furnace test.

Visual observations

During the course of these tests, the main form of monitoring was by visual observation. This was carried out either (a) when samples were found to have failed, and parts were found to have fallen to the bottom of the test rig, or (b) when a target number of cycles was achieved.

The number of samples failing during exposure increased significantly when increasing the target seal surface temperatures from ~1000°C (test 1) to ~1100°C (test 2) and then ~1180°C (test 3). However, the order in which failures were observed remained unchanged:

- fibre mat and metallic spheres seals parted from their base plates
- seals with FeCrAlY based honeycombs (i.e. PM2Hf or Aluchrom YHf) failed just above the braze line (for all the different brazes tested)
- TBC coated/filled seals only failed in the highest temperature tests.

The Haynes 214 honeycomb seals and solid MCrAlY coated seals both survived until the end of the highest temperature test.

Figure 3.13 illustrates the appearance of groups of six samples after exposure in tests 1 and 2. Figure 3.13(a) shows the various different geometries of the conventional honeycomb design tested (small cells, large cells, stretched cells) after 102 cycles (i.e. 306 h) exposure in test 1. Figure 3.13(b) shows a selection of novel seals (fibre mat, TBC coated/filled and metal sphere) after 100 cycles (i.e. 300 h) exposure in test 2.

Cross-sectional microscopic examinations

For each seal type, cross-sections were prepared through samples that had completed their target exposure periods and been removed from the test, or from the longest exposure samples that had been removed due to failure.

Figure 3.14 illustrates the appearance of seals manufactured with honeycombs of Aluchrom YHf, PM2Hf and Haynes 214 after exposure in test 2 (with target seal surface temperatures of ~1100°C). The sample with the PM2Hf honeycomb failed after ~90 cycles, that with the Aluchrom YHf honeycomb after ~190 cycles, and the Haynes 214 was removed after ~300 cycles. Both the PM2Hf and Aluchrom honeycomb samples show the same type of catastrophic oxidation failure just above the honeycomb/braze joint line. In contrast, the sample with the Haynes 214 honeycomb shows only a little surface oxidation and some internal damage on the braze.

Figure 3.15 shows characteristic failures of two TBC coated/filled seal samples. In Fig. 3.15(a) the outer part of the TBC is separating from the rest of the sample by a crack propagating along the top of the metal grid; in addition there is cracking of the TBC parallel to the side of the metal grid and the braze between the metal grid and the substrate is also progressively

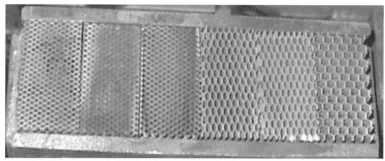

(a) Test 1/sample holder 4; appearance of samples after 102 cycles
(306 hours at target exposure temperature)

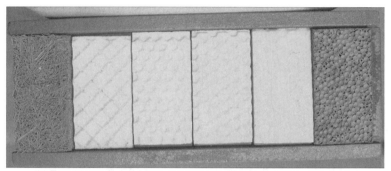

(b) Test 2/sample holder 9; appearance of samples after 100 cycles
(300 hours at target exposure temperature)

3.13 Appearance of sample holders after exposure in ribbon furnace.

oxidising and failing. In Fig. 3.15(b) the TBC has developed cracks parallel to the top of the metal grid and parallel to the grid support legs; in places such cracks led to loss of patches of the outer TBC layer, especially starting at the corner of the samples. It should be noted that failures of TBC coated samples were not observed in tests 1 and 2, only in the highest temperature test (test 3).

3.3.3   General discussion

The findings of the two test series can be summarised as follows:

- Honeycombs alone: at the higher test temperature, PM2000 went into breakaway oxidation, but Haynes 214 did not.
- Seals with conventional honeycomb designs:
  - FeCrAl-RE honeycomb seals:
    Failed close to braze line in highest temperature laboratory furnace test

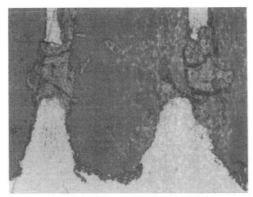

(a) Aluchrom YHf honeycomb after ~190 cycles
(570 hours) at seal face temperature of ~1100°C

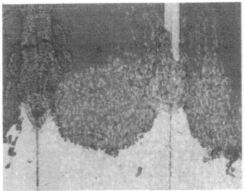

(b) PM2Hf honeycomb after ~90 cycles (270
hours) at seal face temperature of ~1100°C

(c) Haynes 214 honeycomb after ~300 cycles
(900 hours) at seal face temperature of ~1100°C

3.14 Performance of honeycomb seals after ribbon furnace tests.

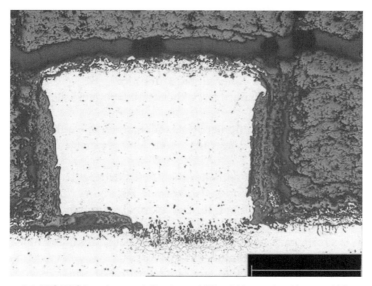

(a) APS TBC/bond coated Aluchrom YHf grid brazed to Haynes 230
substrate exposed for 238 cycles at seal face temperature of ~1180°C

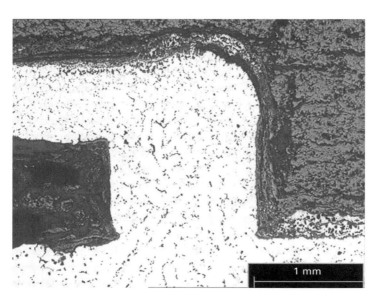

(b) APS TBC/bond coated 247 cast grid and substrate exposed for
175–225 cycles at seal face temperature of ~1180°C

3.15 Appearance of thermal barrier coated metal grid samples after
thermal cycling test.

Failed just above braze line in all three ribbon furnace tests (due to enhanced oxidation)

o   Haynes 214 honeycomb seals:

Pre-oxidised samples performed better than non-pre-oxidised samples (less internal oxidation)

Samples remained intact during all laboratory and ribbon furnace tests

Internal oxidation at all test temperatures – enhanced oxidation above braze line only after long exposures at highest test temperatures

Performed much better than FeCrAl-RE

o   Some samples bent in isothermal laboratory exposures, but all remained flat in ribbon furnace thermal cycling/heat flux exposures

o   Fillers and coatings did not offer any benefits to the seal's oxidation performance, and were generally associated with sample bending, outer layer separation or grid distortion.

• Solid sprayed seals: samples discoloured with exposure – no gross failures observed.

• Novel seal designs

o   Fibre mat and hollow sphere seals: failed at or near braze after short exposure times

o   TBC coated metal grid seals: performed well in ribbon furnace tests 1 and 2, but some failure in test 3, mainly with the outer TBC layers separating from the metal grids.

From the examinations carried out it is evident that the performance of the FeCrAlY materials (PM2000, PM2Hf and Aluchrom YHf) used as honeycombs in the seal structures is not satisfactory for application in high temperature seals in industrial gas turbines. The expected good performance of these honeycomb materials at high temperatures is related to the formation of slow growing α-Al_2O_3 on the alloy surface [4–11]. This is the oxide that is responsible for protecting the alloy substrate against severe oxidation attack at high temperatures; it is slow growing but requires a continuous supply of aluminium and oxygen to maintain its growth. Exposure of these comparatively thin honeycomb foils to oxidising gas at high temperature will, in time, result in the depletion of Al in the matrix to a sufficient extent that catastrophic breakaway oxidation will occur as the matrix level of Al becomes too low to facilitate the repair and/or reformation of the surface oxide. However, the times to sample failure were surprisingly short for these materials at the exposure temperatures used. In addition, samples had been pre-oxidised after manufacture as a means of establishing protective surface oxides prior to exposure to the gaseous environment. Therefore other factors must be important in determining the performance of these materials in the composite seal structures. As the main type of damage that is observed on both materials is typically located just above the braze line, it is believed that this attack is related to the brazing process and/or braze composition and the interdiffusion of elements from the braze and the foils (SEM/EDX mapping has shown extensive interdiffusion of the Ni-based braze material into the Fe-based foils). In the thermal cycling (ribbon furnace) tests, two brazes were included (Nicrobraz 210 and Pd/Ni) as well as two variations on the brazing process (standard vacuum and high vacuum brazing). None of these brazing variations appeared to make much difference to the type of damage observed on the honeycomb foils. In addition, the brazes also suffered significant oxidation (especially internal oxidation) under these conditions. Another type of damage to the honeycomb foils was the oxidation of the tips of the foils; in the ribbon furnace testing this could have been partly as a result of a lack of heat conduction along foils damaged closer to the substrate.

The performance of Haynes 214 honeycomb has been shown to benefit significantly from a pre-oxidation treatment. Similar conventional seal configurations to the FeCrAlY honeycombs, but using Haynes 214 showed that although there were numerous discrete internal oxide particles within the foils, in this case there was only occasional evidence of catastrophic through-section oxidation just above the braze line, even after the longest exposure times. Again, there was significant attack of the Nicrobraz 210, especially in the form of internal oxidation.

The apparently superior resistance of the Haynes 214 honeycomb foils may, at first sight, appear to be contrary to what would be expected, since this alloy has even less aluminium than the FeCrAlY materials tested, i.e. ~4.5% compared to 5.5–5.8%. Therefore, *per se*, the aluminium level should be depleted below a given critical level sooner in the case of the Haynes 214. However, PM2000, Aluchrom YHf and PM2Hf are iron-based alloys whereas Haynes 214 is a nickel-based material. As breakaway oxidation results largely from the very rapid formation of oxides which are more defective/less protective than Al_2O_3, the slower growth of Ni oxides (compared to Fe oxides) will favour the Ni-based alloy. Other factors that may also contribute to the observed difference in behaviour include:

- different strengths of materials (e.g. PM2000 and PM2Hf are significantly stronger than Haynes 214 at elevated temperatures)
- variations in diffusivities in the different matrix structures (ferrite in FeCrAlY materials versus austenite in Haynes 214)
- different fluxing reactions during the high-temperature brazing process
- alloy grain sizes.

These potential factors have not been evaluated or quantified in these tests. However, it is suspected that the role of the braze is critical with interdiffusion between the braze and the honeycomb materials reducing the effective aluminium content remaining in the honeycombs to form a protective oxide; this effect appears to be greater for the iron-based FeCrAlY alloys than the nickel based Haynes 214. The novel fibre mat and hollow sphere designs of seals (which used FeCrAlY materials for these parts) also suffered from a similar type of failure, where the sealant layer parted from the Ni-based substrate material close to the braze line. The development of a better joining method and/or braze material(s) is obviously a topic that needs to be pursued further.

The TBC coated/filled seals performed very well in the two lower temperature thermal cycling tests. However, in the highest temperature test, many failures were observed: in most cases the failures were of the form of the outer TBC layers parting from the rest of the samples along the line of the outer metal grid surfaces. This could be related to both greater stressing from thermal cycling (due to the higher exposure temperature) and the greater oxidation damage to the bond coats and grid structures (also a result of the higher exposure temperature). In addition, it was noted that for samples with grids brazed to the substrates, interdiffusion between the materials occurred along the joints resulting, in some cases, in their separation.

3.4   Conclusions

The performances of a selection of candidate gas turbine seal materials for elevated temperature operation have been studied in simulated and real natural gas combustion environments.

- Laboratory furnace tests have been carried out at 1050, 1100 and 1150°C in simulated combustion gas, using weekly thermal cycles and total exposure times of up to 2520 h. These tests investigated the performance of honeycombs and conventional seal designs using alternative materials with and without fillers.
- A natural gas-fired ribbon furnace has been used to carry out more realistic exposures of seal samples with a heat flux (from rear face cooling) giving seal face temperatures of ~1000, ~1100 and ~1180°C and three-hourly thermal cycles for up to 1050 h. These tests used the most promising candidate materials for conventional and novel seal designs.

The main findings of these exposures were:

- Conventional seal designs/novel materials selections:
  - o FeCrAl-RE (PM2000, Aluchrom YHf and PM2Hf) honeycombs tended to fail just above their braze lines.
  - o Haynes 214 honeycombs performed better than FeCrAl-RE honeycombs with longer times to failure (if observed); failures were also located just above the sample braze lines.

- o   Haynes 214 honeycombs benefited significantly from a pre-oxidation treatment.
- o   Fillers and coatings offered no benefits and instead tended to cause sample failures by distortion or bending.
- o   Solid sprayed seals showed surface oxidation but no failures.
- Novel seal designs:
  - o   Fibre mat and hollow sphere seals failed at/near braze joints.
  - o   TBC coated metal grid seals generally performed well but in the highest temperature tests many samples failed with the outer TBC layer detaching from the rest of the seal; for grids that were brazed to the substrates, there were also instances of braze failures at the highest exposure temperature.
- The results for the seals constructed with honeycombs, fibre mats and metallic spheres highlights the need for development of more appropriate braze material(s) for use of these materials at high temperatures.

This study has shown that further materials development work is required before honeycomb structures can be utilised as higher temperature sealing systems in gas turbines in order to meet the target operating temperatures of $\sim 1100°C$ for periods of at least 24,000 h.

3.5   Acknowledgements

Funding for this work was provided by Siemens Industrial Turbomachinery UK Ltd (formerly Demag Delaval Industrial Turbomachinery Ltd and formerly part of Alstom Power UK Ltd) and the European Union Framework 5 'Growth' Project 'Investigation in Advanced High Temperature Seals' (ADSEALS), project reference number G4RD-CT-2000-00185.

3.6   References

1. C Brown, E Vergheze, D Sporer and R Sellors, Paper 98-GT-565, 43rd ASME Gas Turbine and Aeroengine Tech. Congress, Stockholm, Sweden, ASME International, **1998**.
2. W Smarsly, N Zheng, E Vivo, M Tuffs, K Schreiber, B Defer, C Langlade-Bomba, O Anderson, H Goehler, N Simms and G McColvin, in *Proc. 6th International Charles Parsons Conference*, Maney, **2003**, 625.
3. F H Stott, G C Wood and J Stringer, *Oxidation of Metals*, 44 **1995**, 113.
4. J P Wilber, M J Bennett and J R Nicholls, *Materials at High Temperatures*, 17, **2000**, 125.
5. W J Quadakkers, D Clemens and M J Bennett, in *Microscopy of Oxidation 3*, Eds S Newcomb and J A Little, The Institute of Metals, London, **1997**, 195.
6. R Newton, M J Bennett, J P Wilber, J R Nicholls, D Naumenko, W J Quadakkers, H Al-Badairy, G Tatlock, G Strehl, G Borchardt, A Kolb-Telieps, B Jönsson, A Westerlund, V Guttmann, M Maier and P Beaven, in *Lifetime Modelling of High Temperature*

Corrosion Processes, Eds M. Schütze *et al.*, European Federation of Corrosion (EFC) Publication No. 34, Maney Publishing, London, **2001**, 15.

7. A Kolb-Telieps, U Miller, H Al-Badairy, G Tatlock, D Naumenko, W J Quadakkers, G Strehl, G Borchardt, R Newton, J R Nicholls, M Maier and D Baxter, in *Lifetime Modelling of High Temperature Corrosion Processes*, Eds M. Schütze *et al.*, EFC 34, Maney, **2001**, 123.

8. J R Nicholls, R Newton, M J Bennett, H E Evans, H Al-Badairy, G Tatlock, D Naumenko, W J Quadakkers, G Strehl and G Borchardt, in *Lifetime Modelling of High Temperature Corrosion Processes*, Eds M. Schütze *et al.*, EFC 34, Maney Publishing, London, **2001**, 83.

9. I G Wright, B A Pint, I M Hall and P F Tortorelli, in *Lifetime Modelling of High Temperature Corrosion Processes*, Eds M. Schütze *et al.*, EFC 34, Maney, **2001**, 339.

10. N J Simms, J F Norton, A Encinas-Oropesa, J E Oakey, D Wilshaw and G McColvin, in *Proc. John Stringer Symp. on High Temperature Corrosion*, Eds P Tortorelli *et al.*, ASM, **2003**.

11. J R Nicholls, N J Simms, R Newton and J F Norton, *Materials at High Temperatures*, 20, **2003**, 93.

12. Draft Code of Practice for 'Discontinuous Corrosion Testing in High Temperature Gaseous Atmospheres', EC Project SMT4-CT95-2001, ERA Technology, UK, **2001**.

13. N J Simms, J F Norton, A Encinas-Oropesa and G McColvin, *Materials Science Forum*, Vols 461–464, **2004**, 875–82.

14. O Anderson, *Proc. Conf. on Powder Metallurgy*, 3, **1998**, 33.

15. O Anderson, *Advanced Engineering Materials*, 2, **2000**, 192.

Two calculations concerning the critical aluminium content in Fe-20Cr-5Al alloys

G S T R E H L and G B O R C H A R D T, Institut für
Metallurgie, Germany; P B E A V E N,
GKSS-Forschungszentrum, Germany and
B L E S A G E, Université Paris XI, France

4.1   Introduction

Despite the existence of a number of lifetime prediction models for Fe-20Cr-5Al type alloys that take spallation and regrowth of the oxide into account [1–4], the most rapid and simple estimates of the time to breakaway oxidation (t_B) are still made on the basis of Eq. 4.1:

$$t_B = \frac{1}{k} \left(\frac{V}{A} (C_0 - C_B) \frac{\rho}{v} \right)^{1/n}. \qquad\qquad 4.1$$

This equation balances the available aluminium reservoir, described by the difference between the initial (C_0) and the critical (C_B) aluminium content as well as the volume (V) of the sample with the consumption of aluminium by oxidation. The consumption is represented by the growth constant (k), the corresponding exponent ($1/n$) and, of course, the total surface area (A) of the sample. The alloy density (ρ) and the stoichiometric factor (v) connect the two parts. In [5] this approach has been successfully applied to construct so-called oxidation diagrams from which the lifetime can easily be read, if the temperature and the thickness of the sample are known.

Equation 4.1 implies that two prerequisites are fulfilled: the critical aluminium content has to be known and the aluminium distribution throughout the whole sample has to be uniform, so that the critical aluminium content takes the same value at the centre and at the periphery of the sample.

Within the framework of the LEAFA[1] project, extensive cyclic testing of Fe-20Cr-5Al alloys of different thicknesses from several producers revealed critical aluminium contents ranging from 0 wt.% to 3.5 wt.% over the temperature range 1000–1350°C [6]. This range of values is partly due to the fact that the failure mechanisms for very thin foils and for thicker sheet

---

[1]Life Extension of Alumina Forming Alloys in HT corrosion environments, EC project no. BRPR-CT97-0562.

material are different. As demonstrated in [7] the two mechanisms are intrinsic chemical failure (InCF) for thin foils and mechanically induced chemical failure (MICF) for sheet materials. In thin foils the oxide does not spall and the aluminium is consumed down to a level where chromia formation starts [8–10]. This is the intrinsic chemical failure mechanism. The oxide layer on sheet material spalls upon cycling and thus the bare metal surface is exposed directly to the oxidising environment during reheating. Only if the aluminium content is still above a value, which is called CNOSH (no self healing) in [7] can a new alumina layer be grown to maintain protection and thus extend the life of the component. Especially when the samples are exposed to long cycles, the situation may occur that at the beginning of a new cycle the aluminium content is just above CNOSH, so that during the cycle the oxide layer is adherent to the sample and the aluminium content drops far below CNOSH. When, at the beginning of the next cycle, breakaway oxidation occurs, a critical aluminium level is determined, which can be substantially lower than CNOSH. Thus the scatter in the values for the critical aluminium content may also be caused by different oxidation cycle lengths. It is, of course, also influenced by the general spalling tendency of an alloy, which in turn is governed mainly by its mechanical strength.

Closely related to this discussion is the question of the actual aluminium content at the surface of the sample. Until now, most efforts to measure an aluminium profile in the cross-section of the sample have proved unsuccessful. Only in [11] was the exposure temperature chosen just high enough (1000°C) to allow for α-alumina formation and yet low enough (1080°C) to get aluminium diffusivities which led to the development of a measurable depletion profile. The depletion during the oxidation of a binary alloy near the surface has already been discussed by Wagner [12]. Subsequently, Whittle [13] gave an analytical solution for the depletion profiles in the cross-section of sheets. For the first time it was taken into consideration that the depletion profiles beneath both surfaces of the sheet meet in the middle and that after long exposure times the content of the oxide forming element also drops in the middle of the sheet. The solution is based on the dimensionless function $F_L(x, t)$, where the coordinate x starts at one of the oxide metal interfaces. The original thickness of the sheet is L, and D is the diffusion coefficient of the oxidising element in the alloy.

$$F_L(x, t) = \sum_{n=0}^{\infty} \mathrm{erfc}\left(\frac{nL + x}{2\sqrt{Dt}}\right) + \mathrm{erfc}\left(\frac{(n + 1)L - x}{2\sqrt{Dt}}\right). \qquad 4.2$$

Assuming a parabolic growth law of the form

$$\frac{\Delta m}{A} = (k_p t)^{\frac{1}{2}}, \qquad 4.3$$

the mass gain ($\Delta m/A$) can be transformed into the amount of the oxidised element (n_{Me}^{oxid}) with the help of the molar mass of oxygen (M_O) and the stoichiometric factor (v):

$$\frac{\Delta m}{M_O} v = n_{Me}^{oxid}. \qquad 4.4$$

If the sheet is treated as infinite in two directions, the concentration profile of the element depends on only one coordinate and can be written in the form:

$$c_{Me}(x, t) = c_{Me}^0 - \frac{v}{2 M_O} \left(\frac{\pi k_p}{D} \right)^{\frac{1}{2}} F_L(x, t), \qquad 4.5$$

where c_{Me}^0 is the initial concentration. Such profiles have been found for example in [14] after long-term oxidation experiments on austenitic chromia forming steels.

4.2   The critical aluminium content

The change from regrowth of the alumina scale on Fe-20Cr-5Al type alloys to the onset of breakaway oxidation can also be interpreted as the transition from external to internal oxidation. At the beginning of breakaway oxidation the aluminium content is below CNOSH and a continuous oxide layer can no longer be formed. Thus it can be assumed that internal oxidation of at first aluminium and shortly after that also of chromium starts. The metal surface is denuded in aluminium and chromium and the internal oxides also obstruct the further aluminium supply to the surface, where now the iron starts to react with oxygen. With the formation of fast growing iron oxides the breakaway process is initiated. The following calculations are based on the concepts of internal oxidation as described in [15–18].

During internal oxidation of an alloy, diffusing oxygen meets the component with the most negative enthalpy of formation for the metal oxide concerned. When calculating the fluxes it is assumed that the mole fraction of oxygen at the surface is equal to its solubility in the particular alloy system and that the diffusion profile of the oxide forming element is that of a semi-infinite body with the initial mole fraction of the element at infinity. If the permeability, which is the product of the mole fraction and the diffusion coefficient of the oxide forming element is low compared to that of oxygen, the expansion of the internal oxidation zone is controlled by the diffusion of oxygen into the alloy. If the permeability of the oxide forming element cannot be neglected, the penetration of the internal oxidation zone will be slowed down and enrichment (α) of the oxide forming element in the form of oxide occurs at the reaction front. This enrichment is described in [16] by:

$$\alpha = \frac{2v}{\pi} \left[\frac{X_{Me} D_{Me}}{X_O D_O} \right], \qquad 4.6$$

where v is again the stoichiometric factor. In the case of Al_2O_3 its value is 3/2. At sufficiently high permeability of the oxide forming element, the volume fraction of the oxide is so large that single oxide precipitates no longer form, but rather a closed oxide layer. To calculate this, the enrichment is multiplied by the ratio of the molar volumes of oxide and alloy. In doing so the function h is defined:

$$h = \alpha \frac{V_{MeO_v}^m}{V_{alloy}^m}. \qquad 4.7$$

At a critical value (h^*) of h, which is usually in the range from 0.2 to 0.3, the transition from internal to external oxidation occurs.

In [18] it was shown that the oxidation of ternary alloys is also influenced by the element with the second strongest affinity to oxygen. In the case of the Fe-20Cr-5Al system this is chromium. This second component also helps in capturing oxygen at the surface, stops its inward diffusion and thus the formation of internal oxides. Below this very first oxide skin the oxygen partial pressure drops until the conditions for the selective oxidation of the most reactive element (here aluminium) are given. The whole point is nothing more than calculating the transition between internal and external oxidation, if two elements can take part in the internal oxidation. For this purpose the following approach has been taken:

$$h = \frac{2v}{\pi} \left[\frac{X_{Al} D_{Al} V_{Al_2O_3}^m + X_{Cr} D_{Cr} V_{Cr_2O_3}^m}{X_O D_O V_{Fe\,20\,Cr5Al}^m} \right]. \qquad 4.8$$

The mole fractions of aluminium and chromium, as well as the molar volumes, can easily be determined. The diffusion coefficients have been chosen from the literature as given in Table 4.1. Only the solubility of oxygen in iron has not yet been measured without uncertainty [21], because even small levels of impurities lead to adulteration of the results by oxide formation. Thus in this work another method has been chosen. The critical value h^* has been set to 1/3. From [22] the lowest aluminium contents for the formation of a closed alumina layer on Fe-Al binary alloys have been chosen and the oxygen solubility is then calculated. The result is plotted in Fig. 4.1. The three data points are connected by the best fit straight line $X_O = 778e^{-\frac{155kJ/mol}{RT}}$ Compared

Table 4.1 Data for the determination of the diffusion coefficients, using $D = D_0 e^{-\Delta H_a/RT}$

	D_0 in $\frac{m^2}{s}$	ΔH_a in $\frac{kJ}{mol}$	Reference
D_{Al}	1.8 E-4	228.20	[19]
D_{Cr}	8.5 E-4	250.80	[20]
D_O	0.1 E-4	111.12	[20]

with [23] ($X_O = 1.75 \times 10^{-4}$, 1528°C, δ-Fe) and [21], where a value of $X_O = 2.4 \times 10^{-5}$ for α-Fe without a specification of the temperature is mentioned, the oxygen solubilities in Fig. 4.1 seem to be rather high. Nevertheless, it is not obvious for which type of local alloy composition the oxygen solubility is needed here. Pure iron can be excluded, because at these temperatures a phase transformation to δ-Fe would occur, although pure iron would be closest to the idea of internal oxidation of aluminium and chromium. Other factors such as the detailed diffusion profile or the correction of the diffusion coefficients in iron to those in Fe-20Cr-5Al may well influence these figures. The results presented at the end of the section demonstrate, however, that the extrapolation method chosen here yields satisfactory results.

For reasons of economy, efforts have been undertaken again and again to determine the minimum necessary chromium content to fulfill the above-mentioned condition. The same is true for aluminium, because with increasing aluminium content the alloy becomes more and more brittle and above 6 wt.% hot rolling is no longer possible [24]. In Table 4.2 the results of different investigations that have addressed this question are summarised. It should be noted that in [25] the oxygen partial pressure was 1 bar and thus five times higher than in air.

Using the data now available and Eq. 4.8, the necessary aluminium and chromium contents for the formation of a closed alumina layer on iron based alloys can be calculated. The results are presented in Fig. 4.2 together with the literature values. The good agreement between the calculations and the literature at 1000°C allows for extrapolation to higher temperatures.

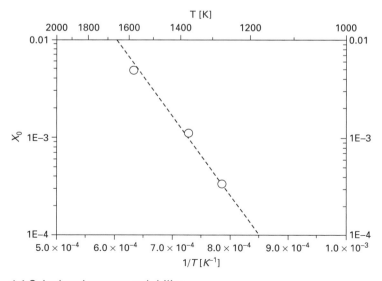

4.1 Calculated oxygen solubility.

Table 4.2 Minimum necessary chromium and aluminium contents in iron-based alloys for the formation of an α-alumina scale

Cr wt.%	Al wt.%	T°C	Reference
0.0	6.0	900–1000	[22]
0.0	9.0	1100	[22]
0.0	11.0	1300	[22]
6.1–12.9	2.0	1000	[26]
10.0	4.0–6.0	950–1050	[25]
16.5	0.5–1.0	1000	[26]

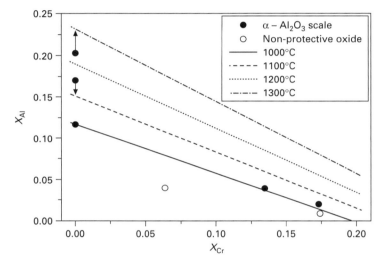

4.2 Minimum necessary chromium and aluminium contents in iron-based alloys for the formation of α-alumina scales. See Table 4.2 for comparison values in the literature.

Values for the critical aluminium content for Fe-20Cr-5Al type alloys can now be read from the calculated curves (Fig. 4.3). At 1200°C the difference between the calculated value, 1.7 wt.%, and the value measured in [5], 1.3 wt.%, is smaller than the differences to the values calculated for 1100°C and 1300°C and thus of an acceptable order of magnitude. Clearly visible is the temperature dependency of the critical aluminium content. From 1000°C to 1350°C C_B covers exactly the range from 0 wt.% to 3.5 wt.% as discussed in [6].

4.3   Three-dimensional Al depletion profile

Within the framework of the LEAFA project one of the authors (Beaven) was intrigued as to why some of the rectangular specimens went into breakaway

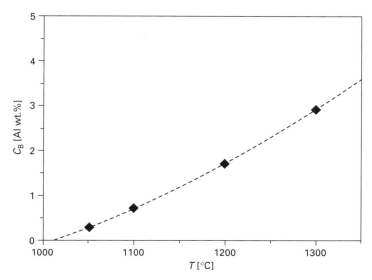

4.3 Critical aluminium content for Fe-20Cr-5Al alloys for the formation of a new alumina layer after scale spallation.

oxidation at the corners well before the replicate samples. Therefore the sample was cut (PM 2000, 20 mm × 10 mm × 0.5 mm) with an initial aluminium content[2] of 40 wt.% in the diagonal direction. The main alloy component profiles were then measured with an electron beam microprobe (Fig. 4.4).

The high consumption of aluminium and chromium in the corner already destroyed by breakaway oxidation is obvious. (The sample concerned was slightly wedge-shaped and this led to premature breakaway oxidation at the right-hand corner.) The influence of this effect even spreads to the centre of the sample. It has been mentioned in, amongst others [27] and [10], that geometries with a high surface to volume ratio are particularly prone to breakaway oxidation, because the necessary aluminium to form a protective scale cannot diffuse as fast as necessary into these regions. In the following, the depletion profile of aluminium for rectangular specimens will be calculated and the potential of the corners to initiate breakaway oxidation will be shown. The border lengths of the sample are l and b. The thickness is d. The origin of the coordinate system is placed in one of the corners of the sample (Fig. 4.5). r is the coordinate on the diagonal of the sample with $z = d/2$. Its

---

[2]The material had been delivered already in the pre-oxidised state. To achieve equal surface conditions for all materials, the initial oxide layer had to be ground off. This reduced the aluminium content to 4.0 wt.%.

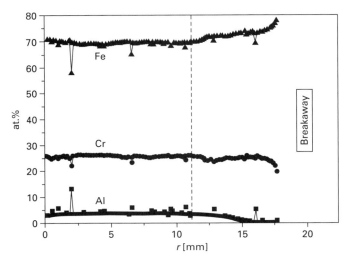

4.4 Profiles of the main alloy constituents of a PM 2000 sample after 920 h of cyclic oxidation at 1200°C in air. The profiles were taken along the diagonal of the sample. The right-hand corner already underwent breakaway oxidation.

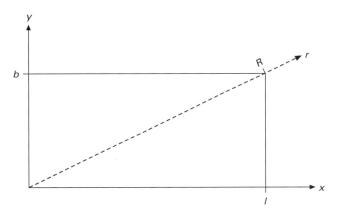

4.5 Coordinate system for the diagonal aluminium profile.

maximum value is $R (= \sqrt{l^2 + b^2})$. From the diagonal coordinate the transformation back to cartesian coordinates is easy.

$$x = r \frac{l}{R},$$ 4.9

$$y = r \frac{b}{R}.$$ 4.10

The diffusion of aluminium in the metallic part of the sample is governed by Fick's second law:

$$\frac{\partial}{\partial t} c(x, y, z, t) = D \, \Delta c(x, y, z, t).$$ 4.11

The initial distribution of the aluminium concentration (c) is assumed to be homogeneous:

$$c(x, y, z, 0) = c_{Al}^0 \text{ for } x \in [0, l], \, y \in [0, b], \, z \in [0, d].$$ 4.12

The initial concentration of aluminium (c_{Al}^0) can easily be calculated with the knowledge of the initial aluminium content (here: $w_{Al} = 4\%$), the molar mass of aluminium (M_{Al}) and the density of PM 2000 ($\rho_{PM2000} = 7.18$ g/cm^3):

$$c_{Al}^0 = \frac{w_{Al} \, \rho_{PM\,2000}}{M_{Al}} = 0.0106 \text{ mol/cm}^3.$$ 4.13

The boundary conditions for all six surfaces can be formulated in a similar way. For $z = 0$ with $x \in [0, l]$ and $y \in [0, b]$, the flux of aluminium through this surface has to be equal to its consumption by oxidation. With the help of Fick's first law as well as Eqs 4.3 and 4.4 it follows that:

$$\frac{\partial}{\partial t} n_{Al}^{oxid} = -\frac{\partial}{\partial t} n_{Al}^{alloy}$$

$$\frac{\partial}{\partial t} \frac{v}{M_O} lb(k_P t)^{\frac{1}{2}} = -j_z \cdot lb$$ 4.14

$$\frac{v}{M_O} \frac{lb}{2} \sqrt{\frac{k_P}{t}} = D \frac{\partial}{\partial z} c(x, y, z, t)\,|_{z=0} \, lb$$

$$\frac{v}{2M_O} \sqrt{\frac{k_P}{t}} = D \frac{\partial}{\partial z} c(x, y, z, t)\,|_{z=0}.$$ 4.15

Because the coordinate system is cartesian, the concentration is separated into a constant and three additive parts:

$$c(x, y, z, t) = c_{Al}^0 + c_x(x, t) + c_y(y, t) + c_z(z, t).$$ 4.16

If each of the functions c_x, c_y and c_z is a solution to Fick's second law in one dimension, then the sum is a solution for three dimensions. With this approach, the boundary conditions separate such that Eq. 4.15 can be solved for each function individually. Because the initial value c_{Al}^0 has already been integrated into the approach, all three functions have to be zero for $t = 0$, which can be realised by the employment of the function F_L (definition see Eq. 4.2). Because

$$\frac{\partial}{\partial x} F_L(x, t)\,|_{x=0} = -\frac{1}{\sqrt{\pi D t}}$$ 4.17

$$\frac{\partial}{\partial x} F_L(x, t)\big|_{x=L} = -\frac{1}{\sqrt{\pi D t}},$$

4.18

all three constants

$$c_x(x, t) = c_{x0} \cdot F_l(x, t)$$

4.19

$$c_y(y, t) = c_{y0} \cdot F_b(y, t)$$

4.20

$$c_z(z, t) = c_{z0} \cdot F_d(z, t)$$

4.21

can be determined from the boundary conditions to be:

$$c_{x0} = c_{y0} = c_{z0} = -\frac{v}{2M_O}\sqrt{\frac{\pi k_P}{D}}.$$

4.22

It has already been shown in [13], that $F_L(x, t)$ fulfills Fick's second law. Hence the solution for the rectangular specimen is:

$$c(x, y, z, t) = c_{Al}^0 - \frac{v}{2M_O}\sqrt{\frac{\pi k_P}{D}} \ (F_l(x, t) + F_b(y, t) + F_d(z, t)).$$

4.23

For the calculation of the aluminium profile in Fig. 4.6 the diffusion coefficient of aluminium in Fe-20Cr-5Al at 1200°C from [28] with $D = 1.10^{-12} \text{m}^2/\text{s}$ was used. Apart from the distortion of the profile caused by the corner already destroyed, the correspondence between calculation and measurement is

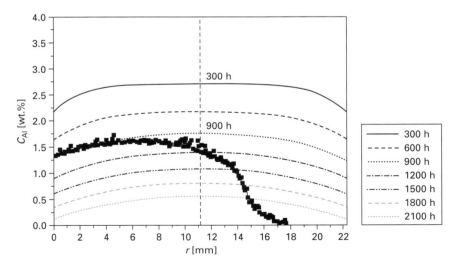

4.6 Calculated aluminium profiles on the diagonal of a rectangular PM 2000 sheet with the dimensions 20 mm × 10 mm × 0.5 mm in comparison with a measured profile (see Fig. 4.4).

satisfactory. With the critical aluminium content of 1.7 wt.% at 1200°C from Fig. 4.3, the breakaway of the sample can easily be explained by the depletion profiles in Fig. 4.6. Perhaps a crack or spall in the oxide scale at the tip of the sample allowed fast growing iron oxides to start the destruction process. Samples made from another Fe-20Cr-5Al alloy, Kanthal AF, have been used to demonstrate that diagonal depletion profiles can also be measured at the beginning of the oxidation. Rectangular specimens with sample dimensions 20 mm × 10 mm × 1 mm were oxidised in 100 h cycles at 1200°C in air. In Fig. 4.7 the profiles measured by EPMA and the calculated profiles are plotted on top of each other.

In Fig. 4.8 the aluminium contents at different places of a rectangular specimen with an initial content of 5 wt.% are compared. Obviously it makes no difference if the aluminium content is measured at the centre or on the main surface (*l-b*) of the sample. Thus the general assumption of a uniform aluminium content across the smallest sample cross-section seems to be justified. A similar behaviour is found for the area (*l-d*) and the border *l*, the area (*b-d*) and the corner *b* as well as for the border *d* and the corner. If the times to the total consumption of aluminium are compared, one obtains 5200 h for the centre of the sample and 4250 h for the corner. This gives a difference of about 1000 h. At an aluminium content of 1.7 wt.% the difference is reduced but the corners still go 600 h earlier into breakaway oxidation than the centre of the sample. This effect cannot be neglected in industrial applications. In Fig. 4.9 contours of equal aluminium content after 1000 h of

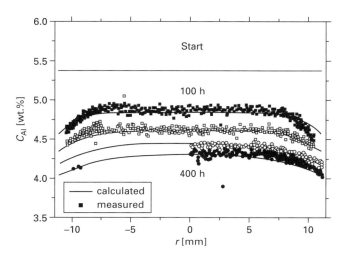

4.7 Calculated and measured diagonal aluminium profiles after oxidation of Kanthal AF samples for 100 h, 200 h, 300 h and 400 h at 1200°C in air. For 300 h and 400 h due to the symmetry only one half of the profile has been measured.

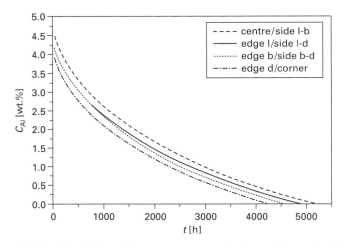

4.8 Calculated aluminium contents at different places of a PM 2000 sample with dimensions 20 mm × 10 mm × 0.5 mm and an initial aluminium content of 5wt.% during oxidation at 1200°C in air.

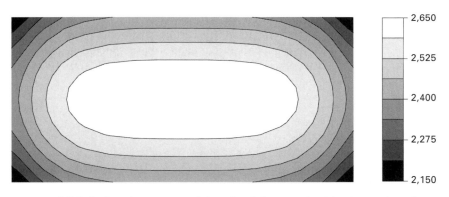

4.9 Calculated contours of the aluminium content in the centre of a PM 2000 sample with the dimensions 20 mm × 10 mm × 0.5 mm after 1000 h of oxidation at 1200°C in air. The aluminium content in the centre is 2.65 wt.% and drops with each greyscale by 0.05 wt.% to 2.15 wt.% in the corners.

oxidation at 1200°C are shown. It is evident that the areas with the lowest aluminium content have the same shape as the corners which are first destroyed by breakaway oxidation in real experiments. By removing the corners parallel to the contour lines, premature destruction of the sample could be avoided. Because these areas are fairly large it is questionable whether this could always be realised in real components.

4.4   Conclusions

A method to calculate the critical aluminium content for breakaway oxidation has been proposed which is based on Wagner's ideas on selective oxidation in ternary alloys. It focusses on the transition between internal and external oxidation, because after spallation of the oxide the bare metal is exposed again to the atmosphere. The calculation shows that the critical aluminium content is temperature dependent and varies from 0 wt.% at about 1000°C to 3 wt.% at 1300°C. This corresponds to the scatter range reported in the literature for the critical aluminium content.

Triggered by the common observation that breakaway oxidation in rectangular specimens very often starts at the corners, an analytical solution for the three-dimensional depletion profile in rectangular samples has been derived for parabolic oxidation kinetics. The solution is based on a one-dimensional function described by Whittle for the depletion of chromium in sheet material. The results are in good agreement with depletion profiles measured across the diagonal of a sample, for both short-term as well as long-term exposures. Thus the method opens up another path for the determination of aluminium diffusion coefficients in Fe-20Cr-5Al alloys.

Both calculations are new tools for better lifetime prediction. While the first gives accurate values for the critical aluminium content, the second allows an estimate of how much sharp corners can reduce the life expectancy of a component.

4.5   Acknowledgments

The careful and accurate preparation and measurement of the diagonal Al depletion profiles by our colleagues at the TU Clausthal, Mr E. Ebeling and Mr K. Herrmann, are highly appreciated.

4.6   References

1. C. E. Lowell, C. A. Barrett, R. W. Palmer, J. V. Auping, and H. B. Probst. COSP: A Computer Model of Cyclic Oxidation. *Oxidation of Metals*, 36(1/2): 81–112, 1991.
2. J. L. Smialek and J. V. Auping. COSP for Windows – Strategies for Rapid Analyses of Cyclic Oxidation Behaviour. *Oxidation of Metals*, 57(5/6): 559–581, 2002.
3. J. R. Nicholls, R. Newton, M. J. Bennett, H. E. Evans, H. Al-Badairy, G. J. Tatlock, D. Naumenko, W. J. Quadakkers, G. Strehl, and G. Borchardt. Development of a Life Prediction Model for the Chemical Failure of FeCrAl(RE) Alloys in Oxidising Environments. In M. Schütze, W. J. Quadakkers, and J. R. Nicholls. (eds), *Lifetime Modelling of High Temperature Corrosion Processes*, pp. 83–106, London, 2001. European Federation of Corrosion, IOM Communications. EFC Publication No. 34.
4. D. Poquillon and D. Monceau. Application of a Simple Statistical Spalling Model for the Analysis of High-Temperature, Cyclic-Oxidation Kinetics Data. *Oxidation of Metals*, 59(3/4): 409–431, 2003.

5. W. J. Quadakkers and K. Bongartz. The Prediciton of Breakaway Oxidation for Alumina Forming ODS Alloys using Oxidation Diagrams. *Materials and Corrosion*, 45: 232–241, 1994.

6. R. Newton, M. J. Bennett, J. P. Wilber, J. R. Nicholls, D. Naumenko, W. J. Quadakkers, H. Al-Badairy, G. Tatlock, G. Strehl, G. Borchardt, A. Kolb-Telieps, B. Jönsson, A. Westerlund, V. Guttmann, M. Maier, and P. Beaven. The Oxidation Lifetime of Commercial FeCrAl(RE) Alloys. In M. Schütze, W. J. Quadakkers and J. R. Nicholls (eds), *Lifetime Modelling of High Temperature Corrosion Processes*, pp. 15–36, London, 2001. European Federation of Corrosion, IOM Communications. EFC Publication No. 34.

7. H. Al-Badairy, G. J. Tatlock, H. E. Evans, G. Strehl, G. Borchardt, R. Newton, M. J. Bennett, J. R. Nicholls, D. Naumenko, and W. J. Quadakkers. Mechanistic Understanding of the Chemical Failure of FeCrAl-(RE) Alloys in Oxidising Environments. In M. Schütze, W. J. Quadakkers, and J. R. Nicholls (eds), *Lifetime Modelling of High Temperature Corrosion Processes*, pp. 50–65, London, 2001. European Federation of Corrosion, IOM Communications. EFC Publication No. 34.

8. G. Strehl, D. Naumenko, H. Al-Badairy, L. M. Rodriguez Lobo, G. Borchardt, G. J. Tatlock, and W. J. Quadakkers. The Effect of Aluminium Depletion on the Oxidation Behaviour of FeCrAl Foils. *Materials at High Temperatures*, 17(1/2): 87–92, 2000. Proceedings of the 4th International Conference on the Microscopy of Oxidation, held at Trinity Hall, Cambridge, 20–22 September 1999.

9. F. H. Stott and N. Hiramatsu. Breakdown of Protective Scales during the Oxidation of Thin Foils of Fe-20Cr-5Al Alloys at High Temperatures. *Materials at High Temperatures*, 17(1/2): 93–99, 2000. Proceedings of the 4th International Conference on the Microscopy of Oxidation, held at Trinity Hall, Cambridge, 20–22 September 1999.

10. G. Borchardt and G. Strehl. On Deviations from Parabolic Growth Kinetics in High Temperature Oxidation. In H. Bode (ed.), *Materials Aspects in Automotive Catalytic Converters*, pp. 106–116, Weinheim, 2002. DGM – Deutsche Gesellschaft für Materialkunde e.V., Wiley-VCH. International Conference 'Materials Aspects in Automotive Catalytic Converters', 3–4 October 2001, Munich, Germany.

11. B. Lesage, L. Maréchal, A. M. Huntz, and R. Molins. Aluminium Depletion in FeCrAl Alloys during Oxidation. *Defect and Diffusion Forum*, 194–199: 1707–1712, 2001.

12. C. Wagner. Theoretical Analysis of the Diffusion Processes Determining the Oxidation Rate of Alloys. *Journal of the Electrochemical Society*, 99(10): 369–380, 1952.

13. D. P. Whittle. The Oxidation of Finite Samples of Heat-resistant Alloys. *Corrosion Science*, 12: 869–872, 1972.

14. H. E. Evans and A. T. Donaldson. Silicon and Chromium Depletion During the Long-Term Oxidation of Thin-Sectioned Austenitic Steel. *Oxidation of Metals*, 50 (5/6): 457–475, 1998.

15. N. Birks and G. H. Meier. *Introduction to High Temperature Oxidation of Metals*. Edward Arnold Ltd, London, 1983.

16. C. Wagner. Reaktionstypen bei der Oxidation von Legierungen. *Zeitschrift für Elektrochemie*, 63(7): 772–782, 1959.

17. G. Böhm and M. Kahlweit. Über die innere Oxidation von Metallegierungen. *Acta Metallurgica*, 12: 641–648, 1964.

18. C. Wagner. Passivity and Inhibition during the Oxidation of Metals at Elevated Temperatures. *Corrosion Science*, 5: 751–764, 1965.

19. I. A. Akimova, V. M. Mironov, and A. V. Pokoyev. Aluminium Diffusion in Iron. *The Physics of Metals and Metallography*, 56(6): 175–177, 1983.
20. O. Madelung (ed). *Landolt-Börnstein, New Series, Group III, Crystal and Solid State Physics*, volume 26. Springer-Verlag, Berlin, 1990.
21. O. Kubaschewski (ed). *Iron – Binary Phase Diagrams*, 2nd edn. Springer Verlag, Berlin, 1982.
22. R. Prescott and M. J. Graham. The Oxidation of Iron-Aluminium Alloys. *Oxidation of Metals*, 38(1/2): 73–87, 1992.
23. K. Schäfer and E. Lax (ed). *Landolt-Börnstein, 6th edn, Volume II, Part 2b, Lösungsgleichgewichte I.* Springer-Verlag, Berlin, 1962.
24. A. Kolb-Telieps, J. Klöwer, A. Heesemann, and F. Faupel. High Temperature Corrosion Resistant FeCrAl Foils. In T. Narida, T. Maruyama, and S. Taniguchi (eds), HTCP 2000, pp. 305–308, 2000. International Symposium on High-Temperature Corrosion and Protection, 17–22 September 2000, Hokkaido, Japan.
25. S. E. Sadique, A. H. Mollah, M. S. Islam, M. M. Ali, M. H. H. Megat, and S. Basri. High-Temperature Oxidation Behaviour of Iron-Chromium-Aluminium Alloys. *Oxidation of Metals*, 54(5/6): 385–400, 2000.
26. J. Davidson, M. Lambertin, and J.-M-Herbelin. Recherche technique acier – Support métallique de catalyseur – Rapport final. Technical Report Contrat n° 7210-MA/313, Commission européenne, 1997.
27. G. Strehl, V. Guttmann, D. Naumenko, A. Kolb-Telieps, G. Borchardt, W. J. Quadakkers, J. Klöwer, P. A. Beaven, and J. R. Nicholls. The Influence of Sample Geometry on the Oxidation and Chemical Failure of FeCrAl(RE) Alloys. In M. Schütze, W. J. Quadakkers, and J. R. Nicholls (eds), *Lifetime Modelling of High Temperature Corrosion Processes*, pp. 107–122, London, 2001. European Federation of Corrosion, IOM Communications. EFC Publication No. 34.
28. A. Heesemann, E. Schmidtke, F. Faupel, A. Kolb-Telieps, and J. Klöwer. Aluminium and Silicon Diffusion in Fe-Cr-Al alloys. *Scripta Materialia*, 40(5): 517–522, 1999.

5

The different role of alloy grain boundaries on the oxidation mechanisms of Cr-containing steels and Ni-base alloys at high temperatures

V B TRINDADE and U KRUPP, University of Applied Science, Germany; P H. E-G WAGENHUBER, S YANG and H-J CHRIST, Universität Siegen, Germany

5.1 Introduction

As a result of a combination of reasonable mechanical properties, efficient corrosion resistance at high temperatures and lower cost compared with other high temperature materials, the Cr-containing steels are mainly used as material for components in power plants such as superheaters and exhaust systems. The surfaces of these tubes are exposed to combustion (outer side) and steam (inner side) atmospheres at temperatures between 400 and 600°C resulting in a time-dependent loss in the tube thickness due to the reaction between gas and metal. It is known that a minimum Cr content of approximately 20 wt.% [1] is needed to establish the formation of a protective, continuous Cr_2O_3 scale on Fe-Cr alloys, which prevents further attack. For steels with lower Cr contents, complex oxide scales composed of hematite (Fe_2O_3), magnetite (Fe_3O_4), spinel ($FeCr_2O_4$), wustite (FeO) and chromia (Cr_2O_3) are formed. In several studies it is assumed that oxides on low-alloy steels grow mainly by outward Fe diffusion [2–4], due to the very small lattice diffusion coefficient of O anions in iron oxides. On the other hand, it was reported [5–8] that inward oxide growth contributes substantially to the overall oxidation process in the temperature range between 500 and 600°C and that the mechanism for inward oxide growth may be attributed to the fast oxygen diffusion along oxide grain boundaries. Considering the relative higher diffusivity along grain boundaries compared to that in bulk, the change in the substrate grain size should considerably influence the oxidation behaviour. Elsewhere [9, 10], the possibility of molecular oxygen permeation through micro-cracks and pores in the oxide scale was mentioned leading to inward oxidation.

The formation of a protective Cr_2O_3 scale is required to avoid degradation by severe corrosion processes for alloys used at high temperatures (up to 1000°C). A number of investigations [11–15] were carried out to improve the understanding of the growth mechanism of Cr_2O_3 scales. The influence of oxide grain size as well as the effect of doping by rare earth elements (e.g.

yttrium or cerium) on the growth kinetics of Cr_2O_3 were carefully investigated and even the diffusion coefficients of chromium and oxygen in the bulk and along grain boundaries of Cr_2O_3 are available [16, 17]. It is well established that Cr_2O_3 scales grow by counter-current diffusion of Cr and O [15, 18, 19].

In addition to understanding the mechanisms of Cr_2O_3 scale growth, it is important to know the minimum bulk concentration of Cr necessary to form a protective scale on the entire surface, in order to prevent oxidation of the base material, e.g. Fe or Ni. Investigations reported a minimum value of Cr in the range between 18 and 20 wt.% [10]. However, this concentration range may depend strongly on the diffusion properties of Cr in the alloy. As a consequence, the grain size of the alloy should play an important role for the supply of Cr to the alloy/oxide interface, since the diffusivity of Cr along grain boundaries is much higher than that in the bulk [20].

Besides the formation of an external oxide scale, internal corrosion may occur under technical service conditions. In some cases, internal corrosion occurs preferentially along grain boundaries (intergranular corrosion). Of course, internal corrosion, particularly intergranular oxidation, is undesirable, since it usually enhances intergranular fracture resulting in premature brittle failure.

Depending on the alloy type (low-Cr steels, high-Cr steels or Ni-base alloys), the alloy grain boundaries play different roles on the oxidation behaviour. In this study emphasis was put not only on the kinetic aspects, but also on the thermodynamics of the oxidation processes in order to establish a mechanism-oriented model for oxidation of different alloys, which is dealt with in Chapter 32.

5.2   Materials and experimental methods

Three low-alloy ferritic steels (c_{Cr} = 0.55–2.29 wt.%), one austenitic steel (c_{Cr} = 17.5 wt.%) and two Ni-base alloys (c_{Cr} = 18.2–19 wt.%) were used. The chemical compositions of these materials are given in Table 5.1.

Table 5.1 Nominal chemical composition (in wt.%) of the materials studied

	C	Cr	Si	Mn	Al	Mo	Ti	Fe	Ni
Steel A	0.076	0.55	0.36	1.01	0.04	–	–	bal.	0.21
Steel B	0.06	1.43	0.22	0.59	0.04	–	–	bal.	0.04
Steel C	0.09	2.29	0.23	0.59	0.01	1.0	–	bal.	0.44
TP 347	0.04	17.50	0.29	1.84	–	–	–	bal.	10.7
Inconel 625 Si	0.01	19.00	1.20	0.07	0.24	8.9	0.28	2.80	bal.
Inconel 718	0.04	18.20	0.40	0.06	0.50	3.0	1.00	18.70	bal.

The grain size was modified by applying a heat treatment at 1050°C in inert gas atmosphere for 2, 12 and 112 h. The grain size was determined using optical microscopy and the mean linear intercept technique was applied. Figure 5.1 illustrates the variation in grain size for the example of Steel C.

The oxidation behaviour of the low-Cr steels and the austenitic steel were investigated for different grain sizes as summarised in Table 5.2. The Ni-base alloys were investigated only in the as-received condition.

Samples with dimensions $10 \times 10 \times 3$ mm^3 were used for thermogravimetric measurements. The samples were ground using SiC paper down to 1200 grid. They were finally cleaned ultrasonically in ethanol prior to oxidation. A hole of 1 mm diameter serves for hanging the samples in the thermobalance by means of a quartz wire. Isothermal and thermal-cycling thermogravimetry was carried out using a SARTORIUS microbalance with a resolution of 10^{-5} g in combination with an alumina chamber and a SiC furnace. The low-alloy steels were oxidized at 550°C, TP347 at 750°C and the Ni-base superalloys at 1000°C. After oxidation the specimens were embedded in epoxy and carefully polished using diamond paste down to 1 μm and cleaned ultrasonically in ethanol. Analysis of the oxide phases and thickness measurements of oxide layers were performed using scanning electron microscopy (SEM) in combination with energy-dispersive X-ray spectroscopy (EDX) and electron back-scattered diffractometry (EBSD). X-ray diffraction (XRD) was employed to analyse the oxidation reaction products.

5.3   Results and discussion

Kinetics of the oxidation process and related microstructural observations of the three classes of alloys are reported in separate sections in order to give a consistent description of the different role of the alloy grain size on the oxidation behaviour of these materials.

5.3.1   Low-alloy ferritic steels

The effect of the alloy grain size on the oxidation kinetics of low-Cr ferritic steels was carefully investigated using thermogravimetric measurements and confirmed by SEM observations. In all cases the oxide scale growth was nearly parabolic, i.e., kinetics can be described by means of a parabolic rate constant (k_p) and plotted against the alloy grain size. As shown in Fig. 5.2, the parabolic rate growth obviously decreases as the alloy grain size increases. Furthermore, the oxidation kinetics decreases as the Cr content increases for alloys with similar grain size.

By means of gold marker experiments it was possible to obtain a better understanding of the effect of alloy grain size on the oxidation mechanism. Prior to oxidation a thin gold layer was sputter-deposited on the sample

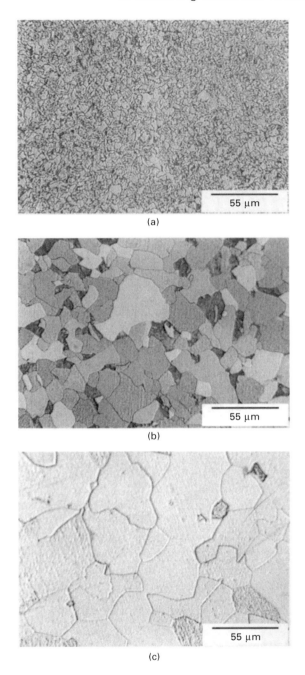

5.1 Microstructure of the low-alloy steel C prior to oxidation with three different grain sizes: (a) 4 μm, (b) 13 μm, (c) 74 μm.

Table 5.2 Alloy grain size (in μm) of the materials studied

	As-received	2 h/1050°C	12 h/1050°C	112 h/1050°C
Steel A	6	24	–	54
Steel B	10	30	60	100
Steel C	4	13	–	74
TP 347	4	11	–	65
Inconel 625	30	–	–	–
Inconel 718	70	–	–	–

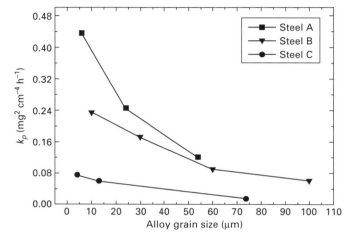

5.2 Parabolic rate constant k_p of the three low-alloy steels oxidised in laboratory air at 550°C for 72 h.

surface. After oxidation, the gold marker was found within the oxide scale (Fig. 5.3a) revealing two mechanism of mass transport: (i) outward transport of Fe cations and (ii) inward transport of O anions. A detailed examination of the inner part of the scale revealed important features about the growth mechanism of the inner scale. By using different techniques (EDS-mapping, XRD and EBSD) the oxide scale structure was characterised. The outer scale consists of Fe_2O_3 at the oxide/gas interface followed by Fe_3O_4. For the inner scale a more complex oxide structure is observed. Iron-chromium spinel ($FeCr_2O_4$), Fe_3O_4 and some amount of Cr_2O_3 were identified. Thermodynamic calculations using the software FactSage (Fig. 5.3b) support these experimental observations.

The growth kinetics of the inner scale can be attributed to the high diffusivity of oxygen along alloy grain boundaries, supporting the experimental observations (Fig. 5.2 and Fig. 5.4). Figure 5.4a documents clearly the effect of the alloy grain size on the thickness of the inner scale and compares this

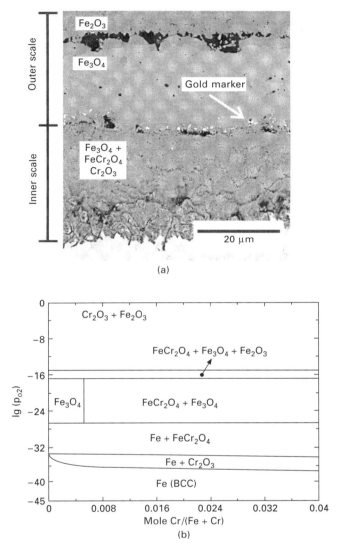

5.3 (a) Microstructural observations of the steel B oxidised in laboratory air at 550°C for 72 h, (b) thermodynamic stability diagram of the system Fe-Cr-O calculated by means of the software FactSage.

with the influence on the growth of the outer scale. As shown in Fig. 5.4b, preferential oxidation takes place along alloy grain boundaries and proceeds into the grain interior determining the progress of the inner scale/substrate interface.

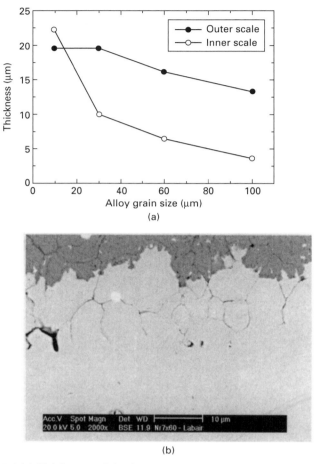

(b)

5.4 (a) Thickness of the inner and outer oxide scales formed on Steel B oxidised in laboratory air at 550°C for a duration of 72 h and (b) intergranular oxidation along the inner scale/substrate interface.

5.3.2   High-alloy austenitic steel

Figure 5.5 shows the thermogravimetrically measured mass gain of TP347 for different grain sizes during exposure at 750°C to laboratory air. The sample with a grain size of 4 µm and 11 µm obey a parabolic rate law. At a grain size of 65 µm, oxidation kinetics is more complex and exhibits a stepwise parabolic behaviour, possibly due to the formation of a multilayer oxide scale and the occurrence of structural defects such as cracks and pores.

On top of the fine-grained specimens a very thin protective Cr_2O_3 scale was formed. However, locally the oxide scale was not totally protective leading to the formation of iron oxide nodules, which grow outward and

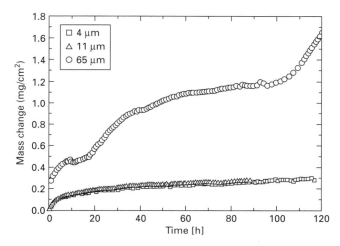

5.5 Thermogravimetrically measured oxidation kinetics of TP347 with different grain sizes, for exposure at 750°C to laboratory air for a duration of 120 h.

inward. Generally, the formation of a protective Cr_2O_3 scale on specimens with a small grain size is favoured by a higher Cr flux from the bulk to the substrate/oxide interface as a consequence of the higher grain boundary density. On the coarse-grained specimen the oxide scale consists of an outer scale of iron oxide ($Fe_2O_3 + Fe_3O_4$) and an inner scale of mixed oxide phases containing Fe, Cr, Mn and Ni, similar to the oxide nodules formed on the fine-grained specimens. The distribution of the elements Cr and Fe (measured by EDS) is shown in Fig. 5.6.

Figure 5.7 shows SEM micrographs of the surface oxide formed during the initial stage of the oxidation process on specimens with different grain sizes. A higher flux of Cr along the alloy grain boundaries leads to the formation of a thicker Cr_2O_3 scale on top of the grain boundaries as compared with the Cr_2O_3 scale formed on the interior of the grains (Fig. 5.7a) on specimens with small grain sizes. On the other hand, in the alloy with the large grain size of $d = 65 \mu m$, the flux of Cr towards the specimen surface is not sufficient to form a continuous and dense Cr_2O_3 scale and consequently to avoid fast oxidation of the base metal (Fe). Figure 5.7b shows the formation of a thicker iron oxide in the interior of the grain and a thinner iron oxide enriched in Cr along the alloy grain boundaries.

From a thermodynamic point of view a rather low concentration of Cr is sufficient to form a Cr_2O_3 scale on Fe-Cr steels. However, the critical minimum concentration is much higher in reality, since the value of the Cr supply to the alloy/oxide interface must be taken into account. As shown in this study, the alloy grain size is an important factor to be considered. As a consequence of the high Cr diffusivity along alloy grain boundaries, fine-grained materials

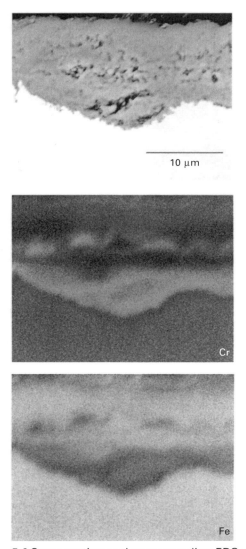

10 µm

5.6 Cross-section and corresponding EDS element mappings (Fe, Cr) of the oxide scale formed on TP347 with a grain size of 65 µm during exposure to laboratory air at 750°C for 120 h.

require a smaller Cr concentration than coarse-grained materials to form a slow growing Cr_2O_3 scale on the entire surface.

5.3.3   Ni-base alloys

The mass gain during isothermal exposure of the two studied Ni-base alloys (Inconel 625Si and Inconel 718) at 1000°C followed the parabolic behaviour

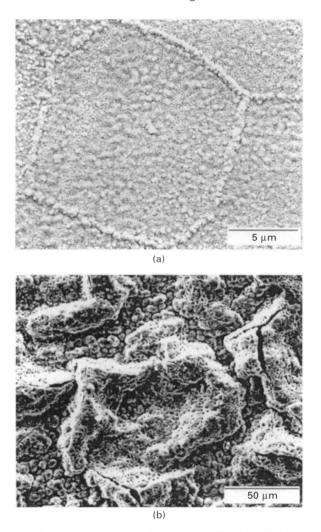

(a)

(b)

5.7 Surface of the scale formed on TP347 at 750°C after 1 h exposure to air: (a) formation of a thick Cr_2O_3 layer along grain boundaries on the fine-grained specimen (grain size = 11 μm) and (b) iron oxide formed on grain interior areas and Cr-enriched oxide formation along the grain boundaries of the coarse-grained specimen (grain size = 65 μm).

(Fig. 5.8). This observation is in agreement with the idea that for both alloys the oxidation processes are controlled by solid state diffusion. The oxidation of Inconel 718 is substantially faster ($k_p = 0.374$ mg^{2}cm^{-4}h^{-1}) than that of Inconel 625Si ($k_p = 0.0825$ mg^{2}cm^{-4}h^{-1}).

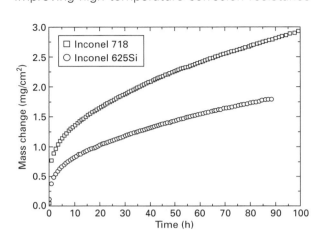

5.8 Thermogravimetrically measured mass gain of Inconel 625Si and Inconel 718 for exposure to laboratory air at 1000°C.

Figure 5.9 shows the oxide scales formed on the Ni-base alloys for the exposure to laboratory air at 1000°C. A continuous external Cr_2O_3 oxide scale is formed on the surface of both alloys. In the case of Inconel 625Si a discontinuous SiO_2 scale is formed underneath the Cr_2O_3 scale (Fig. 5.9a) which is probably responsible for the relatively low oxidation rate. Internal oxidation was observed in both alloys, occurring preferentially along alloy grain boundaries, enhanced by the fast oxygen diffusion along these short-circuit diffusion paths.

5.4   Conclusions

This study revealed that oxide scales grow outward as well as inward at 550°C in steels containing low Cr concentrations. An increased oxidation attack was observed with decreasing grain size due to higher oxygen transport along substrate grain boundaries.

A slow-growing Cr_2O_3 oxide was observed on high-alloy Cr steel (TP347) with small grain size due to the high density of grain boundaries leading to fast outward Cr transport along the substrate grain boundaries. Specimens with larger grain sizes form a complex mixture of Fe, Cr, Mn and Ni oxide. When adding Cr to high-temperature alloys in order to promote the formation of a slow-growing superficial Cr_2O_3, the high diffusivity of Cr along substrate grain boundaries must be considered. Grain boundary diffusion may significantly contribute to the Cr transport toward the substrate/oxide interface in materials with small grain sizes. Therefore, an alloy with a larger grain size requires a higher bulk Cr concentration to establish a protective Cr_2O_3 scale.

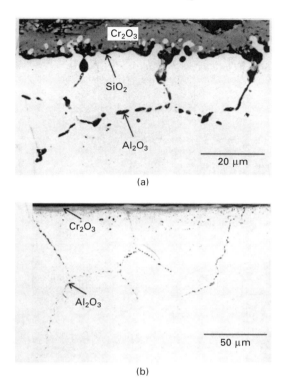

5.9 Oxidation of Ni-base superalloys after exposure to laboratory air at 1000°C: oxide scale and intergranular oxidation zone of (a) Inconel 625Si after 90 h and (b) Inconel 718 after 140 h exposure.

A significant intergranular attack consisting of a Al_2O_3 precipitation was observed for both Ni-base superalloys. Both alloys formed an outer Cr_2O_3 scale. In the case of Inconel 625Si, a partial SiO_2 layer was formed underneath the Cr_2O_3 scale reducing its oxidation kinetics as compared with that of Inconel 718.

5.5   Acknowledgments

This research has been supported by the EU-project OPTICORR and by the Brazilian Research Foundation (CAPES) through a fellowship to one of the authors (V.B. Trindade).

5.6   References

1. J. R. Davis, *Stainless Steels*, ASM Speciality Handbook, ASM, Materials Park, OH, 1994.

2. G. Granaud, R. A. Rapp, *Oxidation of Metals*, 1977, 11, 193.
3. J. E. Castle, P. L. Surman, *The Journal of Physical Chemistry*, 1967, 71, 4255.
4. L. Himmel, R. T. Mehl, C. E. Birchenall, *Journal of Metals*, 1953, 5, 827.
5. S. Matsunaga, T. Homma, *Oxidation of Metals*, 1976, 10, 361.
6. N. Appannagaari, S. Basu, *Journal of Applied Physics*, 1995, 78, 2060.
7. W. Wegener, G. Borchardt, *Oxidation of Metals*, 1991, 36, 339.
8. K. Nakagawa, Y. Matsunaga and T. Yanagisawa, *Materials at High Termperatures*, 2001, 18, 51.
9. R. Bredesen, P. Kofstad, *Oxidation of Metals*, 1990, 34, 361.
10. R. J. Hussey, M. Cohen, *Corrosion Science*, 1971, 11, 713.
11. R. C. Lobb, H. E. Evans, *Metal Science*, 1981, 267.
12. D. R. Baer, M. D. Merz, *Metallurgical Transactions A*, 1980, 11A, 1973.
13. J. S. Dunning, D. E. Alman, J. C. Rawers, *Oxidation of Metals*, 2002, 57, 409.
14. F. J. Pérez, F. Pedraza, C. Sanz, M. P. Hierro, C. Gómez, *Materials and Corrosion*, 2002, 53, 231.
15. S. C. Tsai, A. M. Huntz, C. Dolin, *Materials Science and Engineering A*, 1996, 212, 6.
16. S. N. Basu, G. J. Yurek, *Oxidation of Metals*, 1991, 36, 281.
17. G. J. Yurek, D. Eisen, A. Garrat-Reed, *Metallurgical Transactions A*, 1982, 13A, 473.
18. A. M. Huntz, *Journal of Materials Science Letters*, 1999, 18, 1981.
19. A. F. Smith, *Metal Science*, 1975, 9, 375.
20. I. Kaur, W. Gust, *Fundamentals of Grain and Interphase Boundary Diffusion*, Ziegler Press, Stuttgart, 1988.

Improving metal dusting resistance of Ni-X alloys by an addition of Cu

Y NISHIYAMA, K MORIGUCHI, N OTSUKA,
and T KUDO, Sumitomo Metal Industries Ltd, Japan

6.1   Introduction

Metal dusting, a type of corrosion resulting from catastrophic carburization or graphitization of steels and alloys occurring in carbonaceous atmosphere, is a prominent cause of corrosion damage for high temperature materials used in ammonia, methanol, and syngas plants [1–4]. Metal dusting is often encountered when steels and alloys are exposed at 450–700°C to CO-containing syngas environments where the carbon activity (a_c) is greater than unity, and the oxygen potential (P_{O_2}) is relatively low. For carbon steels, Hochman [5, 6] suggested that the important step in metal dusting is the formation of an unstable cementite (Fe_3C) as an intermediate followed by its decomposition. The detailed reaction steps for carbon steels and low alloy steels are demonstrated by Grabke and his co-workers [7–10].

One of the most common techniques to prevent metal dusting is to form a protective oxide scale on alloy surfaces. Shueler [11] has proposed that theoretical account for effective metal dusting resistance can be calculated by the equivalent equation: $Cr_{eq} = Cr\% + 2 \times Si\% > 22$. Both Cr and Si can form protective oxide scales as Cr_2O_3 and SiO_2, respectively. Schillmoller [12] has modified the equivalent equation in a recent paper. He considered the effect of Al as an Al_2O_3 scale and proposed it as follows: $Cr_{eq} = Cr\% + 3 \times (Si\% + Al\%) > 24$. These scales can act as a barrier against the carbonaceous gas. As long as the protective oxide scale is maintained with no cracks/flaws, pits associated with metal dusting do not appear. From the point of protective oxide-scale integrity, the equivalent equation of resistance to carburization and metal dusting of steels and alloys in terms of alloying elements was used for the selection of conventional and newly developed alloys for application to the metal dusting environments [13, 14]. However, the oxide scales may crack at elevated temperature because of stresses caused by their growth and thermal cycles. These stresses that compress the oxide scale are sufficient to cause its spallation for austenitic stainless steels and nickel-base alloys. Therefore, even with a controlled oxide scale it is difficult to maintain complete protection against metal dusting for a long time. Once the scales cracked, pitting may or may not occur, depending on the competition between the attack of the carbonaceous gas and the healing of the oxide scales. Furthermore, it would depend on the reactivity between the gas and the exposed metal surfaces where the oxide scales are damaged.

Formation of carbon (coking) has been studied in terms of catalytic activity in which carbonaceous gases such as hydrocarbons or CO react with transition metals [15–19]. Figueiredo has suggested that the mechanism of carbon formation from hydrocarbon includes the following steps [18]:

(a) The hydrocarbon that is adsorbed on the metal surface produces chemisorbed carbon atoms by surface reactions.
(b) The chemisorbed carbon atoms can dissolve in and diffuse through the metal, resulting in their precipitation at grain boundaries. Such precipitation causes metal crystallites to be removed from the matrix. The metal can catalyze the filamentous coke formation and transport on top of the growing filaments.
(c) Alternatively, the chemisorbed species may react on the surface to originate an encapsulating film of polycrystalline carbon.

According to step (b), the chemisorption of carbon atoms decomposed from the carbonaceous gases leads to carbon diffusion into the metal, i.e. initiation of metal dusting. This means that metal dusting may be prevented if the reaction between CO and metals is suppressed at their surface.

The objective of the present study is to examine the behavior of both metal dusting and coking for transition metals and Ni-Cu binary alloys exposed in a carbonaceous gas, which simulates the synthetic gas produced in the reforming process. The experimentally observed phenomena are also theoretically discussed from the viewpoint of electronic properties under the chemical reactions of CO with transition-metal surfaces.

6.2   Experimental methods

Transition metals and Ni-Cu binary alloys were tested. The metals of 99.9%Cr, 99.9%Mn, 99.9%Fe, 99.96%Co, 99.97%Ni, and 99.5%Cu (in mass%) were chosen from the fourth periodic transition-metal elements, and both 99.98%Ag and 99.98%Pt were also tested. The binary alloys of Ni-Cu have differing contents of Cu: 1.0, 2.0, 2.9, 4.9, 10.0, 18.3, 49.5, and 69.6 (in mass%). Coupon specimens were cut from sheets or plates, and a small hole of 2mm diameter was drilled in them for support. Specimens of Cr, Mn, and Pt were supported directly by a spot-welded Pt hook. For the binary alloys, button-ingots were made in an arc-melting furnace under an inert argon atmosphere. Coupon specimens were cut from the hot-rolled plates (5 mm thickness) with a solution-heat treatment at 1100°C in air. All of the specimens were mechanically ground with 600-grit emery paper, followed by ultrasonic cleaning in acetone. Metal dusting tests were conducted in a horizontal reaction chamber with a double-layered structure, quartz tube inside and a nickel-base alloy outside. The reaction gas composition of 60%CO, 26%H$_2$, 11.5%CO$_2$, and 2.5%H$_2$O (in vol.%), which gives the carbon activity of 10 and oxygen partial pressure of 4.6×10^{-25} atm at 650°C at equilibrium, was chosen to simulate the actual reforming plants [20]. Corrosion behavior under the test gas condition was predicted from a thermodynamic aspect. Figures 6.1 (a) and (b) show the phase stability diagrams of Fe-Cr-C-O (activity of Cr,

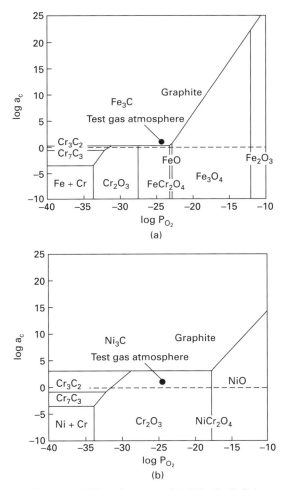

6.1 Phase stability diagram of (a) Fe-Cr-C-O (a_{Cr} = 0.2) system and (b) Ni-Cr-C-O (a_{Cr} = 0.2) system at 650°C. A graphite phase is stable above the broken line shown in each figure. Test gas atmosphere of a 60%CO-26%H$_2$-11.5%CO$_2$-2.5%H$_2$O gas mixture was also plotted.

$a_{Cr} = 0.2$) system and Ni-Cr-C-O ($a_{Cr} = 0.2$) system. The test gas atmosphere is also plotted in these figures. As shown in Fig. 6.1 (a), steels of Fe-Cr form cementite as Fe_3C on their surface. Both scales of spinel-type $FeCr_2O_4$ and Cr_2O_3 might form beneath the Fe_3C. Graphite as a coke can also be deposited on the Fe_3C by a gas phase reaction. For Ni-Cr alloys, the Cr_2O_3 scale can cover their surface, followed by the coke deposition. The Ni_3C cannot form on the surface, even though any Cr carbides can form as internal precipitates under the Cr_2O_3 scale. Water vapor was added by bubbling the test gas in purified water. Dew point of the gas mixture was controlled by heating of water.

The specimens were exposed to the gas mixture at a flow rate of 300 sccm (standard cubic centimeters per minute) at a temperature of 650°C under a pressure of 1 atm. The gas velocity was calculated at 1.6 mm/s at the specimen. Each specimen was suspended on a quartz rack through a Pt hook and then set in the reaction chamber, allowing gas flow parallel to their specimen surfaces. The reaction chamber was first purged with a N_2-5%H_2 gas mixture for an hour at room temperature. After complete replacement of the purge gas with the test gas mixture, the specimens were heated at 650°C for 100 h.

Each specimen was weighed after the exposure test, followed by removing the coke by ultrasonic cleaning in acetone. The treated specimens were weighed and observed with an optical microscope. For transition metals of Cu and Ni, a depth profile of elements was made with glow discharge spectroscopy (GDS, Rigaku System 3860). Metallographic cross-section of the specimen surface was investigated for test specimens by an optical microscope and a scanning electron microscope (FE-SEM, Hitachi S-4100) with energy dispersive X-ray spectroscopy (EDS).

6.3   Theoretical analysis

In order to comprehensively show the chemical dissociation process of CO on metal surfaces, electronic structure calculations have been performed for simple models. We have chosen two methods for the present analyses. The first method is the Discrete Variational Xα (DV-Xα) method, which is the first-principles molecular orbital calculation using Slater's Xα functional for the electron many body term [21]. This method is applied for the electronic structural analyses of CO adsorption on metal surfaces. The second method is the Full-Potential Linear Muffin-Tin Orbital (FP-LMTO) method, which is the first-principles band structure calculation method [22]. The FP-LMTO implementation code of LmtART [23, 24] is used for the calculations of the density of states (DOS) of non-magnetic fcc iron phase. We discuss the electronic structure of transition metal alloys from the rigid band analyses using this DOS. The local density approximation (LDA) parameterized by Vosko *et al.* [25] is used for the present FP-LMTO calculations. The tetrahedron mesh of $8 \times 8 \times 8$ is adopted for the Brillouin zone sampling. In the present work, the magnetic interaction is not considered and the non-spin-polarized calculations are carried out for both methods. In addition, the experimental lattice constants and/or bond length are used for constructing the configuration models analyzed. (The experimental lattice constants of bcc-Fe and fcc-Cu are 2.8664 Å and 3.6146 Å, respectively. The bond length of isolated Cu Molecule is 1.1283 Å.) An approximation of the lattice constant of hypothetical non-magnetic fcc phase of Fe is deduced from the experimental atomic volume in the ground state bcc phase.

6.4   Results

6.4.1   Metal dusting of transition-metals

Transition metals were exposed in CO-H_2-CO_2-H_2O simulated gas atmosphere at 650°C for 100 h. The amounts of coke deposited on their surfaces were measured and are listed in Table 6.1. After the removal of coke, each specimen was weighed, also listed in Table 6.1. Specimens of Cr and Mn had small amounts of coke because they were able to form an oxide scale in the test atmosphere. Except for Cr and Mn, the test metals did not form any oxide scale on the surface. A specimen of Fe had an extremely large amount of coke, over 1400 mg/cm^2, while its mass decreased to –37 mg/cm^2. Specimens of Co and Ni also had coke deposition of 6.55 and 11.95 mg/cm^2, respectively. Their mass changed to a minus value, which includes both mass gain due to carbon ingress and mass loss due to metal wastage, i.e. metal dusting. In contrast, metals of Cu, Ag, and Pt whose masses remained unchanged, had slight amounts of coke deposition on the surface. Neither metal dusting nor coking appeared to occur in the carbonaceous environment. Figure 6.2 shows the appearance of the specimens after the test. Surfaces of Cr and Mn colored dark green and grey, respectively, due to the formation of the oxide scale. For the Fe, Co, and Ni metals, coke covered the whole of their surfaces. In particular, the coke had a layered structure on the Fe specimen, though the outer layer could easily flake away. In contrast, specimens of Cu, Ag, and Pt preserved a metallic surface.

Table 6.1 Amount of coke deposition and mass change of specimens for transition metals exposed in CO-H_2-CO_2-H_2O gas mixture at 650°C for 100 h. Specimens of Cr and Mn formed oxide scale on the surface in the test gas environment

	Cr	Mn	Fe	Co	Ni	Cu	Ag	Pt
Amount of coke in mg/cm^2	0.06	<0.01	1,434	6.55	11.95	0.03	<0.01	<0.01
Mass change in mg/cm^2	0.14	2.22	–37.15	–0.83	–0.15	0	0	0

6.2 Specimen surfaces of transition metals exposed in CO-H_2-CO_2-H_2O gas mixture at 650°C for 100 h.

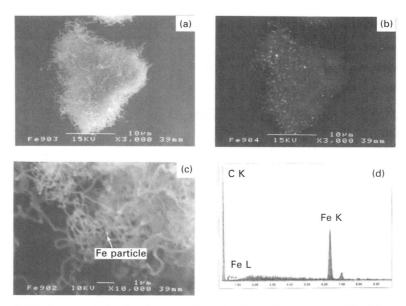

6.3 Coke morphologies deposited on the Fe surface exposed in CO-H_2-CO_2-H_2O gas mixture at 650°C for 100 h. (a) SE image, (b) BS image of (a), (c) Higher magnification of (a), and (d) EDS analysis at a bright spot shown in (b).

Morphologies of coke deposited on the Fe and Ni were observed by FE-SEM and are shown in Figs. 6.3 and 6.4, respectively. Coke on Fe was agglomerated like lint, shown in Fig. 6.3(a). At the same region observed in Fig. 6.3(a), back-scattering electrons imaged many bright spots (Fig. 6.3(b)),

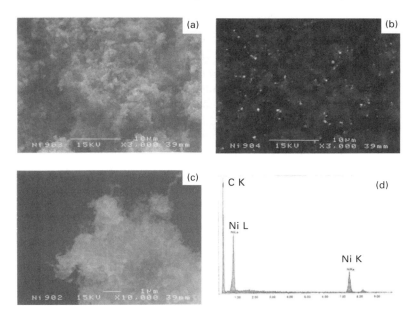

6.4 Coke morphologies deposited on the Ni surface exposed in CO-H_2-CO_2-H_2O gas mixture at 650°C for 100 h. (a) SE image, (b) BS image of (a), (c) Higher magnification of (a), and (d) EDS analysis at a bright spot shown in (b).

suggesting that some metal or carbide particles are involved in the coke. Toh *et al.* [26] reported that the particles could consist of cementite. In this investigation, it is unclear whether the particles consist of metal and/or cementite. A higher-magnification observation showed quite a lot of coke filaments with 0.1–0.2 μm diameter, intricately intertwined with each other (Fig. 6.3(c)). As indicated by the arrow in this figure, a number of nodes studded in the filamentous coke were observed, corresponding to the bright spots imaged by the back-scattering electrons. EDS analysis showed that these nodes consisted of Fe, suggesting the deposited filamentous coke were catalyzed by Fe particles, which were detached from the Fe specimen (Fig. 6.3(d)). The detachment of metals, which is a characteristic phenomenon of metal dusting, was demonstrated by the result of mass loss listed in Table 6.1.

Figure 6.4 shows the result of SEM observations and EDS analysis for coke deposited on Ni. As shown in Fig. 6.4(b), the coke contained some metals. There were fewer imaged bright spots for Ni than there were for Fe. The coke looked filamentous, but the filaments were thinner as compared with the coke on Fe. At the bright spots imaged in Fig. 6.4(b), EDS detected Ni, supporting the theory that it can also act as a catalyst of coke formation (Fig. 6.4(d)).

A cross-section of Ni was observed by SEM, and is shown in Fig. 6.5(a). It had a considerably rough surface where coke (graphite) aligned perpendicular to the base metal like a lamellar structure. As a result, Ni was thinned and then detached from the base metal. The metal wastage progressed uniformly, because it could not form any oxide scales. Carbon ingress in the Ni and Cu, which were exposed in the simulated gas at 650°C for 100 h, was investigated by GDS, shown in Figs 6.5(b) and (c), respectively. For the specimen of Ni, carbon penetrated from the metal surface and reached a depth of approx. 5–10 μm where any nickel carbides was not precipitated. In contrast, carbon did not penetrate into Cu, demonstrating it had neither metal dusting nor coking. The results demonstrated that Cu, Ag, and Pt were useful elements for resisting metal dusting and coking. Because of the absence of protective oxide scales as a barrier against CO gas, those metals were concluded to have less reactivity with a carbonaceous gas such as CO.

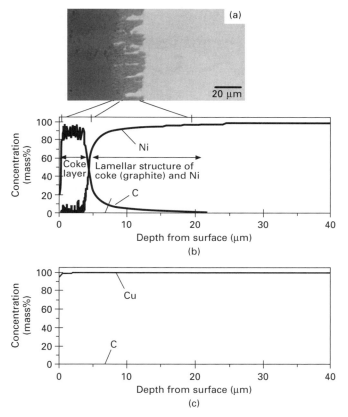

6.5 SEM observation and GDS depth profiles of C and alloying elements for (a) and (b) Ni and (c) Cu, exposed in CO-H_2-CO_2-H_2O gas mixture at 650°C for 100 h.

6.4.2   Metal dusting of Ni-Cu binary alloys

Among the useful transition metals identified above, Cu might be the most practical element to consider as an alloy for high temperature materials. For Ni-Cu binary alloys, therefore, the effect of Cu was investigated on metal dusting. The binary alloys with differing amounts of Cu were exposed in the $CO-H_2-CO_2-H_2O$ gas mixture at 650°C. After 100 h exposure, the amounts of coke deposition and mass changes of the specimens were plotted against Cu content, shown in Figs 6.6 and 6.7, respectively. The results for pure Ni and Cu were also plotted in each figure. Alloys with a small addition of Cu had a large amount of coke deposited on their surface. The coke deposition was decreased drastically with increasing Cu content, and reduced to almost zero for Ni-Cu alloys of 20% and more Cu. In accordance with the coking

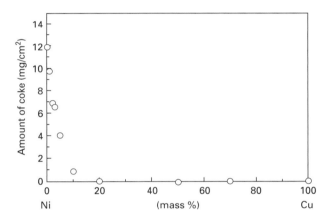

6.6 Amount of coke deposition for Ni-x%Cu binary alloys exposed in $CO-H_2-CO_2-H_2O$ gas mixture at 650°C for 100 h.

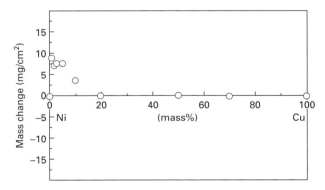

6.7 Mass change of specimens for Ni-x%Cu binary alloys exposed in $CO-H_2-CO_2-H_2O$ gas mixture at 650°C for 100 h.

behavior, alloys containing low Cu, but not pure Ni, increased their masses due to the carbon ingress. Specimens of 10%Cu showed less increase in mass compared with the low Cu alloys. For specimens of 20% and more Cu, their mass remained unchanged even after 100 h exposure. Their surfaces without removal of coke are shown in Fig. 6.8. A large amount of coke with filamentous morphology was formed over the specimens of 1.0% and 2.0%Cu. Coke deposition decreased with increasing Cu content, and then materially alleviated for Ni-10.0%Cu alloy. Alloys of 18.3% and 69.5%Cu, which showed no mass change in Fig. 6.7, had little coke formation on their surfaces.

Figure 6.9 shows metallurgical cross-sections of Ni alloys with 2.0%, 10.0%, and 49.5%Cu after removal of coke, following the exposure test at 650°C for 100 h. Nickel alloy with 2.0%Cu showed a considerably rough surface with a lot of crystalline fragment collapse, resulting in detachment from the base metal. The metal wastage progressed uniformly: neither cracks nor pitting occurred at the surface, because the alloy could not form any oxide scales. Alloy of Ni-10.0%Cu also showed a rough surface, though the detailed morphology of metal wastage was different from that for 2.0%Cu alloy. For Ni-49.5%Cu alloy, in contrast, its surface was extremely smooth without metal wastage. From these results, Ni alloys with less than 20%Cu demonstrated coke formation and metal dusting under the condition of carbonaceous gas. These levels of Cu content clearly cannot suppress the reactivity of metal with CO gas at the metal surface where no oxide scale is formed. Therefore, 20% and more Cu in Ni alloy is needed to inactivate the surface reaction with CO gas, and results in lowering of metal dusting attack in the synthetic gas environment tested.

6.5   Discussion

6.5.1   Initial stage of metal dusting

Nickel alloys containing Cu, even though they did not have any oxide scales to protect against CO gas, were demonstrated to improve the metal dusting

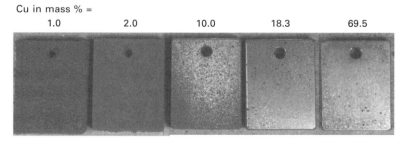

6.8 Specimen surfaces of Ni-Cu binary alloys exposed in $CO-H_2-CO_2-H_2O$ gas mixture at 650°C for 100 h.

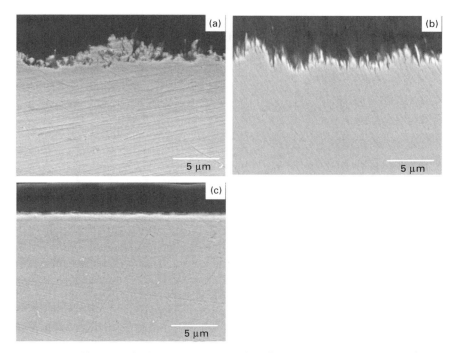

6.9 Metallurgical cross-sections of Ni-Cu binary alloys, (a) 2.0% Cu, (b) 10.0% Cu, and (c) 49.5% Cu, exposed in $CO-H_2-CO_2-H_2O$ gas mixture at 650°C for 100 h.

resistance, and thereby to inhibit the coke formation catalyzed by the detached metal particles. The amount of Cu for significant improvement needed to be almost 20% under the test gas condition. Since the effect of Cu may be related to the reduction of the reactivity with CO gas, an understanding of elementary reaction steps for the initial stage of metal dusting is important. Figure 6.10 shows the possible steps for carbon to penetrate into the base metal, illustrated at the atomic level. Under carbonaceous gas atmospheres, CO molecules collide on the metal surface, and part of them adsorb on it (reaction step I). Each atom of carbon and oxygen, while still binding together as molecular CO, forms a chemical union with the metal atom at the metal surface. The chemisorbed CO molecule is dissociated into carbon and oxygen, because the binding energy of C-O is weakened (reaction step II). The continuous carbon atom penetrates into the base metal (reaction step III). In the carbonaceous gas, the sequence continues with a charging of carbon into the base metal, and as a result, metal dusting occurs on the whole of the metal surface. Of great importance for preventing metal dusting is to suppress the dissociation of CO molecules adsorbed on the metal surface in the reaction steps mentioned above.

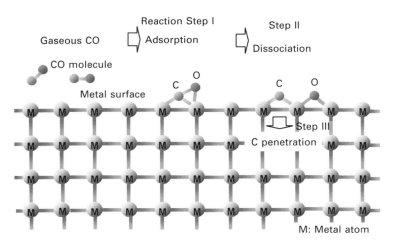

6.10 Schematic representation of metal/CO gas reaction on the metal surface at the atomic level.

For a given transition metal series, it is experimentally known that the tendency of CO dissociation becomes greater, when the metallic elements of the substrate are located further to the left side in periodic table [27]. This dissociative tendency is assumed to be strongly related to the results of the present metal dusting experiments, while the protective properties against metal dusting found in the specimens of Cr, Si, and Al originate in another mechanism, the formation of oxide. In addition, Grabke has recently reported that the adsorption of sulfur on pure Fe surfaces induces protective properties against metal dusting [28]. It is interesting to note that the adsorption of sulfur atoms on Fe(100) surfaces has also been experimentally detected to prevent the dissociative adsorption of CO [29]. This is because those adsorption sites of CO [30], C, O [31], and S [32] on Fe(100) surfaces are identical with the four-fold hollow sites, and CO molecules lose their dissociative reaction fields with a sulfur monolayer. These experimental results suggest that an understanding of atomistic properties of CO adsorption on transition metal surfaces is the key ingredient for metal dusting suppression technologies.

6.5.2   Dissociative and non-dissociative adsorption of CO at metal surfaces

In metal dusting phenomena, the most basic question for the CO adsorption on metal surfaces concerns the difference between the dissociative and non-dissociative properties. While some controversial phenomena still exist, especially in platinum systems [33], the model widely accepted for the question so far is the donation and back-donation mechanism proposed by Blyholder [34]. In this section, we describe the Blyholder mechanism for the dissociative and non-dissociative adsorption properties of CO at metal surfaces using the electronic structures calculated by the DV-Xα method.

On bcc-Fe(100) surface, a CO molecule is experimentally known to occupy the four-fold hollow site, where the molecular axis is tilted along the vertical direction of the surface [30]. In the present work, however, the electronic structures of non-tilted states are calculated for simplicity. The equilibrium distance between Fe(100) surface and CO molecule in this configuration is theoretically reported to be r(Fe-CO) = 0.57Å [35]. In contrast, it is experimentally understood that a CO molecule is vertically adsorbed at the one-fold on-top site on fcc-Cu(100) surface, and is reported to be r(Cu-C) = 1.85 Å [36]. For both cases, it is known that a carbon atom in CO is adsorbed with the metallic surface. This is because the chemically active frontier orbital of the CO molecule, consisting of the highest occupied molecular orbital (HOMO: CO-5σ) and the lowest unoccupied molecular orbital (LUMO: CO-2π), are constructed from the carbon states much larger than the oxygen ones. The LUMO of the CO molecule has an anti-bonding character, where the occupation of electrons weakens the interatomic C-O bond.

Figure 6.11 shows the electronic density of states (DOS) for the cluster models which mimic the adsorption of CO on bcc-Fe(100) and fcc-Cu(100) surfaces. These figures are the partial DOSs projected to 3d states of metal elements and to total ones of CO molecules, as a function of distance between metal surface and CO. In dissociative adsorption systems such as CO/Fe, the DOSs indicate that a strong hybridization between metallic states and CO ones occurs as the CO molecule approaches the metallic surface, in which the anti-bonding CO-2π orbital is partially occupied by electrons. In contrast, in the case of non-dissociative adsorption systems such as CO/Cu, the CO-2π state is still located above the Fermi level even when the CO molecule approaches the equilibrium adsorption configuration, which means that the anti-bonding CO-2π orbital is not occupied by electrons. This is the Blyholder mechanism [34] for the dissociative and non-dissociative adsorption properties of CO at metal surfaces, in which the CO-metal bonding properties are characterized arising from electron transfer of the CO-5σ orbital (HOMO) to unoccupied metal states accompanied by back-donation from occupied metal states to the CO-2π orbital (LUMO). Since the electronic sites around the Fermi level in transition metals are mainly composed of the d-band states, the energetic difference between the CO-2π state and the metallic d-band states is essentially important in CO dissociative reactions on transition-metal surfaces.

Next, we focus on the general trend of electronic structure of transition metals. It is known that the electronic structure of transition metal alloys can be described by means of the so-called 'rigid band model' as a first approximation, which states that the electronic band structure is unchanged upon substitutional alloying and the electronic structure is simply described

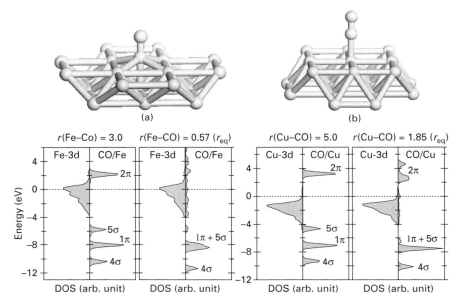

6.11 The electronic density of states (DOS) for the cluster models which mimic the adsorption of CO (a) on bcc-Fe(100) and (b) fcc-Cu(100) surfaces as a function of distances between CO and metal surfaces, calculated by the DV-Xα method. r(M-CO) denotes the distance between CO and metal (M) surfaces in Å. The partial DOSs projected to 3d states of metal elements (left panels) and to total ones of CO molecule (right panels) are shown. The Fermi level in each configuration is defined as zero.

by filling the DOS with the associated electrons [37]. Figure 6.12 exemplifies the Fermi level of 3d transition metals as a function of valence electrons per atom within the rigid band model using the total density of states (DOS) of non-magnetic fcc-Fe, which is calculated using the FP-LMTO method. In this figure, the zero energy is set to the bottom of the s-state band, i.e. the Γ_1 eigenvalue. For a given transition metal series, the Fermi level shifts to the upper side by filling the d-band with electrons. This means that the unoccupied d-states decrease as valence electrons increase. Considering the Blyholder mechanism mentioned above, the interaction between the CO and the transition metal surfaces is predicted to weaken as the Fermi level of the system rises. This electronic structural view suggests that the upper shifting of Fermi level in the d-band state due to electron donor elements is one of the key factors for the appearance of metal dusting resistance in transition metal alloys. In the previous section, we have reported that it is necessary to add copper of 20% or more in Ni base alloys to completely prevent metal dusting phenomena without an oxide scale. Based on the discussion here, the complete filling of the d-band with electrons is assumed to be necessary to prevent metal dusting

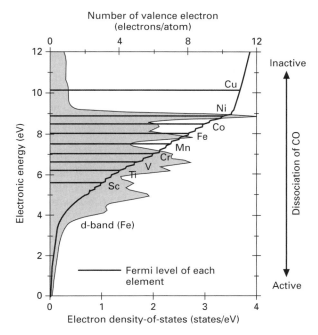

6.12 Calculated Fermi level as a function of valence electrons in the 3d transition metals within the rigid band model using the DOS of non-magnetic fcc-Fe. The zero energy is set to the bottom of the s-state band, i.e. the Γ_1 eigenvalue.

phenomena without oxide scales, and this preventing effect is called 'surfactant-mediated suppression'.

Finally, the surface segregation phenomena is briefly discussed that can assist the metal dusting suppression. In binary Ni-Cu alloys, a strong surface segregation has been observed experimentally, which indicates a strong preference of the Cu impurity for a surface site rather than a bulk site [38]. These segregation phenomena have also been addressed theoretically from the viewpoint of electronic theory by Ruban *et al.* [39], in which impurity atoms segregated on vicinal surfaces lower the surface energies and work as 'surfactants' in transition metal alloys. From the experimental results reported so far, the surface of the Ni-Cu alloys including Cu of 20% or more is predicted to be mainly composed of Cu atoms [38]. Therefore, the surface segregation of Cu at chemically active sites may possibly assist the metal dusting suppression in the present experiments. The experimental confirmation of this 'segregation effect' for the metal dusting suppression is currently underway in our group. In view of the surface segregation, it may be possible to predict the variation of metal dusting rate with copper content.

6.6   Conclusions

♦ Preliminary metal dusting tests using transition metals suggest that noble metals such as Cu, Ag, and Pt have good resistance against metal dusting. There is neither coke formation nor change in specimen mass after exposure in a simulated synthetic gas at 650°C for 100 h. For Ni-Cu binary alloys, which do not form any protective oxide scales, Cu content of almost 20% is needed for prevention of metal dusting.

♦ Metal dusting resistance of copper, which acts as a 'surfactant-mediated suppression' effect, is discussed from the viewpoint of atomistic interaction of CO with transition metal surfaces using electronic structure analyses. Our experimental and theoretical results suggest that the upper shifting of Fermi level in the d-state band due to electron donor elements, also strongly enhanced by their surface segregation, is one of the key factors for the appearance of metal dusting resistance in transition metal alloys.

6.7   References

1. F.A. Prange, *Corrosion* 15, 12 (1959), 619t.
2. F. Eberle and R.D. Wylie, *Corrosion* 15, 12 (1959), 622t.
3. W.B. Hoyt and R.H. Caughey, *Corrosion* 15, 12 (1959), 627t.
4. H.J. de Bruyn and M.L. Holland, 'Materials experience in a methane reforming plant', CORROSION/98, paper no.429, (Houston, TX: NACE, 1999).
5. R.F. Hochman and J.H. Burson, 'The fundamentals of metal dusting', *API Division of Refining Proc.*, 46 (1966), 331.
6. R.F. Hochman, 'Basic studies of metal dusting deterioration ("metal dusting") in carbonaceous environments at elevated temperatures', *Proc. of the 4th International Congress on Metallic Corrosion*, NACE (1972), 258.
7. H.J. Grabke, R. Krajak, and Müller-Lorenz, *Werkst. Korros.* 44 (1993), 89.
8. J.C. Nava Paz and H.J. Grabke, *Oxid. Met.* 39, 5/6 (1993), 437.
9. H.J. Grabke, R. Krajak, and J.C. Nava Paz, *Corros. Sci.* 35, 5–8 (1993), 1141.
10. H.J. Grabke, C.B. Brancho-Troconis, and E.M. Müller-Lorenz, *Mater. Corros.* 45 (1994) 215.
11. R.C. Schueler, *Hydrocarbon Process* (1972), 73.
12. C.M. Schillmoller, *Chem. Eng.* 93, 1 (1986), 83.
13. J. Klöwer, H.J. Grabke, E.M. Müller-Lorenz, and D.C. Agarwal, 'Metal Dusting and Carburization Resistance of Nickel-base Alloys', *CORROSION/97*, paper no. 0139, (Houston, TX: NACE, 1997).
14. B.A. Baker, G.D. Smith, V.W. Hartmann, and L.E. Shoemaker, 'Nickel-base Material Solutions to Metal Dusting Problems', *CORROSION/2002*, paper no. 2394, (Houston, TX: NACE, 2002).
15. D.L. Trimm, *Catal. Rev. Sci. Eng.* 16 (1977), 155.
16. I. Alstrup, *J. Catal.* 109 (1988), 241.
17. I. Alstrup, M.T. Tavares, C.A. Bernardo, O. Sørensen, and J.R. Rostrup-Nielsen, *Mater. Corros.* 49 (1998), 367.
18. J. L. Figueiredo, *Mater. Corros.* 49 (1998), 373.
19. M.T. Tavares, I. Alstrup, and C.A.A. Bernardo, *Mater. Corros.* 50 (1999), 681.

20. Y. Nishiyama, N. Otsuka, T. Kudo, and O. Miyahara, 'Metal Dusting of Nickel-Base Alloys in Simulated Syngas Mixtures', *CORROSION/2003*, paper no. 3471, (Houston, TX: NACE, 2003).

21. H. Adachi, M. Tsukada, and C. Satoko, *J. Phys. Soc. Jpn.* 45 (1978), 875.

22. M. Methfessel, *Phys. Rev. B* 38 (1988) 1537; M. Methfessel, C.O. Rodriguez, and O. K. Andersen, *Phys. Rev. B* 40 (1989), 2009.

23. S.Y. Savrasov, *Phys. Rev. B* 54, (1996) 16470.

24. The LmtART code is available from the WWW site of Department of Physics, New Jersey Institute of Technology (http://physics.njit.edu/~mindlab/index.html).

25. S.H. Vosko, K. Wilk and N. Nusair, *Can. J. Phys.* 58 (1980), 1200.

26. C. Toh, D.J. Young and P.R. Muroe, *Oxid. Met.* 58, 1/2 (2002), 1.

27. W. Andreoni and C.M. Varma, *Phys. Rev. B* 23 (1981), 437.

28. H.J. Grabke, *Mater. High Temp.* 17 (2000), 483.

29. D.W. Moon, D.J. Dwyer, and S. L. Bernasek, *Surf. Sci.* 163 (1985), 215.

30. R.S. Saiki *et al.*, *Phys. Rev. Lett.* 63 (1989), 283.

31. F. Jona *et al.*, *Phys. Rev. Lett.* 40 (1978), 1466.

32. K.O. Legg, F. Jona, D.W. Jepsen, and P.M. Marcus, *Surf. Sci.* 66 (1977), 25.

33. S. Ohnishi and N. Watari, *Phys. Rev. B* 49 (1994) 14619, and references therein.

34. G. Blyholder, *J. Phys. Chem.* 68 (1964), 2772.

35. D.C. Sorescu, D.L. Thompson, M.M. Hurley, and C.F. Chabalowski, *Phys. Rev. B* 66 (2002) 35416.

36. S. Andersson and J.B. Pendry, *Phys. Rev. Lett.* 43 (1979), 363.

37. O.K. Andersen *et al.*, *Physica B*, 86–88 (1977), 249.

38. M.J. Kelley and V. Ponec, *Prog. Surf. Sci.* 11 (1981), 139 and references therein.

39. A.V. Ruban, H.L. Skriver, and J.K. Nørskov, *A Phys. Rev. B* 59 (1999) 15990.

Part II

Surface treatment

7

Effects of minor additions and impurities on oxidation behaviour of FeCrAl alloys in the development of novel surface coatings compositions (SMILER)

V K O C H U B E Y, Forschungszentrum Jülich GmbH, Germany;
H A L - B A D A I R Y and G T A T L O C K, University of
Liverpool, UK; J L E - C O Z E, Ecole Nationale Supérieure des
Mines de Saint-Etienne, France and D N A U M E N K O and
W J Q U A D A K K E R S, Forschungszentrum Jülich,
GmbH, Germany

7.1   Introduction

Due to the formation of a protective alumina scale, FeCrAl alloys exhibit excellent oxidation resistance at temperatures up to 1300°C. However, the lifetime of the FeCrAl alloys may be limited by oxidation because of depletion of the alloy Al reservoir, which is consumed by the scale growth and re-healing in the case of scale spalling [1]. The growth rate and adherence of the alumina scale depend to a large extent on minor alloying constituents including reactive elements (e.g. Y, Zr, Hf, etc.) and impurities (e.g. S, C, etc.). Extensive research has been performed into the respective effects over the last 30 years and many of the mechanisms are still under debate [2–5]. It has, however, been widely accepted that the lifetime extension of FeCrAl components requires optimisation of the minor alloy composition. Studies aiming at lifetime extension have been carried out in a number of research programmes in the European Union and the United States [6, 7].

Unequivocal interpretation of the minor element effects using commercial FeCrAl materials is frequently difficult to derive because the alloys contain several reactive elements and a number of impurities, whereby their concentrations may vary between different alloy batches [8]. Therefore, within the framework of the EU-funded BRITE-EURAM programme LEAFA, a series of high purity model alloys were procured, whereby different minor elements were added in a systematic way to an FeCrAlY-base alloy. The alloy compositions were selected to study the effects of single additions (impurities). Important data regarding the effects of P, V, Ca, Ti and Zr on the oxidation behaviour of FeCrAlY alloys were obtained with test times up to 27000 h, as reported elsewhere [9, 10]. In a further EU-funded programme, SMILER, the studies on the model alloys have been extended mainly to account for the effects of multiple RE additions and to get more insight into the effect of several common impurities (C, Si). The results of the model alloy studies in the SMILER project are intended to be transferred to the development of overlay coatings for improvement of the cyclic oxidation performance of commercial, oxide dispersion strengthened (ODS) FeCrAl alloys.

In the present chapter, a summary of the effects of different alloying additions to the base FeCrAlY composition will be demonstrated on the basis of long-term oxidation data. The mechanisms responsible for these effects will be discussed, using results obtained with specimens after shorter exposure times using a number of analytical techniques. In addition, implementation of the main findings on the model alloys for coating applications will be considered.

7.2   Experimental methods

The model alloys with a base composition (in wt%) Fe-20Cr-5Al-0.05Y (Table 7.1) were prepared by melting and solidification in a cold crucible under pure Ar. The ingots were forged and hot rolled down to 1 mm at 1000°C. The final annealing treatment was 1 h at 1000°C in pure Ar followed by water quenching. The chemical compositions of the sheets analysed by ICP-MS are summarised in Table 7.2.

Specimens 20×10 mm in size were machined from the alloys sheets. The specimens were ground using 1200 grit SiC-paper and cleaned in a detergent prior to oxidation exposure. The discontinuous oxidation tests were performed in a resistance-heated tube furnace in laboratory air at 1200°C. The cycle duration was 92 h at the oxidation temperature and 4 h still air cooling at room temperature, respectively. For collecting the oxide spalls the specimens were oxidised in alumina crucibles. The mass of the crucibles containing the specimens (gross mass change) and of the crucibles without the specimens (spall mass change) were measured at cooling intervals. In addition to the discontinuous oxidation tests, thermogravimetric experiments were performed

Table 7.1 Nominal compositions of the studied model alloys

MRef (Fe-20Cr-5Al-0.05Y; wt%)
+ Zr (300 ppm)
+ C (500 ppm)
+ Hf (250 ppm)
+ Si (5000 ppm)
+ Zr + Hf (300 ppm of each)
+ Zr + Ti (300 ppm of each)
+ Hf + Ti (300 ppm of each)

Table 7.2 Chemical analyses (ICP-MS) of the studied model alloy sheets (wt%)

	MRef	+Zr	+C	+Hf	+Si	+Zr+Hf	+Zr+Ti	+Hf+Ti
Cr	19.9	19.7	19.60	19.40	19.32	19.28	19.78	19.53
Al	5.0	4.86	5.10	5.00	5.01	5.06	4.97	5.01
C	0.0092	0.009	0.053	0.0115	0.0114	0.0131	0.0138	0.0137
S	< 0.001	< 0.001	0.0007	0.0005	0.0005	0.0005	0.0006	0.0005
O	< 0.001	< 0.001	0.0003	0.0004	0.0004	0.0006	0.0005	0.0007
N	< 0.001	< 0.001	0.0006	0.0006	0.0006	0.0004	0.0006	0.0004
P	0.0086	0.0078	0.0086	0.0097	0.0087	0.0076	0.0051	0.0092
Y	0.051	0.049	0.046	0.045	0.046	0.047	0.043	0.046
Ti	0.0015	0.0012	–	–	–	–	0.032	0.031
Hf	< 0.005	< 0.005	–	0.031	–	0.030	–	0.029
Zr	< 0.001	0.029	–	–	–	0.032	0.031	–
Si	0.002	0.0018	–	–	0.496	–	–	–

in a SETARAM® thermobalance and to study the effect of cooling rate selected samples were furnace cooled and quenched in liquid nitrogen. The oxidised specimens were characterised mainly as metallographic cross-sections using scanning electron microscopy/energy dispersive X-ray analysis (SEM/EDX). In several cases the samples were etched in HNO_3-solution to reveal the important microstructural features of the bulk alloy.

7.3   Results and discussion

Figure 7.1 shows the mass change data of the studied model alloys obtained during 7000 h discontinuous oxidation at 1200°C. The higher initial (up to ca 500 h exposure) oxidation rates of the Zr-containing alloys ('+Zr'; '+Zr+Hf' and '+Zr+Ti'), as compared to those of the alloys without Zr, were not found to deteriorate the long-term oxidation behaviour. Moreover, the lifetime of 1 mm thick alloy '+Zr' appeared to be superior to that of 'MRef' (27000 h vs 20000 h respectively) [11]. This superior behaviour of alloy '+Zr' could be mainly attributed to a better scale adherence compared to that of 'MRef' (Fig. 7.1 and reference [11]). Another obvious observation, which could be made from the mass change data in Fig. 7.1 was the very poor scale adherence on alloy '+C' (500 ppm carbon) resulting in early failure of this material.

Extensive SEM studies performed on the prevailing materials after the discontinuous oxidation revealed that in respect to extent as well as the mode of scale spallation, the studied alloys could be divided into two groups, i.e. those containing Zr and/or Hf compared to 'MRef', '+Si', '+C' (Table 7.3). In the former case a 'cohesive mode' of spallation, which occurred within the scale prevailed, whereas in the latter case the scales mainly spalled at the oxide/metal interface, i.e. in an 'adhesive mode' (Fig. 7.2).

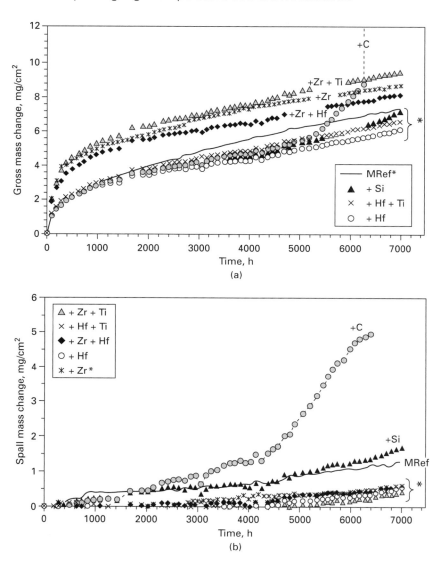

7.1 Mass change data of the studied model alloys during discontinuous (92 h heating/4 h cooling) air oxidation at 1200°C: (a) gross mass change; (b) mass change of spalled oxide.

An extensive description of the scale characterisation of all studied alloys falls outside the scope of this chapter. Therefore, in the following the mechanistic studies performed on these two groups of materials will be demonstrated mainly using alloys '+C' and '+Zr' as examples.

Table 7.3 Microscopic observations of the oxide spallation mode of the studied alloys after 3100 h air oxidation at 1200°C

Alloy	Mode of spallation	
	Adhesive	Cohesive
'MRef'	✓	
'+C'	✓	
'+Si'	✓	
'+Zr'		✓
'+Zr+Hf'		✓
'+Hf'		✓
'+Hf+Ti'		✓

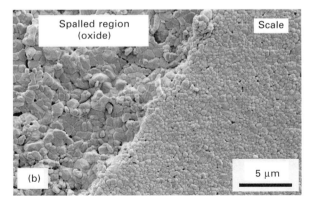

7.2 SEM images after 3100 h discontinuous oxidation at 1200°C showing different modes of oxide spallation: (a) alloy '+C' – adhesive mode; (b) '+Zr+Ti' – cohesive mode.

7.4   Alloys with adhesive mode of scale spallation: 'MRef', '+C'

A comparison of the macro-photographs taken from the breakaway specimens in Fig. 7.3 reveals the occurrence of massive, adhesive type scale spallation from the high carbon alloy (in agreement with the oxidation mass changes in Fig. 7.1b). In contrast, on the reference alloy with 90 ppm carbon the spallation is mainly restricted to macro-cracks occurring across the scale. The protrusions growing outwards through the alumina scale could be observed on both materials, although the locations and the morphology of the protrusions differed slightly. The protrusions on alloy '+C' started to grow much earlier (after *ca* 2300 h) than those on 'MRef' (5000 h). The formation of these local protrusions of alumina and Fe-based oxides (termed 'broccolis'), was previously [12] proposed to be related to the formation of Cr carbides in combination with Y aluminates at the alloy grain boundaries near the scale/metal interface. It was suggested that the reason for the 'broccoli effect' is oxidation of the Cr carbides, exposed to the atmosphere after scale cracking/spallation.

Figure 7.4 shows the alloy microstructure of alloy 'MRef' in the non-oxidised condition with clear Cr carbide formation at the alloy grain boundaries. It is important to note that the phase contrast in backscattered electron images between the alloy and the carbides is very poor and without chemical etching it was hardly possible to detect the carbide formation and distribution. In the high-carbon alloy ('+C'; Fig. 7.5) the Cr carbide formation not only occurs at the alloy grain boundaries, as in 'MRef', but also within the alloy grains. The amount of carbides formed in the alloy matrix appeared to be promoted by decreasing the cooling rate as illustrated in Fig. 7.6. This result indicates that the carbides are primarily formed during cooling, which is supported by

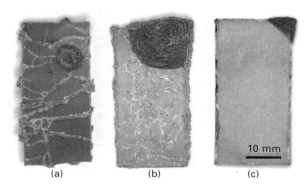

(a)   (b)   (c)

7.3 Macro-photographs of three model alloys (20 × 10 × 1 mm in size) after discontinuous air oxidation till breakaway at 1200°C: (a) alloy 'MRef' for 20800 h; (b) alloy '+C' for 6400 h; and (c) alloy '+Zr' for 27200 h.

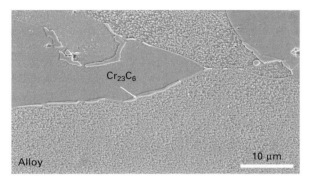

7.4 SEM of etched cross-section of alloy 'MRef' in as-received condition illustrating formation of Cr carbides at the alloy grain boundaries.

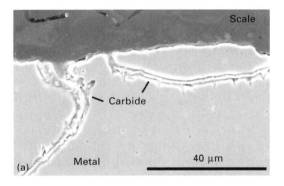

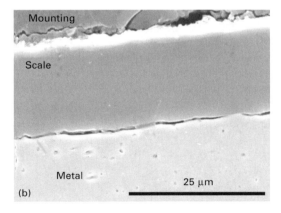

7.5 SEM of etched cross-section of alloy '+C' after 3100 h discontinuous oxidation at 1200°C showing chromium carbide formation in the metal (a) and crack propagating along scale/metal interface (b).

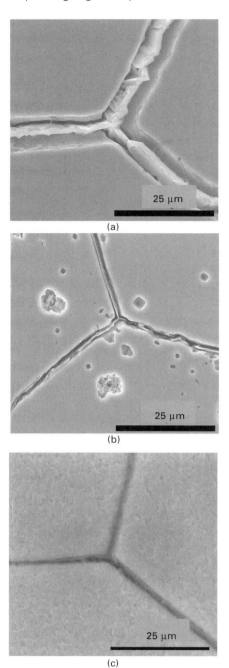

(a)

(b)

(c)

7.6 SEM images of chemically etched alloy '+C' after 20 h annealing at 1200°C in air: (a) furnace cooled; (b) air cooled; (c) liquid nitrogen quenched.

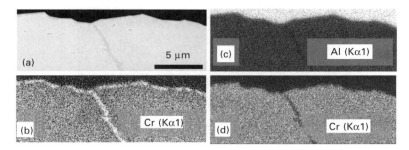

7.7 SEM backscattered electron image near scale/alloy interface (a) and respective EDX-mappings for Cr (b), Al (c) and Fe (d) of alloy '+C' after 1000 h discontinuous oxidation at 1200°C.

thermodynamic calculations [13]. In addition to alloy matrix precipitation, the carbides in alloy '+C' could locally be found at the scale/metal interface of the chemically etched specimens, but they could also be verified on non-etched cross-sections by EDX-mapping (Fig. 7.7). Such interfacial carbide formation was not observed in the case of alloy 'MRef' with the lower C content. Hence, it is suggested that the interfacial formation of a brittle carbide with low strength is responsible for the extensive oxide scale spallation observed after long-term oxidation of alloy '+C'. The deteriorated scale adherence in combination with the broccoli effect is responsible for the much earlier breakaway failure of alloy '+C' compared to that of alloy 'MRef' (6400 h vs 20800 h respectively for 1 mm thick specimens). It is important to note that the measured Al content in the breakaway specimen of alloy '+C' was also higher than that in alloy 'MRef' (1.5wt.% vs. 0.1 wt.% respectively).

7.5   Alloys with cohesive mode of scale spallation: '+Zr', '+Hf'

Compared to the reference alloy, the Zr-containing material revealed up to breakaway no broccoli effect and only minor macroscopically visible spallation at the specimen corners and edges (Fig. 7.3), in agreement with the mass change data in Fig. 7.1. The suppression of the broccoli effect by Zr has previously been related to tying up the carbon impurity into more stable carbide [12]. It should be noted that no Cr carbide formation was detected in the alloy '+Zr'. The microstructures of the scales after 1000 h oxidation are shown in Fig. 7.8. Compared to alloy 'MRef' the alumina scale on alloy '+Zr' is thicker and contains zirconia inclusions and porosity in the outer part. Studies of the scale grain structure were performed using electron back-scattered diffraction (EBSD). The studies revealed (Fig. 7.9) that the alumina

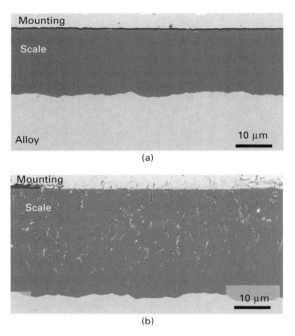

7.8 SEM images of the scales formed on alloy 'MRef' (a) and alloy '+Zr' after 1000 h discontinuous air oxidation at 1200°C.

scale on alloy 'MRef' has a columnar microstructure, typical for Y-containing FeCrAl alloys. In contrast in the outer part of the scale on alloy '+Zr' small, equi-axed grains with a much higher density of grain boundaries were observed compared to the scale on alloy 'MRef'. Nearer to the scale/metal interface the oxide grain size on alloy '+Zr' increases and the scale microstructure becomes similar to that of 'MRef'. Based on the observations of the oxide microstructure, the initially enhanced growth rate of the oxide scale on alloy '+Zr' (Fig. 7.1) can be explained by a higher density of oxygen transport paths, i.e. oxide grain boundaries, porosity and zirconia particles. However, such a microstructure is expected to be more resistant against macro-fracture, e.g. by a crack blunting mechanism as commonly observed in construction ceramics [14, 15]. The latter argument can be used to explain the increased resistance to spalling of the scale on alloy '+Zr' as well as the different, i.e. 'cohesive' mode of spallation compared to alloys 'MRef' and especially '+C', of which the alumina scales fail in adhesive mode. Indeed, it has been shown [16] that the scale through cracking is the necessary prerequisite for the scale spallation during cooling. Hence, if the scale through cracking is hindered, it should lead to an enhanced scale resistance to spallation.

The absence of the zirconia particles and the different oxide microstructure in the inner part of the scale (Figs 7.8 and 7.9) lead to the conclusion that the

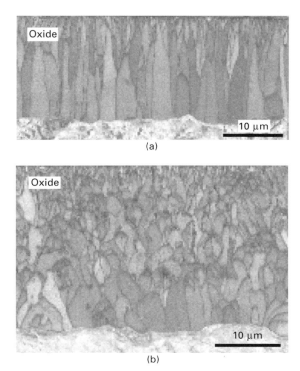

(a)

(b)

7.9 EBSD images of the scales formed on alloy 'MRef' (a) and alloy '+Zr' after 1000 h discontinuous air oxidation at 1200°C (cf. Fig. 7.8).

effect of Zr on the scale microstructure not only depends on the alloy Zr content but also on the Zr reservoir in the prevailing specimen. The relatively small Zr content of 300 ppm wt. seems to be depleted from the 1 mm thick specimen already within 1000 h oxidation at 1200°C. In order to verify the assumption that the Zr reservoir is equally important for the oxidation behaviour as the alloy Zr concentration, thermogravimetric experiments at 1300°C were performed with specimens of alloy '+Zr' of different thickness (Fig. 7.10). It can be seen that the thicker 1 mm specimen shows initially a reproducible higher oxidation rate than the 0.3 mm specimen with its factor of three lower Zr reservoir. After depletion of the Zr reservoir, the oxidation rates of both specimens exhibit similar values. The depletion of the Zr reservoir is expected to occur especially rapidly in thin FeCrAl components (coatings) during exposure at high temperatures and hence it should be considered when optimising their compositions.

The effect of Hf, which also belongs to the group IVa of the periodic system is to some extent different to that of Zr. Although both Zr and Hf improve the scale adherence, a Hf addition of 0.03 wt.% to the FeCrAlY base alloy does not lead to an increase of the oxide growth rate, which was

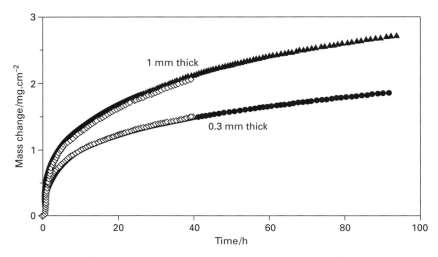

7.10 Thermogravimetric data measured during isothermal oxidation for 40 h and 100 h in Ar-20%O_2 at 1300°C on 0.3 and 1 mm thick specimens of alloy '+Zr'.

found for the materials containing 0.03 wt.% of Zr (Fig. 7.1). Microstructural studies reveal that Hf together with Y becomes incorporated into the oxide scale thereby forming tiny precipitates at the oxide grain boundaries (Fig. 7.11). At the outer scale surface, 0.2 to 1 µm large hafnia precipitates could be observed indicating the occurrence of Hf transport across the alumina scale (Fig. 7.12). However, based on the data available at the moment, Hf is far less rapidly incorporated into the alumina scale compared to Zr, which is confirmed by the fact that the oxidation rate of alloy '+Hf' is not initially increased. Therefore, it is unlikely that the scales on the Hf-containing alloys exhibit the same amount of the short circuit paths for oxygen transport as the scales formed on the Zr-containing alloys.

7.6   Implications for alloy/coating development

Based on the results of the studies described in this chapter, several model alloys were selected as coating material to improve the oxidation resistance of high creep resistant, FeCrAl ODS materials. The coating compositions were similar to those of alloys '+Zr', '+Zr+Hf' and '+Zr+Ti' with lower carbon content than that normally present in commercial FeCrAl alloys. It has been anticipated that a FeCrAl coating with a low creep strength may contribute to the relaxation of the thermal stresses in the scale and thus to better scale adherence during cyclic oxidation. Poor oxide stress relaxation by metal creep [16] has been often reported to increase oxide scale spallation and thus decrease the lifetime of ODS alloys, such as PM2000 or Kanthal APM [17]. In Fig. 7.13(a) a magnetron-sputtered coating on the commercial

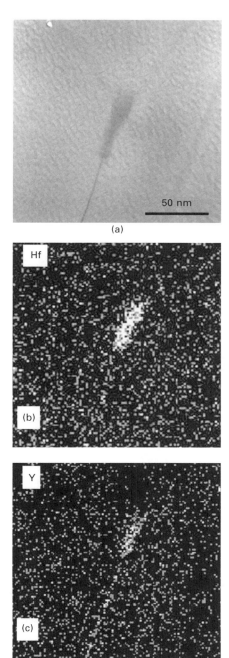

7.11 STEM bright field image (a) and corresponding elemental maps for Hf (b) and Y (c). Oxide grain boundary on alloy '+Hf' after 3100 h air oxidation at 1200°C.

7.12 SEM plan view image showing alumina surface scale containing hafnia precipitates after oxidation of alloy '+Hf' for 3100 h at 1200°C in air.

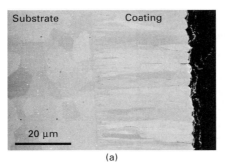

(a)

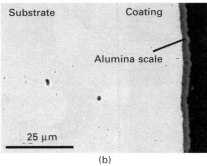

(b)

7.13 Sputter-coating '+Zr+Hf' on commercial, high strength alloy Kanthal APMT substrate in as-coated condition (a) and after 100 h cyclic oxidation (20 h cycles) at 1200°C (b).

alloy Kanthal APMT is shown. During cyclic oxidation at 1200°C the formed alumina scales appeared to possess excellent adherence to the coating (7.13(b)). The optimisation of the coating parameters is on-going and the results will be published in a future paper.

7.7   Conclusions

Increasing the carbon content of an FeCrAlY alloy from 90 to 500 ppm wt. results in more extensive spallation of the alumina scale, more extensive formation of local oxide protrusions ('broccoli effect') and a higher critical Al content, at which breakaway occurs. The effects induced by a high carbon content of 500 ppm lead to shortening of the alloy oxidation limited lifetime by a factor of three to four compared to that of the alloy with 90 ppm carbon. It is suggested that the negative effects of carbon can be related to extensive Cr carbide formation within the alloy (grain boundaries) as well as at the scale/metal interface. An important finding was that the extent and morphology of the carbide precipitates depends substantially on cooling rate. Further studies are necessary to elucidate the effect of cooling rate on the lifetime of high-C alloy.

Minor alloying additions of Zr prevent the negative effects of carbon apparently by tying up the carbon impurity into more stable carbides. Furthermore, Zr is rapidly incorporated into the alumina scale thereby modifying its structure by reducing the oxide grain size as well as by formation of zirconia precipitates and nano-porosity. The Zr-imparted modifications to the scale structure lead to an enhanced oxidation rate compared to the Zr-free FeCrAlY composition. However, the alumina scale structure modified by Zr incorporation shows an excellent resistance against spallation under thermal cycling conditions. It is important to note that the rapid Zr incorporation into the alumina scale may lead to an early exhaustion of the Zr reservoir and thus to a loss of the Zr effect on scale microstructure after longer exposure times, especially in the case of thin specimens.

Addition of Hf has a similar effect to that of Zr with respect to increasing the scale resistance to spallation and suppressing the 'broccoli effect'. However, the effect of Hf on the scale microstructure is somewhat different from that of Zr resulting in slower oxidation kinetics.

The results of the studies with the model alloys have been used for optimisation of the overlay coating compositions for high strength, commercial ODS FeCrAl alloys.

7.8   Acknowledgment

This work has been supported by the European Commission under the 5th Framework Programme – Competitive and Sustainable Growth (Project SMILER, Contract No. G5RD-CT-2001-00530).

7.9   References

1. W.J. Quadakkers and M.J. Bennett, Oxidation induced lifetime limits of thin walled, iron based, alumina forming, oxide dispersion strengthened alloy components, *Materials Science and Technology*, 10, 126–131 (1994).

2. D.P. Whittle and J. Stringer, Improvements in high temeperature oxidation resistance by additions of reactive elements or oxide dispersions, *Philosophical Transactions of the Royal Society in London*, A295, 309–329 (1980).

3. R. Prescott and M.J. Graham, The formation of aluminium oxide on high-temperature alloys, *Oxidation of Metals*, 38, 3/4, 233–254 (1992).

4. B.A. Pint, Experimental observations in support of the dynamic segregation theory to explain the reactive element effect, *Oxidation of Metals*, 45, 1/2, 1–31 (1996).

5. P.Y. Hou, Compositions at Al_2O_3/FeCrAl interfaces after high temperature oxidation, *Materials and Corrosion*, 51, 329–337 (2000).

6. J.R. Nicholls, Life Extension of alumina forming alloys – background, objectives and achievements of the BRITE/EURAM Programme LEAFA, in *Lifetime Modelling of High Temperature Corrosion Processes*, EFC Monograph No. 34, The Institute of Materials, London, 3–15 (2001).

7. B.A. Pint, Optimization of reactive-element additions to improve oxidation performance of alumina-forming alloys, *J. Am. Ceram. Soc.*, 86, 4, 686–695 (2003).

8. W.J. Quadakkers, D. Naumenko, L. Singheiser, H.J. Penkalla, A.K. Tyagi and A. Czyrska-Filemonowicz, Batch to batch variations in oxidation behaviour of alumina forming Fe-based alloys, *Materials and Corrosion*, 51, 350–357 (2000).

9. D. Naumenko, W.J. Quadakkers, V. Guttmann, *et al.*, Critical role of minor elemental constituents on the life time oxidation behaviour of FeCrAl-RE alloys, in *Lifetime Modelling of High Temperature Corrosion Processes*, EFC Monograph No. 34, The Institute of Materials, London (2001).

10. H. Al-Badairy, D. Naumenko, J. Le-Coze, G.J. Tatlock and W.J. Quadakkers, Significance for minor alloying additions and impurities on alumina scale growth and adherence on FeCrAl alloys, *Materials at High Temperatures*, 20, 3, 405–412 (2003).

11. V. Kochubey, Effect of Ti, Hf and Zr addition and impurity elements on the oxidation limited lifetime of thick-and thin walled FeCrAly-components, PhD Thesis, University of Bochum, Germany, p 73 (2005).

12. D. Naumenko, J. Le-Coze, E. Wessel, W. Fischer and W.J. Quadakkers, Effect of trace amounts of carbon and nitrogen on the high temperature oxidation resistance of high purity FeCrAl alloys, *Materials Transactions*, 43, 2, 168–172 (2002).

13. V. Kochubey, H. Al Badairy, J. Le Coze, D. Naumenko, G. Tatlock, E. Wessel, W.J. Quadakkers, Effect of Carbon Content on the Oxidation Behaviour of FeCrAly Alloys in the Temperature Range 1200–1300°C, Materials at High Temperatures, 22(3/4), p. 297–302 (2005).

14. W.D. Kingery, H.K. Bowen and D.R. Uhlmann, *Introduction to Ceramics*, J. Wiley and Sons, New York (1976).

15. W.H. Tuan, R.Z. Chen, T.C. Wang, C.H. Cheng and P.S. Kuo, Mechanical properties of Al_2O_3/ZrO_2 composites, *Journal of the European Ceramic Society*, 22, 2827–2833 (2002).

16. H.E. Evans, Stress effects in high temperature oxidation of metals, *International Materials Reviews*, 40, 1, 1–40 (1995).

17. J.P. Wilber, M.J. Bennett and J.R. Nicholls, The effect of thermal cycling on the mechanical failure of alumina scales formed on commercial FeCrAl-RE alloys, in *Cyclic Oxidation of High Temperature Materials*, EFC monograph No. 27, The Institute of Materials, London, 133–147 (1999).

8

Lifetime extension of FeCrAlRE alloys in air. Potential role of enhanced Al reservoir and surface pre-treatment (SMILER)

J R NICHOLLS, M J BENNETT and N J SIMMS, Cranfield University, UK; H HATTENDORF, Thyssen Krupp VDM GmbH, Germany; D BRITTON and A J SMITH, Diffusion Alloys Ltd, UK; W J QUADAKKERS, D NAUMENKO and V KOCHUBEY, Forschungszentrum Jülich GmbH, Germany; R FORDHAM and R BACHORCZYK, JRC Petten, The Netherlands and D GOOSSENS, Bekaert Technology Centre, Belgium

8.1 Introduction

Focused technological advances are essential to control/eliminate critical domestic and industrial emissions to combat global warming. Potential approaches include exhaust gas catalytic converters and improved industrial process efficiency, inevitably based on higher temperature operations. Foremost among metallic alloys for such applications are the FeCrAlRE steels (where RE is a reactive element), primarily because their outstanding corrosion resistance is based on protective alumina scales. Intensive research during the last decade or so, worldwide, and most especially in Europe in a succession of initiatives (COST 501/WP2 and the CEC BRITE/EURAM programmes IMPROVE [1] and LEAFA [2]), has generated lifetime corrosion data for commercial FeCrAlRE alloys exposed to most industrial relevant environments across potential operational and envisaged fault temperature ranges [e.g. 3–5]. A framework now exists also on scientific understanding underlying the growth, mechanical failure and ultimate chemical failure of protective alumina scales [e.g. 6]. Combination of data and understanding has enabled development of a life prediction model [7, 8]. However, with this considerable knowledge base it is clear that the existing commercial FeCrAlRE alloys cannot satisfy completely all the envisaged technological demands for emissions control and improved industrial process operations, in particular necessitating longer life-times, thinner metallic component sections, etc. Investigation of potential approaches to achieve such objectives has been undertaken in the most recent European research SMILER project.

Identification of such potential approaches stems from understanding of the oxidation behaviour of FeCrAlRE alloys [3–8]. As mentioned, their corrosion resistance is afforded by the formation and growth of an α-alumina scale, which concurrently depletes the alloy aluminium reservoir. The oxidation induced alloy lifetime is governed by chemical failure of the protective scale and ultimately to non-protective/breakaway attack, essentially in two stages [3, 5], as shown schematically in Fig. 8.1. The first occurs when the alloy Al activity falls below a critical value the $[Al]_{crit}$ is unable to sustain the protective alumina scale. Both the aluminium consumption rate and $[Al]_{crit}$ value can be influenced detrimentally by the occurrence of prior scale mechanical failure, by cracking and spallation, with subsequent scale rehealing on re-exposure to the oxidant. For all situations, however, the second chemical failure stage derives from oxidation of iron leading eventually to complete metal consumption. Under certain circumstances, during the intervening period a Cr_2O_3 sub-scale can form beneath the outer-alumina layer and afford what has been termed 'psuedo protection' [9]. Its duration depends on the alloy composition, section thickness and, in general, decreases markedly with increasing temperature. To date for practical purposes and modelling, therefore, lifetimes have been defined as the onset of the first chemical failure stage.

The most obvious approach to extend lifetimes would be to increase the available alloy aluminium reservoir. Most commercial FeCrAlRE alloys, fabricated by either the conventional melting, powder metallurgical or mechanical alloying routes, contain between 5 and 6 w/o aluminium. Using experimental oxide dispersion strengthened (ODS) alloys with Al contents

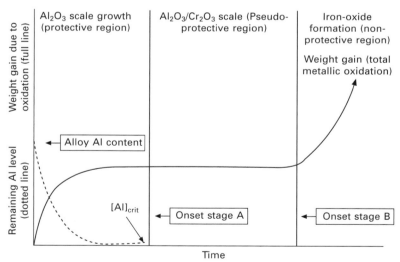

Stage A – Alloy Al level falls below $[Al]_{crit}$: Cr oxidised to form a Cr_2O_3 sub-scale.
Stage B – Iron oxidation initiated : Breakaway/Non-protective oxidation.

8.1 Schematic representation of the oxidation induced chemical failure of FeCrAlRE alloys.

ranging between 3.1 and 7.1 w/o, the times to breakaway, in air, at 1200–1300°C increased markedly with increasing Al content [10, 11]. Commercially it has not proved possible to increase homogeneous Al levels to 7 w/o or above by conventional processing [12], as decreasing ductility with increasing Al content causes severe metallurgical processing problems. However, early in this project ThyssenKrupp VDM increased the Al content of their standard wrought alloy Aluchrom YHf, and the resulting alloy Aluchrom YHfAl (containing 6 w/o Al) has been included in the evaluation test programme. An alternative way forward to enrichment could be by Al surface deposition and subsequent diffusion into the substrate. Previous studies [9, 13] using electron beam Al vapour deposition have confirmed the viability of this approach. The SMILER project has assessed two procedures to enrich the alloy Al content to 7–8 w/o, firstly, by hot-dipping or cladding to form Aluchrom P, and secondly by aluminising the commercial alloys Aluchrom YHf, Aluchrom YHfAl and Kanthal AF. Although the initial development of Aluchrom P has been reported [12], the unique aluminising process optimisation work was undertaken entirely within the SMILER project.

Although the use of FeCrAlRE alloy foils as a metallic substrate for vehicle exhaust catalysts was advanced [14] as early at 1978, this potential has only really been pursued vigorously within the last decade. The initial Harwell work was based on envisaged operating temperatures of $\geq 1100°C$. However, currently perceived catalyst temperatures are much lower, down to $\sim 800°C$ [15]. The main focus of most corrosion testing of FeCrAlRE alloys has been directed to performance at elevated temperatures ($\geq 1000°C$), where the protective scale is α-Al_2O_3. By comparison, only limited previous research has been directed to the oxidation behaviour of these alloys over the lower temperature range [16–19]. This has been augmented by further extensive testing of 30–50 μm thick commercial FeCrAlRE alloy foils in the SMILER project at 800°, 850°, 900° and 950°C, which has involved substantially longer test exposures than in the earlier studies [20, 21]. At these temperatures oxidation involves the formation of transitional aluminas, produced by outward cation growth until an inner protective α-Al_2O_3 layer is nucleated. With increasing temperature the oxidation rate increased through a maximum at around 850–900°C before increasing again progressively with increasing temperature with α-Al_2O_3 formation by inward anion movement. For the low temperature range (i.e. 800–950°C) surface protection could be provided best by pre-oxidation to preform an α-Al_2O_3 scale prior to service, as has been examined previously [22]. A novel, controlled gas annealing treatment has been developed by the SMILER project [23], which minimised depletion of the alloy Al prior to service. Assessment of the effectiveness of gas annealing upon the long-term oxidation behaviour of FeCrAlRE alloys in air has been undertaken in the current study. As with evaluation of the potential of an increased alloy Al reservoir, experiments have been undertaken not only on sheet and foil coupons but also on near real-life components, including fibre mats and model catalytic converter supports.

8.2   Experimental methods

8.2.1   Alloys

The commercial FeCrAlRE alloys tested were Kanthal AF, Aluchrom YHf and Aluchrom YHfAl. These alloys, prime products of the project industrial partners, Kanthal AB and ThyssenKrupp VDM, were all manufactured by the conventional route, involving ingot melting, hot rolling and cold rolling with intermediate heat treatments. The final product forms were 1 mm thick sheet and 50 μm thick foils, from which standard 2 cm × 1 cm coupons were cut for oxidation testing. In addition, to confirm that results obtained from these standard corrosion coupons were relevant technologically, two types of model components were manufactured from the Aluchrom YHf and Aluchrom YHfAl foils, namely interwoven mats of 20 × 50 μm fibres by Bekaert and a model catalyst support (39 mm o.d., 71.5 mm long) by ThyssenKrupp VDM. The composition of these alloys (Table 8.1) was determined by chemical analyses (mainly using atomic absorption spectroscopy (AAS), inductively coupled plasma mass spectroscopy (ICPMS), glow discharge mass spectroscopy (GDMS), colorimery and oxidation (C and S) or melting (O and N)).

8.2.2   Surface modification to increase alloy
Al reservoir

As mentioned, the conventional production route for the FeCrAlRE alloys limits their aluminium content to around 5–6 w/o because they tend to embrittle during hot rolling. Two procedures have been examined to enrich the Al levels up to 8 w/o Al: by fabrication of Aluchrom P (ThyssenKrupp VDM) and by aluminising (Diffusion Alloys Ltd). Since the hot rolling stage is easier for alloys with a lower Al content, Aluchrom P manufacture commences with FeCrAl strip, with a reduced Al content and additions of Y, Hf and Zr, produced by the conventional route; hot rolling followed by cold rolling to 0.6–1.4 mm, roll bonding with Al or hot dipping in Al + 10 mass% Si, cold rolling to 50 μm thickness and finally, a diffusion anneal. The analysed chemical composition of Aluchrom P is also given in Table 8.1, while EPMA elemental line scans across a half foil cross-section (Fig. 8.2) demonstrate the compositional uniformity. Aluchrom P was tested only as a model catalyst support.

Aluminising of Aluchrom YHf, Aluchrom YHfAl and Kanthal AF corrosion coupons specimens of both alloy forms (50 μm thick foil and 1 mm thick

Table 8.1 Chemical compositions of commercial alloys

Alloy	wt.%		ppm														
	Al	Cr	Mn	Si	Y	Hf	Zr	Ti	Mg	V	C	S	O	N	P
Kanthal AF	5.2	21.0	610	1900	340	3.1	580	940	17	200	280	1.5	<10	150	140
Aluchrom YHf	5.6	20.0	2250	2900	480	400	320	99	85	600	220	1.3	<10	40	130
Aluchrom YHfAl	6.1	20.6	2000	2700	600	400	530	100	80	790	240	6.3	14	24	110
Aluchrom P	7.6	20.2	1900	3300	500	300	400	–	90	800	250	20	–	50	–

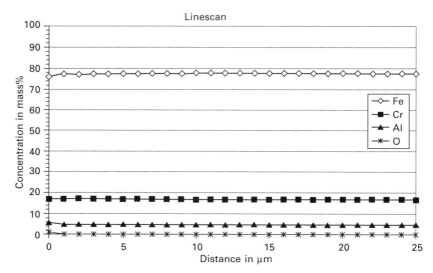

8.2 Elemental line scans after diffusion annealing Aluchrom P
(ThyssenKrupp VDM).

sheet) and also of Aluchrom YHf and Aluchrom YHfAl model catalyst supports
and Aluchrom YHf as fibre mats was undertaken by Diffusion Alloys Ltd.
For this purpose new procedures had to be developed and optimised, involving
surface coating and subsequent either integrated or separate diffusion treatment.
The process utilised varied with product form. Variable parameters included
reactant vapour or pack composition, coating temperature and time and finally,
diffusion temperature and time. The level of aluminium enrichment could be
controlled to either 7 or 8 w/o as desired, and the distributions across the
foils and sheets were effectively uniform (Fig. 8.3), as were those across the
component foils of a model catalyst support (Fig. 8.4).

8.2.3   Surface modification to preform an α-Al$_2$O$_3$ protective scale

Gas annealing of specimens (foil, sheet, fibre mat and model catalyst supports)
either in the as-fabricated condition or following aluminising, was carried
out at FZJ Jülich by a standard procedure developed in the SMILER project
[23] involving 2 h exposure at 1100–1200°C to argon containing 4% H$_2$ and
2% H$_2$O (p_{H_2O} and P$_{o_2}$ were controlled by an ice bed at –75°C). The resulting
mass gains were measured and ranged between 0.11 and 0.26 mg/ cm^2 (i.e.
equivalent Al$_2$O$_3$ layer thickness was 0.55–1.30 μm) and these were added to
the subsequent gravimetric measurements following air oxidation, and
subsequent figures show mass gain as a function of exposure time.

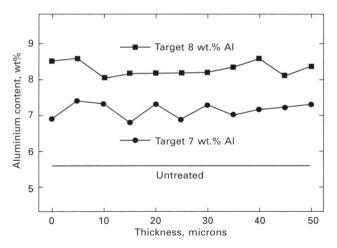

8.3 Al distribution across a 50 μm thick Kanthal AF foil cross-section following aluminising (Diffusion Alloys Ltd).

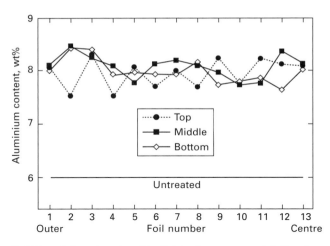

8.4 Aluminium levels of individual component foils of a model catalyst support fabricated from Aluchrom YHfAl following aluminising (Diffusion Alloys Ltd).

8.2.4   Oxidation testing

Testing was undertaken, as far as practicable, following TESTCORR procedures (the European Guidelines for High Temperature Corrosion Testing). Some limited exposures (typically ~100 h duration) were undertaken isothermally, with continuous gravimetric monitoring, using controlled atmosphere microbalances. However, the majority of exposures were discontinuous, in which standard experimental sized specimens (i.e. 2 cm × 1 cm coupons of foil, sheet and fibre mat) were contained in alumina crucibles during oxidation. These, together with model catalysts, were located within the furnace reaction vessel, in flowing laboratory air, at several temperatures (800°, 850°, 900°, 950°, 1100°, 1200° and 1300°C). In these discontinuous tests the cycles involved 1–100 h exposure at temperature, followed by either furnace or air cooling to room temperature. The extents of oxidation and of spallation, where this occurred, were measured gravimetrically. Testing was discontinued either after 3000 h total oxidation or breakaway onset, whichever was the shorter.

Following exposure, a range of conventional surface analytical techniques (macro/optical, and scanning electron microscopy, X-ray diffraction, EDX/EPMA analysis) were deployed to examine the oxidation scales formed.

8.3   Results

For clarity the oxidation data for as-fabricated and surface treated experimental corrosion coupons will be described first and will be followed by presentation of the corresponding results obtained on surface treated near real-service components. The laboratory responsible for each test series is given in the relevant figure caption. In all figures the onset of breakaway attack is denoted as B/O.

8.4   Air oxidation of foil and sheet coupons

8.4.1   Comparison of the oxidation behaviour of Aluchrom YHfAl and Aluchrom YHf in air at 800–1300°C

The extents of oxidation (expressed as the increase in gross mass gains with time) of 50 μm alloy foils, in air, at 800°, 850°, 900° and 950°C are shown in Figs 8.5–8.8, respectively. Aspects of the Aluchrom YHf and Kanthal AF data have been described already [20]. During exposure times out to 3000 h, or breakaway if earlier, the oxidation characteristics were similar at 800°, 900° and 950°C. The oxidation rates were fastest initially and associated with the formation of a transitional alumina scale, which then decreased with time due first, to the transformation of the transitional alumina to or formation of α-alumina and later, to the formation of a chromia sub-scale (the samples then had a green appearance) (i.e. onset of pseudo-protection (Fig. 8.1)). At 850°C, the oxidation kinetics differed from those observed at 800°C or 900°C and above. The oxidation kinetics were essentially linear until either the surface colour changed from grey to green (onset of breakaway) or a α-alumina layer had completely formed. The proportion of transitional alumina within the scales decreased from 800°C to 900°C following 3000 h exposure.

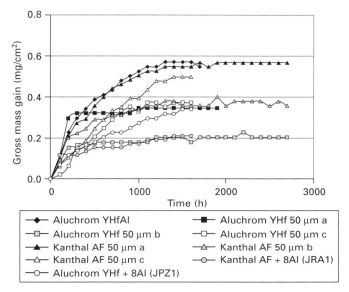

8.5 Air oxidation at 800°C of as-fabricated Aluchrom YHfAl, aluminised and as-fabricated Aluchrom YHf and Kanthal AF, 50 μm thick foils (Cranfield University).

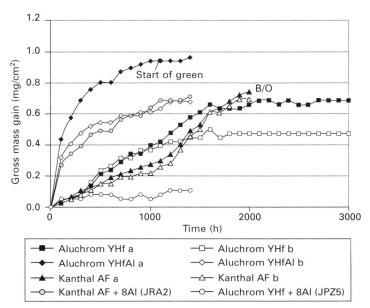

8.6 Air oxidation at 850°C of as-fabricated Aluchrom YHfAl, aluminised and as-fabricated Aluchrom YHf and Kanthal AF, 50 μm thick foils (Cranfield University).

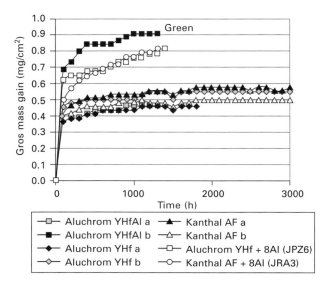

8.7 Air oxidation at 900°C of as-fabricated Aluchrom YHfAl, aluminised and as-fabricated Aluchrom YHf and Kanthal AF, 50 μm thick foils (Cranfield University).

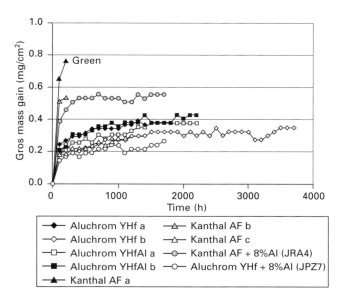

8.8 Air oxidation at 950°C of as-fabricated Aluchrom YHfAl, aluminised and as-fabricated Aluchrom YHf and Kanthal AF, 50 μm thick foils (Cranfield University).

At 950°C the scale was completely α-alumina after 1400 h. This balance between the rate of transitional alumina growth and its conversion to α-alumina led to a peak rate of oxygen uptake at 850°C. Thus, oxidation rates were slower at 800°C, reaching a maximum at 850–900°C decreased at 900–950°C before increasing again as the rate of α-alumina growth increased with temperature.

Aluchrom YHfAl had a higher Al content (6.0 wt.%) than Aluchrom YHf (5.5 wt.%). During the exposure of 50 μm thick foils at 800–900°C (Figs 8.5–8.7) throughout the exposure at least one Aluchrom YHfAl specimen oxidised more rapidly than that of Aluchrom YHf, due primarily to a faster initial oxidation rate. At both 850° and 900°C, at a mass gain comparable with that calculation for exhaustion of the alloy Al reservoir (0.93–0.97 mg/cm^2), the colour of these samples changed from grey to green. However, in contrast, at both temperatures, a duplicate Aluchrom YHfAl specimen exhibited improved behaviour, with an initial fast oxidation rate being reduced abruptly to a slower value before alloy Al exhaustion. (Figs 8.6 and 8.7) This pattern was observed also for the oxidation of both Aluchrom YHfAl and Aluchrom YHf, 50 μm thick foils, in air, at 950°C (Fig. 8.8) resulting in α-Al$_2$O$_3$ scale formation and growth throughout the exposure duration (3000 h).

The oxidation behaviours of the two alloys, in the form of 1 mm thick sheet, were compared at 1200°C and 1300°C and are shown in Figs 8.9 and 8.10, respectively. The Aluchrom YHf data were derived in the previous LEAFA programme [3]. Throughout the exposures, at both temperatures, Aluchrom YHfAl oxidised faster than Aluchrom YHf, the enhanced attack

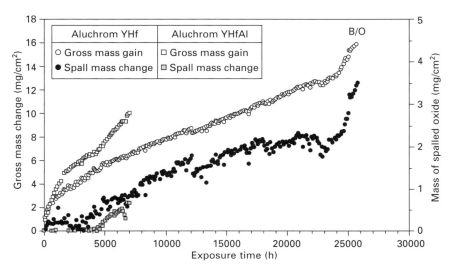

8.9 Comparison between the oxidation behaviour of Aluchrom YHfAl and Aluchrom YHf, 1 mm thick sheet, in air, at 1200°C (FZJ Jülich).

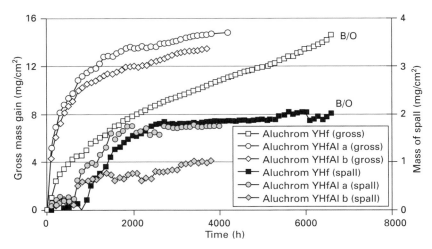

8.10 Comparison between the oxidation behaviour of Aluchrom YHfAl and Aluchrom YHf, 1 mm sheet, in air at 1300°C (Cranfield University).

occurred again principally during the initial exposure period. Breakaway attack of Aluchrom YHf resulted from Al depletion at a gross mass gain of ~15mg/cm² after 26000 h at 1200°C and a shorter time (~6600 h) at 1300°C. Unfortunately, the exposure durations have been too short for Aluchrom YHfAl breakaway oxidation initiation, thereby negating consideration of the possible role of the higher Al content in this alloy compared with Aluchrom YHf.

At both temperatures oxide spallation was initiated when the scales exceeded critical thicknesses, being thicker on Aluchrom YHfAl than on Aluchrom YHf. Thereafter, after each successive thermal cycle spall accumulated at comparable rates on both alloys, at each temperature, with the rate being greater at the higher temperature (1300°C).

8.4.2   Influence of increasing the alloy Al reservoir by aluminising

The Al contents of both 50 μm thick foils and 1 mm sheets of both Aluchrom YHf and Kanthal AF were enriched by CVD aluminising to both 7% and 8%. The Al distributions throughout the cross-sections were remarkably uniform (e.g. Fig. 8.3). The foils were oxidised at 800–1300°C, while sheets were tested only at the two higher temperatures (1200° and 1300°C). Increasing the Al content of Kanthal AF to 8% was beneficial at 800°C (Fig. 8.5), as it was also at 850–950°C (Figs 8.6–8.8). Although the initial rate was increased, subsequently it flattened off such that within these exposure durations (~1500 h) breakaway was not observed, as occurred on the as-fabricated alloy at 850° and 950°C. Turning to Aluchrom YHf, increasing the Al content by aluminising had no significant influence on the oxidation behaviour at 800°C (Fig. 8.5), whereas at 850°C (Fig. 8.6) and 950°C (Fig. 8.8) it reduced the oxidation rate throughout exposure. In contrast at 900°C (Fig. 8.7), although the initial attack was increased the rate flattened subsequently. In no case was breakaway observed during the exposures of the aluminised foils within this temperature range. In confirmation the mass gains on exposure completion were significantly lower than those for complete oxidation of the available Al in the 8 wt.% aluminised alloys (1.1–1.4 mg cm^2).

Although aluminising to provide both 7% and 8% Al contents did not affect the oxidation kinetics in air, at 1100°C of both Aluchrom YHf and Kanthal AF, 50 μm thick foils, it did extend dramatically the respective times to breakaway attack (Fig. 8.11, Kanthal AF; Fig. 8.12, Aluchrom YHf). This was reflected by a corresponding increase in the mass gain at breakaway. The lifetimes, given in Table 8.2, increased progressively with increasing alloy Al content, i.e. from the nominal 5 wt.% in the as-fabricated alloys through 7 wt.% and a maximum of 8 wt.% in those aluminised.

Aluminising (to provide both 7 wt.% and 8 wt.% Al contents), in contrast, had no significant positive influence on the oxidation behaviour of Aluchrom YHf and Kanthal AF 50 μm foils in air at 1200°C (Fig. 8.13) and 1300°C. The kinetics were similar and oxidation rapidly led to breakaway, at shorter times at 1300°C than at 1200°C.

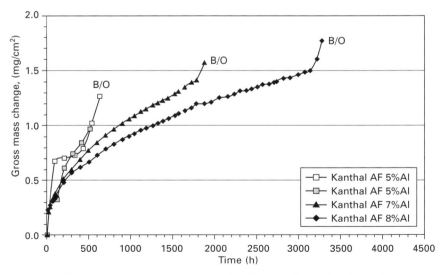

8.11 Increasing lifetime of 50 μm thick Kanthal AF foils during air oxidation at 1100°C with increasing alloy Al content from 5 wt.% to 8 wt.% (JRC, Petten).

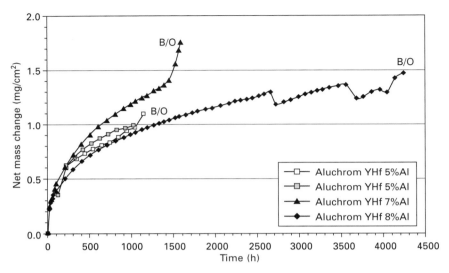

8.12 Increasing lifetime of 50 μm thick Aluchrom YHf foils during air oxidation at 1100°C with increasing alloy Al content from 5 wt.% to 8 wt.% (JRC, Petten).

Table 8.2 Lifetime increase of Aluchrom YHf and Kanthal AF during air oxidation at 1100°C with increasing alloy Al content

Alloy	Al content (wt.%)	Lifetime (h)	Alloy	Al Content (wt.%)	Lifetime (h)
Aluchrom	5	1050	Kanthal	5	400
YHf	7	1400	AF	7	1700
	8	4050		8	3100

In the corresponding tests on aluminised (8 wt.% Al), 1 mm thick, Aluchrom YHf and Kanthal AF sheets, the aluminised alloys oxidised at slightly faster rates than the as-fabricated alloys at 1200°C (Fig. 8.14). However, although the exposures were 3000 h, these were much shorter than the time needed to establish whether the increased Al alloy content affected the time to breakaway, e.g. the lifetime of 1 mm Kanthal AF sheet at 1200°C was 13,000 h [3]. At 1300°C, in contrast, the increased Al content reduced the oxidation rate of Kanthal AF (Fig. 8.16), so that whereas the lifetime of the as-fabricated alloy was 1300 h [3], chemical failure of the aluminised Kanthal AF did not occur during the current exposure (3000 h). Aluminising Aluchrom YHf to 8 wt.% Al had no influence on the oxidation rate (Fig. 8.16). However, at both temperatures aluminising enhanced the spallation propensity of both alloys (Figs 8.15 and 8.17).

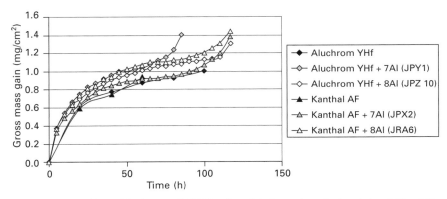

8.13 Air oxidation at 1200°C of as-fabricated and aluminised (7 wt.% and 8 wt.%) Aluchrom YHf and Kanthal AF, 50 μm thick foils (Cranfield University).

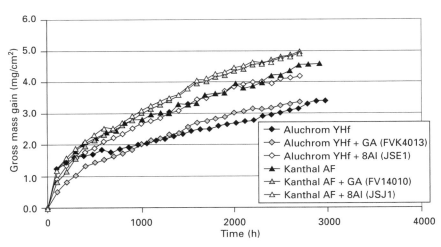

8.14 Air oxidation at 1200°C of as-fabricated, gas annealed (GA) and aluminised (8 wt.% Al) Aluchrom YHf and Kanthal AF, 1 mm sheet (Cranfield University).

8.4.3 Potential of gas annealing to preform an α-Al$_2$O$_3$ scale

The development at FZJ Jülich of a gas annealing treatment for FeCrAlRE alloys is described in detail in [23]. The twofold objectives of such a treatment in Ar + H$_2$ mixtures were: (a) the formation of a stable, thin, α-Al$_2$O$_3$ layer to prevent growth of transitional metastable aluminas at low temperatures, and (b) the removal of deleterious non-metallic alloy impurities (such as C, S, etc.). Pre-oxidation was undertaken primarily at 1200°C. Raman fluorescence spectrometry was used to establish the oxide phase composition, which

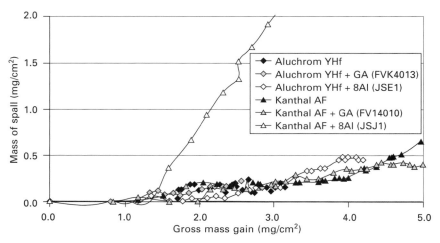

8.15 Spallation during air oxidation at 1200°C of as-fabricated, gas annealed (GA) and aluminised (8 wt.% Al) Aluchrom YHf and Kanthal AF, 1 mm sheet (Cranfield University).

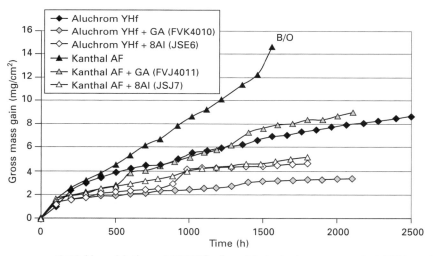

8.16 Air oxidation at 1300°C of as-fabricated, gas annealed (GA) and aluminised (8 wt.% Al) Aluchrom YHf and Kanthal AF, 1 mm sheet (Cranfield University).

depended on the water vapour content and/or oxygen partial pressure. At lower water vapour contents (dew points of –75°C and –40°C) the scales consisted mainly of theta alumina, whereas at higher water vapour contents (dew point 20°C) only α alumina was observed, which should afford the greatest protection. Chemical analyses demonstrated that this gas annealing treatment also reduced the S content of Aluchrom YHf from 60–1100 ppm to

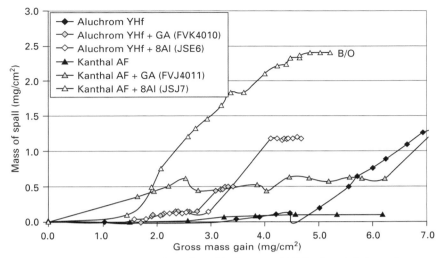

8.17 Spallation during air oxidation at 1300°C of as-fabricated, gas annealed (GA) and aluminised (8 wt.% Al) Aluchrom YHf and Kanthal AF, 1 mm sheet (Cranfield University).

<10 ppm but did not affect the C content. The standard gas treatment adopted was 2 h annealing in Ar + 4% H_2 + 2% H_2O at 1200°C during which the mass gains ranged between 0.1 and 0.25 mg cm^2, such that the preformed α-Al_2O_3 layer thickness was between 0.55 and 1.30 µm. The treatment, therefore, satisfied another major criterion that the Al consumption during gas annealing was minimised to ensure a sufficient remaining Al reservoir for service operation.

The effectiveness of these gas annealing treatments in reducing the mass change on 50 µm thick Aluchrom YHf foils during successive 24 h and 48 h exposures in air, at 900°C, is shown in Fig. 8.18, with the standard treatment being the most effective. A similar result was also obtained for 50 µm thick Kanthal AF foil. At 1100°C gas annealing reduced the air oxidation rate of 1 mm thick Aluchrom YHf sheet after exposure for over 4300 h (Fig. 8.19) and a similar behaviour was also observed for Kanthal AF. It should be noted that in figures such as Fig. 8.19, the gross mass gains shown include those arising during this surface treatment. Gas annealing had no significant effect on either the oxidation rate or the spallation behaviour of 1 mm sheets of both alloys at 1200°C in tests at Cranfield University (Figs 8.14 and 8.15). Concurrent exposures of similar specimens at this temperature at FZJ Jülich were in broad agreement, although gas annealing slightly reduced the oxidation rate of Aluchrom YHf throughout their 2000 h exposure. At 1300°C gas annealing proved remarkably beneficial in reducing the respective oxidation rates of the two alloys (Fig. 8.16); the corresponding role on spallation, as yet, is not so clearly defined (Fig. 8.17).

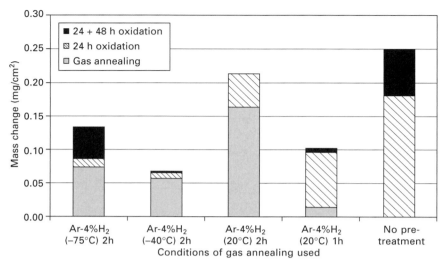

8.18 Effect of gas annealing at 1200°C on the subsequent air oxidation at 900°C of 50 μm thick Aluchrom YHfAl foils (FZJ, Jülich).

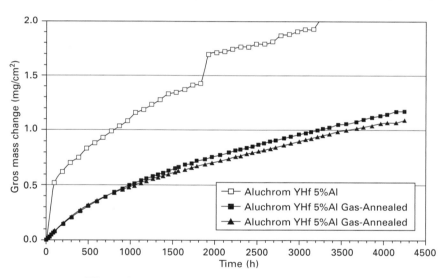

8.19 Effect of gas annealing upon the air oxidation at 1100°C of Aluchrom YHf, 1 mm sheet (JRC Petten and VDM).

8.4.4   Combined effect of aluminising and gas annealing

Although only a single series of isothermal air oxidation tests were undertaken on 50 μm thick Aluchrom YHf and Kanthal AF foils at 900°C, these demonstrated unequivocally that in combination, aluminising to 8 wt.% and gas annealing, inhibited completely any further oxidation, at least during the early stages of oxidation at 900°C (e.g. Fig. 8.20 for Aluchrom YHf). In contrast, as described in Section 8.4.2, 8 wt.% aluminised foils oxidised initially at a considerably faster rate than the as-fabricated alloy but then the rate progressively decreased such that at mass gains of ~0.3 mg cm^2 it had effectively flattened off. In contrast, the as-fabricated alloy continued oxidising throughout the exposure but at a diminishing rate.

8.5   Air oxidation of near real-service components

8.5.1   Fibre mats

Metal fibres shaved from a bulk alloy, such as Aluchrom YHf, were woven into mats by Bekaert SA. The Aluchrom YHf fibres' cross-sectional dimensions were typically 20×50 μm. During air oxidation, the mats became embrittled and breakaway occurred at times. These oxidation effects increased with decreasing temperature over the range 950 to 800°C (Fig. 8.21). Such lifetimes proved unacceptable technologically because the mats became embrittled. The effectiveness of the gas annealing pretreatment in extending lives in air oxidation to in excess of 2000 h at 900°C is shown in Fig. 8.22. The untreated fibre mat oxidised rapidly and even after 20 h it had turned green, confirming complete Al consumption from the fibres. In contrast, over the same exposure

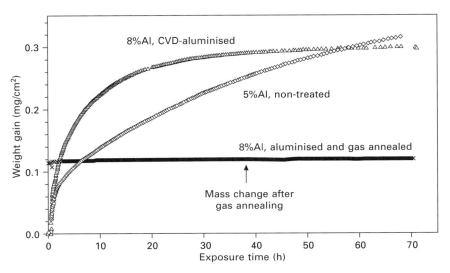

8.20 Combined effect of aluminising and gas annealing upon the oxidation in air at 900°C of a 50 μm thick Aluchrom YHf foil (FZJ, Jülich).

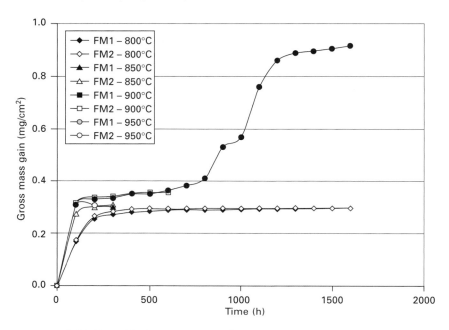

8.21 Air oxidation at 800–950°C of Aluchrom YHf fibre mats (Cranfield University).

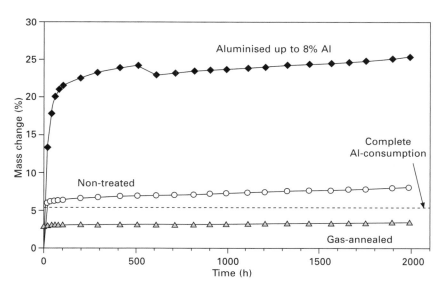

8.22 Effect of gas annealing in air oxidation at 900°C of Aluchrom YHfAl fibre mat (FZJ, Jülich).

period, the mass change of the gas annealed fibre mat increased only slightly above that associated with the surface treatment at 1200°C and at the end of the exposure the mat was grey in colour and was still flexible. Further oxidation tests at 1100°C revealed that although CVD aluminising (8 wt.%) did not prolong the mat lifetime, gas annealing was beneficial at least for up to 200 h. The surface treatment examined in these tests at 900° and 1100°C was carried out at 1200°C and as a consequence ~ 50% of the Al available in the fibre mats was consumed to preform the α-Al$_2$O$_3$ protective layer. Further recent work (to be published) has demonstrated that gas annealing at 1100°C reduced by one-third the Al utilisation during this process. Nevertheless, the thinner α-Al$_2$O$_3$, thereby preformed, proved to be equally as effective as the thicker layer produced by annealing at 1200°C.

8.5.2   Model FeCrAlRE catalyst supports

Model catalyst supports (39 mm dia. × 71.5 mm long) were manufactured by ThyssenKrupp VDM from 50 μm thick foils of Aluchrom YHfAl and of Aluchrom P. Air oxidation of untreated and surface treated (gas annealed, aluminised (8 wt.%) and gas annealed + 8 wt.% aluminised) Aluchrom YHfAl and Aluchrom P model catalyst supports was undertaken at 900° and 1100°C, with 100 h thermocycles.

Throughout 2000 h cumulative exposure at 900°C the mass gains on all surface treated and untreated model catalyst supports increased slowly, essentially at a decreasing rate with increasing time. None of the model catalysts underwent breakaway oxidation, confirmed by calculation that the oxygen uptake at 2000 h was less than half that for total Al consumption. After 2000 h the extents of oxidation (including that during gas annealing) decreased in the order: gas annealed + aluminised, untreated, aluminised, gas annealed Aluchrom YHfAl and finally, Aluchrom P (Fig. 8.23).

The relevant effectiveness of the various surface treatments was more clearly delineated in the corresponding oxidation exposures, in air, at 1100°C. In these tests the protective oxidation region and the two stages of chemical failure (shown schematically in Fig. 8.1) were established clearly from the mass gain versus time graphs and were backed by optical microscopy of the appearance of the catalyst supports after each successive oxidation thermocycle. In the protective region the catalyst support colour was grey, then when the level of Al in the alloy fell below [Al]$_{crit}$, the colour changed to green due to chromia subscale formation and then the end of this pseudo-protection region was marked by colour change to black, at first in striations, then at the edges indicating iron-oxide formation. This Stage B failure onset was defined also by a rapid rate of mass gain.

The respective oxidation graphs are separated into two figures. Figure 8.24 compares the behaviour of model catalyst supports fabricated from

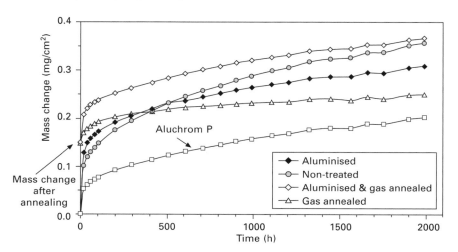

8.23 Effect on air oxidation at 900°C on model catalyst supports fabricated from Aluchrom P and untreated and surface treated Aluchrom YHfAl (FZJ, Jülich).

Aluchrom YHf and Aluchrom YHfAl, together with those of aluminised (8 wt.%) Aluchrom YHfAl and Aluchrom P. In Fig. 8.25 corresponding oxidation plots examine the effectiveness of the other surface treatments: gas annealing and gas annealing + aluminising (8 wt.%) Aluchrom YHfAl. All the Aluchrom YHf Al foils were produced from the same heat.

The protective oxidation region oxide grew without spalling and obeyed a power law time dependence $\Delta m = kt^n$, with the growth rate parameters k, n. Additionally the characteristic time $t' = (10/k)^{1/n}$ to reach a mass gain of 10 g/m^2 is a useful parameter of catalyst support life [24, 25]. For all samples this function was fitted to the alumina forming region (from 100 h up to 1000 h before the first appearance of other oxide or an obvious abrupt change in oxidation rate). The curves derived are shown in Figs 8.24 and 8.25. The respective growth parameters are collated in Table 8.3. No oxide spallation occurred up to breakaway onset.

An assessment of Figs 8.24 and 8.25, in association with the optical images throughout the oxidation exposures, has enabled definition of the times (given in Table 8.4) for appearance of green areas (i.e. chemical failure Stage A onset) and of black areas (i.e. chemical failure Stage B onset), which with one possible exception, that of aluminised Aluchrom YHfAl, agreed well with the time for rapid mass gain. The pseudo-protection region duration lasted between the onsets of Stage A and Stage B failure. Finally, the $[Al]_{crit}$ values at Stage A failure onset were calculated from the mass gains at that juncture. The values for five samples, Aluchrom YHf, aluminised, gas annealed and as-fabricated Aluchrom YHfAl and Aluchrom P were at or below 0.5

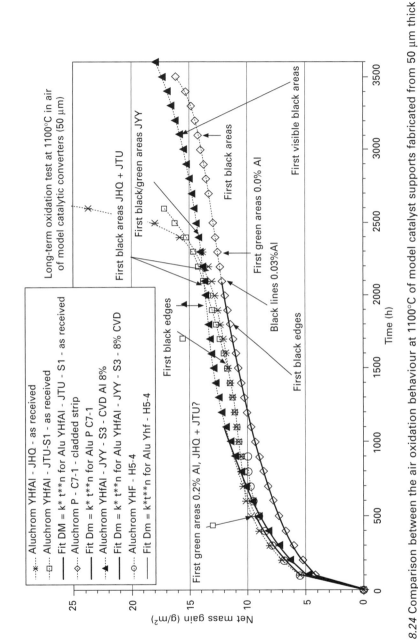

Net mass gain (g/m²)

Time (h)

Long-term oxidation test at 1100°C in air of model catalytic converters (50 μm)

--✳-- Aluchrom YHfAl - JHQ - as received
--☐-- Aluchrom YHfAl - JTU-S1 - as received
—— Fit DM = k* t**n for Alu YHfAl - JTU - S1 - as received
--◇-- Aluchrom P - C7-1 - cladded strip
—— Fit Dm = k* t**n for Alu P C7-1
--▲-- Aluchrom YHfAl - JYY - S3 - CVD Al 8%
—— Fit Dm = k* t**n for Alu YHfAl - JYY - S3 - 8% CVD
--◉-- Aluchrom YHF - H5-4
—— Fit Dm = k*t**n for Alu Yhf - H5-4

First green areas 0.2% Al, JHQ + JTU?

First black edges

First black areas JHQ + JTU

First black/green areas JYY

First black areas

First green areas 0.0% Al

Black lines 0.03%Al

First black edges

First visible black areas

8.24 Comparison between the air oxidation behaviour at 1100°C of model catalyst supports fabricated from 50 μm thick foils of Aluchrom YHf, Aluchrom YHfAl, aluminised (8 w/o) Aluchrom YHfAl and Aluchrom P (ThyssenKrupp VDM).

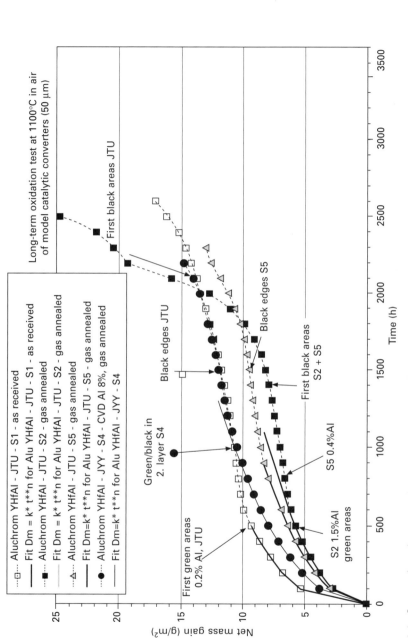

Long-term oxidation test at 1100°C in air of model catalytic converters (50 μm)

Legend:
- ···□··· Aluchrom YHfAl - JTU - S1 - as received
- —— Fit Dm = k* t**n for Alu YHfAl - JTU - S1 - as received
- ···■··· Aluchrom YHfAl - JTU - S2 - gas annealed
- —— Fit Dm = k* t**n for Alu YHfAl - JTU - S2 - gas annealed
- ···△··· Aluchrom YHfAl - JTU - S5 - gas annealed
- —— Fit Dm=k* t**n for Alu YHfAl - JTU - S5 - gas annealed
- ···●··· Aluchrom YHfAl - JYY - S4 - CVD Al 8%, gas annealed
- —— Fit Dm=k* t**n for Alu YHfAl - JYY - S4

Net mass gain (g/m²)

Time (h)

First green areas 0.2% Al, JTU

Green/black in 2. layer S4

Black edges JTU

Black edges JTU

Black edges S5

First black areas S2 + S5

S5 0.4%Al

S2 1.5%Al green areas

First black areas JTU

8.25 Comparison between the air oxidation behaviour at 1100°C of model catalyst supports fabricated from 50 μm thick foils of Aluchrom YHfAl and the influence of two surface treatments (gas annealing and gas annealing + aluminising) (ThyssenKrupp VDM).

Table 8.3 Growth rate parameters of the fit of the power law $\Delta m = kt^n$ to the oxidation curves

Aluchrom, sample number	Surface treatment	Period of fit (h)	k g/m^2/(h^n)	n	t'(h)
YHf, H5-4	As received	100–300	1.23	0.33	636
YHfAl, JTU-S1	As received	100–400	1.10	0.34	626
YHfAl, JYY-S3	Aluminised to 8 wt.%	100–1100	0.84	0.38	681
P, C7-1	Cladded to 7.6 wt.% Al	100–2000	0.80	0.35	1215
YHfAl, JTU-S2	Gas annealed (Ar + 4% H$_2$ + 2% H$_2$O for 2 h at 1200°C)	100–400	0.32	0.46	1664
YHfAl, JTU-S5	Gas annealed (Ar + 4% H$_2$ + 2% H$_2$O for 2 h at 1200°C)	100–700	0.34	0.47	1327
YHfAl, JYY-S4	Aluminised to 8 wt.%, then gas annealed (Ar + 4% H$_2$ + 2% H$_2$O for 2 h at 1200°C)	100–900	0.49	0.45	864

Table 8.4 Effectiveness of surface treatments to prolong the lifetimes of Aluchrom FeCrAlRE alloy model catalyst supports during air oxidation at 1100°C

Aluchrom, sample number	Surface treatment	Time to appearance in (h)			Duration of pseudo protection (h)	[Al]$_{crit}$ at first green in (mass%)
		Green areas	Black areas	Rapid mass gain		
YHf, H5-4	As received	400			1700	0.5
YHfAl, JTU-S1	As received	500	2100	2200	1800	0.2
YHfAl, JYY-S3	Aluminised to 8 wt.% Al	1300	2400	3100–3600	1000	0.0
P, C7-1	Cladded 7.6 wt.% Al	2300–2100	3100	3300		0.0
YHfAl, JTU-S2	Gas annealed Ar + 4% H$_2$ + 2% H$_2$O for 2 h 1200°C	500	1400	1600	1100	1.5
YHfAl, JTU-S5	Gas annealed Ar + 4% H$_2$ + 2% H$_2$O for 2 h 1200°C	800	1400	1800	1100	0.4
YHfAl, JYY-S4	Aluminised 8 wt.% Al, gas annealed Ar + 4% H$_2$ + 2% H$_2$O for 2 h 1200°C	1000		2000	1000	1.3

wt.%. The remaining values for a repeat gas annealed and gas annealed + aluminised Aluchrom YHfAl were 1.5 and 1.3 wt.%, respectively.

The effectiveness of the various surface treatments is apparent from detailed considerations of these data (Figs 8.24 and 8.25, Table 8.3 and 8.4). The two as-fabricated Aluchrom YHfAl model catalyst supports behaved in a similar manner. Increasing the Al alloy content by fabrication (i.e. Aluchrom YHfAl compared with Aluchrom YHf) marginally increased (by 25%) the lifetime to breakaway. In contrast, increasing the Al reservoir to 8 wt.%Al by aluminising/plating increased substantially the times to breakaway (Stage A) and also the times to onset of non-protective attack (Stage B) by factors ranging between 3 and 6 times. Aluminising also maintained the duration of pseudo-protection, while it was reduced by up to two fold for all other surface treatments. The initial oxidation rates during the protective scale region were lowest on Aluchrom P and gas annealed Aluchrom YHfAl.

8.6   Discussion

Based on current understanding of temperature dependence, the mechanisms of scale growth and scale compositions and also of the two-stage, oxidation induced chemical failure process of FeCrAlRE alloys, the development and evaluation of two potential procedures to increase alloy lifetimes have been major objectives of the CEC SMILER project. The majority of testing has been undertaken in laboratory air, as described in this chapter but has been substantiated by concurrent evaluation in other simulated relevant industrial environments [26]. The procedures involved first increasing the alloy Al reservoir, and second gas annealing prior to service to pre-form a stable α-Al_2O_3 protective scale and also to remove possible deleterious non-metallic alloy impurities, such as sulphur.

With regard to increasing the available Al reservoir, three approaches were studied based on Al enrichment during standard metallurgical fabrication, by aluminising and by plating/cladding. ThyssenKrupp VDM produced by the conventional metallurgical fabrication processing operations, Aluchrom YHfAl (Al content ~6 wt.%) having ~ 1 wt.% more Al than Aluchrom YHf. During air oxidation at 800° and 900°C (Figs 8.5 and 8.7) of duplicate 50 μm thick foil specimens, the effect of the higher Al content proved variable. On one specimen at least, at each temperature this proved to beneficial, yet on an identical specimen oxidised concurrently it was detrimental. The higher Al content, as observed with the Kanthal AF, which has a similar Al content, could favour the nucleation and faster growth of transitional aluminas by outward cation movement. Oxidation then continued either by this mode until alloy Al exhaustion led to Stage A failure onset or was arrested by the subsequent nucleation of a more stable α-Al_2O_3 inner layer, which slowed growth by a change to inward anion movement. This is entirely consistent with current mechanistic understanding of the oxidation behaviour of FeCrAlRE alloys at these temperatures. Variability, particularly regarding α-Al_2O_3 layer nucleation/development, undoubtedly underlay, therefore, the observed differences in Aluchrom YHfAl behaviour at 800° and 900°C. At higher temperatures, 950°C (Fig. 8.8), 1200°C (Fig. 8.9) and 1300°C (Fig. 8.10), Aluchrom YHfAl oxidised slightly faster than Aluchrom YHf. The enhanced attack emanated mainly during the initial exposure stage, but thereafter, rates on the two alloys were comparable. Spallation initiated at 1200° and 1300°C at a thicker critical scale thickness than on Aluchrom YHf, which could be attributed to the two alloys having different creep properties. Since Aluchrom YHfAl is now a ThyssenKrupp VDM standard alloy, it was used for the fabrication of all the near real-service components tested in the second phase of this study.

Increasing the alloy aluminium content of both Aluchrom YHf and Kanthal AF, to 8 wt.% maximum proved extremely beneficial to the oxidation behaviour of 50 μm thick foils at 800–1100°C (Figs 8.5–8.8, 8.11, 8.12, 8.20). Although, in certain instances at the lower temperature 800–900°C, an increase to 8 wt.% Al caused a faster initial oxidation rate compared with those on the as-fabricated alloys, the rate abruptly flattened with inner α-Al_2O_3 layer nucleation/development and this subsequently provided continuing protection through the remaining exposure period (1500–3000 h). In no case was breakaway oxidation of aluminised alloys observed at 800–950°C, as had occurred on some as-fabricated alloys (e.g. Kanthal AF, at 850° and 950°C). One point of interest, for which there does not seem to be a clear explanation, but could elicite speculation, is why over this temperature range aluminised Aluchrom YHf resulting in an 8 wt.% Al content was uniformly beneficial, whereas enhancement to 6 wt.% Al in Aluchrom YHfAl had a variable influence. At 1100°C (Figs 8.11 and 8.12) the successive increase in Al alloy content of 50 μm thick foils from 5 wt.% (nominal) through 7 to 8 wt.% did not affect the oxidation kinetics but did increase progressively the mass gain for onset of breakaway and, thereby, the time at which this occurred. The foil lifetimes (Table 8.2) were increased by factors of 4 (Aluchrom YHf) and 8 (Kanthal AF) following aluminisation to 8% Al. At higher temperatures 1200°C (Fig. 8.13) and 1300°C any potential benefit of aluminising 50 μm thick alloy foils was negated by the progressively higher rates of Al consumption by oxidation. Likewise for the corresponding 1 mm thick alloy sheets it did not prove possible to ascertain what was the possible influence of aluminising, as in this case, the test duration (3000 h) was shorter than the times to breakaway determined previously [3] for as-fabricated alloys (e.g. 13,000 h for Kanthal AF). At 1300°C aluminising reduced the oxidation rates of both alloys and within the exposure period aluminising certainly increased the lifetime of 1 mm thick Kanthal AF at this temperature. A possible detrimental consequence of aluminising both alloys, at both oxidation temperatures, was the enhancement of spallation propensity (Figs 8.15 and 8.17), which could derive from concurrent changes in alloy creep properties. This is an aspect needing further study.

The gas annealing treatment (2 h pre-oxidation in Argon + 4% H_2 + 2% H_2O at 1200°C) developed in the SMILER project by FZJ Jülich [23] pre-formed an α-Al_2O_3 protective surface layer and concurrently significantly reduced the alloy S content. It should be anticipated that this surface treatment would be particularly effective at lower temperatures where transitional aluminas could be formed, at least initially or throughout exposures to failure, but that its role would diminish with increasing oxidation temperature where oxidation was governed normally by α-Al_2O_3 formation. In general the results confirmed expectations. Oxidation of 50 μm thick foils of both alloys was reduced dramatically at 900°C (Fig. 8.18), while even at 1100°C (Fig. 8.19) gas annealing appeared beneficial, although it may be significant that the control and gas annealed foils were from different alloy manufacturing batches. Bearing in mind this same factor, the 1200°C, 1 mm sheet, results (Figs 8.14 and 8.15) conform with prediction but what was surprising, however, was the apparent beneficial influence of gas annealing on the oxidation behaviour of 1 mm sheets of both alloys at 1300°C (Fig. 8.16). A speculative view could be this may be linked with the S level reduction in the alloys, which could be proved by further research.

The final series of experiments based on standard corrosion test coupons examined the potential role of aluminising, followed by gas annealing, upon the air oxidation of 50 μm thick foils of both alloys at 900°C. The combined treatment completely inhibited any further oxidation during the test duration (72 h) (Fig. 8.20).

Overall, the standard corrosion coupon studies clearly defined both the potential viability and range of applicability of the two approaches developed by the SMILER project to increase FeCrAlRE alloy lifetimes, namely an increased Al reservoir and surface pre-treatment by gas annealing. These observations not only confirmed but substantially extended the results of previous studies on these two approaches [12, 13, 22].

To provide further confirmation, additional assessment was undertaken using two technologically important near real-life component geometries: Aluchrom YHfAl alloy fibre mats for gas heater burners and Aluchrom YHfAl supports for automobile emission control catalysts. Gas annealing proved remarkably effective in extending the lifetimes of the fibre mats at 900°C. This surface treatment caused a dramatic oxidation rate reduction, such that the component lifetime would have been substantially longer than the test duration (2000 h) (Fig. 8.22), which in itself represented a 100-fold improvement over the measured lifetimes of an untreated mat. The performance of gas annealing should be expected to decrease with increasing temperature, as was witnessed by the tests at 1100°C. Further work, therefore, will be required at intermediate temperatures to define the envelope of applicability for normal and fault operating temperatures. It is conceivable also that a combination of aluminising followed by gas annealing could be even more effective and needs also to be studied. Finally, with regard to the gas annealing conditions to be employed, it has been mentioned already that the Al consumed by this process can be reduced by carrying out this operation at 1100° rather than 1200°C. Further optimisation may be needed, with regard to gas composition and velocity for each application, to ensure uniformity of surface chemistry changes throughout components.

Turning finally to the potential of the surface treatments to extend the lifetimes of FeCrAlRE alloy model catalyst supports at 900°C (Fig. 8.23), none failed chemically within the 2000 h test duration, so it is not possible to define which treatment provides superior lifetime. However, some indication can be provided by the relative extents of oxidation (including those resulting from gas annealing) at test completion. These decreased in the order aluminised + gas annealing, untreated, aluminised, gas annealed Aluchrom YHfAl, with Aluchrom P being the lowest. In a corresponding test series, in air, at 1100°C extensive gravimetric measurements, backed by optical microscopy, followed all stages of chemical failure (Fig. 8.1) from the protective region, through the pseudo-protection region to non-protective attack of as-fabricated and surface treated Aluchrom YHfAl and Aluchrom P model catalyst supports. The main conclusion to emerge was that increasing the Al reservoir to 8 wt.% Al, either by aluminising or by cladding/plating, was the most effective treatment. Times to breakaway (chemical failure Stage A onset) and also of chemical failure Stage B onset were increased by factors ranging between × 3 and × 6, while maintaining also the duration of pseudo-protection. As might have been expected for this temperature, gas annealing, even in combination with aluminising, proved less effective.

8.7   Conclusions

- Two novel surface modification procedures, increasing the alloy Al reservoir and gas annealing (2 h in Ar + 4% H_2 + 2% H_2O at 1100–1200°C) have been developed by the CEC funded SMILER project with the objective of extending the lifetime of relevant FeCrAlRE alloy components across the industrial operating temperature range.
- The effectiveness, during air oxidation, at 800–1300°C, of these procedures has been established by the present study using standard corrosion coupons of 50 μm thick foils and 1 mm thick sheet of the commercial alloys Aluchrom YHf, Aluchrom YHfAl and Kanthal AF.
- Gas annealing, resulting in α-Al_2O_3 protective scale preformation, proved most effective at low temperatures around 900°C, and even more so,

when combined with prior aluminising. Benefit from gas annealing diminished with increasing temperature.

- Increasing the alloy Al reservoir to 7 wt.% and 8 wt.% Al progressively, and substantially, increased alloy foil lifetimes at 1100°C but prove ineffective at 1200°C because the faster inherent alloy oxidation rate dominated behaviour. At 900°C, although increase in the alloy Al reservoir increased the initial oxidation rate, it then abruptly slowed down so that aluminising proved effective, at least throughout the test duration (typically 3000 h).

- The potential of these procedures was confirmed using near real-life components. Gas annealing dramatically increased the lifetimes of Aluchrom YHfAl fibre mats, in air, at 900°C. In contrast, tests at this temperature on the possible relative benefits of the surface modification of Aluchrom YHfAl model catalyst supports proved inconclusive, as none had failed during the exposure period (2000 h). However, corresponding tests at 1100°C clearly established that increasing the Al reservoir to ~ 8 wt.% Al either by aluminising or by cladding/plating extended lifetimes by technologically important factors of between × 3 and × 6, and also maintained the duration of pseudo-protection.

8.8   Acknowledgements

The authors are most grateful for the significant contributions from all SMILER project partners to this study – a real team effort. We wish also to acknowledge the support and effort from other staff members in our respective companies, research institutes and universities. Finally, this work could not have been undertaken without the financial support of the European Commission under Contract G5RD-CT-2001-00530 and the advice received from the Commission Scientific Officer.

8.9   References

1. BRITE/EURAM IMPROVE programme, Project No. BE7972 (1994), Final Report (1997), CEC Brussels.
2. J. R. Nicholls and M. J. Bennett, *Lifetime Modelling of High Temperature Corrosion Processes* (Eds M. Schütze, W. J. Quadakkers and J. R. Nicholls), European Federation of Corrosion Publication No. 34, Institute of Materials, London (2001), 3–14.
3. R. Newton, M. J. Bennett, J. P. Wilber, J. R. Nicholls, D. Naumenko, W. J. Quadakkers, H. Al-Badairy, G. Tatlock, G. Strehl, G. Borchardt, A. Kolb-Telieps, B. Jonsson, A. Westerlund, V. Guttmann, M. Maier and P. Beaven, (Eds M. Schütze, W. J. Quadakkers and J. R. Nicholls), European Federation of Corrosion Publication No. 34, Institute of Materials, London (2001), 15–36.
4. G. Strehl, H. Hattendorf, A. Kolb-Telieps, R. Newton, R. J. Fordham and G. Borchardt, *Mater. High Temp.*, **20** (2003), 339–346.

5. M. J. Bennett, R. Newton and J. R. Nicholls, *Mater. High Temp.*, **20** (2003), 347–356.
6. H. Al-Badiary, G. J. Tatlock, H. E. Evans, G. Strehl, G. Borchardt, R. Newton, M. J. Bennett, J. R. Nicholls, D. Naumenko and W. J. Quadakkers, Life-time Modelling of High Temperature Corrosion Processes (Eds M. Schütze, W. J. Quadakkers and J. R. Nicholls), European Federation of Corrosion Publication, No. 34, Institute of Materials, London (2001), 50–65.
7. J. R. Nicholls, R. Newton, M. J. Bennett, H. E. Evans, H. Al-Badairy, G. J. Tatlock, D. Naumenko, W. J. Quadakkers, G. Strehl and G. Borchardt, (Eds M. Schütze, W. J. Quadakkers and J. R. Nicholls), European Federation of Corrosion Publication No. 34, Institute of Materials, London (2001), 83–106.
8. J. R. Nicholls, M. J. Bennett and R. Newton, *Mater. High Temp.*, **20** (2003), 429–438.
9. A. Andoh. S. Taniguchi and T. Shibata, *Oxid. Met.*, **46** (1996) 481.
10. M. J. Bennett, *Solid State Phenomena*, **41**, (1995), 235–252.
11. W. J. Quadakkers, D. Clements and M. J. Bennett, *Microscopy of Oxidation – 3* (Eds S. B. Newcomb and J. A. Little), The Institute of Materials, London (1997), 195–206.
12. H. Hattendorf, A. Kolb-Telieps, T. H. Strunskus, V. Zaporojchenko and F. Faupel, *Lifetime Modelling of High Temperature Corrosion Processes* (Eds M. Schütze, W. J. Quadakkers and J. R. Nicholls), European Federation of Corrosion Publications No. 34, The Institute of Materials, London (2001), 135–147.
13. S. Taniguchi, T. Shibata and A. Andoh, *Metal-Supported Automotive Catalytic Converters* (Ed. H. Bode), Werkstoff-Informationsgesellshaft, Frankfurt, Germany (1997), 179–189.
14. J. E. Antill, J. B. Warburton and R. F. A. Carney, UKAEA, Harwell Laboratory Report, AERE-G1045 (1978).
15. *Metal Supported Automotive Catalytic Converters* (Ed. H. Bode), MACC 1997, (1997) and MACC 2001 (2002). Werkstoff-Informationsgesellshaft, Frankfurt, Germany.
16. R. Molins, A. Germidis and E. Andrieu, *Microscopy of Oxidation – 3* (Eds S. B. Newcomb and J. A. Little), Institute of Materials, London (1997), 3–11.
17. P. T. Moseley, K. R. Hyde, B. A. Bellamy and G. Tappin, *Corros. Sci.*, **24** (1984), 547.
18. H. E. Kadiri, R. Molins and Y. Bienvenu and M Hortstemeyer, *Materials Science Forum*, **461–464**, 1107–1116 (2004).
19. R. Molins and A. M. Huntz, *Materials Science Forum*, **461–464**, 29–36 (2004).
20. M. J. Bennett, R. Newton, J. R. Nicholls, G. J. Tatlock, H. Al-Badairy and A. Galerie, *Materials Science Forum*, **461–464**, 463–472 (2004).
21. D. Naumenko, W. J. Quadakkers, A. Galerie, Y. Wouters and S. Jourdain, *Mater. High Temp.*, **20** (2003), 287–293.
22. S. Taniguchi, A. Andoh and T. Shibata, *Metal-Supported Automotive Catalytic Converters* (Ed. H. Bode), Werkstoff-Informationsgesellshaft, Frankfurt, Germany (2001), p 83–92.
23. E. N'Dah, A. Galerie, Y. Wouters, D. Goossens D. Naumenko, V. Kochubey and W. J. Quadakkers, *Materials and Corrosion*, **56**, 843–847 (2005).
24. W. J. Quadakkers and K. Bongartz, *Werkstoff und Korrosion*, **45** (1994), 222–241.
25. W. J. Quadakkers, D. Naumenko, W. Wessel, V. Kochubey and L. Singheiser, *Oxid. Met.*, **61** (2004), 17–37.
26. G. J. Tatlock, H. Al-Badairy, R. Bachorczyk-Nagy and R. Fordham, *Materials and Corrosion*, **56**, 867–873 (2005).

9

Effect of gas composition and contaminants on the lifetime of surface-treated FeCrAlRE alloys (SMILER)

G J TATLOCK and H AL-BADAIRY, University of Liverpool, UK and R BACHORCZYK-NAGY and R FORDHAM, JRC Petten, The Netherlands

9.1   Introduction

In many industrially developed nations, up to 3% of gross national product is consumed by corrosion-related wastage [1]. Among the corrosion-resistant alloys, Fe-Cr-Al alloys are some of the most oxidation resistant materials available for use in high temperature applications, above 1000°C, where harsh and highly corrosive environments exist. Such applications include heating elements for furnaces, burner elements, hot gas filters and, most recently, widespread use in the car industry as a catalyst support for catalytic converters. The superior oxidation resistance of these alloys is due to the formation of a highly stable, compact and slow growing α-Al_2O_3 scale when oxidised above 1000°C [2]. Despite the excellent stability of this scale, the lifespan of Fe-Cr-Al alloys can be significantly reduced during oxidation by the depletion of the scale-forming element, aluminium, in the substrate. Scale growth, spallation and re-growth of the protective scale can accelerate the reduction of the aluminium reservoir and ultimately the formation of a non-protective Fe-rich oxide, signalling the end of the service lifetime of the Fe-Cr-Al alloy [3–6]. The critical Al level required to sustain a protective alumina scale is about 2 wt.% [7, 8].

In practice, the alloys are used in many different oxidising environments, which can also influence the lifespan of the alloys. For example, Janakiraman *et al.* [9] found that water vapour increases the spallation rate of alumina scales formed on FeCrAl alloys and consequently a more rapid degradation of the alloys results. Others showed that the presence of water vapour causes the oxidation of all alloying elements during the initial stages of oxidation resulting in the formation of an alumina scale which is less protective due to the presence of iron and chromium oxides [10]. Also it was reported that localised spallation of alumina scales formed on FeCrAl alloys that were oxidised in synthetic exhaust gas ($N_2+CO_2+H_2O$) was caused by H_2O in the atmosphere [11].

The resistance of the protective alumina scale to spallation and decohesion can be improved greatly by the addition of reactive elements such as Y, Zr, Hf, Ti, La, Ce, Nb, V, etc., thus prolonging the lifespan of the Fe-Cr-Al alloys. Many researchers in the field of high temperature oxidation have now extensively studied Fe-Cr-Al alloys [12–18]. For example, a major investigation has been carried out regarding the role of La, Ce, Y, Ti, V, Nb and Ta on the alumina scales formed on Fe-Cr-Al alloys oxidised at temperatures up to 1000°C [19]. It was concluded that elements such as Y, Ce, La and Ti which form sulphides more stable than Al_2S_3 improved scale adhesion by preventing the segregation of S to the scale/metal interface, whereas elements such as V and Nb which form sulphides less stable than Al_2S_3 had no beneficial effects on scale adhesion. Other researchers have reported that incorporating Zr, Hf and La into Fe-Cr-Al alloys reduced the growth rate of the scale and enhanced its bonding with the substrate [20].

Another approach is to apply coatings to a FeCrAl substrate. Hence, plasma spraying as a pre-treatment technique for the surface of a FeCrAl metal substrate for motorcycle exhaust catalysis has been attempted [21], while further protection and increases in lifetime have been sought by applying a thermal barrier coating system [22, 23].

In this chapter, the research has been focussed on two areas. Firstly, the role of single and multiple additions of reactive elements to FeCrAl alloys under a range of oxidising conditions from laboratory air to 60 vol.% H_2O in synthetic air to simulated automotive exhaust gases. Secondly, to study the effects of modifying the surfaces of the alloys by sputter coating an oxidation resistant, weak coating onto a mechanically strong substrate where the oxide scale would normally fail by repeated spallation under conditions of thermal cycling. This appears to be a feasible technique to prolong the lifetime of alloys used in complex atmospheres in high temperature industrial applications.

9.2   Experimental methods

9.2.1   Materials

The cyclic oxidation experiments were performed on commercial alloys (Table 9.1) and model alloys of well-defined composition* (Table 9.2). Most of the samples were in the form of 10 mm × 20 mm coupons at least 1 mm thick, or 50 μm thick foils.

9.2.2   Testing atmospheres

Four different atmospheres were used for the bulk of the tests. Laboratory air, in which the moisture content was estimated to be approximately 1.5

* Prepared by Armines St. Etienne, France.

Table 9.1 Typical compositions of commercial Fe-Cr-Al alloys

Material	Composition (wt.%)									
	Fe	Cr	Al	Y	Zr	Hf	Ti	C	S	N
Kanthal AF	Base	21	5.0	0.03	0.05	<0.01	0.09	0.03	0.001	0.010
Kanthal APMT	Base	20–23	5.0	0.10	0.05	0.10	0.02	0.03	0.002	0.050
Aluchrom YHf	Base	20	5.5	0.05	0.05	0.04	<0.01	0.02	0.002	0.005
Aluchrom YHfAl	Base	21	6.0	0.06	0.05	0.04	<0.01	0.03	0.002	0.005

Table 9.2 Chemical composition (wt.%) of model Fe-Cr-Al alloys

Alloy	M8	M9	M10	M11	M12	M13	M14
Fe	Base	Base	Base	Base	Base	Base	Base
Cr	19.70	19.53	19.56	19.55	19.70	19.58	19.90
Al	4.83	4.85	4.83	4.85	4.83	4.89	4.91
C	0.053	0.0115	0.0114	0.0131	0.0138	0.0137	0.0128
S	0.0007	0.0005	0.0005	0.0005	0.0006	0.0005	0.0006
O	0.0003	0.0004	0.0004	0.0006	0.0005	0.0007	0.0006
N	0.0006	0.0006	0.0006	0.0004	0.0006	0.0004	0.0004
P	0.0086	0.0097	0.0087	0.0076	0.0051	0.0092	0.0087
Y	0.036	0.033	0.036	0.033	0.030	0.034	–
La	–	–	–	–	–	–	0.025
Ti	–	–	–	–	0.032	0.031	–
Hf	–	0.031	–	0.030	–	0.029	–
Zr	–	–	–	0.032	0.031	–	–
Si	–	–	0.496	–	–	–	–

vol.% H_2O, moist air with a moisture content of 10 vol.% H_2O, highly humid synthetic air with a moisture content of 60 vol.% H_2O, and simulated exhaust gas. The 10 vol.% H_2O atmosphere was produced by bubbling air through a water bath held at 60°C but to produce the 60 vol.% H_2O atmosphere, a water bath at 92°C and a gas pressure of 1.3 bars was needed, while the simulated exhaust gas used a gas mixture of N_2 + 0.23% C_3H_8 + 4.2% CO + 9.2% H_2O + 10.2% CO_2. In each case 100 h cycles were used for oxidation testing at temperatures between 1100°C and 1300°C. In the 10 vol.% H_2O test the samples were withdrawn from the furnace into laboratory air each cycle, while at 60 vol.% H_2O the samples were heated and cooled in argon between cycles.

9.2.3   Sample preparation and characterisation

Prior to testing the 1 mm coupons were cleaned and degreased using standard procedures [7]. Samples were mounted in crucibles during oxidation testing, in order to catch any spalled oxide generated, and the gross and net mass gains were recorded periodically between testing cycles. After exposure, plan view and cross-section scanning electron microscopy (SEM) images were recorded and the various oxidation products and depleted substrate components identified by energy dispersive X-ray analysis (EDX).

9.3   Results

9.3.1   Mixed gas corrosion

Typical mass gain curves in 10 vol.% H_2O and 60 vol.% H_2O are shown in Figs 9.1 and 9.2. Only one sample was used for each 10 vol.% H_2O measurement, but the 60 vol.% H_2O test results are the average of three samples of each alloy. Although the detailed behaviour is slightly different as the water vapour content is changed, the overall picture shows that an increase in water vapour content has only a small effect on the overall mass gain, with the model alloys rich in Zr showing the highest initial gains along with the commercial alloy YHfAl in both atmospheres. Another interesting feature of the data is that some of the alloys show very little mass gain in the 10 vol.% H_2O atmosphere during the early stages of oxidation; and this point will be returned to later in the discussion. Comparisons of the gross mass gains in the high Zr (samples M11 and M12) and Hf+Ti alloys (M13) measured in this work with previous results from KFA Jülich [24], measured in laboratory air and a simulated exhaust gas, show remarkably good agreement between the data measured in air in three laboratories under different conditions of moisture content (Fig. 9.3), while the mass gains in simulated exhaust gas are lower in every case. This could be explained by the lower partial pressure of oxygen available in the simulated exhaust gas.

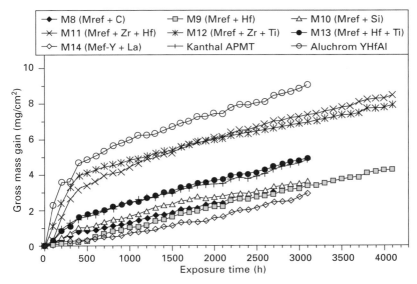

◆ M8 (Mref + C)	◻ M9 (Mref + Hf)	△ M10 (Mref + Si)
✕ M11 (Mref + Zr + Hf)	✳ M12 (Mref + Zr + Ti)	● M13 (Mref + Hf + Ti)
◇ M14 (Mef-Y + La)	┼ Kanthal APMT	○ Aluchrom YHfAl

9.1 Gross specific mass gain in air + 10 vol.% H_2O oxidised at 1200°C (100 h/cycle), 1 mm thick samples.

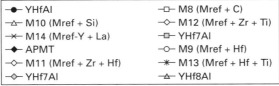

● YHfAl	◻ M8 (Mref + C)
△ M10 (Mref + Si)	◇ M12 (Mref + Zr + Ti)
✕ M14 (Mref-Y + La)	◻ YHf7Al
◆ APMT	○ M9 (Mref + Hf)
◇ M11 (Mref + Zr + Hf)	✳ M13 (Mref + Hf + Ti)
◇ YHf7Al	△ YHf8Al

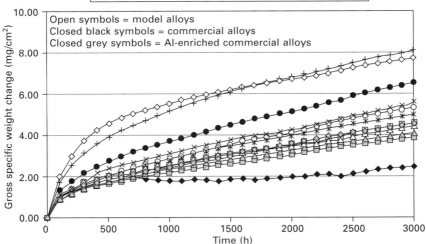

9.2 Gross specific mass change vs exposure time in air + 60 vol.% H_2O at 1200°C for the model alloys in comparison with two commercial alloys and 3 Al-enriched commercial alloys.

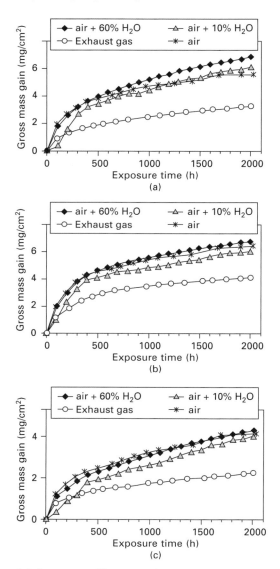

9.3 Gross specific mass change vs exposure time in various atmospheres for (a) alloy M11 (Y+Hf+Zr), (b) alloy M12 (Y+Ti+Zr) and (c) alloy M13 (Y+Ti+Hf).

Comparison of the surface morphologies of the samples after exposure showed two different microstructures when viewed under a low power optical microscope. In some samples the oxide scale had spalled randomly across the whole surface of the coupon, in small discrete regions, while in other samples the surface was criss-crossed with a series of cracks leaving the surface reminiscent of crazy paving. (Fig. 9.4). On closer examination in the SEM, backed up by EDX analysis, it was clear that the samples that had spalled usually failed at the metal/oxide interface as illustrated, for example, in Fig. 9.5. On the crazy paving samples, however, the only areas showing any spallation were those where several cracks intersected at triple points. At these junctions the scales appeared to have failed in a cohesive manner, within the oxide, and the underlying substrate was still protected from further attack by a (thinner) layer of oxide (Fig. 9.6). Many voids and defects were also present in these scales, which were usually formed on the alloys where the mass gain data indicated a rapidly growing initial oxide.

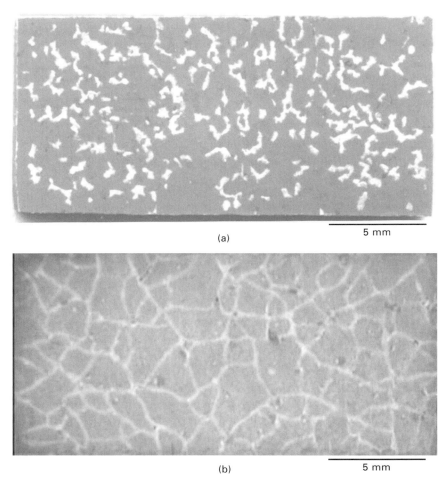

(a)   5 mm

(b)   5 mm

9.4 Low magnification optical images of sample coupons after oxidation at 1200°C in air + 10 vol.% H_2O for 400 h: (a) alloy M14 (+La), (b) alloy M12 (Y+Ti+Zr).

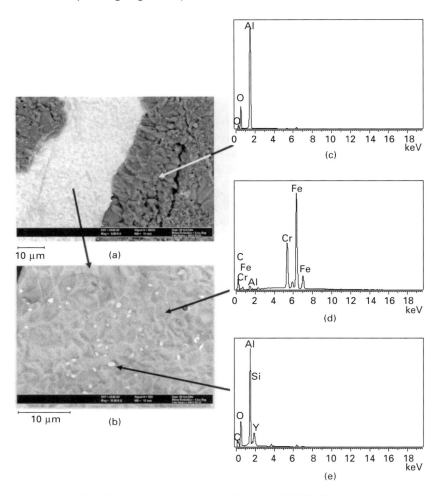

9.5 Spalled region of oxide on Kanthal APMT after oxidation at 1200°C in air + 60 vol.% H_2O for 600 h: (a) and (b) show the spalled interface while (c), (d) and (e) are EDX spectra taken from the oxide, the region under the spalled oxide and particles left on this surface, respectively.

The Zr-containing alloys were always found to have higher oxidation rates than the alloys without this addition. This includes the commercial material Aluchrom YHfAl and the model alloys containing Y+Zr+Hf (M11) and Y+Zr+Ti (M12). The alloys with Zr additions and the one containing Y+Hf+Ti (M13) showed good resistance to scale spalling under temperature cycling conditions during long-term cyclic oxidation tests. On alloys with the above additions the scales spall without exposing the metal substrate. In contrast, on the other alloys studied, such as APMT, or the model alloy containing only La, scale spallation at the scale/metal interface prevailed.

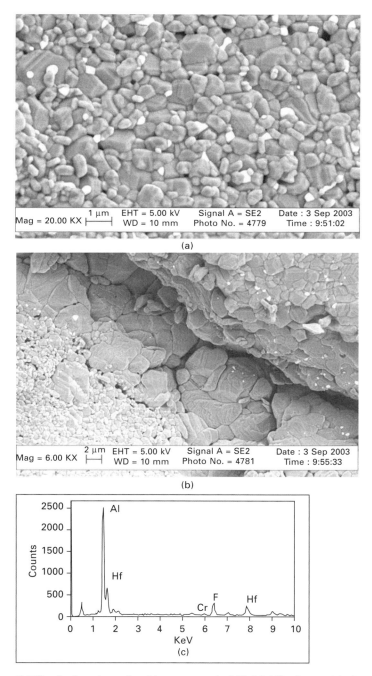

9.6 Spalled region of oxide on sample M9 (Y+Hf) after oxidation at 1200°C in air + 10 vol.% H_2O for 4100 h: (a) the outermost layer of the oxide and (b) a spalled region. (c) is an EDX spectrum taken from one of the white particles shown in (a).

Studies of the growth kinetics of alloys containing Zr have shown that Zr rapidly diffuses into the alumina scale [25]. Upon incorporation it segregates to the oxide grain boundaries, and subsequently leads to the formation of zirconia precipitates and/or pores [17]. The latter findings apparently explain the observed initial higher oxidation rates of the Zr-containing model alloys. It might be, however, that in the commercial alloy Aluchrom YHfAl the effect of Zr is different to that of the model alloys, because of higher carbon content in the material. In the case of alloys that failed at the metal oxide interface, yttria-rich particles were often found on the fracture surface (Fig. 9.5).

9.3.2   Surface-treated alloys

The behaviour of surface-treated alloys was also monitored in this study. PVD methods were used to add a weak oxidation resistant coating to a strong substrate, which would otherwise be much less able to accommodated any growth stresses during cyclic oxidation testing. In this case the combination chosen to test the hypothesis was a coating of model alloy M11 containing Y+Zr+Hf on a Kanthal APMT substrate. A cross-section through the coating deposited by PVD at Jülich is shown in Fig. 9.7. It should be emphasised that this is a trial coating and the coating conditions have not been fully optimised. The cross-section after a 100 h oxidation test (20 h cycles) in laboratory air

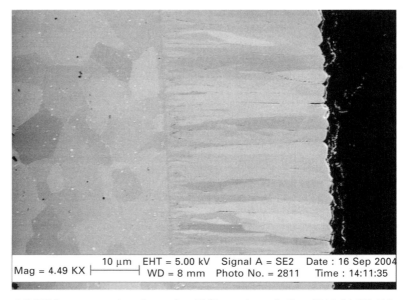

9.7 SEM cross-section through a PVD coating of alloy M11 (Y+Hf+Zr) on an APMT substrate. The columnar nature of the grains in the coating are clearly visible.

at 1200°C is shown in Fig. 9.8, where it can be seen that an adherent alumina scale has formed on the surface of the coating with only a small amount of internal attack. The mass gain results after 600 h of testing of the same coating in 60 vol.% water vapour are compared with the behaviour of the bulk alloy in Fig. 9.9 and it would appear that, under oxidation, the coating is behaving in a similar way to the bulk alloy.

9.4   Discussion

9.4.1   Mixed gas corrosion

Very little work has been done on the synergistic effects of the additions of multiple reactive elements to Fe-Cr-Al alloys even though the commercial variants of these routinely contain multiple additions. In this study a range of model alloys was tested in a variety of atmospheres so that any trends would become apparent. Although there are some trends in behaviour between atmospheres, the major effects appeared to be due to the different compositions of the alloys, which could be roughly divided into two categories: those in which the oxide spalled randomly across the surface and those where the oxide cracked, giving rise primarily to spallation at the junctions where the cracks intersected. Oxides which grew rapidly, initially, tended to have a

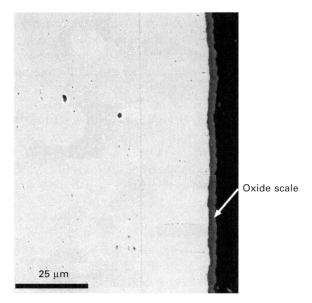

Oxide scale

25 μm

9.8 Electron backscattered image of a cross-section through a PVD coating of alloy M11 (Y+Hf+Zr) on an APMT substrate after oxidation for 100 h at 1200°C in air (20 h/cycle).

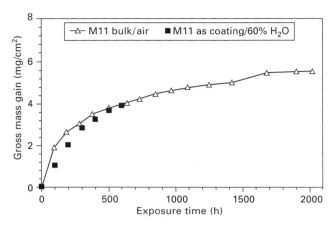

9.9 Gross specific mass gain change vs exposure time in air + 60 vol.% H₂O at 1200°C for the bulk alloy M11 compared with APMT coated with M11.

more open porous structure, especially near the surface, and the cracking was generally confined to failure within the scales, leaving a layer of (thinner) oxide still protecting the underlying substrate. This was true particularly for those alloys, which contained Zr, which has been reported to be particularly mobile through aluminium oxides and form particles of Zr-rich oxide within the outer scale [25]. The more slowly growing oxides, which tended to spall in a more random manner, nearly always failed at the metal oxide interface, leaving an imprint of the oxide grains on the metal surface. This suggests that most of the oxide spalled on cooling rather than at high temperatures, when a new oxide would have started to form. The formation of chromium-rich carbides at the metal oxide interface may also have a major influence on the failure mechanism, in the case of the high carbon alloy, M8 [26]. Although the separation into two categories is not a rigid classification, it does form a useful summary of the behaviour of the samples as shown in Table 9.3. The one exception is alloy M9, which does not fit neatly into either category, but spans across them both, exhibiting slow initial oxide growth and less spallation.

When considering the influence of the environment on the oxidation behaviour, it has already been shown that in the case of the Zr-rich alloys, which show a high initial mass gain, the behaviour is almost independent of the water content of the atmosphere (Fig. 9.3). This is thought to be due to the overriding influence of the oxidation of Zr during the early stages of exposure. The growth and transformation of transient aluminas into α-alumina will also contribute to the mass gain in the early stages, but at the high temperatures used in the present experiments, the transient stage will be well within the first 100 h oxidation cycle. It would therefore appear that the

Table 9.3 Summary of the overall oxidation behaviour

Sample	Initial growth rate		Spallation		Type of failure	
	Slow	Fast	Random	Network	Adhesive	Cohesive
M8	◆		◆		◆	
M10	◆		◆		◆	
M14	◆		◆		◆	
APMT	◆		◆		◆	
M9	◆			◆		◆
M11		◆		◆		◆
M12		◆		◆		◆
M13		◆		◆		◆
YHfAl		◆		◆		◆

*Prepared by Armines St. Etienne, France.

reactions involving Zr dominate the measured mass gain, and that these reactions are insensitive to the water content of the atmosphere.

In the case of the remaining alloys with relatively low weight gains, however, comparison of the initial weight gain curves for the alloys containing La but no Y (M14), for example, or the alloy M9, containing Y and Hf only, shows that, although the results in laboratory air and 60 vol.% H_2O are similar, those for 10 vol.% H_2O are not. The initially high weight gain appears to be lower in a 10 vol.% H_2O atmosphere, although the longer-term oxide growth rate is similar in all cases. There could be two explanations for this: either the water vapour atmosphere is affecting the rate of formation of the transient oxides and their conversion to α-alumina, or there was some other interaction within the furnace during the first few hours of testing. The use of gaseous pre-treatments to minimise the growth of transient aluminas has been reported elsewhere [27], although in a different atmosphere from the one used here.

However, if the results are anomalous due, for example, to the use of new furnace components (which are always baked overnight before use, as a matter of course), then repeat experiments in well-used crucibles and furnaces should settle the matter. Hence the tests were repeated for the first few hundred hours with old crucibles. The new results in 10 vol.% H_2O were then indistinguishable from the laboratory air data measured in Jülich and the 60 vol.% H_2O data recorded in Petten. This in itself confirms the reproducibility of the procedures used in the different laboratories, but also indicates that, although water vapour may have a very small effect on the formation and transformation of transient aluminas, this appears to be less than the experimental scatter of the results for a given alloy over the whole moisture range from 1.5 vol.% H_2O to 60 vol.% H_2O.

9.4.2   Surface-treated alloys

Although FeCrAl alloys produced by mechanical alloying and/or containing oxide dispersions have improved mechanical properties at high temperature, the high strength of the alloys can sometimes be a problem during oxidation. The oxide scales have a greater tendency to crack and spall than those formed on weaker substrates where any oxide growth stresses or stresses built up during thermal cycling can be relieved by the creep of the substrate. In an attempt to combine the advantages of high substrate strength with good oxidation resistance, it was therefore decided to sputter a weak oxidation resistant coating on to a strong substrate, as shown in Fig. 9.7. The results of oxidation shown in Figs 9.8 and 9.9, do indeed illustrate that we can in principle achieve the advantages of a weak coating on a strong substrate. They also demonstrate that the oxidation mechanisms for the coating and the bulk alloy from which it was derived are essentially the same. Work is now in progress to optimise the coating deposition conditions so as to eliminate any porosity in the coating and improve the substrate-coating interface.

9.5   Conclusions

- Multiple additions of minor elements to Fe-Cr-Al alloys affect the failure mechanisms of the oxide scales. There are two broad categories, cohesive failure after rapid initial growth and adhesive failure following slow scale growth.
- Alloys containing Zr show a large initial mass gain dominated by reactions involving Zr; and the water vapour content of the oxidising environment appears to have little effect.
- In non-Zr containing alloys, a lower overall weight gain is observed, but once again the water content of the atmosphere has little effect on the oxidation behaviour at this temperature.
- The application of a weak oxidation resistant coating onto a strong substrate by PVD appears to change the oxidation mode to that of the bulk alloy, which has been applied as a coating.

9.6   Acknowledgements

This work has been supported by the European Commission under the 5th Framework Programme – Competitive and Sustainable Growth (Project acronym SMILER, Contract No. G5RD-CT-2001-00530). The authors gratefully acknowledge the production of the model alloys by Dr J. Le Coze, Ecole des Mines, St Etienne and the contributions of Dr Dmitry Naumenko and Dr J. Quadakkers at Forschungszentrum, Jülich, for the experiments in laboratory air and simulated exhaust gas.

9.7   References

1. J. C. Scully, *The Fundamentals of Corrosion*, 3rd edn, Pergamon Press (1990).
2. G. C. Wood and F. H. Stott, Proc. Int. Conf. on High Temperature Corrosion NACE-6 (March 1981, San Diego California, USA), 227–250, (1983).
3. W. J. Quadakkers and M. J. Bennett, *Materials Science and Technology*, **10**, 126–131 (1994).
4. H. Al-Badairy and G. J. Tatlock, *Oxidation of Metals*, **53**, 157–170 (2000).
5. H. E. Evans and J. R. Nicholls, Lifetime Modelling of High Temperature Corrosion Processes, Proceedings of European Federation of Corrosion Publication No. 34, 37–48 (2001).
6. J. Klower, *Mater. Corros.*, **49**, 758–763 (1998).
7. H. Al-Badairy and G. J. Tatlock, *Materials at High Temperatures*, **18**, 101–106 (2001).
8. H. Al-Badairy and G. J. Tatlock and M. J. Bennett, *Materials at High Temperatures*, **17**, 101–107 (2000).
9. R. Janakiraman, G. H. Meier and F. S. Pettit, *Cyclic Oxidation of High Temperature Materials* (Eds M. Schutze and W. J. Quadakkers), EFC Publication No. **27**, 38–66 (1999).
10. H. Buscail, S. Heinze, Ph. Dufour and J. P. Larpin, *Oxidation of Metals*, **47**, 445–464 (1997).
11. D. R. Sigler, *Oxidation of Metals*, **40**, 295–312 (1993).
12. G. C. Wood, *Oxidation of Metals*, **2**, 11–57 (1970).
13. F. H. Stott, G. C. Wood and J. Stringer, *Oxidation of Metals*, **44**, 11–145 (1995).
14. P. Y. Hou and J. Stringer, *Materials Science and Engineering*, **A202**, 1–10 1995.
14. K. Messaoudi, Doctor thesis, University Paris XI, Orsay France (1997).
16. V. K Tolpygo, *Oxidation of Metals*, **51**, 449–477 (1999).
17. B. A. Pint, *Oxidation of Metals*, **45**, 1–37 (1996).
18. W. J. Quadakkers, J. Nicholls, D. Naumenko, J. Wilber and L Singheiser, *Materials Aspects in Automotive Catalytic Converters* (Ed. by H. Bode), 93–105 (2001).
19. D. R. Sigler, *Oxidation of Metals*, **32**, 337–355 (1989).
20. T. Biegun, M. Danielewski and Z. Skrzypek, *Oxidation of Metals*, **38**, 207 (1992).
21. P. Hou, *Journal of the Electrochemical Society*, **139**, 1119–1126 (1992).
22. E. Sommer, S. G. Terry, W. Sigle, C. Mennicke, T. Gemming, G. H. Meier, C. G. Levi and M. Ruhle, *Materials Science Forum*, **369–372**, 671–678 (2001).
23. B. A. Pint, *Journal of Thermal Spray Technology*, **9**, 198–203 (2000).
24. D. Naumenko, V. Kochubey and W. J. Quadakkers, private communication.
25. D. Naumenko, V. Kochubey, J. Le-Coze, L. Niewolak, L. Singheiser, W. J. Quadakkers, *Materiaux and Techniques*, **7–9** 63–69 (2003).
26. V. Kochubey, D. Naumenko, E. Wessel, J. Le Coze, L. Singheiser, W. J. Quadakkers, H. Al-Badairy and G. J. Tatlock, *Materials Letters* **60**, 1654–1658 (2006).
27. E. N'Dah, A. Galerie, Y. Wouters, D. Goosens, D. Naumenko, V. Kochubey and W. J. Quadakkers Materials and Corrosion, **56**, 843–847 (2005).

Development of novel diffusion coatings for
9–12% Cr ferritic-martensitic steels
(SUNASPO)

V R O H R and M S C H Ü T Z E, Karl-Winnacker-Institut der
DECHEMA e.V., Germany; E F O R T U N A and
D N T S I P A S, Aristotle University of Thessaloniki,
Greece and A M I L E W S K A and F J P É R E Z,
Universidad Complutense Madrid, Spain

10.1   Introduction

9–12% Cr ferritic-martensitic steels are on the borderline of the protective
situation with respect to oxidation corrosion at 650°C [1, 2]. Austenitic
steels and nickel-based alloys are potential candidates for increasing the
steam temperatures and thus the efficiency of power plants. However, they
are more expensive, have higher coefficients of thermal expansion and show
lower heat conductivity, which are the significant drawbacks for their use as
heat exchanger tubes.

Nonetheless, at least from the mechanical point of view, the ferritic-
martensitic steels have been designed for steam service temperatures up to
650°C [3]. Therefore, if the corrosion resistance of ferritic-martensitic steels
is enhanced by a suitable coating, the service temperature of these steels
could be increased up to their mechanical limits. Among the currently available
coating processes, the widely used spray coatings can provide a protection
adapted to each type of corrosion environment at high temperature [4].

Also, the Al-slurry coatings produced by Aguero and Muelas [5] showed
promising behaviour with regard to corrosion protection of 9% Cr steels in
water vapour. Nevertheless, neither the spraying nor the slurry techniques
seem to be suitable for coating the inside of a heat exchanger tube, which can
be several kilometres in recently commissioned plants.

Vapour deposition processes present a certain potential with respect to
their applicability to power plant components. This chapter concentrates on
addressing the development of chemical vapour deposition (CVD) coatings
on 9–12% ferritic-martensitic steels. Two processes have been investigated:
the in-situ CVD pack cementation process and the fluidised bed CVD (FBCVD)
process.

The high temperature diffusion coating processes generally consist of
enriching the substrate surface with elements such as Al, Cr or Si that are
expected to form a protective oxide scale in aggressive environments [6].

The CVD processes rely on the formation and diffusion of gaseous species, usually halides, at high temperatures. For 9–12% Cr steels, this temperature has to be kept below about 700°C because the rearrangement of lath martensite with high dislocation density into equiaxed cells is enhanced above 700°C, as observed for instance on heat affected zones after welding [7]. High coating temperatures would thus lead to a degradation of the mechanical properties of the 9–12% Cr steels.

Thermodynamic calculations showed that for the same temperature and the same stoichiometric conditions, the partial pressure of Al-halide is higher than that of Si or Cr-halides [8]. It is therefore expected that deposition of Al is more suitable for the development of a low temperature coating process.

Very recently, the pack cementation technique has been extended below 700°C by Xiang and Datta on low alloy steels [9], while our group has been concentrating on 9–12% Cr steels [10]. These reports show that Al diffusion is fast enough for forming aluminide coatings of suitable thickness in a reasonable coating time. They also showed that the coating thickness may be increased by a mechanical surface treatment prior to the coating process leading to an enhancement of the diffusion rate by increasing the number of rapid diffusion paths.

The FBCVD method has been developed in the last 15 years. Extensive work has been carried out on austenitic steels and nickel-based alloys. Pérez *et al.* even developed an Al-Si codiffusion coating at temperatures as low as 600°C on the austenitic steel AISI 304 [11]. However, the applicability of the FBCVD process to 9–12% Cr steels seems to be unexplored.

Furthermore, it has been shown that boron improves the oxidation properties of aluminide coatings [12]. Boron can be deposited by both pack cementation and FBCVD processes and previous studies showed that on carbon steels, pack cementation leads to textured coatings which is not the case for the FBCVD method [13]. Boron was also codiffused with Al by pack cementation on low alloy steels [14]. Unless the coating temperature is decreased, these conditions cannot be used for 9–12% Cr steels.

The aim of this study is to present the microstructures of the coatings developed during the first two years of the SUNASPO project. These include single element aluminides as well as silicon and boron modified aluminides developed at low temperature by pack cementation or FBCVD on 9–12% Cr steels. Also, first attempts at the combination of the heat treatment with the coating process in order to allow the diffusion of Cr are reported.

10.2   Experimental methods

10.2.1 Substrate materials

The composition (in wt.%) of P91 is reported in Table 10.1. Samples were machined to the dimensions of $20 \times 9.5 \times 3$ mm. Samples to be coated were

Table 10.1 Composition of the substrate alloys analysed at FZ Jülich, Germany

Material	Fe	C	Si	Cr	Al	Ni	Mn	Mo	Nb	V	N	Other elements
HCM12A	Bal.	0.07	0.25	12.5	0.01	0.34	0.54	0.36	0.08	0.21	0.06	W: 1.89, Cu: 0.85
P91	Bal.	0.09	0.40	9.2	0.01	0.38	0.50	0.90	0.06	0.22	0.05	Co: 0.01

Table 10.2 Experimental conditions for CVD-FBR coatings

Coating	Material	Stage	Time (h)	Temperature (°C)
Al	P91	1	2	550
Si	P91/HCM12A	1	0.5	490
Al/Si	P91/HCM12A	1° (Al)	1	550
		2° (Si)	0.5	490

either glass bead blasted in the conditions described elsewhere [15] or ground with 120 grit SiC paper. They were degreased in acetone prior to the coating process.

The sample preparation after the coating process included electroplating with nickel, cross-sectioning and polishing up to 1 μm diamond paste. X-ray diffraction (XRD), light microscopy, scanning electron microscopy with energy dispersive spectrometer (SEM/EDS) and electron probe microanalysis (EPMA) were used for the coating characterisation.

10.2.2 Pack cementation coating process

For the pack cementation process, the specimens were put into an alumina retort, surrounded by a powder mixture, composed of a source supplier (Al, Cr, Si or B_4C) called the masteralloy, an activator (NH_4Cl) and an inert filler (Al_2O_3). Afterwards, the retort was placed in a tubular furnace and heated to the coating temperature for a certain time in inert or reducing environment, i.e. Ar or Ar-10%H_2. The detailed experimental conditions are summarised in Table 10.2.

10.2.3 Fluidised bed chemical vapour deposition

The fluidised bed was composed of aluminium, silicon or aluminium/silicon powder. The particle size was less than 200 μm to obtain a homogeneous bed and to improve the conditions of heat and mass transfer. The bed was then fluidised with an increasing flow rate of argon during the heating-up period.

A schematic diagram of the CVD-FBR system used in this study is shown in reference [16]. The reactor was made of quartz with an internal diameter of 5 cm. It was heated inductively by an external furnace.

- For the aluminum or silicon deposition, the reactor was filled with a treatment agent, consisting of an aluminum donor powder (Al 99.5% and grain size ≤ 200 μm) or silicon donor powder (Si 99.5% and grain size ≤ 40 μm).
- For the aluminium/silicon coatings the fluidised bed was made of a mixture of silicon donor powder (grain size about 40 μm) and aluminium powder (Al 99.5% and grain size ≤ 200 μm).

The mixture was fluidised by Ar gas (99.999% purity), and hydrogen chloride (HCl) was used as an activator of the process. The $Ar_{(g)}$, $HCl_{(g)}$ and $H_{2(g)}$ gas mixture entered the CVD-FB with ratio of 5/1 to 10/1 of HCl/H_2; the deposition conditions are shown in Table 10.3. For the codeposition process, the thermodynamic calculations were performed with the Thermocalc software [17] as shown in Fig. 10.1.

10.3   Experimental results

10.3.1   Single element Al coating on P91

Figure 10.2 gives the microstructure and composition of the aluminide obtained by FBCVD at 550°C on P91. The coating is composed of 10 μm Fe_2Al_5 and probably an interdiffusion zone of less than 1 μm, which is visible by SEM on the back scattered electron image. The EDX analysis gives 5 at.% Al for this interdiffusion zone but it has to be kept in mind that due to its very low thickness some information may arise from the Fe_2Al_5 phase. The coating itself is quite irregular in thickness and shows some cracks.

The coating shown in Fig. 10.3 is an aluminide, too. It was obtained by pack cementation with an Al masteralloy and an NH_4Cl activator at 650°C for 6 h. The coating is 50 μm thick with Fe_2Al_5 phase. However, it appears to be more homogeneous in thickness than the previous FBCVD coating and does not show any cracks. The EPMA analyses reveal that there is not so much of an interdiffusion zone underneath the Fe_2Al_5 phase. This coating

Table 10.3 Experimental conditions for FBCVD coatings

Coating	Material	Stage	Time (h)	Temperature (°C)
Al	P91	1	2	550
Si	P91	1	0.5	490
Al/Si	P91	1° (Al)	1	550
		2° (Si)	0.5	490

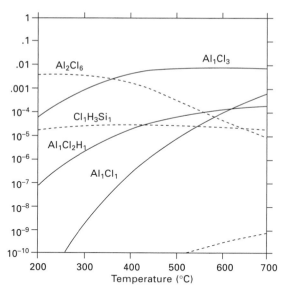

10.1 Thermodynamic calculation for FBCVD Al-Si co-deposition.

also contains Cr-Si rich precipitates, both elements coming from the substrate steel.

Aluminide coatings with lower aluminium contents were then developed. The coating temperature was kept at 715°C (for 6 h) and some changes in the pack composition were carried out in order to reduce the Al activity in the gas phase.

By using the more stable activator $AlCl_3$ instead of NH_4Cl at 715°C, the coating phase was shifted into the range of FeAl with up to 53 at.% Al (Fig. 10.4). The first layer of the coating is quite porous, the second layer is much denser with a composition varying from 45 at.% Al to 26 at.% on the first 5 μm. The aluminium enrichment in the ferritic matrix reaches a depth of around 30 μm underneath the surface and leads to the formation of aluminium nitrides appearing as black needles.

The same coating phase can be obtained at 650°C by replacing the pure Al masteralloy by a powder of Cr_5Al_8 intermetallic compound. The coating phase is less than 5 μm thick and shows some thickness irregularities. Its Al concentration is in the same range as in the previous case with 53 at.% (Fig. 10.5)

10.3.2 Single element Si coating on P91

The single element Si coatings are shown in Figs 10.6 and 10.7 for FBCVD and pack cementation, respectively. With the thickness of the enriched layer

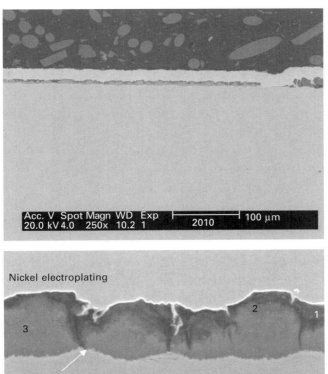

10.2 FBCVD aluminide developed on P91 in 2 h at 550°C.

being about 10 µm for the pack coating and 3–4 µm for the FBCVD, both coatings show a certain porosity. The Si content is 18 at.% for FBCVD and 27 at.% for the pack coating. The first concentration corresponds to the DO_3-Fe_3Si phase concluding from the Fe-Si binary phase diagram [18], whereas the second falls into the range between Fe_3Si and ε-FeSi.

10.3.3 Codiffusion Al-Si coating on P91

Figures 10.8 and 10.9 show the Al-Si codiffusion coatings developed by pack cementation and FBCVD, respectively. On P91, both processes lead to the formation of Fe_2Al_5 as the main coating phase, an FeAl interdiffusion

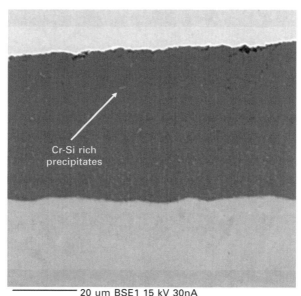

20 um BSE1 15 kV 30nA

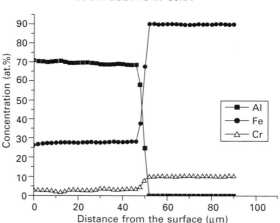

10.3 Aluminide obtained by pack cementation on P91 at 650°C for 6 h in Ar-10%H$_2$. Pack composition (wt.%): 10 Al, 1 NH$_4$Cl, 89 Al$_2$O$_3$. Composition profile analysed by EPMA.

zone and black needles underneath in the ferritic matrix (Fig. 10.8). Both coatings show thickness irregularities and in both cases there is no significant amount of Si enrichment in the Fe$_2$Al$_5$ phase. However, Cr-Si precipitates are detected in the coating. These coatings are different in thickness, 30–40 μm for the pack coating and 2–8 μm for FBCVD, which could be due to the difference in process temperatures. The pack cementation coating also shows some cracks.

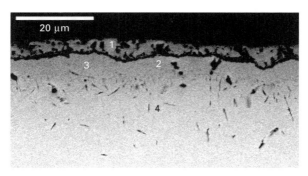

10.4 SEM micrograph and EDX analyses of P91 aluminised at 715°C for 6 h in Ar. Pack composition (wt.%): 5 Al, 2 AlCl$_3$, 93 Al$_2$O$_3$.

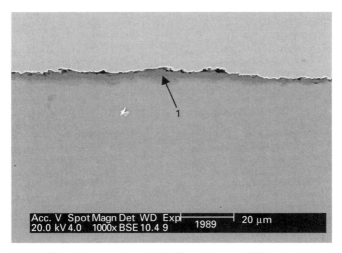

10.5 SEM micrograph and EDX analyses of P91 aluminised at 650°C for 6 h in Ar-10% H$_2$. Pack composition (wt.%): 20 Cr$_5$Al$_8$, 2 NH$_4$Cl, 58 Al$_2$O$_3$.

10.3.4 Codiffusion Al-B coating on P91

Al-B co-deposition was carried out with pure Al as Al donor and B$_4$C as boron donor. The results presented in Fig. 10.10 show the formation of a 100 µm thick coating composed of Fe$_2$Al$_5$ with a thin FeAl layer in the interdiffusion zone. This coating contains some pores and cracks. No significant effect of the boron addition could be observed.

On the other hand, the same type of deposition was carried out by a fluoride activated pack. Besides the replacement of the NH$_4$Cl activator by KBF$_4$, the pure Al was replaced by Na$_3$AlF$_6$ that could also act as an activator. The coating is shown in Fig 10.11. It consists of a 20 µm dense layer. The Al

10.6 FBCVD silicide on P91 and HCM12A developed at 490°C for 0.5 h.

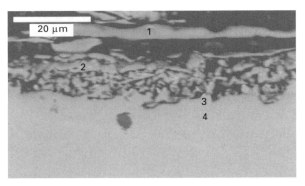

10.7 SEM micrograph and EDX analyses of P91 siliconised at 715°C for 6 h in Ar. Pack composition (wt.%): 10 Si, 2 NH_4Cl, 88 Al_2O_3.

concentration lies below 1 at.% and the X-ray diffraction revealed the presence of iron borides: $(Fe, Cr)_2B$ and $(Fe, Cr)B$.

10.3.5 Two-step Cr + Al coating on P91

The Cr deposition by a chloride activated pack cementation process requires high temperatures. In order to enable this kind of treatment on P91, the coating process was combined with the heat treatment of the steel. For 9–

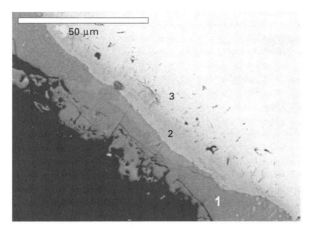

10.8 SEM micrograph and EDX analyses of P91 silicoaluminised at 715°C for 6 h in Ar. Pack composition (wt.%): 40 (0.92 Si + 0.08 Al), 2 NH$_4$Cl, 58 Al$_2$O$_3$.

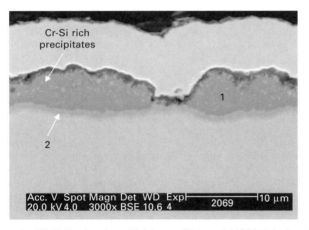

10.9 FBCVD aluminosilicide on P91 and HCM12A developed at 550°C for 1 h and at 490°C for 0.5 h.

12% Cr steels, this consists of an austenitisation at 1000–1100°C, followed by a rapid cooling in air for the martensite formation. A second tempering treatment is generally applied afterwards at 730–790°C [19, 20].

Our investigations concentrated on a Cr deposition at 1000°C for 2 h followed by rapid cooling in Ar. The deposition of Al in a second step was then carried out at 650°C for 1 h.

The coating structure is given in Fig. 10.12. It consists of a 10 μm Al-rich layer (Al$_4$Cr) followed by a Cr-rich layer, that contains up to 70 at.% Cr. Underneath, no Al diffusion could be detected but the Cr enrichment goes

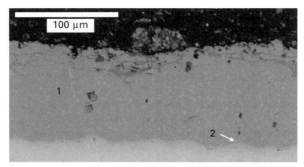

10.10 SEM micrograph of P91 boroaluminised at 715°C for 6 h in Ar. Pack composition (wt.%): 17 (0.3 Al + 0.7 B₄C), 2 NH₄Cl, 81 Al₂O₃.

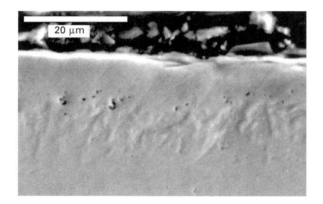

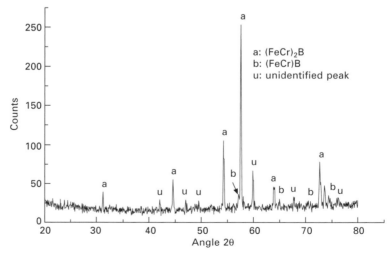

10.11 SEM micrograph and XRD analyses of P91 boroaluminised at 715°C for 6 h in Ar. Pack composition (wt.%): 10 B₄C, 5 KBF₄, 5 Na₃AlF₆, 80 Al₂O₃.

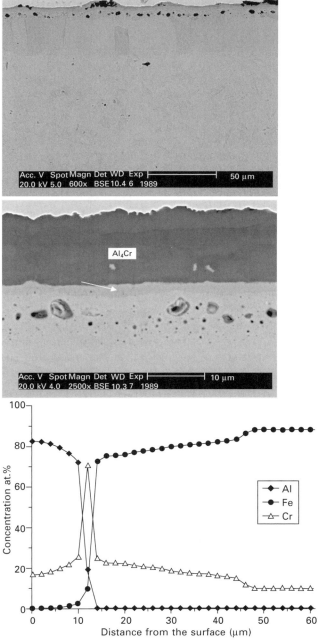

10.12 Two-step Cr+Al pack cementation coating on P91. Step 1: Cr deposition at 1000°C for 2 h in Ar followed by rapid cooling (Pack composition in wt.%: 30 Cr, 5NH₄Cl, 65 Al₂O₃). Step 2: Al deposition at 650°C for 1 h in Ar-10%H₂ (Pack composition in wt.%: 20 Al, 1 NH₄Cl, 79 Al₂O₃).

further into the ferritic matrix, down to 45 μm below the surface. Occasionally, some pores appear in the changeover region between the intermetallic phases and the ferritic matrix. These pores certainly arise from the Kirkendall effect during the Cr diffusion and were observed already after the first step of Cr deposition [10].

10.4   Discussion

10.4.1 Single element diffusion coatings

The aluminium diffusion coatings developed at low temperatures by both pack cementation and FBCVD are mainly composed of Fe_2Al_5. The preferential formation of this phase has certainly to be attributed to its thermodynamic stability: the enthalpy of formation of Fe_2Al_5 is about three times lower than for the other iron aluminides [21]. This phase has shown promising results with regard to corrosion resistance in water vapour [5] and in simulated coal firing environment [22] since it forms a protective aluminium oxide scale. Moreover, it was shown that Fe_2Al_5 transforms into FeAl due to further interdiffusion at high temperature.

The structure and the mechanical properties of Fe_2Al_5 were investigated by Hirose *et al.* [23]. Although it shows a significant plasticity above 600°C, this phase has poorer mechanical properties than the widely studied lower Al containing phases such as B2-FeAl or DO_3-Fe_3Al. For instance, the yield strength at 650°C published by Hirose *et al.* is about two times lower than that for FeAl or Fe_3Al [24]. One possibility for transforming Fe_2Al_5 into lower aluminium-containing iron aluminides would involve heat treating the Fe_2Al_5 coated steel. However, previous results suggest that several hundred hours might be required for a complete transformation of a 30 to 50 μm Fe_2Al_5 coating into B2-FeAl [22].

For this reason, part of the present work aimed at reducing the Al concentration of the pack cementation coatings in order to obtain the low aluminium containing iron aluminides in one single step. This cannot be achieved by using a pure metallic Al donor together with the unstable NH_4Cl activator. One of these two reactants has to be replaced in order to reduce the Al chloride activity during the pack cementation process. Both the replacement of the NH_4Cl activator by the more stable $AlCl_3$ and the replacement of the pure Al source by the less Al active Cr_5Al_8 alloy lead to the formation of FeAl, with about 53 at.% Al. Nevertheless, the coating thickness is not yet sufficient for long-term high temperature applications on power plant components. Higher coating temperatures or longer process durations are certainly required in order to get thicker coatings. However, higher coating temperatures could only be achieved by combining the coating process with the heat treatment of the steel. Therefore the heat treatment duration dictated by the steel properties has to be respected.

Dense Si diffusion coatings seem difficult to obtain by both of the processes investigated. The possible formation of iron dichlorides, which competes with the formation of silicon chloride [8], may be an explanation for the porous coatings obtained. Furthermore, the potential application of silica-forming alloys in power plants needs to be verified in detail, since silica might dissolve in pressurised steam at high temperature.

10.4.2 Al-Si or Al-B codiffusion

Our experience shows that as long as pure Al is used as donor for the aluminium deposition, the coating obtained is composed of Fe_2Al_5. The co-element, Si or B, then enters the coating in very low amounts, usually under 1%. According to the Al-Fe-Si ternary phase diagram at 600°C [25] the solubility of Si in Fe_2Al_5 reaches around 6 at.%. At least for Fe_2Al_5 on P91, our concentration is, thus, significantly below the solubility limit. The presence of Si in the form of Si-Cr-rich precipitates does not allow the conclusion that Si enrichment took place, since this kind of phase also was observed on single phase aluminides. Moreover, a comparison of the silicon content between the precipitates obtained after Al deposition and Al-Si co-deposition is impossible using the microanalytical techniques in the present investigations because of their very low dimensions.

For FBCVD Al-Si co-deposition, thermodynamic calculations performed with the Thermocalc software are shown in Fig. 10.1. Comparing the partial pressures of SiH_3Cl with the most reactive aluminium chloride AlCl helps in optimising the deposition conditions. It can be seen that during the first treatment at 550°C, the partial pressure of AlCl is comparable to SiH_3Cl that allows the deposition of Al. However, during the second treatment at 490°C, the partial pressure of AlCl is about ten times lower than that of SiH_3Cl. Our results thus suggest that either this ratio is not enough for depositing a significant amount of silicon or the aluminium deposition due to $AlCl_3$ and Al_2Cl_6 has to be taken into account.

Furthermore, the observation of Cr-rich precipitates in Fe_2Al_5 suggests that this phase is saturated in Cr. The solubility of Cr in Fe_2Al_5 in the range of 550–715°C has not been explored yet, but at 1000°C the limit lies at about 5–8 at.% Cr depending on the Al concentration [26]. None of the Fe_2Al_5 coatings reported here contains so much Cr in solution, suggesting that the solubility must be lower at the temperatures investigated.

Boroaluminide coatings on low carbon and low alloy steel have been found to be protective against steam oxidation at 650°C for periods up to 168 h [12]. This protection was due to the formation of a protective Al_2O_3 scale resting on an internal Fe_2B layer. It seems that this Fe_2B layer between the substrate and the external Al_2O_3 layer acts as a diffusion barrier. In the boro-aluminide coatings on P91 where $Fe(Cr)B_2$ was observed, the Al enrichment was less than 1 at.%, and therefore this proposed mechanism is not applicable.

Attempts are under way, using different pack cementation compositions, to obtain boroaluminide coatings containing both Fe_2B and ~15 at.% Al.

10.4.3 Role of the aluminium diffusion rate

Al diffuses quite rapidly in ferritic steels. This is the explanation for the formation of suitable coating thicknesses in reasonable coating durations even at low temperatures, for instance a 50 µm coating grows in 6 h at temperatures as low as 650°C. Thus, it is expected that during service at almost the same temperature, the aluminium loss due to interdiffusion between the coating and the substrate will contribute to a significant degree to the Al reservoir degradation. For this reason, even though higher coating temperatures would allow the development of the mechanically more favourable iron-aluminides (e.g. FeAl or Fe_3Al) with lower Al content, they would certainly degrade earlier in service at high temperature. Furthermore, the effect of a too strong Al enrichment of the substrate steel has not yet been elucidated but if Al precipitates form with nitrogen from the substrate as observed by Aguero and Muelas [5], this will probably change the creep properties of the steel, whose strengthening is based on cabonitrides formation.

10.4.4 Combination of the coating process with the heat treatment of the steels

The combination of the coating process with the conventional heat treatment of 9–12%Cr steels aims at taking benefit from coating temperatures higher than 650°C. Although higher coating temperatures would enable the direct development of low Al containing coatings due to the higher diffusion rates of the substrate elements, this strategy was not employed. Indeed it is expected that single aluminide coatings might degrade early in service, due to interdiffusion between the coating and the substrate, as mentioned earlier.

Higher temperatures are particularly required for Cr deposition but our aim is not to develop single element Cr coatings, since the environments encountered in power plants usually contain some water vapour, in which chromium oxide is known to form volatile species [27]. The Cr enrichment is rather the background for the development of the two-step Cr+Al coatings, where the Cr-rich layer prevents the Al from diffusing further into the substrate in service conditions. The aluminium thus stays close to the surface for a longer time, where it enables the formation of a protective alumina scale. This effect, observed earlier by Rosado and Schütze [28] on other types of materials, is expected again with the Cr+Al coating developed in the present work.

10.5   Conclusion

Pack cementation as well as fluidised bed chemical vapour deposition have been extended to temperatures below 715°C. Homogeneous diffusion aluminide Fe_2Al_5 coatings of up to 50 μm thick have been obtained by pack cementation. FBCVD has shown a higher success for the Al-Si co-deposition. Further improvement is, though, required for the Si deposition as well as the pack cementation codiffusion at low temperatures.

A two-step Cr+Al coating was developed by combining the pack cementation process with the heat treatment of P91 steel. The Cr-rich interlayer is expected to act as an interdiffusion barrier between the Al-rich reservoir and the substrate.

From our investigations two general routes for the protection of 9–12%Cr steels can be highlighted. The first solution consists in applying a thin coating, with a thickness below 10 μm. This implies a small Al reservoir that would rapidly degrade especially by interdiffusion. However, the Al amount being limited, the Al enrichment of the substrate will have limited effect on the microstructure of the 9–12% Cr steel in the subcoating zone. As a consequence, the coating will have little influence on the mechanical properties of the steel. The present investigations showed that FBCVD is better adapted to the application of thin coatings.

The second solution consists in applying a thicker coating, with a thickness above 10 μm. It is then to be expected that a high amount of Al enters the substrate due to interdiffusion as previously described. However, this can be avoided by using the Cr interdiffusion barrier. The higher Al reservoir would thus remain in the subsurface zone and achieve a long-term corrosion protection. Up to now, pack cementation has proved a greater success for the deposition of thick coatings.

10.6   Acknowledgements

This work was supported by the SUNASPO, European Commission funded RTN project No. HPRN-CT-2001-00201, which is gratefully acknowledeged. The authors thank the Forschungszentrum Jülich (Gerrmany) for providing the samples. Also, we would like to express our deep appreciation to the referee of this chapter for his enriching suggestions.

10.7   References

1. J.P.T. Vossen, P. Gawenda, K. Rahts, M. Röhrig, M. Schorr, M. Schütze, *Materials at High Temperature*, **14**, 4 (1997) 387–401.

2. M. Thiele, H. Teichmann, W. Schwarz, W.J. Quadakkers, *VGB Kraftwerkstechnik*, **77**, 2 (1997) 135–140.

3. R. Viswanathan, W. T. Bakker, *IJPGC2000-15049 Proceedings of 2000 International Joint Power Generation Conference*, Miami Beach, Florida, July 23–26 (2000).

4. L. Pawlowski, *The Science and Engineering of Thermal Spray Coatings*, John Wiley & Sons, New York (1995).

5. A. Aguero, R. Muelas, *Materials Science Forum*, **461–464** (2004) 957–966.

6. R. Mevrel, *Material Science and Engineering*, **A120** (1989) 13–24.

7. F. Arav, H.J.M. Lentferink, *Proceedings of the Fifth International Conference on Creep of Materials*, Lake Buena Vista, Florida, USA, 17–21 May (1992).

8. S.C. Kung, R.A. Rapp, *Oxidation of Metals*, **32**, 1–2 (1989) 89–109.

9. Z.D. Xiang, P.K. Datta, *Surface and Coatings Technology*, **184** (2004) 108–115.

10. V. Rohr, M. Schütze, *Materials Science Forum*, **461–464** (2004) 401–408.

11. F.J. Pérez, J.A. Trilleros, M.P. Hierro, A. Milewska, M.C. Carpintero, F.J. Bolívar, *Materials Science Forum*, **461–464** (2004) 313–320.

12. N.E. Maragoudakis, S.D. Tsipa, G. Stergoindis, H. Omar, D.N. Tsipas, Boro-aluminide coatings for protection against high temperature steam oxidation, *Proc. of Corrosion science in the 21st century, UMIST* (2003) to be published in *Electronic Journal of Corrosion Science and Engineering*, **6**, http://www.jcse.org/.

13. K.G. Anthymidis, N. Maragoudakis, G. Stergioudis, O. Haidar, D.N. Tsipas, *Materials Letters*, **57** (2003) 2399–2403.

14. N.E. Maragoudakis, G. Stergioudis, H. Omar, H. Paulidou, D.N. Tsipas, *Materials Letters*, **53** (2002) 406–410.

15. V. Rohr, M. Schütze, *Surface Engineering*, **20**, 4 (2004) 266–274.

16. F.J. Pérez, M.P Hierro, F. Pedraza, C. Gomez, M.C Carpintero, *Surface and Coating Technology*, **120–121** (1999), 151.

17. Thermocalc Software AB. Version P. Copyright 1995–2003. Foundation of Computational Thermodynamics, Stockholm, Sweden.

18. T.B. Massalski, *Binary Alloy Phase Diagrams*, American Society for Metals, (1986).

19. G. Gupta, B. Alexandreanu, G.S. Was, *Metallurgical and Materials Transactions A*, **35A** (2004) 717–719.

20. H. Kaest, *VGB Kraftwerkstechnik*, **70**, 12 (1990) 1050–1053.

21. S. C. Deevi, V. K. Sikka, *Intermetallics*, **4** (1996) 357–375.

22. V. Rohr, A. Donchev, M. Schütze, A. Milewska, F.J. Pérez, Diffusion coatings for the high temperature corrosion protection of 9–12% Cr steels, *Proc of the EUROCORR Conference in Nice (CD) EFC event no 266, paper no 237* (2004).

23. S. Hirose, T. Itoh, M. Makita, S. Fujue, S. Arai, K. Sasaki, H. Saka, *Intermetallics*, **11** (2003) 633–642.

24. K. Vedula, Chapter 9 in: Westbrook J.H., Fleischer P.L. (eds), *Intermetallic Compounds*, John Wiley, New York (1995).

25. G. Petzow, G. Effenberg, *Ternary Alloys, A Comprehensive Compendium of Evaluated Constitutional Data and Phase Diagrams*, vols 1–4, VCH Weinheim, (1991).

26. M. Palm, *Journal of Alloys and Compounds*, **252** (1997) 192–200.

27. E.J. Opila, *Materials Science Forum*, **461–464** (2004) 765–774.

28. C. Rosado, M. Schütze, *Materials and Corrosion*, **54**, 11 (2003) 831–853.

11

Steam oxidation and its potential effects on creep strength of power station materials (SUNASPO)

M S C H Ü T Z E and V R O H R, Karl-Winnacker-Institut der
DECHEMA e.V., Germany and L N I E T O H I E R R O,
P J E N N I S and W J Q U A D A K K E R S,
Forschungszentrum Jülich GmbH, Germany

11.1   Introduction

The development of steam power plants towards higher thermal efficiencies is currently an important issue in the power generation industry, as increasing the efficiency would reduce their environmental impact. The most effective means for increasing the thermal efficiency is to raise the steam temperature. However, the operating conditions of the plant components become more severe in terms of creep rupture strength and oxidation resistance of the materials. Therefore materials with improved creep strength and corrosion resistance are required [1, 2].

The aim of this research is to investigate the oxidation resulting from exposures in steam and in combustion gas and how this may affect the creep resistance of the materials. For that, the oxidation behaviour of some different commercial materials will be described. Considering the material loss due to corrosion, the interaction between creep and corrosion could be studied by separating both effects [3]. Thus creep tests in air of pre-oxidised specimens have been carried out; specimens were pre-oxidised in different atmospheres and at different temperatures.

11.2   Steam oxidation tests

Five different materials were selected for these studies: two ferritic-martensitic steels (P91 and HCM12A), two austenitic steels (1.4910 and Alloy 800) and the nickel base Alloy 617. Their chemical compositions are listed in Table 11.1.

From these materials $20 \times 10 \times 2$ mm specimens were machined and ground to 1200 grit surface finish. The samples were exposed in Ar-50%H_2O, at 550, 600 and 650°C for 250 and 1000 h. To measure the corrosion rate the weight gain was used. In case of the 1000 h exposure, the samples were cooled to room temperature every 250 h for weight measurement. Oxide scale characterisation after exposure was carried out using optical

Table 11.1 Chemical composition of the selected materials (wt.%)

Material	Fe	Ni	Co	Cr	Si	Mn	Al	Ti	Mo	V	W	C
P91	88.2	0.38	0.01	9.2	0.40	0.5	0.01	0.01	0.9	0.22	–	0.089
HCM12A	83.5	0.34	–	12.5	0.25	0.54	0.01	–	0.36	0.21	1.9	0.071
1.4910	66.8	12.5	–	16.7	0.27	1.3	–	0.32	2.2	–	–	0.019
Alloy 800	45.1	34.2	–	20.1	0.14	1.0	0.35	0.49	–	–	–	0.061
IN 617	1.2	54.8	11.1	22.4	0.05	0.05	1.1	–	9.0	–	–	0.053

metallography, scanning electron microscopy (SEM), X-ray diffraction and Raman spectroscopy.

The weight gain of the materials after 250 h is shown in Fig. 11.1. The highest oxidation rates were obtained for the lowest chromium content steel P91, whereas the high chromium alloys, Alloy 800 and Alloy 617 exhibited the lowest weight gains. Intermediate values were obtained for the HCM12A and 1.4910 steels.

The oxide scale morphology of P91 after 250 h exposure did not substantially change between the three different temperatures. In all the cases it consisted of two layers (Fig 11.2). At 550°C there are two layers: Fe_2O_3, Fe_3O_4 and spinel rich in Cr. The external layer was grown over the original surface of the sample and possessed the spinel structure. By SEM/EDX it was found that there were no other metallic elements except iron within this layer, therefore its nature is magnetite Fe_3O_4. On the outside of the scale some Fe_2O_3 inclusions were found as well. The presence of these inclusions is related to the formation of voids within the scale; these voids decrease in size and number with increasing temperature [4, 5]. Beneath the original surface of the sample a second layer appeared. Its structure was also spinel, but in contrast to the externally growing scale, it contains dissolved chromium. Underneath the internal layer some internal oxidation precipitates consisting of FeO and Cr_2O_3 were formed.

The scale morphology described above was also found after the exposure of the HCM12A at 550°C. At higher temperatures an inhomogeneous attack

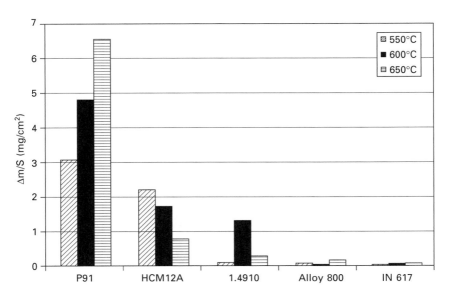

11.1 Weight change of materials after 250 h exposure at three temperatures 550, 600 and 650°C in Ar-50%H_2O.

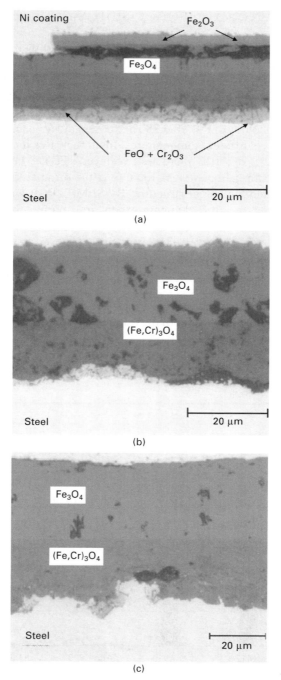

11.2 Cross-section of P91 after 250 h exposure in Ar-50%H$_2$O at
(a) 550°C, (b) 600°C and (c) 650°C.

was observed. On some parts of the surface, oxide nodules were formed, whereas a thin Cr_2O_3 layer covered the rest of the surface. The nature of the oxide nodules was similar to the oxide scale that was formed on the P91 steel (Fig. 11.3). It was observed that increasing the temperature decreased the size of nodules. This is in agreement with the temperature dependence of the weight gain shown in Fig. 11.1, at high temperatures the weight gain is lower than at low temperatures. The austenitic steel 1.4910 exhibited similar oxidation behaviour at temperatures above 600°C. However, at 550°C it formed a thin Cr_2O_3 scale (Fig. 11.4) similar to the scales observed on Alloy 800 and Alloy 617.

After 1000 h exposure, the oxide scale morphology of the different materials did not significantly change. As with 250 h exposure, the highest oxidation

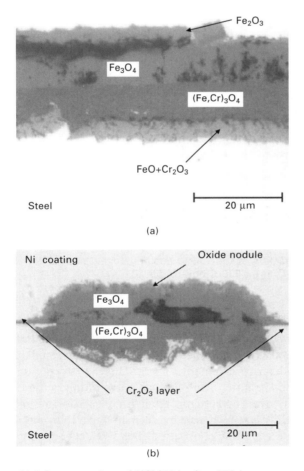

11.3 Cross-section of HCM12A after 250 h exposure in Ar-50%H_2O at (a) 550°C and (b) 650°C.

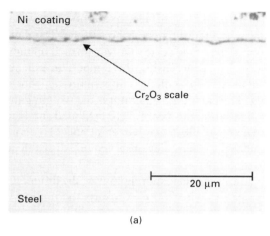

(a)

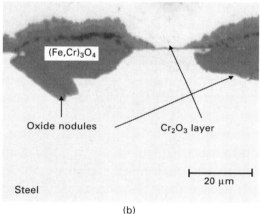

(b)

11.4 Cross-section of the 1.4910 austenitic steel after 250 h exposure in Ar-50%H_2O at (a) 550°C and (b) 650°C.

rates were obtained for the P91 ferritic-martensitic steel at 600°C and 650°C (Figs 11.5, 11.6). In both cases the oxide scale had spalled. The initiation of the spallation occurred as a gap formed at the interface between the inner Fe-Cr spinel and the external magnetite layer. This gap forms due to the accumulation of voids resulting from the fast outward migration of cations [6], separates the two layers and after a certain time leads to the spallation of the external layer (Fig. 11.7). It can be seen in Figs 11.5 and 11.6 that spallation of the oxide scale on P91 occurred after 500 h at both temperatures, and spallation also occurred in case of 1.4910 austenitic steel after the same period of time at 600°C. At 650°C no spallation was observed for this material. The HCM12A steel showed intermediate values of the weight change and no

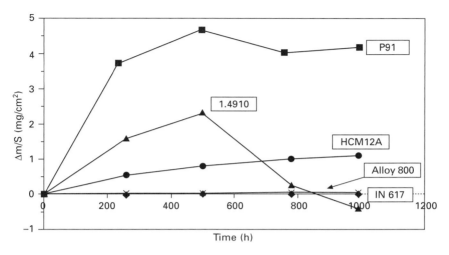

11.5 Weight change of materials during exposure at 600°C in Ar-50%H_2O.

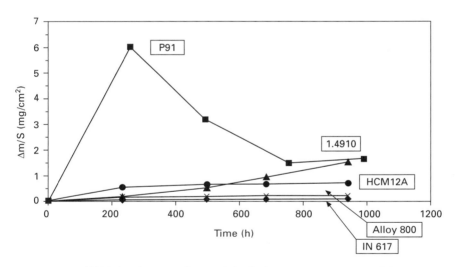

11.6 Weight change of materials during exposure at 650°C in Ar-50%H_2O.

evidence of spallation. At both temperatures the weight change tended to be nearly constant after 250 h and the morphology of the oxide scale was similar to that obtained after short-term exposures. The lowest weight change values were found for the high chromium materials, Alloy 800 and Alloy 617. These two materials formed thin Cr_2O_3 scales at both temperatures.

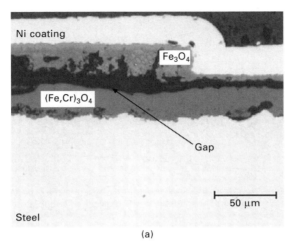

(a)

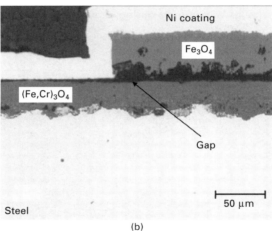

(b)

11.7 Cross-section of P91 after 1000 h exposure in Ar-50%H$_2$O at (a) 600°C and (b) 650°C.

11.3   Model alloys

In order to study in more detail the oxidation behaviour of chromium containing alloys in steam, similar oxidation tests were carried out using Fe-Cr model alloys with systematic variations in their chemical composition from 9 to 16%Cr wt. All the model alloys were manufactured by induction melting, and exposed for 250 h in simulated steam at 550, 600 and 650°C. Figure 11.8 shows the weight gain of the model alloys as a function of their chromium contents. It can be seen that in this range of composition a transition from protective to non-protective behaviour takes place. Low chromium model alloys formed thick and porous multilayered scales, which correspond to the highest weight change values in Fig. 11.8. The nature of these oxide scales was very similar to the scale observed for the P91 (Fig. 11.9(a)). Compared to the P91 the model alloys contained a larger volume fraction of voids, especially in the inner spinel layer. As a result of void coagulation, continuous gaps and cracks were formed, which led to spallation. These gaps usually appeared at the original surface of the sample or at the interface between the inner spinel layer and the zone where the internal oxidation precipitates of FeO and Cr_2O_3 were located. It has been reported that pores, voids and cracks play an important role in the growing mechanism of oxide scales in steam [7, 8].

Increasing the chromium content to above 12%, a certain reduction in the weight change was observed. The oxide scale became more similar to the one observed for the steel HCM12A (Fig. 11.9(b)). The inhomogeneous attack which was described above appeared on the alloys with 13, 14 or 15%Cr. In contrast to what was observed in the case of the steel HCM12A, none of the model alloys exhibited the same temperature dependence of the weight gain. In all the model alloys the weight change increased with increasing temperature, as well as the critical chromium content to form a protective oxide scale. This can be seen in Fig. 11.10, where the temperature dependence is compared for the two commercial steels P91 and HCM12A and the Fe-Cr model alloys with the same chromium contents. The weight gain of the commercial steels was always lower than those obtained from the model

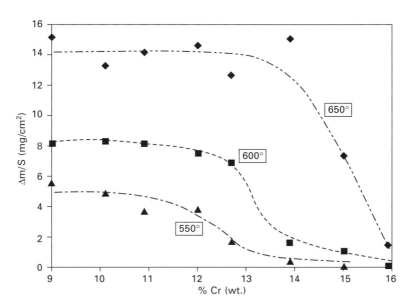

11.8 Weight gain of the Fe-Cr model alloys after 250 h exposure in Ar-50%H_2O vs their chromium content.

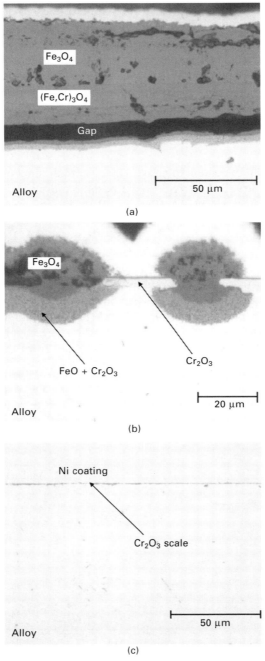

11.9 Cross-section of several Fe-Cr model alloys after 250 h exposure in Ar-50%H_2O at 600°C: (a) Fe-9%Cr, (b) Fe-14%Cr and (c) Fe-16%Cr.

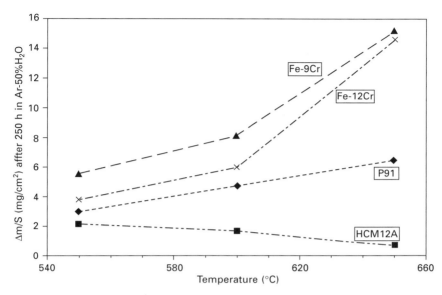

11.10 Temperature dependence of Fe-9Cr and Fe-12Cr model alloys compared to the commercial steel P91 and HCM12A.

alloys, and it increased with increasing temperature in the case of the P91 and the model alloys, whereas it decreased only for the HCM12A. Therefore, the anomalous temperature dependence of the HCM12A cannot be attributed only to the presence of chromium in the alloy. It has been reported for similar steels, which showed this kind of temperature dependence, that several factors might have an effect on it, such as the presence of other alloying elements, grain size, microstructure, etc. [9]. Model alloys with high chromium contents formed thin Cr_2O_3 scales after exposures at the three temperatures (Fig. 11.9(c)).

11.4   Effect of oxidation on creep behaviour

The formation of thick oxide scales has several consequences for the service behaviour of components:

1. The loss of material reduces the available load-bearing cross-section, so that as oxidation proceeds, the stresses acting on the component increase. The magnitude of the effect will, of course, depend on the initial wall thickness; for thick-walled components, such as steam headers (30–50 mm wall thickness), a 1 mm loss will not be very significant, but a similar material loss in a thin-walled tube of say 5 mm wall thickness would have grave repercussions for the service life [10]. Figure 11.11 illustrates this for a P92 pipe (300 mm outer diameter and 40 mm wall

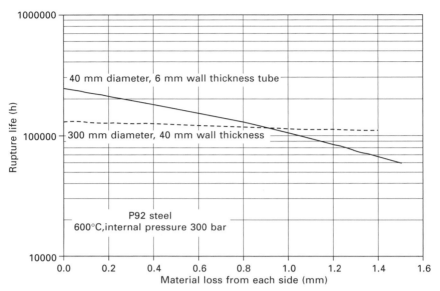

11.11 Rupture life reduction for two P92 components, a pipe of 300 mm diameter and wall thickness 40 mm and a tube of 40 mm diameter, wall thickness 6 mm.

thickness) and a P92 tube (40 mm outer diameter and 6 mm wall thickness) both exposed at an internal pressure of 300 bar. The life of the thick-walled pipe is hardly affected by a material loss of 1.5 mm, whereas for the thin-walled tube, the service life is reduced from 230,000 h to 60,000 h.

2. The thermal insulation effect of a thick oxide scale which reduces the heat transfer across the component wall can lead to overheating. A small increase in component operating temperature will reduce the stress rupture life considerably, as shown in Fig. 11.12. A relatively thin oxide scale leads to a dramatic reduction in rupture life due to the increase in temperature.

3. If the oxides spall during service, there is the danger of tube blockage as the spalled oxide accumulates inside the tubes, resulting again in overheating. If oxide particles are carried with the steam into the turbine, erosion of the blades may occur.

4. The matrix becomes depleted in those elements that are selectively oxidised to form the thick oxide scales, leading to microstructural changes and associated potential effects on the long-term strength. This effect has been investigated in this work for Alloy 800 and P92.

11.5   Creep tests on oxidised Alloy 800 and P92

The aim behind these experiments was to ascertain whether the depletion of Cr in the regions immediately below the oxide scale has any influence on the

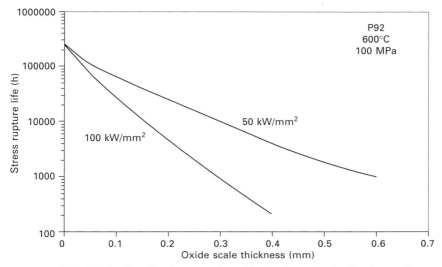

11.12 Reduction in stress rupture life for P92 due to the thermal insulating effect of an oxide scale for two heat transfer conditions, 50 and 100 kW/mm². The temperature increase leads for a given stress to a reduction in rupture life.

creep resistance. The test specimens were of 3 mm gauge diameter to represent dimensions close to typical boiler tubes. The specimens were tested in three conditions.

1. As received.
2. The as-received material was heat treated for 1000 h at 800°C and 650°C for Alloy 800 and P92, respectively. Following the heat treatment the specimens were machined. In this way, the effect of thermal ageing on the creep behaviour could be assessed.
3. Specimens were machined from the initial material, and then exposed to oxidising conditions in Ar-50%H$_2$O or in flue gas simulating coal firing (same temperature and durations as for the heat treatments in 2.). The flue gas environment is composed of: 14% CO$_2$, 10% H$_2$O, 1% O$_2$, 0.1% SO$_2$, 0.01% HCl, bal. N$_2$. The results reveal the influence on creep resistance of oxidation and of any microstructural changes caused by thermal ageing and by selective oxidation of Cr.

Creep tests have been carried out in laboratory air at 800°C/25 MPa and 650°C/120 MPa for Alloy 800 and P92, respectively. The dimensions of all samples were measured by light microscopy in order to determine their length and especially their exact diameter before creep test. This enabled a precise adjustment of the load at 120 ± 0.1 MPa (respectively 25 ± 0.1 MPa) although the specimen diameters could vary slightly from one sample to another. During the creep tests, the strain was recorded vs time. The creep

Table 11.2 Creep results for P92 at 650°C/120 MPa

Type of sample	Time to 1% creep strain (h)	Time to 2% creep strain (h)	Minimum creep rate (s^{-1})
As received	210	1160	4.5×10^{-9}
Heat treated	49	463	6×10^{-9}
Oxidised in Ar-50%H$_2$O	17	384	5.5×10^{-9}

tests were stopped when the strain exceeded 12% for Alloy 800 and 2% for P92. As shown in Table 11.2, for P92 the creep strength decreased from the as-received specimen to the corroded in flue gas in the following order: As received > Heat treated > Ar-50%H$_2$O. It first of all appears that the thermal ageing at 650°C for 1000 h affects the microstructure of P92, resulting in an increase in the creep rate.

Comparing the heat-treated specimen with the one oxidised in Ar-50%H$_2$O shows the influence of the Cr depleted zone and the oxide scale, whose thickness is included in the measurement of the diameter before test. It appears that the pre-oxidation reduces the time to 1 or 2% creep strain. A possible explanation for this faster 1 and 2% creep is the fact that the diameter measured just before the creep test includes the thickness of the oxide. By this means, the bearing cross-section has been overestimated in the case of the oxidised specimen. Nevertheless, the minimum creep rate stays more or less the same. This latter finding suggests that the properties of the bulk metal, which are dictating the creep rate, are not affected by the Ar-50%H$_2$O treatment. Final interpretation will, however, be possible only after more detailed metallographic investigations.

Concerning Alloy 800, the creep curves show that the thermal ageing at 800°C decreases the creep strength of the alloy (Fig. 11.13). The pre-corroded specimen shows a secondary creep rate which is about twice that of the aged specimen. However, the pre-corroded specimen rapidly changes from secondary to ternary creep with a higher deformation rate. This can be correlated with the presence of cracks after creep resulting from the pre-corrosion treatment. As shown in Fig. 11.14, after 1000 h at 800°C in flue gas, oxidised metal grain boundaries were found under the protective chromia layer. After creep, cracks can be observed on the surface of the pre-corroded specimen but not on the others, and the cracks appear to be deeper and larger for the pre-corroded specimen (Fig. 11.15). These observations suggest that during the creep test, the cracks initiated at internally oxidised grain boundaries leading to a reduction of the load-bearing cross-section. This effect had been observed for Alloy 800 and described in detail by Bruch *et al.* [11]. As a consequence, a higher creep rate and an earlier transition from secondary to ternary creep were found for the pre-corroded specimen.

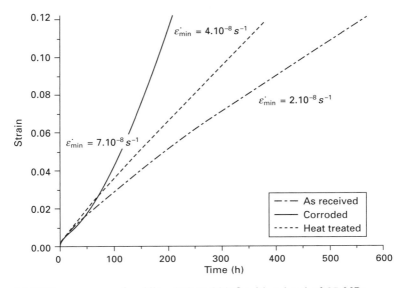

11.13 Creep curves for Alloy 800 at 800°C with a load of 25 MPa.

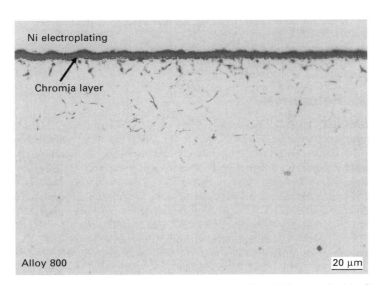

11.14 Cross-section by light microscopy. Alloy 800 corroded in flue gas for 1000 h at 800°C.

11.6   Acknowledgement

L. Nieto Hierro and V. Rohr acknowledge the financial support provided through the European Community's Human Potential Programme 'SUNASPO' under contract HPRN-CT-2001-00201.

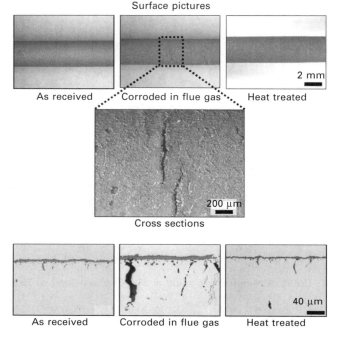

Surface pictures

As received   Corroded in flue gas   Heat treated

2 mm

200 μm

Cross sections

As received   Corroded in flue gas   Heat treated

40 μm

11.15 Light microscopy pictures of Alloy 800 after creep test at 800°C for 1000 h.

11.7   References

1. D. Allen, J. Oakey and B. Scarling, The New COST Action 522 – Power Generation in the 21st Century: Ultra Efficient, Low Emission Plant, in *Materials for Advanced Power Engineering 1998*, (eds J. Lecomte-Beckers, F. Schubert, P.J. Ennis), Forschungszentrum Jülich, Energy Technology Series, Vol. 5, 1998, Part III, 1825–1839.

2. R. Blum, J. Hald, in *Materials for Advanced Power Engineering 2002*, (eds) J. Lecomte-Beckers, M. Carton, F. Schubert and P.J. Ennis), Forschungszentrum Jülich, Energy Technology Series, Vol. 21, 2002, Part II, 1009–1016.

3. P.J. Ennis, W.J. Quadakkers and H. Schuster, *Journal de Physique IV*, Colloque C9, supplement au *Journal de Physique III*, Vol. 3, December 1993, 979–986.

4. P.J. Ennis, Y. Wouters and W.J. Quadakkers, The effects of oxidation on the service life of 9–12% chromium steels, in *Advanced Heat Resistant Steels for Power Generation* (eds R Viswanathan and J Nutting), Institute of Materials Book No. 708, IOM Communications Ltd, 1999, 457–467.

5. R.J. Ehlers and W.J. Quadakkers, Oxidation von ferritischen 9–12% Cr-Stählen in wasserdampfhaltigen Atmosphären bei 550 bis 650°C. PhD Thesis, 2001. Published as Report of the Research Centre Jülich, Jül-3883, June 2001.

6. J. Zurek, L. Nieto Hierro, P.J. Ennis, L. Singheiser and W.J. Quadakkers, EPRI/DOE Fourth International Conference on Advances in Materials Technology for Fossil Power Plants, Hilton Head Island, South Carolina, October 26–28, 2004.

7. A. Rahmel, J. Tobolski, Einfluss von Wasserdampf und Kohlendioxid auf die Oxidation von Eisen in Sauerstoff bei hohen Temperaturen, *Corrosion Science, Vol.* 5, 333 (1965).

8. J. Zurek, L. Nieto Hierro, J. Piron-Abellan, L. Niewolak, L. Singheiser and W.J. Quadakkers, Effect of alloying additions in ferritic 9–12%Cr steels on the temperature dependence of the steam oxidation resistance. 6th International Symposium on High-Temperature Corrosion and Protection of Materials HTCPM-2004, Les Embiez, May 16–21, 2004.

9. J. Zurek, E Wessel, L. Niewolak, F. Schimtz, T.-U. Kern, L. Singheiser, W.J. Quadakkers, Anomalous temperature dependence of oxidation kinetics during steam oxidation of ferritic steels in the temperature range 550°C–650°C, *Corrosion Science* Vol. 46, No. 9, 2301–2317.

10. P.J. Ennis, W.J. Quadakkers and J. Zurek, Steam oxidation of chromium steels and its implications for the service life of components, Proceedings of International Conference on High Temperature Plant Integrity and Life Extension, April 4–16, 2004, Robinson College, Cambridge, UK (eds A. Fleming, D. G. Robertson, I. A. Shibli), published by European Technology Development, UK.

11. U. Bruch, K. Döhle, S. Pütz, A. Rahmel, M. Schütze, K.D. Schuhmacher, Proc. of the National Research Council of Canada, Vol. 3, 1984, 325–329.

12

The erosion-corrosion resistance of uncoated and aluminized 12% chromium ferritic steels under fluidized-bed conditions at elevated temperature (SUNASPO)

E HUTTUNEN-SAARIVIRTA, Institute of Materials Science, Finland, S KALIDAKIS and F H STOTT, Corrosion and Protection Centre, UMIST UK and V ROHR and M SCHÜTZE, Karl-Winnacker-Institut der DECHEMA e.V., Germany

12.1 Introduction

Many industrial countries have agreed to decrease the release of greenhouse gases, including carbon dioxide (CO_2), according to the Kyoto Protocol, while the consumption of energy is strongly increasing worldwide. One step towards lower gas emission levels is to increase the efficiency of existing energy production processes, such as coal, biowaste and biomass combustion; however, this requires operating temperatures that are higher than currently used. The development of new materials for components such as heat exchangers is key for the development of power plants that can operate at higher temperatures. Such materials must have acceptable creep strength, formability, weldability, corrosion resistance and erosion-corrosion resistance under the new operating conditions.

Fluidized-bed combustion offers a versatile means of burning a range of fuels, including low-grade fuels, and is characterized by remarkably low adverse emissions and improved efficiency compared to other existing combustion techniques. However, heat exchanger tubes in fluidized-bed combustors are generally exposed to a relatively stable oxidizing atmosphere and to mechanical impact by bed particles; together with elevated temperatures, these lead to more severe erosion-corrosion problems than those encountered in more traditional fuel-burning processes [1]. Therefore, improved high-temperature erosion-corrosion resistance becomes crucial for materials that can be used in fluidized-bed conditions at temperatures above those currently in use. This necessitates a better understanding of material erosion-corrosion interactions at high temperatures.

This chapter aims to provide an insight into the erosion-corrosion behaviour of uncoated and aluminized 12% chromium ferritic steels. Such steels are candidate materials for heat exchangers in future power plants. They belong to the so-called third generation ferritic steels for power plants, with improved weldability and creep strength compared to prior ferritic steel versions [2]. Very few erosion-corrosion studies, however, have been reported on the behaviour of such steels under fluidized-bed conditions at elevated temperatures. Aluminization, in turn, is a potential surface treatment procedure for these low-alloy steels because of the resulting excellent oxidation resistance at high temperatures, as well as good wear resistance [3] and good erosion-corrosion resistance under oxidizing conditions at elevated temperatures [4]. However, there have been no fundamental studies of the erosion-corrosion interactions of aluminized steels under fluidized-bed combustion conditions. The present research is intended to determine the interactions between erosion and corrosion processes under several fluidized-bed conditions for such uncoated and coated ferritic steels that are candidate materials for heat exchangers in future power plants.

12.2   Experimental methods

12.2.1 Materials

The materials under study are uncoated and aluminized 12% chromium steels. The 12% chromium steel reported in this chapter was a ferritic alloy steel HCM12A, prepared by Sumitomo, Japan, of nominal composition: 12.22 wt.% Cr, 1.88 wt.% W, 0.86 wt.% Cu, 0.53 wt.% Mn, 0.36 wt.% Mo, 0.35 wt.% Ni, 0.31 wt.% Si, 0.20 wt.% V, 0.10 wt.% C, 0.052 wt.% N, 0.05 wt.% Nb, 0.014 wt.% P, 0.001 wt.% S, 0.0008 wt.% Al and Fe (balance). This is a so-called third generation ferritic steel for power plants, with improved weldability and creep strength compared to prior ferritic steel versions. It has a duplex microstructure of tempered martensite and δ-ferrite [2]. The uncoated steel samples were ground to 320 grit finish using SiC grinding paper. Prior to the experiments, the samples were washed ultrasonically in acetone and then in ethanol.

Some of the steel samples were further prepared for aluminizing by grinding to 120 grit surface finish using SiC paper and cleaning ultrasonically in ethanol for 10 minutes. Aluminizing was carried out by employing a pack cementation process. The powder mixture contained 10 wt.% aluminium powder, 1 wt.% ammonium chloride activator and balance (89 wt.%) alumina as an inert filler. The substrate samples were placed in the mixture in an alumina crucible covered with an alumina cap. Then, the crucible was inserted into a furnace and held at 650°C for 6 h in a flowing Ar-10% H_2 atmosphere. After the pack cementation treatment, the pack was cooled to room temperature. The samples were removed and ultrasonically cleaned in ethanol to remove any loosely embedded pack material.

12.2.2 Erosion-corrosion tests

The erosion-corrosion experiments were carried out in a fluidized-bed erosion-oxidation apparatus, shown in Fig. 12.1. This consists of a fluidized bed containing 40 vol.% particles, a heating system and a specimen holder assembly. The heating system has three independent heating zones, which were set at the same temperature in these tests; air is preheated before it enters the bed of particles, the bed is heated by the bed heater while the temperature of the specimen is maintained by the body heater. During operation, air is pumped through the spiral flow tube in the preheater, enabling it to attain the necessary acceleration as well as the desired temperature before entering the bed of particles. After fluidizing the particles, the air enters the coolers and cyclones and, finally, recirculates through the system. The specimen arm is rotated through the fluidized bed by a drive motor. The specimen arm is designed to allow the attached samples to be completely immersed in the bed at the lowest point, then to move through the space above the bed. The specimens, of size 35 mm × 7 mm × 4.5 mm, were fastened in the two holders at the ends of the specimen arm assembly. The back side of each specimen was protected from the environment by the sample holder, so only the front of the sample was subjected to erosion-corrosion. By adjusting the rotational speed of the drive motor, the linear speed of the samples relative to bed particles could be set at the desired value.

In this study, silica sand particles of average diameter of 200 μm (Fig. 12.2) were used as bed particles. The tests were conducted using two impact angles, namely 30° and 90°, in the temperature range from 550°C to 700°C. The drive motor rotational speed was maintained at 932 rpm, giving linear speeds from 7.0 m s^{-1} to 9.5 m s^{-1} along the sample length. The test duration was 50 h for the uncoated steel and 200 h for the aluminized steel specimens.

12.2.3 Characterization methods

Thickness changes for the samples following the tests were determined by measuring the cross-section of the sample, using a micrometer, at twelve locations along the specimen length. Each location gave a particular linear speed between the specimen at that location and the erodent particles. To ensure that the readings were accurate, the thickness of the specimen was measured at five different points across the specimen at each location and ten measurements were taken at each point. The average values and standard deviations for the 50 measurements at the location for each linear speed were then calculated. Measurements were made before and after the tests, enabling the average thickness change to be determined at each speed.

The sample surfaces were examined by scanning electron microscopy (SEM) using an Amray 1810 microscope, equipped with an ISIS energy

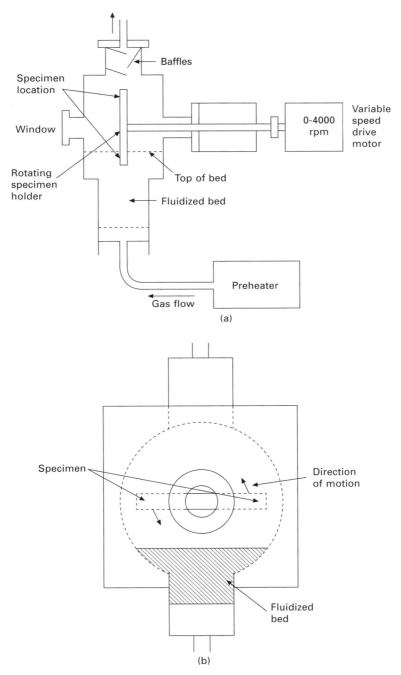

12.1 Schematic presentation of the erosion-oxidation testing device: (a) side elevation, (b) front evation.

12.2 Silica sand particles used as erodent in the erosion-oxidation tests.

dispersive spectrometer (EDS) system. In addition to the specimen surfaces, the aluminized steel specimens were studied by SEM in cross-section, following etching in 4% nital for 20–30s, to reveal the coating thickness and microstructure. The aluminized steel specimens were also examined by X-ray diffractometry (XRD), using a Philips PW 3710 X-ray diffractometer with CuK_α radiation, to determine the phase structure.

12.3   Results

12.3.1 Uncoated 12% chromium steel

Figure 12.3 shows the thickness changes for the uncoated 12% chromium steel specimens as a function of speed at several temperatures. At 550°C (Fig. 12.3(a)) and impact angle of 30°, an average thickness loss of 2 μm was measured. Although there was no close correlation between the thickness change and speed, in general, thickness losses were observed at speeds below 7.6 m s^{-1}, while slight thickness gains were more typical at higher speeds. The maximum thickness loss (33 μm) was recorded at 7 m s^{-1} while the maximum thickness gain (13 μm) was observed at 8.8 m s^{-1}. At an impact angle of 90°, a relatively constant thickness loss of 20 to 30 μm was observed at nearly all speeds, giving an average value of 21 μm.

At 600°C (Fig. 12.3(b)), the steel experienced thickness losses at all speeds for a 30° impact angle, giving an average loss value of 38 μm. At speeds below 8 m s^{-1}, the thickness loss remained at between about 20 and 30 μm, while at higher speeds, the thickness loss increased with increase in speed, finally reaching a value of 91 μm at a speed of 9.2 m s^{-1}. At an impact angle of 90°, there was less correlation between thickness change and speed. The average change was a loss of 7 μm, with a maximum loss value of 23

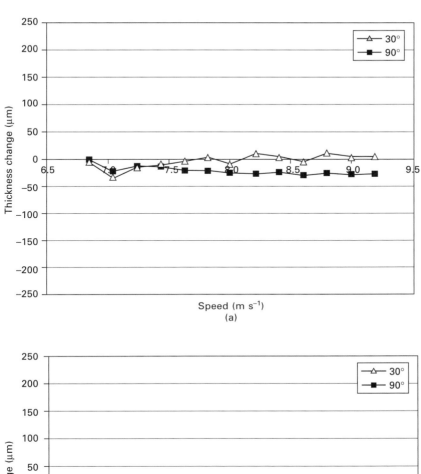

(a)

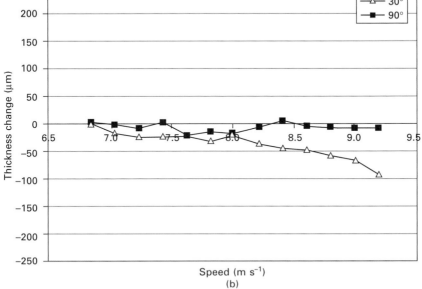

(b)

12.3 Thickness changes for uncoated 12% chromium steel samples as a function of speed at various temperatures: (a) 550°C, (b) 600°C, (c) 650°C, (d) 700°C.

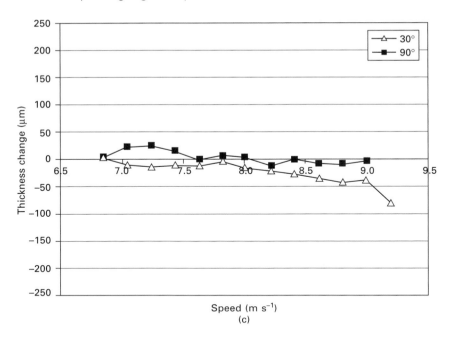

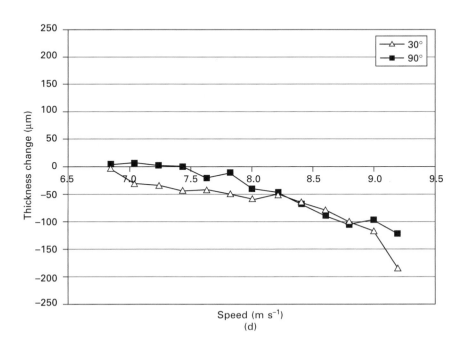

12.3 Continued

µm. Indeed, at speeds of 7.4 m s^{-1} and 8.4 m s^{-1}, small thickness gains, to a maximum value of 5 µm, were recorded.

Similar to the results at 600°C, the uncoated steel experienced thickness losses at every speed for an impact angle of 30° at 650°C (Fig. 12.3(c)), with an average value of 24 µm. Moreover, the thickness loss remained relatively constant, at about 10 µm, for speeds below 7.8 m s^{-1}, but at higher speeds, it increased with increase in speed, approaching 80 µm at 9.2 m s^{-1}. For an impact angle of 90° at 650°C, the average thickness change was a small gain of 4 µm. In general, there were gains of 15 to 25 µm at speeds below 7.6 m s^{-1}; thereafter, at higher speeds, the steel first experienced small thickness gains of a few microns, at speeds from 7.6 m s^{-1} to 8.2 m s^{-1} and small thickness losses at higher speeds, up to a maximum loss of 12 µm.

At 700°C (Fig. 12.3(d)), the steel sample experienced thickness losses at every speed, for an impact angle of 30°, with an average loss of 66 µm. There was a general trend for material loss to increase with increase in speed, approaching a value of 184 µm at the highest speed. For an impact angle of 90°, the specimen experienced slight thickness gains of 3 to 6 µm at speeds up to 7.2 m s^{-1}. However, at speeds above 7.2 m s^{-1}, noteworthy thickness losses were observed, the value increasing with increase in speed. The greatest thickness loss, 121 µm, was experienced at the highest speed. The average thickness loss over the full range of speeds was 46 µm.

The thickness changes obtained for the steel at speeds of 7 m s^{-1}, 8 m s^{-1} and 9 m s^{-1} as a function of temperature for both impact angles are shown in Fig. 12.4. For an impact angle of 30° (Fig. 12.4(a)), the temperature dependency of thickness change varied with the speed. Thus, at 7 m s^{-1}, while thickness losses were observed at all temperatures, these decreased with increase in temperature from 550°C to 650°C, but increased slightly at 700°C. At 8 m s^{-1}, there was a progressive increase in thickness loss with increase in temperature while, at 9 m s^{-1}, apart from the value at 650°C, the trend of increasing thickness loss with increase in temperature was more pronounced, with a small thickness gain at 550°C but a thickness loss of over 100 µm at 700°C.

At an impact angle of 90°, the thickness changes followed a similar trend with temperature at all three speeds (Fig. 12.4(b)). Thus, small thickness losses were recorded in each case at 550°C, with these losses decreasing with increase in temperature to 650°C, with almost no change in thickness, or for 7 m s^{-1}, a small increase, being recorded at 650°C. However, at 700°C, this trend was reversed, with relatively large thickness losses being observed at 8 m s^{-1} and 9 m s^{-1} and a smaller thickness gain than at 650°C for a speed of 7 m s^{-1}. It should be noted that the largest thickness losses occurred at the highest temperature (700°C) and the highest speed (9 m s^{-1}) for both impact angles. Moreover, in general, the thickness losses were smaller for an impact angle of 90° than for one of 30°C.

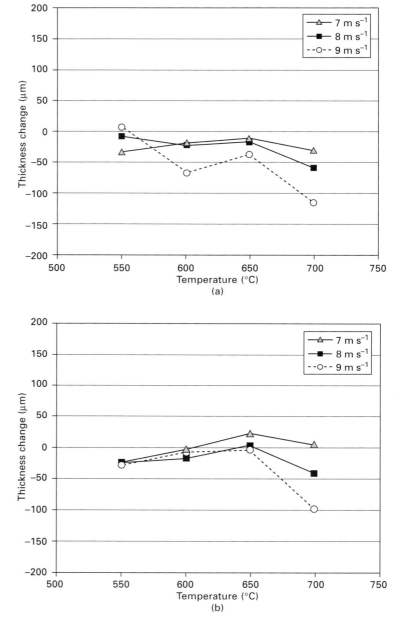

12.4 Thickness changes for uncoated 12% chromium steel as a function of temperature at various impact angles: (a) 30°, (b) 90°.

The microstructural examination of the damaged steel surfaces (Fig. 12.5) revealed the presence of an oxide scale comprising iron oxides and chromium oxides after exposure under almost all conditions, although at 550°C,

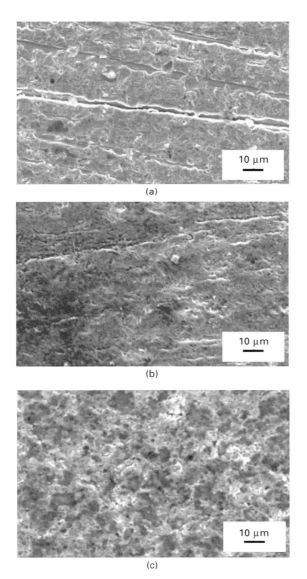

12.5 SEM photographs of the surfaces of uncoated 12% chromium steel samples after the erosion-corrosion tests under various erosion-corrosion conditions: (a) 7 m s⁻¹, 550°C, 30°, (b) 9 m s⁻¹, 550°C, 30° (c) 7 m s⁻¹, 550°C, 90°, (d) 8 m s⁻¹, 600°C, 90°, (e) 9 m s⁻¹, 650°C, 90°, (f) 9 m s⁻¹, 700°C, 90°.

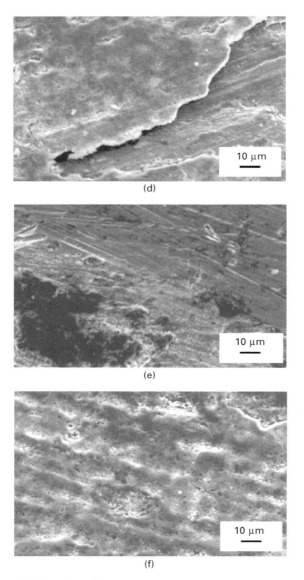

12.5 Continued

characteristics of the scales were somewhat different from those at higher temperatures. For an impact angle of 30° at 550°C, the oxide scale developed at 7 m s^{-1} (Fig. 12.5(a)) was not completely adherent to the substrate and did not cover the whole surface. With increases in speed and temperature, the scale was more uniform (Fig. 12.5(b)) and there was some evidence of spalling and subsequent healing, and, at 700°C, of crack and crater formation.

For an impact angle of 90°, at all speeds, the scale was uniform at 550°C (Fig. 12.5(c)). However, at higher temperatures, of 600°C and above, severe spalling was observed to have occurred. Thus, at 600°C, only the outermost parts of the scale had spalled (Fig. 12.5(d)), while, at 650°C, the scale had spalled completely in some areas and no oxide was detected on the surface at these locations (Fig. 12.5(e)). At 700°C, the oxide scale was denser and more consolidated than those at lower temperatures and contained cracks and craters (Fig. 12.5(f)). Such surface features have previously been observed and associated with high rates of spalling [5]. It is worth noting that, at both impact angles, residues of the silica particles were occasionally observed in the surfaces after impact at the lower speeds, while, at the higher speeds, there was considerably more silica embedded in the oxide scale.

12.3.2 Aluminized 12% chromium steel

Figure 12.6 shows typical SEM micrographs and XRD analyses of the aluminium diffusion coating formed in the pack cementation process. It consisted of relatively equiaxed columnar grains (Figs 12.6(a) and (b)); the thickness ranged from 31 µm to 40 µm, with a mean value of 36 µm. The average composition of the coating was Al 75.7 at.%, Cr 3.9 at.% and Fe 20.4 at.%, and it appears to consist of a single phase; XRD data (Fig. 12.6(c)) indicated the presence of the Al_5Fe_2 phase only.

Figure 12.7 shows the thickness changes for the coated steel specimens as a function of speed for the four temperatures. At 550°C (Fig. 12.7(a)) and an impact angle of 30°, thickness gains were observed for all speeds, with an average value of 13 µm. There was no apparent trend with changes in speed; the values varied between 6 µm and 23 µm. For an impact angle of 90°, in contrast, the specimens experienced thickness losses at all speeds, with an average loss of 79 µm. Also, the loss increased in an almost linear manner with increase in speed. The maximum loss, 124 µm, was observed at the highest speed.

Similarly, at 600°C (Fig. 12.7(b)), the thickness change did not show a correlation with speed for an impact angle of 30°; it stayed relatively constant, varying between a moderate thickness loss, maximum 7 µm, and a moderate thickness gain, maximum 8 µm. The average thickness change was a loss of 1 µm. For an impact angle of 90°, thickness losses were recorded at all speeds, with an average value of 40 µm. Although the loss increased with increase in speed, the trend was less pronounced than at 550°C and the maximum loss was only 53 µm.

At 650°C (Fig. 12.7(c)), the observed thickness changes did not follow similar tendencies as at lower temperatures. At this temperature, a pronounced thickness loss was recorded at all speeds for an impact angle of 30°, with an average value of 109 µm. Moreover, the loss increased in an almost linear

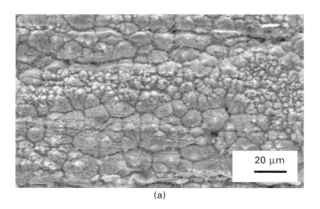

(a)

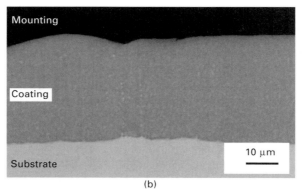

(b)

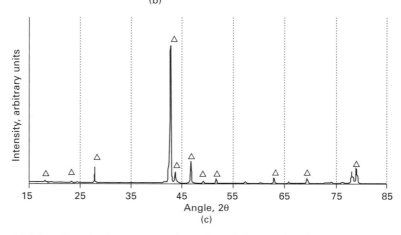

(c)

12.6 Results of microstructural studies of the coating formed during the pack aluminization of 12% chromium steel. (a) SEM photograph, showing the surface of the coating. (b) SEM back-scattered electron photograph, showing a cross-sectional view of the coating. (c) XRD spectrum from the pack aluminized 12% chromium steel. $\triangle$ refers to the orthorhombic Al_5Fe_2 phase.

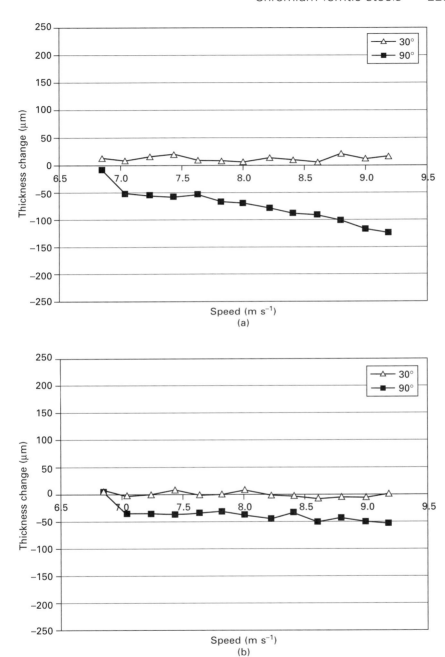

12.7 Thickness changes for aluminized 12% chromium steel samples as a function of speed at various temperatures: (a) 550°C, (b) 600°C, (c) 650°C, (d) 700°C.

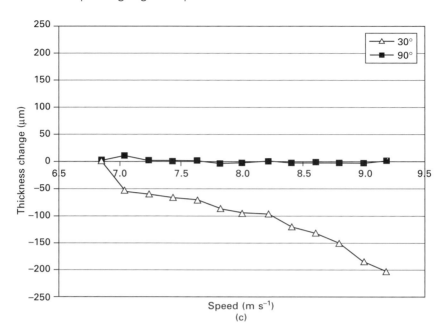

Speed (m s⁻¹)
(c)

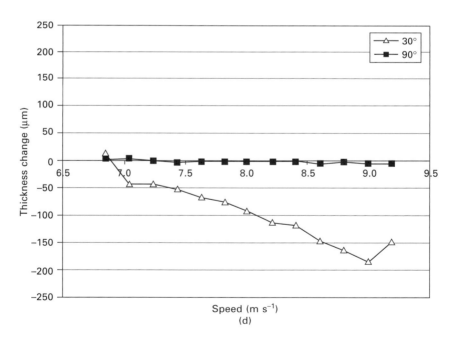

Speed (m s⁻¹)
(d)

12.7 Continued

manner with increase in speed. The maximum loss, 201 µm, was accordingly experienced at the highest speed. Conversely, for an impact angle of 90°, there was no dependence of thickness change on the speed. Indeed, the thickness changes were very small, ranging from a slight thickness loss, maximum 5 µm, to a slight thickness gain, maximum 9 µm, with an average value of a 1 µm loss.

At 700°C (Fig. 12.7(d)), the results were almost identical to those observed at 650°C. For an impact angle of 30°, the aluminized steel experienced substantial thickness losses, with an average of 103 µm; again, the magnitude of the thickness loss increased in a relatively linear manner with increase in speed, reaching a maximum value of 183 µm at a speed of 9 m s^{-1}. However, the observed loss was slightly smaller at 9.2 m s^{-1} than that at 9 m s^{-1}, being of a similar value to that recorded at 8.6 m s^{-1}. For an impact angle of 90°, the thickness changes were very small and independent of speed. They varied from 3 µm thickness gain to 5 µm thickness loss, with the average value being a loss of 2 µm.

Figure 12.8 shows the thickness changes at speeds of 7 m s^{-1}, 8 m s^{-1} and 9 m s^{-1} as a function of temperature for both impact angles. For an impact angle of 30° (Fig. 12.8(a)), the smallest thickness changes occurred at 550°C. For all three speeds, the change was slightly positive. However, with increase in temperature, the change became a thickness loss, the value increasing with increase in speed and temperature from 550°C to 650°C. Thus, the largest loss was experienced at 650°C and the highest speed. Moreover, at 700°C, the thickness losses were very similar to those at 650°C (for a speed of 9 m s^{-1}) or even slightly less (for speeds of 7 m s^{-1} and 8 m s^{-1}).

For an impact angle of 90° (Fig. 12.8(b)), the trend of thickness change with temperature was very different. At 550°C, the largest thickness change, a thickness loss, was experienced, with the greatest thickness loss taking place for the highest speed. Also, the extent of thickness loss decreased with increase in temperature, becoming almost zero (for speeds of 8 m s^{-1} and 9 m s^{-1}) or even a small thickness gain (for 7 m s^{-1}). It is worth noting that, at both impact angles, a change in the trend of thickness change with temperature occurs between 600°C and 650°C.

Examination of the specimens after the tests indicated that there was no build-up of aluminium oxide scale on the coating surface at either 550°C or 600°C, because EDS analyses did not show a presence of oxygen. At these temperatures, for an impact angle of 30°, the coating surfaces were slightly deformed and cracked during the tests. The deformation was mainly concentrated on the thicker areas (Figure 12.9(a)) and the topmost parts of the columnar grains of the coatings (Fig. 12.9(b)). At the lowest speeds, residues of silica sand particles were detected (Fig. 12.9(c)). In contrast, for an impact angle of 90° at 550°C and 600°C, the coating formed during aluminization was either partially (Fig. 12.9(d)) or totally (Figs 12.9(e) and

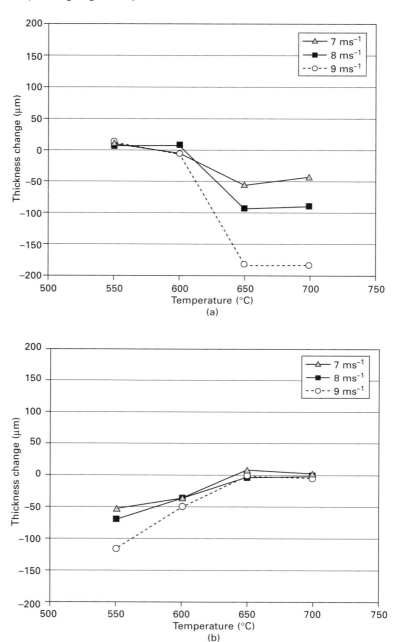

12.8 Thickness changes for aluminized 12% chromium steel as a function of temperature at various impact angles: (a) 30°, (b) 90°.

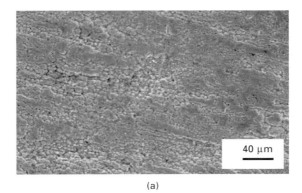

(a)

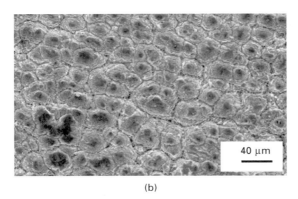

(b)

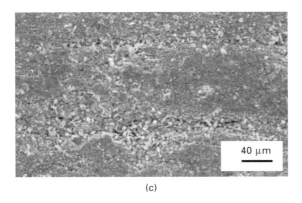

(c)

12.9 SEM photographs of the surfaces of aluminized 12% chromium samples after the erosion-corrosion tests at 550°C and 600°C under various erosion-corrosion conditions: (a) 9 m s^{-1}, 550°C, 30°, (b) 9 m s^{-1}, 600°C, 30°, (c) 7 m s^{-1}, 550°C, 30°, (d) 7 m s^{-1}, 600°C, 90°, (e) 7 m s^{-1}, 550°C, 90°, (f) 9 m s^{-1}, 600°C, 90°.

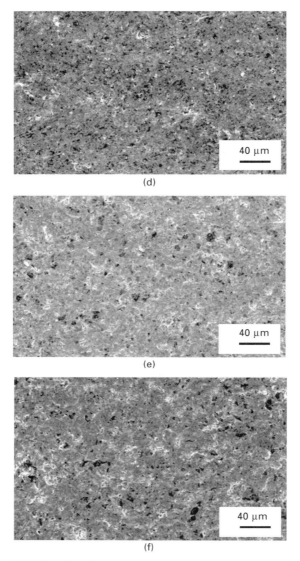

12.9 Continued

(f) lost during the test. At 550°C for all the speeds, the coating was lost completely, while, at 600°C, complete loss of coating was observed at speeds greater than 7.4 m s⁻¹, but, at lower speeds, some residual coating was retained. In areas that were no longer covered by residual coating, a porous, uniform and adherent oxide scale developed on the exposed steel surface.

Examination of the exposed specimens after the tests at 650°C and 700°C revealed a considerable oxidation of the surfaces. For an impact angle of

30°, where the coating was lost completely during the test to reveal the underlying steel, the surface was covered by an oxide scale (Figs 12.10(a) and (b)). In addition, SEM studies showed the presence of residual silica sand particles either on the surface (Fig. 12.10(a)) or embedded in the oxide scale (Fig. 12.10(b)). The surface features of these specimens indicated the presence of several oxide layers. In contrast, at an impact angle of 90°, where little thickness change had been observed, an adherent oxide scale was observed on the surface of the aluminium-rich coating (Figs 12.10(c) and (d)). The scale was more pronounced on the edges of the columnar grains of the coating. Furthermore, no residual silica sand particles were detected on the oxidized surfaces, although some silicon was detected within the scale.

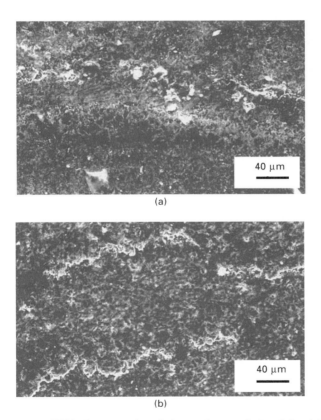

(a)

(b)

12.10 SEM photographs of the surfaces of aluminized 12% chromium steel samples after the erosion-corrosion tests at 650°C and 700°C under various erosion-corrosion conditions: (a) 9 m s^{-1}, 650°C, 30°, (b) 7 m s^{-1}, 700°C, 30°, (c) 9 m s^{-1}, 650°C, 90°, (d) 9 m s^{-1}, 700°C, 90°.

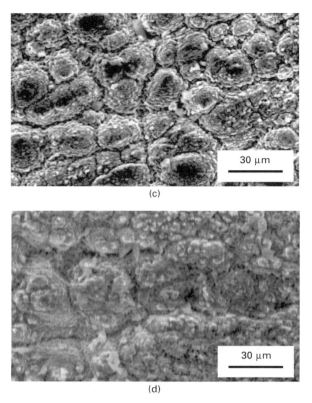

(c)

(d)

12.10 Continued

12.4   Discussion

12.4.1 Uncoated 12% chromium steel

The results reported in this study suggest that there are changes in the erosion-corrosion behaviour of the uncoated 12% steel between 550°C and 600°C for both 30° and 90° impact angles. At 550°C, the thickness changes are almost independent of speed and are relatively small for the 30° impact angle tests. The trends are similar for the 90° tests, although the somewhat larger thickness losses recorded are more consistent with brittle behaviour than ductile behaviour at this temperature. For the smaller impact angle, although a thin oxide scale was observed on part of the exposed surface, it was not complete and did not cover all areas, particularly at the lower speeds. However, it was more complete and adherent at the higher speeds, consistent with the observed small increases in thickness. Similar scales were observed for 90° impact tests, although the measured thickness losses are consistent with loss of material by scale spallation.

At temperatures of 600°C and above, oxides clearly play a greater role in the erosion-corrosion behaviour than at 550°C. Moreover, the steel shows an angle-dependence that is more typical of a ductile material than a brittle material; i.e. the extent of material loss at 30° is greater than at 90°. However, as discussed, the main mode of damage is formation and loss of oxide scale, so the extent of damage is influenced by both the properties of the scale and those of the scale/substrate interface and not just those of the surface of the specimen.

At 600°C, 650°C and 700°C for a 30° impact angle, the main trends are for the extent of thickness loss to increase with increase in speed, particularly at speeds above 8 m s^{-1}, while for a 90° impact angle, the thickness changes are relatively small for all speeds at 600°C and 650°C but increase significantly with increase in speed at 700°C (Figs 12.3(b)–(d)). Previous research has reported similar effects of speed on the extents of spallation of scale during erosion-corrosion tests and suggested that higher speeds favour densification of the oxide scale [5], as observed in the present work, as well as in earlier work [6]. Overall, the extent of damage is related to the ease of loss of scale by spallation (favoured by thicker scales, and, thus, higher temperatures), the stresses imported by the contacting particles (favoured by higher speeds and influenced by impact angle) and the coherency of the scales.

Although there has been little published research on the erosion-corrosion behaviour of 12% chromium steels, several papers have reported on the behaviour of 9% chromium steels [5, 7–10]. These have shown that the oxide scales present on the surfaces after such tests at 30° impact angle have very different morphologies from those at 90° impact angle. In the former case, the primary mechanism of scale loss is cracking and chipping, while, in the latter case, it is spallation. Although such studies have involved higher speeds (several tens of m s^{-1}) than in the present research, there are similar differences in morphologies between the scales present after tests at 30° and those at 90° impact angle.

From the results of this study, it is unclear why the extent of damage is greater for 30° impact angle tests than for 90° impact angle tests, particularly at 600°C and 650°C. A possible explanation is that the scales are thinner and more resistant to spallation by 90° angle impacts at the lower temperatures, but are susceptible to damage by a cracking and chipping process by 30° angle impacts at all temperatures. SEM studies, however, show that oxide scales formed by 30° angle impacts tend to be denser than those formed by 90° angle impacts (Figs 12.5(b) and (c)), providing an alternative and a more plausible explanation. Moreover, it is apparent that, in both cases, the prevailing erosion-corrosion mode for the steel under all conditions of this study is, essentially, an oxidation-affected erosion mechanism [11–13].

12.4.2 Aluminized 12% chromium steel

It is apparent from the thickness change data that there are significant differences in the erosion-corrosion modes between particle impacts at 30° and at 90° for the coated steel specimens. Moreover, these differences are influenced markedly by temperature. Thus, at 550°C and 600°C, the specimens underwent very little damage for 30° impacts, irrespective of speeds, while more significant thickness losses were recorded for 90° impacts, with the extent of damage increasing with increase in speed, particularly at 550°C. Conversely, at 650°C and 700°C, the specimens underwent very little damage for 90° impacts, irrespective of speed, while significant thickness losses were recorded for 30° impacts, with the extent of damage increasing rapidly with increase in speed.

The results at 550°C and 600°C are consistent with a mode of erosion-corrosion damage that follows brittle behaviour, i.e. the extent of material loss is greater for 90° angle impacts than for 30° angle impacts and increases with increase in speed. This is in agreement with the results of Hirose *et al.* [14], who studied the deformation behaviour of Al_5Fe_2 intermetallic compound in compression at 550°C to 800°C. Below 500°C, the phase is perfectly brittle with no observable plastic deformation. However, the results of Hirose *et al.* demonstrate clearly that the Al_5Fe_2 phase shows plasticity at temperatures above 600°C, with the extent of plasticity increasing rapidly with increase in temperature and decrease in strain rate.

At elevated temperatures in oxidizing gases, a slow growing and adherent alumina scale is expected to form on the surface of iron aluminides [15]. However, this would be very thin at relatively low temperatures, such as 550°C. Indeed, following the tests at this temperature, for 30° impact angle where very little damage is observed, EDS analyses did not show the presence of oxygen on the surface, confirming that the residual oxide film was too thin to be detected by this technique. Thus, it is possible that oxidation does not have a significant role in the erosion-corrosion process at this temperature. At 90°, the coating is removed completely during the tests, with the thickness loss increasing with increase in speed, probably by the well-documented brittle erosion mode. In each case, it is not until the coating has been removed completely, to expose the substrate, that oxidation starts to play a more significant role. For the tests at 600°C, the results are similar, although here the higher temperature and, thus, faster rate of oxidation, affect the damage process, particularly for 90° impacts, leading to smaller thickness losses than at 550°C.

There are two main factors that are responsible for the changes in erosion-corrosion behaviour with increase in temperature, particularly from 600°C to 650°C. Firstly, as discussed by Hirose *et al.* [14], the Al_5Fe_2 phase shows significant plasticity at temperatures above 600°C. Secondly, the rate of oxidation increases rapidly with increase in temperature, so oxidation plays a more significant role at higher temperatures. Thus, at 650°C and 700°C, the thickness change data have a dependence on impact angle that is closer to that for 'ductile' behaviour than that for 'brittle' behaviour, with large material losses (that increase with increase in speed) at 30° impact angle but little damage at 90° impact angle. Moreover, particularly at the higher speeds, the coating is lost completely during the 30° impact angle tests, so much of the later damage involves loss of the steel substrate, in a similar manner to that described earlier. The precise relative roles of the coating, substrate and oxide scales at these higher temperatures during tests for 30° angle impacts are unclear since the coating was lost completely. However, in situations where material is lost at such high rates, and the growth rates of the alumina-rich scale are relatively low, even at 700°C, it is probable that much of the damage was due to erosion of the intermetallic phase, rather than erosion of the oxide scale. For tests at 90° impact angles, where the more plastic intermetallic phase is able to resist the particle impacts, oxidation can play a greater role and an adherent, relatively impact-resistant oxide scale is able to develop and be retained on the coating surface, as observed in practice (Figs 12.10(c) and (d)). Although not investigated here, as the amount of oxide scale was relatively abundant, it is probable that the primary erosion-corrosion mode is erosion-enhanced oxidation (type I) [12, 13]. This mode requires oxidation to take place rapidly enough for a continuous and protective oxide scale to develop.

12.5 Conclusions

Erosion-corrosion resistance of uncoated and pack-aluminized ferritic 12% chromium steels was studied under fluidized-bed conditions in a fluidized-bed erosion-corrosion test rig. The erosion-corrosion experiments were conducted in the temperature range from 550°C to 700°C using silica sand particles as an erodent. The speed between the sample surfaces and the silica sand particles was varied from 7.0 m s^{-1} to 9.5 m s^{-1} and the impact angles of interest were 30° and 90°. The main results can be summarized as follows:

- The erosion-corrosion behaviour of uncoated 12% steel changes at temperatures above 550°C. At 550°C, the extent of material loss is greater for tests involving particle impacts at 90° than those involving impacts at 30°, while, at temperatures above 550°C, the trend is reversed and the extent of material loss is generally greater at 30° than at 90°. This change in the angle-dependence with temperature is related to the characteristics, i.e. thickness, uniformity and density, of the formed oxide scales.
- The prevailing erosion-corrosion mode for the uncoated 12% chromium

steel under all the studied erosion-corrosion conditions is oxidation-affected erosion. Spallation of oxide scale is the primary mode of material wastage.

- Material wastage of uncoated 12% chromium steel increases with increases in temperature and speed, because these changes favour the development of more uniform, thicker and denser oxide scales and greater impact stresses on the oxide scales. The highest average thickness loss for both impact angles (30° and 90°) occurs at 700°C.

- The erosion-corrosion behaviour of the aluminized 12% chromium steel changes in the temperature range from 600°C to 650°C. This is due to a shift in the erosion behaviour of the coating from 'brittle' to 'ductile', and to a more rapid oxide scale build-up at temperatures above 600°C.

- For an impact angle of 30° and at 550°C and 600°C, the prevailing erosion-corrosion process for aluminized steel is oxidation-affected erosion. For an impact angle of 30° at 650°C and 700°C and for an impact angle of 90° at 550°C and 600°C, the utilized coating thickness is insufficient to allow determination of the primary erosion-corrosion process for the coating, because the coating was totally worn away during the test, to reveal the substrate material. At 650°C and 700°C, the primary erosion-corrosion mode for 90° angle impacts is erosion-enhanced oxidation.

- A single-phased aluminized Al_5Fe_2 coating, of average thickness 36 μm, provides some erosion-corrosion protection to the steel for a shallow impact angle at 550°C and 600°C and for a steep impact angle at 700°C.

12.6   Acknowledgements

The authors wish to express their gratitude to the European Commission for the funding allocated to the study through a Research Training Network: SUNASPO. In addition, the Academy of Finland is acknowledged for its financial support to E. H.-S.

12.7   References

1. K. Natesan, *Surf. and Coatings* Tech. **1993**, *56*, 185–197.
2. R. Viswanathan, W.T. Bakker, presented at International Joint Power Generation Conference 2000, Miami Beach, Florida, USA, July 23–26, **2000**. 22 pp.
3. S.C. Deevi, V.K. Sikka, C.T. Liu, *Prog. in Mater. Sci.* **1997**, *42*, 177–192.
4. M.A. Uusitalo, P.M.J. Vuoristo, T.A. Mäntylä, *Wear* **2002**, *252*, 586–594.
5. A. Levy, Y.-F. Man, *Wear* **1986**, *111*, 135–159.
6. F.H. Stott, S.W. Green, G.C. Wood, *Mater. Sci. and Engng.* **1989**, *A121*, 611–617.
7. A. Levy, Y.-F. Man, *Wear* **1986**, *111*, 161–172.
8. A. Levy, B.-Q. Wang, Y.-F. Man, N. Jee, *Wear* **1989**, *131*, 85–103.
9. A. Levy, E. Slamovich, N. Jee, *Wear* **1986**, *110*, 117–149.
10. A. Levy, Y.-F. Man, *Wear* **1989**, *131*, 39–51.

11. C.T. Kang, F.S. Pettit, N. Birks, *Met. Trans.* **1987**, *18A*, 1785–1803.
12. D.M. Rishel, F.S. Pettit, N. Birks, *Mater. Sci. and Engng.* **1991**, *A143*, 197–211.
13. M.M. Stack, S. Lekatos, F.H. Stott, *Tribology Int.* **1995**, *28*, 445–451.
14. S. Hirose, T. Itoh, M. Makita, S. Fujii, S. Arai, K. Sasaki, H. Saka, *Intermetallics* **2003**, *11*, 633–642.
15. K. Natesan, *Mater. Sci. and Engng.* **1998**, *A258*, 126–134.

13

Silicon surface treatment via CVD of inner surface of steel pipes

K BERRETH, K MAILE and A LYUTOVICH,
University of Stuttgart, Germany

13.1   Introduction

Increased Si content at the surface of steel leads to a beneficial effect improving the anti-oxidation behaviour. The surface protection is achieved by several mechanisms that take place at the same time, including the formation of silicon oxide films, increased Cr diffusion from the bulk to the surface, formation of phases such as γ-Fe, Cr_2O_3, SiO_2, Si_xFe_y, passivation of the surface by oxidation and surface diffusion. As mentioned in references [1–4], silicon seems to retard breakaway in the presence of water vapour in the environment and may facilitate Cr rediffusion from the bulk which would help repassivation observed after breakaway. As a consequence the level of the Cr reservoir possibly may be kept lower than for Si-free steels. At least, a continuous silica layer is not the reason for the improved oxidation behaviour. The positive effect of silicon seems to stabilise at values above 0.5% Si. A possible reason for the influence of silicon seems to be that silicon enhances the diffusion of Cr in the metal matrix.

Formation of mesoscopic structures by the CVD (chemical vapour deposition) of silane (SiH_4) on Fe was investigated by Rebhan et al. [5, 6, 7]. The morphology of the resulting surface can be summarized by the following five categories:

1. smooth, flat surface
2. dendrites
3. octagonal islands
4. porous surface layer
5. cracked surface.

All these different types of mesoscopic structures can be achieved by varying pressure within about one order of magnitude (0.02–0.5 mbar) and time within less than two orders of magnitude (20–1000 s). These findings have a number of possible applications, e.g. reduction of the adhesion of liquids to the surface by silane CVD films. Klam et al. [8] used a gas mixture of Ar,

SiH_4, $SiCl_4$ and H_2. The experiments were carried out at $T\alpha\gamma = 912°C$ (the temperature of the phase transition from body centred (bcc) γ-Fe to face-centred cubic (fcc) γ-Fe). This led to a complex mechanism of weight loss by the formation of gaseous $FeCl_2$ due to the presence of $SiCl_4$ and the formation of solid Fe_3Si crystals. Sanjurio et al. [9] showed that silicides are formed even below 750°C, results consistent with those of Cabrera et al. [10], who reported that the Fe catalysed SiH_4 decomposition, with the limiting step being the diffusion of Si in Fe. During investigation of the formation of silicon diffusion alloys from their reaction with silane [11], a method for surface modification of metals using silane-hydrogen reactive atmospheres was developed, which resulted in the formation of a silicon diffusion coating. The silicon was confined mainly in the near surface of the metal and imparted high-temperature oxidation protection. The kinetics of the surface reaction between silane and the metal substrate, as well as the behaviour of the coating in oxidizing environments at high temperatures, were studied by a gravimetric technique. With diffusion of hydrogen in and out of the steel lattice, hydrogen contents could be achieved in accordance with the given standard and were not higher than those before the proposed surface modifying process.

The current study was focussed on martensitic 9–11%Cr steels and concept advanced by the Materials Testing Institute (MPA), University of Stuttgart is based on the idea that, by CVD, a modifying surface process with the diffusion of Si, together with Cr, would lead to an improved anti-oxidation effect at high temperatures in the range of steam parameters for power plants operating at $\geq 600°C$. The new concept here is to achieve an improved oxidation resistance for Cr steels using Si from $SiCl_4$ at high temperatures, given that $SiCl_4$ reduces at high temperature during the first cycle of standard heat treatment, which is required to achieve the appropriate martensite structure.

13.2   Experimental methods

In principle, thermal CVD processes could be adjusted to the standard heat treatment conditions of the tube material, including cleaning and chemical reduction. The architecture of the coating should be a ductile material coherent with the tube material, showing improved oxidation resistance. The thickness of the diffusion coating with elevated Si content can be controlled by the process time [12]. The plastic deformation capability of the ferritic layer with lower hardness is better than that of the martensitic tube material. The automated CVD process leads to reproducible results [12–15], and a schematic representation is given in Fig. 13.1.

At the start of the process the oxygen in the reactor has to be evacuated and then replaced by introduction of hydrogen atmosphere and heating starts to the coating temperature. First, cleaning in the hydrogen atmosphere occurs

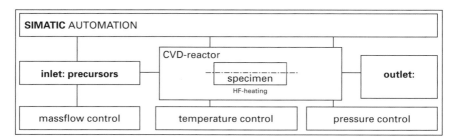

13.1 Schematic representation of the CVD process.

and hydrogen as carrier gas saturated with $SiCl_4$ flows into the reactor. The $SiCl_4$ is reduced and free Si is absorbed at the surface of the heated tube. Then Si diffuses into the steel and distributed in the lattice. The remaining chlorine reacts with hydrogen to form HCl and can be removed. After this the hydrogen is removed to the original amount given in the material standard. Then the standard annealing follows primarily under nitrogen atmosphere.

The Si is deposited by automated CVD with HF heating in a horizontal 'cold wall' reactor chamber under atmospheric pressure for easy implementation in industry. For lower gas consumption and to stimulate the chemical reactions the coating process can be at low pressure, where the chemical reactivity is usually higher. Up-scaling from a laboratory scale or bench scale unit to industrial application is based on a totally computer-controlled laboratory process. Total automation eases transferability and integration with other industrial processes.

For our study, different base materials of various forms and sizes, including tubes with typical production scales, were used (Fig. 13.2).

The inner surface of tubes of modern creep resistant 9–11%Cr steels have been modified using chemical vapour deposition (CVD) and chemical surface treatment. The reaction of $SiCl_4$ and H_2 with the steel was used to produce silicon, which diffused in the steel at a processing temperature of 1050°C. Due to the increased Si content the surface layer showed a modified microstructure with improved oxidation behaviour. The excess of Si was converted to Si_3N_4 using NH_3 [15]. Si or codiffusion for Al-Si coatings were developed at 600°C [16]. The physical and chemical properties of the modified surfaces were investigated and characterized using SEM, EDX electron microprobe and X-ray diffraction. In this layer Fe was substituted by Si with a maximum concentration of ~ 5%; however, the Cr content was almost constant. The deformation capacity of the layer was proven by means of high temperature tensile tests. SEM investigation of the surface of the specimen yielded up to 3% showing no cracking. Ageing and creep tests showed no disadvantages with respect to the creep behaviour. First, oxidation tests with atmospheres of dry air, 10% water, and an artificial combustion gas mixture

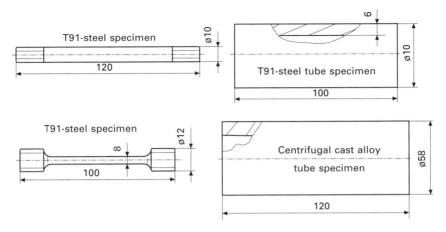

13.2 Different specimens for experimental investigations (dimensions are in mm).

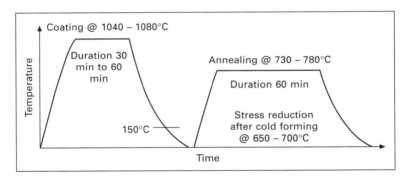

13.3 Standard heat treatment, e.g. for 9Cr steel T91.

show the positive protective influence of the Si modified surface. Owing to the adjustment of the temperature–time–course of the coating process to the requirements of the standard heat treatment of the material, the optimized characteristics (creep strength) have not been changed. The limiting diffusion of Si in Fe is lower at phase transition temperature than at the high temperature of the standard heat treatment (see Fig. 13.3). To achieve the phase transformation temperature is not a trivial problem and takes time to steady state. Fe and Cr catalyses the reduction process of Si from $SiCl_4$ so that $SiCl_4$ reduces significantly below 1150°C. This has a beneficial environmental effect. The anti-oxidation effect for steam conditions and the effect on centrifugal cast tubes for anti-coking are under investigation.

13.3   Results and discussion

A cross-section of a modified surface is shown in Fig. 13.4. The schematic of the modifying process is shown in Fig. 13.5 and Figs 13.6 and 13.7 show the iron carbon diagram for austenitic T91 at 1050°C and 0.1%C, and the iron-silicon phase diagram with phase transformation from austenitic to ferrite steel lattice with growing Si content, respectively.

It can be seen in Fig. 13.8 that the silicon content is increased in the ferrite modified zone to about 5% and drops in the transition zone. Interesting is the chromium content which does not change significantly from the original 9%. The supposed reason therefore is increased diffusion of chromium with increased silicon content. The chromium reservoir is the base material. The values of the iron content balance to 100%.

Figure 13.9 shows the modified inner surface of a tube specimen, which indicates that the silicon content can be increased under the original scale of the production process of the tube and must not be machined before coating.

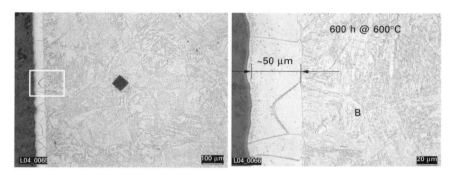

13.4 Ferritic modified steel surface with improved oxidation resistance.

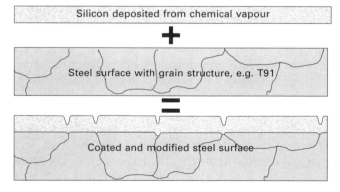

13.5 Schematic of surface modification via diffusion of silicon.

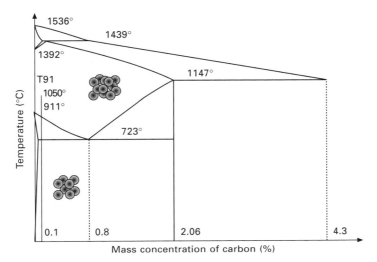

13.6 Iron carbon diagram, with austenitic T91 at 1050°C and 0.1% C; crystal lattice.

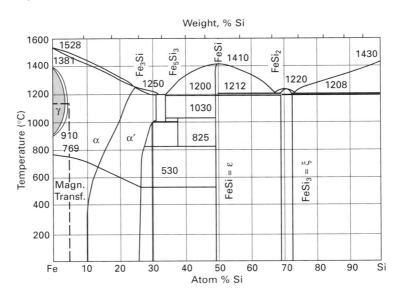

13.7 Iron-silicon phase diagram with phase transformation from austenitic to ferrite steel lattice with growing Si content from 0% to ~ 5% Si, γ-Fe section [17].

A beneficial effect improving the oxidation and corrosion resistance occurs even when the Si content is not sufficient to lead to a phase transition to ferrite, lower than ~ 5%, since the Si content is more than in the base tube material.

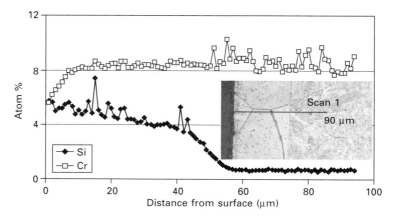

13.8 Si and Cr content; electron probe line-scans across a cross-section of a modified surface after 600 h annealing at 600°C.

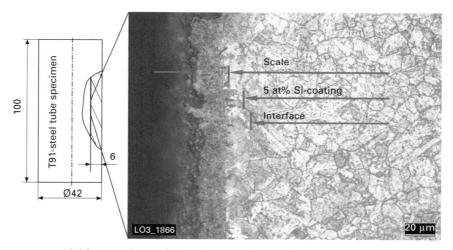

13.9 Inner tube surface with corrosion scale.

Figure 13.10 shows the microstructure before ageing. It gives some EDX measurements which show that the Si content is not constant in the modified surface but significantly higher than in the tube base material.

Figures 13.11 and 13.12 give the XRD pattern before and after modification. The patterns look different. The XRD patterns of uncoated and coated T91 steel in comparison to each other show different results and have to be further investigated. Main peaks at about 68° and 167° remain visible.

Figure 13.13 shows that the hardness in the surface region is lower than in the martensitic steel. Tensile tests at 600°C with 1%, 2% and 3% show no cracks in the surface due to the ductile behaviour of the modified ferrite zone

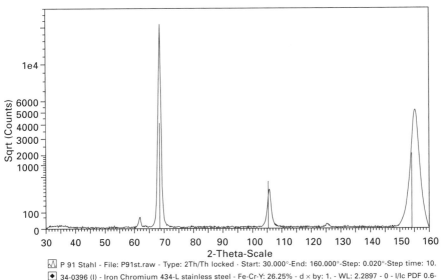

Element	Si	Cr	Fe	Mo	C	V
Measurement	Atom %	Atom %	Atom %	Atom %	Atom %	Atom %
1 Grain boundary	5.86	8.80	75.30	6.18	0.80	3.06
2 Grain boundary	4.26	9.37	80.91	2.80	0.93	1.74
3 Grain boundary	12.96	7.63	49.64	24.63	2.02	3.12
4 Substrate	0.70	10.09	86.92	0.57	0.88	0.84
5 Coating	4.49	10.04	81.67	2.04	0.88	0.87
6 Coating	12.33	8.92	51.27	25.85	1.63	0.00
7 Coating	6.28	9.45	65.41	6.40	1.50	3.44
8 Coating	5.91	8.07	73.66	8.73	1.37	1.05

13.10 EDX results on a cross-section of a modified surface.

M P 91 Stahl - File: P91st.raw - Type: 2Th/Th locked · Start: 30.000°-End: 160.000°-Step: 0.020°-Step time: 10.
◆ 34-0396 (I) - Iron Chromium 434-L stainless steel - Fe-Cr-Y: 26.25% - d × by: 1. - WL: 2.2897 - 0 - I/Ic PDF 0.6-

13.11 XRD diffraction pattern for the surface of a T91 tensile test specimen.

preserving the surface. Tensile testing at elevated temperatures 200°C, 300°C, 400°C, 500°C, 600°C, 700°C were performed, which indicated that there was only a minor influence on the strength of the base material.

For the investigation of the oxidation and corrosion resistance several tests were performed. As can be seen in Fig. 13.14, the modified surface

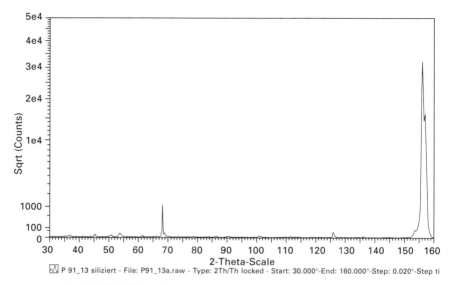

13.12 XRD diffraction pattern for a modified surface of a T91 tensile test specimen.

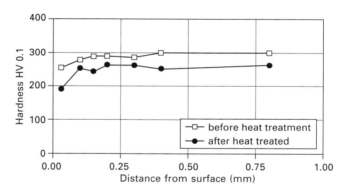

13.13 Hardness measurements.

does not show significant oxidation attack. The optical microscopic pictures in Fig. 13.14 are all made from the same position, which was found with help of a hardness indent. The hardness indent was deeper than the thickness of the modified surface. The surroundings of the hardness indents does show corrosion.

Figure 13.15 shows the results of gravimetric investigations of specimens aged for periods of 200 h at different atmospheres at 600°C. The oxidation resistance of specimens with silicon modified surface are superior to those of unmodified tube specimens (Fig. 13.16). Because the tube specimen therefore were treated without machining away the original scale at the inner surface,

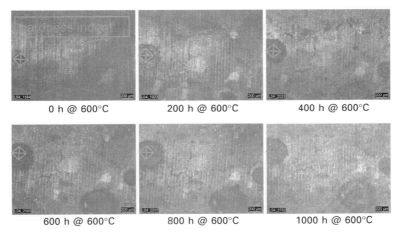

| 0 h @ 600°C | 200 h @ 600°C | 400 h @ 600°C |
| 600 h @ 600°C | 800 h @ 600°C | 1000 h @ 600°C |

13.14 Thermo-cyclic oxidation tests up to 1000 h at 600°C with 200 h cycles. Comparison of oxidation test results (tubes, round bar specimens).

and without treating the outer surface, these specimens show more weight gain in the oxidation test in comparison to a round bar specimen with overall surface treatment. With increased Si content this protecting effect occurs already below 5% without phase transformation to a ferrite layer. An additional effect could be achieved if the Si content is increased above ~5% Si. In this case the lattice of the steel transfers to the ferritic phase at coating temperature. This ferritic phase does not change to martensite during the cooling and annealing that usually follow. The ferrite phase has lower hardness and better ductility than the martensite phase of the base material. To enhance the diffusion of Si into the Fe lattice, the tube therefore can have maximum temperature at the inner surface. As Si is not an appropriate alloying element for creep resistant steels (causes embrittlement, welding difficulties) it is advantageous to have a thin surface region containing more Si, which provides oxidation and crack resistance without any adhesion problems. If this zone is thin compared with the tube thickness it does not affect the basic properties (creep strength, crack initiation) of the base metal.

Within this concept the oxidized surface of typical tubes must not even be machined. It will be cleaned during the CVD process with hydrogen at high temperature. The hydrogen that diffuses into the steel has to be removed by effusion, for example under vacuum, keeping the high temperature to avoid embrittlement. In principle silane can be used as precursor, but since this compound is toxic and pyrophoric, $SiCl_4$ is easier to handle as the coating precursor. With this precursor also the removal of chlorine from the steel surface has to be achieved during the process, to avoid the detrimental effect of chlorine-induced corrosion. The need to improve the oxidation resistance

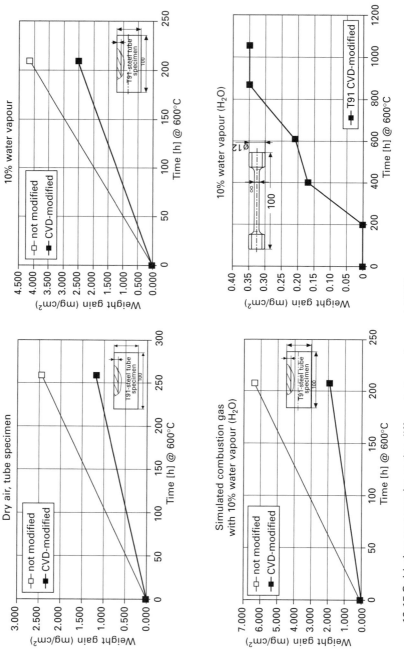

13.15 Oxidation test results under different atmospheres at 600°C.

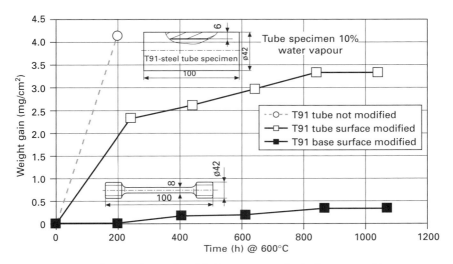

13.16 Comparison of oxidation test results (tubes, round bar specimens).

occurs with modern martensitic 9–11%Cr steels at service temperatures around 600 to 650°C, therefore advanced surface treatment developments are necessary. The surface modifying process is not a coating process resulting in the formation of an additional layer. The strength of the material of the tubes is unchanged as the optimized microstructure of the steel is maintained, since diffusion is applied during a one-step process within the heat treatment. The coating process temperatures therefore must be in accordance with the respective thermal heat treatment of the specific steel alloy. In addition the fabricability must be sustained, so that the coating must be ductile for unavoidable bending of long tubes and for weldability. Also the use of hard particles in the coating should be completely avoided, as they may cause damage at the turbine blades following the tubes of heat exchangers of power plants. The main goal is the improvement of the inner surfaces of long tubes. The outer surface is of additional interest, but seems easier to be cleaned and coated. One result is that the scale must not be machined to clean and to prepare the surface for coating.

13.4   Conclusion

Experimental investigations modifying the inner surface of tubes of 9% Cr steel via CVD with silicon containing precursors have been performed to provide improved protection against high temperature oxidation in steam atmosphere.

A surface modifying CVD coating technology, taking into account the specific requirements of material heat treatment, has been successfully developed as a one-step process. By means of this technology Si diffusion occurs in a limited zone at the inner surface of tubes. The physical and chemical properties of the modified surface were investigated using optical microscopy, SEM, EDX electron probe line scans and XRD. These revealed that the diffusion layers were formed with a ferritic structure. Fe atoms were substituted by Si with a maximum concentration of ~ 5% Si. The hydrogen concentration was in the standard range and not higher than before the process. The surface modified tubes kept their weldability. The ferrite modified surface showed more ductility and less hardness than the original martensite T91 steel. The strength of the substrate was not influenced, but oxidation behaviour was improved.

13.5   Acknowledgment

The authors performed their work at MPA University of Stuttgart, and they wish to gratefully acknowledge support from the AVIF, Project A176.

13.6   References

1. M. Schütze, *Korrosion und Korrosionsschutz*, Wiley-VCH, Weinheim, **2001**, 687–700.
2. F. Dettenwanger, M. Schorr, J. Ellrich, T. Weber, and M. Schütze, *Lifetime Modelling of High Temperature Corrosion Processes* (eds M. Schütze, W. J. Quadakkers, and J. R. Nicholls), Maney Publ., London, **2001**, 206–219.
3. M. Schütze, M. Schorr, D. P. Renusch A. Donchev and J. P. T. Vossen, *Materials Research*, **2004**, 7, 1, 111–123.
4. J. P. T. Vossen, P. Gawenda, K. Raths, M. Röhrig, M. Schorr and M. Schütze, *Materials at High Temperature*, **1997**, 14, 387–401.
5. M. Rebhan, M. Rohwerder and M. Stratmann, *Chem. Vap. Deposition*, **2002**, 8, 6.
6. M. Rebhan *et al.*, *Applied Surface Science*, **2001**, 178, 194–200.
7. M. Rebhan, M. Rohwerder and M. Stratmann, *Applied Surface Science*, **1999**, 140, 99–105.
8. C. Klam, J. P. Millet. H. Mazille and J. M. Gras, *J. Mater. Sci.*, **1991**, 26, 4945.
9. A. Sanjurio, B. Wood, K. Lau and G. Krishnan, *Scr. Metall. Mater.*, **1994**, 31, 1019.
10. A. L. Cabrera, J. F. Kirner and R. Pirantozzi, *J. Mater. Res.*, **1990**, 5, 74.
11. A. L. Cabrera and J. F. Kirner, *Surface and Coatings Technology*, **1989**, 39–40, 43–51.
12. K. Maile, K. Berreth, A. Lyutovich, *VGB-Fachtagung*, Dortmund, March, **2004.**
13. K. Maile, K. Berreth, A. Lyutovich, G. Zies, R. Weiss and T. Perova, *Materials Week*, Munich, **2003**, mw2002_374.pdf, available on CD-Rom.
14. E. Roos, K. Maile, K. Berreth, A. Lyutovich, R. Weiss and T. Perova, and A. Moore, *Surface and Coating Technology*, **2004,** 180–181C, 465–469.
15. K. Maile, K. Berreth, and A. Lyutovich, FGM2004, Belgien, Leuven, *Materials Science Forum*, **2005**, 492–493, 347–352.
16. V. Rohr, M. Schütze, E. Fortuna, D.N. Tsipas, A. Milewska and F.J. Pérez, EFC-Event No 275, **2004**, book of abstracts 15.
17. W. G. Moffat, *The Handbook of Binary Phase Diagrams*, General Electric, Vol. 2, **1978–97**.

14

Performance of thermal barrier coatings on γ-TiAl

R B R A U N, German Aerospace Centre (DLR), Germany;
C L E Y E N S, Technical University of Brandenburg at Cottbus,
Germany and M F R Ö H L I C H, German Aerospace
Centre (DLR), Germany

14.1   Introduction

Because of their attractive properties, γ-TiAl-based alloys are considered for
high temperature applications in the aerospace and automotive industries [1,
2]. They possess low density, high stiffness and yield strength, and good
creep resistance up to high temperatures. These advantages make them attractive
as structural materials in the temperature range between 700 and 900°C.
They offer the potential to replace steels and nickel-based alloys presently
used. In the automotive industry, titanium aluminides have proven their
suitability as construction materials for parts in future engine applications
[3]. These alloys meet the requirements for exhaust engine valves and
turbocharger wheels. In aeroengines, γ-TiAl alloys were successfully tested
as low-pressure turbine blades [4]. They are also considered for use in high
temperature components such as nozzle components in high-speed civil aircraft
for noise attenuation [5].

A drawback of gamma titanium aluminides, which limits their wide
application, is their poor oxidation resistance at temperatures above
approximately 750°C. This poor oxidation resistance results from the formation
of a non-protective oxide scale consisting of a heterogeneous mixture of
alumina and titania on high temperature exposure [6, 7]. The oxidation
behaviour can be improved by alloying with high amounts of niobium [8].
However, to increase the oxidation resistance to a sufficiently high level,
protective coatings probably have to be used [9]. A novel approach is the use
of thermal barrier coatings (TBCs) on γ-TiAl, which are commonly applied
to nickel-based superalloys [10–12]. In combination with component cooling,
this allows reduction of the metal surface temperatures. The aim of the
present work was to study the performance of thermal barrier coatings on
gamma titanium aluminides, which were coated with different types of oxidation
protection layers.

14.2   Experimental methods

The substrate used was extruded gamma titanium aluminide alloy Ti-45Al-8Nb. The material was supplied by GKSS Research Centre in Germany. The composition of the alloy determined by EDX analysis was 45.4Ti-43.6Al-11.0Nb (at.%). Disk-shaped specimens were machined with a diameter of 15 mm and 1 mm in thickness. A specimen holder made of conventional titanium alloys was welded to the sample for handling during coating. The surfaces of the specimens were ground with SiC paper up to 2500 grit. After polishing and cleaning, the specimens were coated with different protective layers.

Four types of bond coat layers were applied: Al_2O_3, TiAlCrYN, $TiAl_3$ and $TiAl_2$ coatings. Alumina layers were produced by physical vapour deposition carried out by a job coater. Monolithic TiAlCrYN coatings were deposited using a combined cathodic arc/unbalanced magnetron sputtering technique. Details of the coating process, which was performed at Sheffield Hallam University, UK, are given elsewhere [13]. The nitride coated specimens were pre-oxidised for 100 h at 750°C. Diffusion coatings were produced using the pack-cementation process and annealing treatments. In a first step, the surface region of specimens was aluminised to $TiAl_3$ by burying in a powder mixture consisting of aluminium, ammonium chloride and alumina. Part of the specimens were subsequently annealed at 910°C. This high temperature exposure resulted in the formation of a $TiAl_2$ layer due to interdiffusion between the $TiAl_3$ coating and the substrate. The aluminising process and the interdiffusion treatment were performed at Dechema in Frankfurt, Germany. Details are described elsewhere [14]. Specimens with both aluminide coatings were pre-oxidised for 10 h at 900°C. As reference, uncoated specimens which were pre-oxidised at 750°C in air for 100 h were included in this study.

Thermal barrier coatings of 7 wt.% partially yttria stabilised zirconia were deposited on the various pre-treated specimens using a 150 kW electron-beam physical vapour deposition (EB-PVD) coater installed at DLR. The thickness of the TBCs was about 170 μm. While a substrate temperature of 1000°C is usually used during ceramic topcoat deposition with nickel-based superalloys, the deposition temperature was reduced in the case of gamma titanium aluminides. Figure 14.1 shows side and top views of the EB-PVD thermal barrier coating on γ-TiAl alloys. The TBC exhibited a columnar microstructure with multiple featherarms. The threefold symmetry suggests that the columns grew in <111> direction. Compared to thermal barrier coatings on Ni-based superalloys, those on γ-TiAl alloys had a lower density, and their columnar microstructure was less pronounced.

The oxidation behaviour of the specimens with environmental and thermal protective coatings was determined under cyclic oxidation conditions at 900°C in air. One cycle consisted of 1 h at temperature and 10 min cooling down to

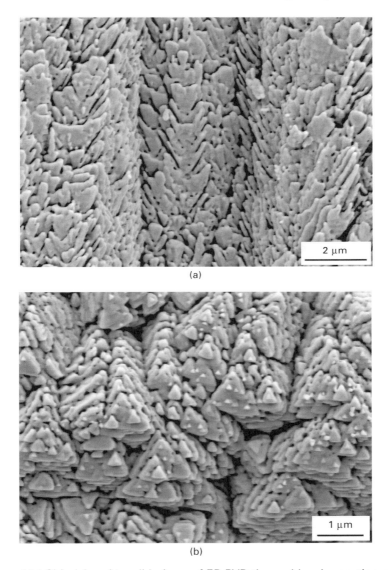

(a)

(b)

14.1 Side (a) and top (b) views of EB-PVD thermal barrier coating on gamma titanium aluminide alloy Ti-45Al-8Nb.

about 70°C. Furthermore, quasi-isothermal oxidation tests were performed at 900°C in air. In these tests, the specimens were removed from the furnace and weighed once every week. Prior to oxidation testing, the specimen holder was removed leaving a small unprotected area of remnant holder material. This area represented spots of preferential oxidation due to the poor oxidation resistance of conventional titanium alloys at 900°C. After oxidation, the microstructure of the specimens was examined using a scanning electron microscope with EDX detector.

14.3   Results and discussion

Figure 14.2 shows the mass change vs time and number of cycles for pre-oxidised specimens with TBC, which were exposed at 900°C in air under quasi-isothermal and cyclic oxidation conditions, respectively. Under quasi-isothermal conditions, spallation was observed after 840 h during cooling for weighing. When cyclically tested, the TBC spalled off partly after 700 cycles. Figure 14.3 shows micrographs of a pre-oxidised specimen with TBC which was oxidised for 980 cycles at 900°C. A thick oxide scale formed underneath the TBC, exhibiting a layered structure (Fig. 14.3(a)). Failure occurred in the porous titania-rich layer, whereas the TBC was well adhering to the oxide scale formed during pre-oxidation (Fig. 14.3(b)). SEM micrographs and EDX analysis revealed an outer layer of alumina interspersed with titania, and an adjoining layer of porous titanium dioxide containing small amounts of Nb- and Al-oxides. As shown in Fig. 14.4(a), the lower part of the oxide scale exhibited a layered structure with bands rich in niobium (bright porous layer) and adjoining dark layers consisting mainly of alumina. Between the oxide scale and the substrate a complex transition zone formed. In this transition zone, alumina particles and an intermetallic phase rich in niobium were found as well as a discontinuous nitride layer (Fig. 14.4(b)). This

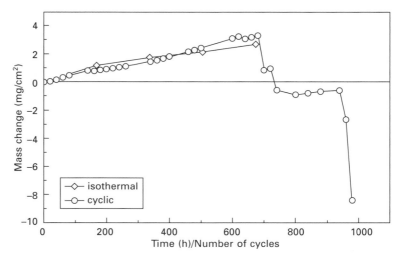

14.2 Mass change vs time and number of cycles of pre-oxidised Ti-45Al-8Nb specimens with TBC during exposure to air at 900°C under quasi-isothermal and cyclic oxidation conditions, respectively.

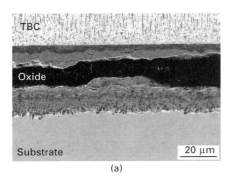
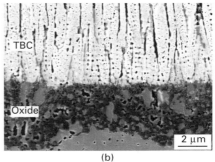

14.3 Scanning electron micrographs of a pre-oxidised Ti-45Al-8Nb specimen with TBC, which was oxidised at 900°C in air for 980 cycles, showing the thermally grown oxide scale (a) and the transition region between TBC and oxide scale (b).

nitride layer consisted of two different nitrides, as confirmed by EDX analysis. Adjacent to the substrate, Ti_2AlN was observed followed by TiN. Titanium nitride was oxidised with increasing oxygen partial pressure further away from the scale-substrate interface [15]. The porous bright spots were rich in niobium, indicating that the niobium-containing intermetallic phases were oxidised behind the nitride layer. The subsurface zone of the Ti-45Al-8Nb alloy was depleted in titanium and enriched in aluminium.

The adhesion of the TBC on specimens coated with alumina was weak, indicating that the PVD alumina coatings were not optimised for this particular type of testing. The thermal barrier coatings spalled off on removing the specimen holder from the sample. Flakes of the alumina coatings were observed on the surfaces of these specimens. Cross-sectional examination revealed a thin nitride layer (Fig. 14.5). The composition measured by EDX analysis was 22.3Ti-25.6Al-52.1N (at.%), indicating that the layer consisted of TiN and AlN. Spallation of the TBC was probably associated with this first stage of oxidation of the substrate. A specimen which was quasi-isothermally oxidised with specimen holder failed after 336 h of exposure during cooling.

TiAlCrY-nitride layers, monolithically grown or with superlattice structure, provide an excellent oxidation protection for γ-TiAl alloys in air at 750°C [11, 16]. Analysis of specimens coated with monolithically grown TiAlCrYN after 3000 h of isothermal oxidation at 750°C revealed that approximately half of the nitride layer was oxidised, but oxygen was not detected in and beyond the remaining nitride layer. Figure 14.6 shows the performance of thermal barrier coatings on specimens coated with nitride under quasi-isothermal and cyclic oxidation conditions at 900°C. When isothermally oxidised, the TBC spalled off after 504 h during cooling for weighing. Under cyclic oxidation conditions, partial spallation occurred after 160 cycles. Figure 14.7(a) shows the cross-section of a specimen coated with nitride which was

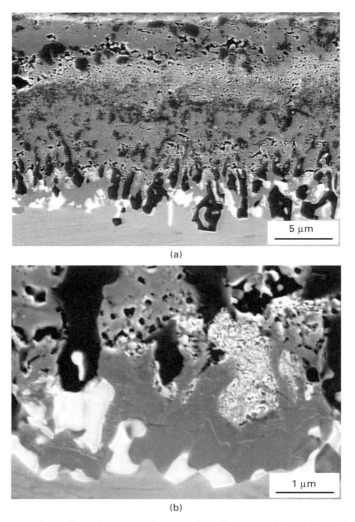

(a)

(b)

14.4 Scanning electron micrographs of a pre-oxidised Ti-45Al-8Nb specimen with TBC, which was oxidised at 900°C in air for 980 cycles, showing the inner part of the oxide scale (a) and the transition zone between oxide scale and substrate (b).

pre-oxidised at 750°C for 100 h. A mixed oxide scale formed on the surface consisting of titanium, aluminium and chromium oxides (Fig. 14.7(b)). The thickness of the nitride layer was about 2 μm. After oxidation at 900°C for 504 h, the nitride layer was completely oxidised (Fig. 14.8). Subsequently, a thick oxide scale formed similar to that found with pre-oxidised specimens (Fig. 14.3, 14.4). Again, a good adhesion of the TBC was observed on the oxide scale formed during pre-oxidation of the nitride layer.

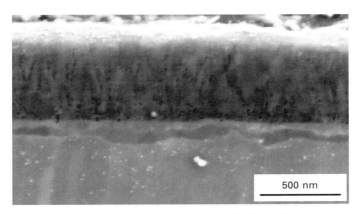

14.5 Scanning electron micrograph of a Ti-45Al-8Nb specimen, which was coated with PVD alumina and TBC, showing a nitride layer on the substrate surface.

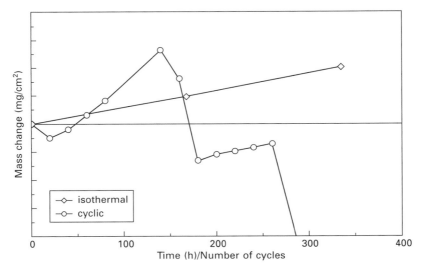

14.6 Mass change vs time and number of cycles of Ti-45Al-8Nb specimens coated with a TiAlCrYN layer and TBC during exposure to air at 900°C under quasi-isothermal and cyclic oxidation conditions, respectively.

Figure 14.9 shows the oxidation behaviour of Ti-45Al-8Nb specimens coated with different aluminide layers and TBCs under cyclic and quasi-isothermal oxidation conditions. For comparison, mass change data of pre-oxidised specimens with TBC are included. No spallation of the thermal barrier coatings was observed after 1000 cycles at 900°C for specimens coated with the two types of aluminides investigated. Drops in mass change

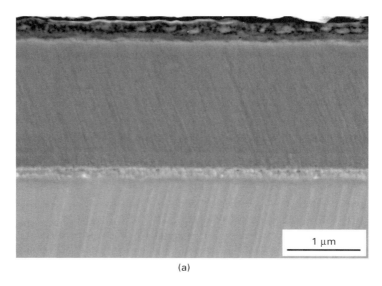

(a)

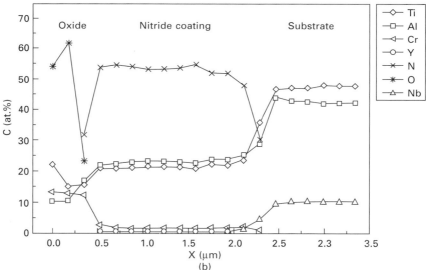

X (μm)

(b)

14.7 Scanning electron micrograph (a) and EDX analysis (b) of a Ti-45Al-8Nb specimen coated with a TiAlCrYN layer, which was pre-oxidised at 750°C in air for 100 h.

observed at about 400 cycles were not associated with failure of the TBC but with removal of the oxidised remnants of the specimen holders (Fig. 14.9(a)). Compared with the specimen which was only pre-oxidised, both aluminide coatings provided a significant reduction of mass gain. This was most evident for the sample coated with TiAl$_3$ after 400 cycles when the oxidised remnants

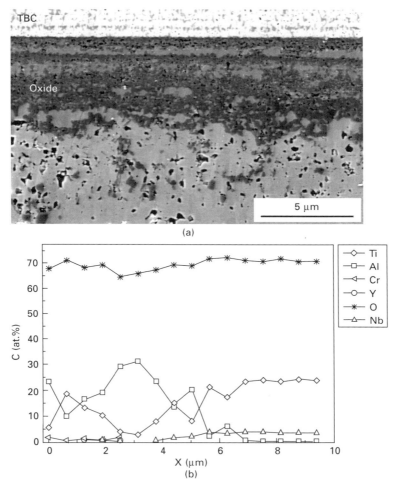

14.8 Scanning electron micrograph (a) and EDX analysis (b) of a Ti-45Al-8Nb specimen coated with a TiAlCrYN layer and TBC, which was oxidised at 900°C in air for 504 h under quasi-isothermal conditions.

of holder material were removed. The subsequent mass change was very low. The minor weight loss of specimens during the very early stages of exposure was associated with release of water vapour taken up by the TBC from natural air humidity during storage before being tested. Similarly, no spallation was found for both aluminide coatings after 3000 h at 900°C under isothermal oxidation conditions. In the latter oxidation tests, remaining holder material was not removed. Again, the mass gain was considerably lower for the specimens coated with aluminides, in particular with a TiAl$_3$ layer than for a specimen which was only pre-oxidised.

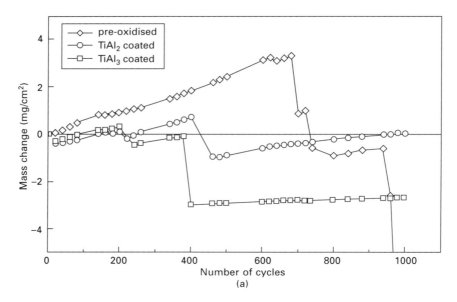

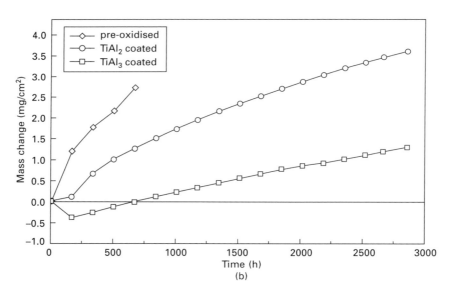

14.9 Mass change vs number of cycles (a) and time (b) of Ti45Al-8Nb specimens coated with aluminides and TBC, which were exposed to air at 900°C under cyclic and quasi-isothermal oxidation conditions, respectively. Data for non-coated, pre-oxidised specimens with TBC are included.

A continuous compact alumina layer with a thickness of about 1 μm was found on the surface of specimens coated with TiAl₃ and oxidised at 900°C for 1000 cycles (Fig. 14.10(a)). The thermal barrier coating exhibited an excellent adhesion to the alumina layer, which was probably formed during the pre-oxidation heat treatment [10]. In the subsurface zone beneath the alumina layer, a large number of cracks were observed as a result of the brittleness of the TiAl₃ diffusion layer [17] (Fig. 14.10(b)). Figure 14.11 shows the elemental profiles across the subsurface zone. The concentrations of Ti and Al beneath the oxide scale did not correspond to the chemical composition of the aluminides TiAl₃ or TiAl₂. The amount of aluminium was

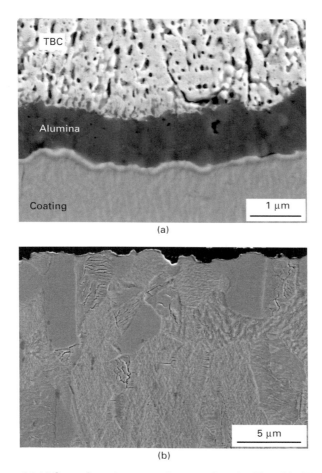

(a)

(b)

14.10 Scanning electron micrographs of a Ti-45Al-8Nb specimen coated with TiAl₃ aluminide and TBC, which was oxidised at 900°C in air for 1000 cycles: (a) continuous alumina layer on the coating surface, (b) subsurface region.

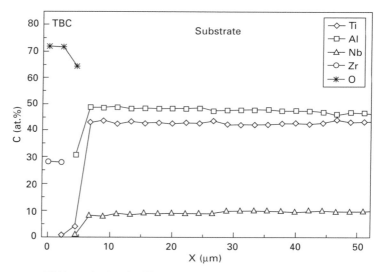

14.11 EDX analysis of a Ti-45Al-8Nb specimen coated with TiAl₃ aluminide and TBC, which was oxidised at 900°C in air for 1000 cycles.

lower, i.e. ~50 at.%, decreasing continuously to the substrate level with increasing distance from the interface between alumina scale and coating. Obviously, the TiAl₃ layer was entirely consumed by interdiffusion between coating and substrate. In the initial stages of oxidation of aluminised γ-TiAl, an intermediate TiAl₂ layer formed between the TiAl₃ coating and the γ-TiAl alloy [14]. However, in the present study, the TiAl₂ phase was not observed due to substantial interdiffusion during long-term oxidation. Locally, small regions were found in the matrix beneath the alumina layer consisting of two different phases which were rich in aluminium. Their compositions were close to TiAl₃ and TiAl₂. The local occurrence of these phases might be associated with variation in the TiAl₃ coating thickness due to specimen preparation prior to pre-oxidation and TBC deposition. Beneath the alumina scale, neither oxygen nor nitrogen were detected. Thus, aluminising the substrate surface to TiAl₃ seems to be an effective coating process to improve the oxidation behaviour of γ-TiAl alloys.

Figure 14.12 shows elemental profiles across the subsurface layer of a specimen coated with TiAl₂ in the as-received, polished condition. With further distance from the surface, the concentrations of Al and Ti continuously decreased and increased, respectively, to the substrate levels. The chemical composition at the top of the coating did not correspond to the aluminide TiAl₂. The original TiAl₂ layer might have been removed on grinding and polishing the specimens after the pack cementation process. It is also possible that annealing applied to convert the TiAl₃ phase to TiAl₂ was too severe,

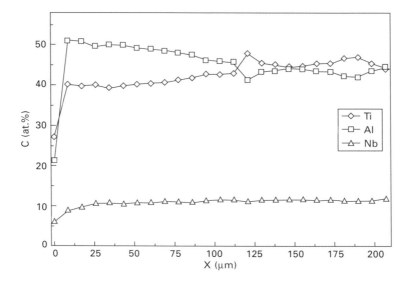

14.12 Elemental profiles across the subsurface layer of a Ti-45Al-8Nb specimen coated with $TiAl_2$ aluminide in the as-received, polished condition.

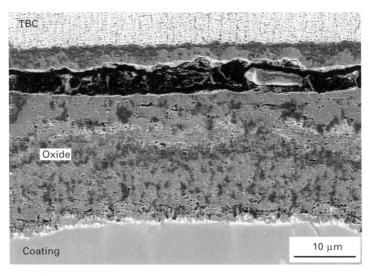

14.13 Scanning electron micrograph of a Ti-45Al-8Nb specimen coated with $TiAl_2$ aluminide and TBC, which was oxidised at 900°C in air for 1000 cycles.

causing the dissolution of the latter phase by interdiffusion with the Ti-45Al-8Nb substrate. Pre-oxidation of these specimens resulted in a mixed oxide scale including formation of nitrides at the scale-metal interface. A continuous protective alumina layer was not formed. On the contrary, stable Al_2O_3 formation was found on a $TiAl_2$ layer when oxidised at 900°C, suggesting this phase as attractive protective coating for γ-TiAl alloys [14]. Good adhesion of the TBC was found again on the oxide scale formed during pre-oxidation. Oxidation of the specimens with TBC at 900°C for 1000 cycles resulted in the formation of a thick oxide scale (Fig. 14.13). However, it should be noted that spallation of the TBC was not observed after 1000 cycles. Probably, failure would have occurred if specimens had been tested for longer exposure times. Nevertheless, thermal barrier coatings on specimens coated with a supposed $TiAl_2$ layer exhibited a longer lifetime than specimens which were only pre-oxidised. The higher amount of aluminium in the surface zone of the former specimens might retard the formation of a porous layer which predominantly consists of TiO_2. Although spallation of the ceramic coating did not occur during the maximum exposure time period, oxide scale damage, mainly in the titania-rich zone of the oxide scale, was already visible in the metallographic cross-section as demonstrated by cracking shown in Fig. 14.13.

14.4   Conclusions

- EB-PVD partially yttria stabilised zirconia coatings offer a very promising concept to reduce service temperatures on γ-TiAl components.
- Thermal barrier coatings revealed excellent adhesion to oxide scales consisting predominantly of alumina.
- Failure of the thermal barrier coatings was associated with cracking of the oxide scale in the porous titania-rich layer.
- Monolithic TiAlCrYN layers exhibited poor oxidation resistance at 900°C.
- $TiAl_3$ aluminide coating provided an excellent oxidation protection associated with the formation of a continuous alumina surface layer.
- $TiAl_2$ aluminide coating did not form a protective alumina layer. The intended surface chemistry of the coating corresponding to the $TiAl_2$ phase was not achieved.

14.5   Acknowledgments

TiAlCrYN coatings were provided by P.Eh. Hovsepian at Sheffield Hallam University, UK, which is gratefully acknowledged. The authors would like to thank D. Renusch from the Karl Winnacker Institute of Dechema in Frankfurt for aluminising the specimens.

14.6   References

1. H. Clemens, H. Kestler, *Advanced Engineering Materials*, **2000**, *2*, 551.
2. H. Clemens, F. Appel, A. Bartels, H. Baur, R. Gerling, V. Güther and K. Kestler, *Ti-2003, Science and Technology*, Volume IV, G. Lütjering, J. Albrecht, Eds., Wiley-VCH Verlag, Weinheim, **2004**, 2123.
3. H. Baur, D.B. Wortberg, *Ti-2003, Science and Technology*, Volume V, G. Lütjering, J. Albrecht, Eds., Wiley-VCH Verlag, Weinheim, **2004**, 3411.
4. W. Smarsly, H. Baur, G. Glitz, H. Clemens, T. Khan and M. Thomas, *Structural Intermetallics 2001*, K.J. Hemker *et al.*, Eds., The Minerals, Metals & Materials Society, Warrendale, **2001**, 25.
5. P.A. Bartolotta and D.L. Krause, *Gamma Titanium Aluminides 1999*, Y.-W. Kim, D.M. Dimiduk, M.H. Loretto, Eds., The Minerals, Metals & Materials Society, Warrendale, **1999**, 3.
6. S. Becker, A. Rahmel, M. Schorr and M. Schütze, *Oxidation of Metals*, **1992**, *38*, 425.
7. A. Rahmel, W.J. Quadakkers and M. Schütze, *Materials and Corrosion*, **1995**, *46*, 271.
8. Y.-W. Kim, *Niobium for High Temperature Applications*, Y.-W. Kim, T. Carneiro, Eds., The Minerals, Metals & Materials Society, Warrendale, **2004**, 125.
9. L. Niewolak, V. Shemet, E. Wessel, A. Gil, L. Singheiser and W.J. Quadakkers, *Structural Intermetallics 2001*, K.J. Hemker *et al.*, Eds., The Minerals, Metals & Materials Society, Warrendale, **2001**, 535.
10. V. Gauthier, F. Dettenwanger and M. Schütze, *Intermetallics*, **2002**, *10*, 667.
11. C. Leyens, R. Braun, P.E. Hovsepian and W.-D. Münz, *Gamma Titanium Aluminides 2003*, Y.-W. Kim, H. Clemens, A.H. Rosenberger, Eds., The Minerals, Metals & Materials Society, Warrendale, **2003**, 551.
12. C. Leyens and R. Braun, *Materials Science Forum*, **2004**, *461–464*, 223.
13. C. Leyens, M. Peters, P.E. Hovsepian, D.B. Lewis, Q. Luo and W.-D. Münz, *Surface and Coatings Technology*, **2002**, *155*, 103.
14. V. Gauthier, F. Dettenwanger, M. Schütze, V. Shemet and W.J. Quadakkers, *Oxidation of Metals*, **2003**, *59*, 233.
15. F. Dettenwanger, E. Schumann, M. Rühle, J. Rakowski and G.M. Meier, *Oxidation of Metals*, **1998**, *50*, 269.
16. C. Leyens and R. Braun, *Niobium for High Temperature Applications*, Y.-W. Kim, T. Carneiro, Eds., The Minerals, Metals & Materials Society, Warrendale, **2004**, 239.
17. J.L. Smialek, *Corrosion Science*, **1993**, *35*, 1199.

Looking for surface treatments with potential industrial use in high temperature oxidation conditions

G BONNET, J M BROSSARD and J BALMAIN,
University of La Rochelle, France

15.1 Introduction

One of the most common problems with metallic materials working at high temperatures is their chemical degradation, due to the presence of aggressive gaseous environments, particularly containing oxygen. A lot of research has been carried out to ameliorate some properties of alloys but their resistance to high temperature oxidation is still inadequate and additional surface treatments or coatings are required. Numerous methods have been employed to modify alloy surfaces, such as PVD techniques (e.g. ion implantation [1–3], magnetron sputtering [4]), CVD (e.g. PECVD [5], MOCVD [6, 7], pyrosol method [8], pack cementation [9–11], CVD-FBR [12]), or chemical ones (e.g. sol/gel [13,14]). Most of these techniques are very appropriate for a laboratory use but cannot easily be transferred to an industrial environment, for example because of the equipment (PVD, CVD) or chemical precursor (MOCVD) costs, or because of the difficulty in controlling the deposit quality (sol/gel).

The purpose of this work was to introduce a reactive element at the surface of a NiCr alloy by an easy and cheap method, potentially usable in industry. From this point of view, yttrium-containing films were electrodeposited on Ni20Cr1.5Si coupons, from a water- ethanol solution of yttrium nitrate. The effect of current density was investigated to determine coating conditions of a thin (thickness < 1 μm), homogeneous and adherent films. Then, a thermal treatment allowed the conversion of the deposited films into Y_2O_3 coatings. Coated samples were also tested in high temperature oxidation conditions.

15.2 Experimental methods

All samples were polished before coating or high temperature oxidation. NiCr (Ni-20wt.%Cr-1.5wt.%Si) alloy coupons were cut from a rod, approximately 12 mm in diameter and 1.5 mm in thickness. A 1 mm diameter hole was drilled near the edge of each sample. Specimens were then abraded from 600 grit SiC to 3 μm diamond paste and ultrasonically cleaned with ethanol. Sample characterisation was performed by SEM, supplied with an EDS analysis device, XPS, SIMS and AFM.

15.2.1 Electrodeposition of yttrium-containing thin films

Cathodic electrochemical deposition was achieved using a conventional three electrode cell including a NiCr substrate as cathode, a platinum grid counter electrode and a saturated calomel electrode (SCE) as reference. The electrochemical bath was a 0.01 M mixed ethyl alcohol-water solution (1:1 volume ratio) of commercial $Y(NO_3)_3$, $6H_2O$ (99.99% purity) as previously discussed [15]. Deposits were obtained in galvanostatic mode at room temperature without stirring. Deposition conditions to obtain thin (<1 μm), uniform and adherent films were determined by varying the current density and deposition time. As-deposited samples were washed with ethanol, dried in air for two days and then heated for 1 h at 600°C and then 15 min at 950°C (heating rate: 10°C/min) under argon gas atmosphere before high temperature experiment.

The amount of deposited yttrium vs applied current densities was determined using inductively coupled plasma-optical emission spectroscopy (ICP-OES). Concentration profiles were measured by secondary ion mass spectrometry (SIMS) on the coated samples before and after heat treatment, using a 10 keV O^{2+} primary ion source. The yttrium depth profiles were established with $^{89}Y^+$ signals. The chemical state of the as-deposited and the thermally treated film was also investigated by X-ray photoelectron spectroscopy (XPS) using an Mg Kα X-ray source, without sputtering the surface.

15.2.2 High temperature thermal treatments and isothermal oxidation tests

All high temperature experiments were carried out in a thermobalance, under an inert argon gas (Ar type U: 99.995 vol.% Ar, N_2 < 20 ppm, H_2 < 10 ppm, CH_4 < 5 ppm, H_2O < 5 ppm) or synthetic air (80 vol.% N_2; 20 vol.% O_2) atmosphere, at atmospheric pressure. After heat treatment under inert atmosphere, argon was replaced by reconstituted air which marked the beginning of the oxidation of the blank NiCr specimens. For coated specimens, the described thermal treatment steps have to be included in the temperature increase. Blank and coated NiCr samples were oxidised for 90 h at 850°C.

15.3   Results

15.3.1 Determination of deposition conditions

The electrochemical mechanism forming the yttrium-containing films follows two distinct steps [16]. In the first step, hydroxyl ions are electrochemically generated by reduction of O_2, H_2O and of NO_3^- in the case of nitrate containing bath, depending on the applied current density (galvanostatic mode) [17]. In the second step, hydroxyl ions produced at the surface lead to a local increase of pH producing hydrolysis and precipitation of metallic species ($Y_{(aq)}^{3+}$, $Y(OH)_{(aq)}^{2+}$, $Y_2(OH)_{(aq)}^{4+}$).

The as-deposited films are transparent and have a gel-like appearance particularly at high current densities and long deposition time, which indicates that films formed in these conditions are widely moisturised. After drying, the XRD analysis demonstrated that films were not $Y(OH)_3$, as expected, but rather a hydrated $Y_2(OH)_{6-x}(NO_3)_x$ [18].

The amounts of yttrium, determined by ICP-OES, deposited at different current densities for duration time fixed at 900 s, show that local rise in pH responsible for precipitation is effective even for low current densities (Fig. 15.1). The deposited yttrium content increases with current density and reaches a maximum at high current density ($|j| = 0.8$ and 1 mA.cm^{-2}) when a strong H_2 bubbling is observed. After drying for a few minutes in laboratory air, the deposits obtained at low current density ($|j| = 0.2$ and 0.4 mA.cm^{-2}) are

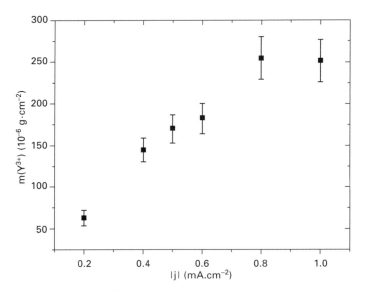

15.1 Amount of Y^{3+} deposited per square centimetre vs current density for a deposition time fixed at 900 s.

transparent with rainbow interference colours whereas, at higher current density ($|j| > 0.6$ mA.cm^{-2}), deposits are non uniform and appear cracked to the naked eye (Fig. 15.2). These observations can be associated to the hydroxide

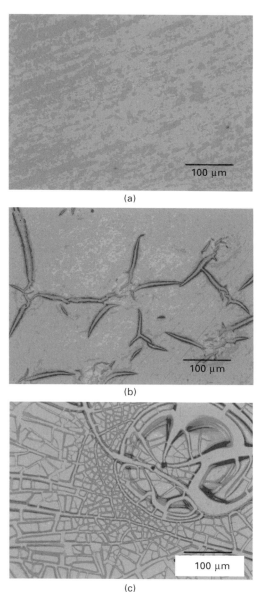

(a)

(b)

(c)

15.2 Surface morphologies of yttrium hydroxynitrate films, after drying in ambient air, electrodeposited for 900 s at different current densities: (a) $|j| = 0.2$ mA.cm^{-3}, (b) $|j| = 0.6$ mA.cm^{-2}, (c) $|j| = 0.8$ mA.cm^{-2}.

films thickness, in agreement with the relation proposed by Lu *et al.* [19]. Uniform, continuous and uncracked hydroxide films are estimated to be thinner than 0.2 μm. Thickness of uniform, non-continuous and hair-like cracked hydroxide films would be comprised between 0.2 and 2 μm. Finally, non-uniform, non-continuous and mud-like cracked films would be thicker than 2 μm.

These results are in agreement with those of Chaim *et al.* [20], who showed that increase of current density for fixed deposition time accelerated the deposition rate and so increased the deposit thickness. Nevertheless, the H_2 gas bubbles produced at high current density lead to a local stirring and hamper the increase of pH close to the surface and limit the film formation. The mud crack can be attributed to H_2 bubbling but also, in the case of thick films, to the evaporation rate of H_2O from the gel surface which is faster than the diffusion rate of H_2O from the gel bulk to the surface layer. These phenomena lead to stress on the gel surface due to shrinkage of the surface which is less moisturised than the bulk gel [21].

Effect of deposition time on film morphology was then investigated for a current density fixed at $|j| = 0.2$ mA.cm^{-2} to avoid H_2 bubbling taking place at higher current densities. Previous work [15] already revealed the existence of an acceleration of deposition rate after 500 s of deposition time. The observations of the deposits formed at different deposition times (Fig. 15.3) show that films obtained for durations shorter than 500 s are thin, transparent with blue interference, adherent and relatively uniform. Circular islands observed are attributed to substrate impurities or to surface defects leading to a local heterogeneous precipitation. For longer times ($t > 500$ s), the deposits consist of agglomerated flakes growing rapidly on the adherent film formed before. When duration increases to 2100 s, films are non-continuous, have a hair-like cracked morphology and an increase in the width of mud crack is observed when increasing deposition time. As described by Yeng [21], the greater the film thickness is, the less the diffusion rate of H_2O in the bulk gel can keep up with the evaporation rate of H_2O from the gel surface to the air. This leads to large stresses and to an increase in the width of mud crack. Nevertheless, the thin film initially formed ($t < 500$ s) seems to be crack free and adherent to the substrate. These observations are also consistent with the mechanism proposed by Chaim *et al.* [20] corresponding to the formation of a thin dense layer during a first stage, then recovered by a relatively porous and thicker one growing during the second stage. Thus, conditions required to obtain thin, uniform, and adherent film are a low current density (fixed at $|j| = 0.2$ mA.cm^{-2}) and short deposition time (fixed at 300 s). These conditions lead to a film thickness around 40 nm, as indicated by the SIMS yttrium profile in Fig. 15.4(a). The XPS analyses (Fig. 15.4(b)) show the Y 3d core level signal consists in two peaks due to high bonding state of Y 3d$^{3/2}$ and to low binding state of Y 3d$^{5/2}$. The area ratio of Y 3d$^{5/2}$

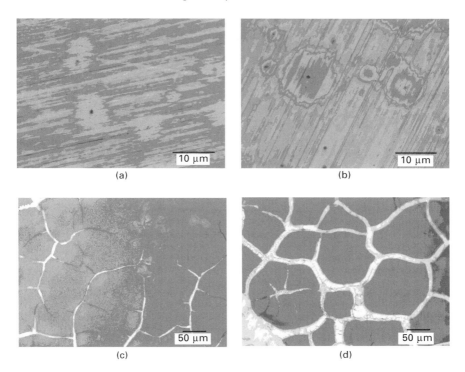

15.3 Surface morphologies of yttrium hydroxynitrate films, after drying in ambient air, electrodeposited at |j| = 0.2 mA.cm^{-2} for different times: (a) 300 s, (b) 1000 s, (c) 2100 s, (d) 3600 s.

to Y 3d$^{3/2}$ is close to the theoretical ratio of 1.5 and the binding energies of Y 3d$^{3/2}$ (159.7 eV) and Y 3d$^{5/2}$ (157.6 eV) are in agreement with the binding energies found in the case of a Y-OH binding. These results confirm that the electrodeposited film is an hydroxide as analysed by XRD [18].

15.3.2 Heat treatment and high temperature behaviour of uncoated or Y$_2$O$_3$-coated NiCr alloy

Deposited films were heat treated to promote dehydration and adhesion to substrate prior to investigating their effects on high temperature oxidation of the chromia-forming substrate Ni20Cr. AFM showed that the conversion heat treatment did not significantly affect the morphology of the deposit (Fig. 15.5).

SIMS analysis of a coated sample after heat treatment at 950°C (Fig. 15.6(a)) reveals a gradual distribution of yttrium to a thicker depth than before heat treatment. The chromium distribution also reveals an increase in

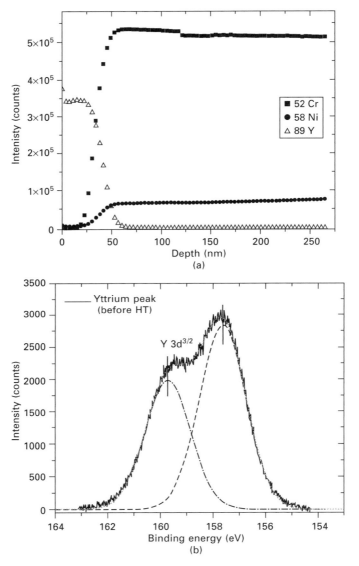

15.4 Surface analysis performed on Ni20Cr coated by yttrium hydroxynitrate hydrate ($|j| = 0.2$ mA.cm^{-2}, $t = 300$ s), after drying in air and before heat treatment: (a) SIMS depth profiles, and (b) XPS spectra.

the Cr content towards the gas/coating interface. In addition, the Y 3d core level signal (Fig. 15.6(b)) obtained on this sample always consists in two peaks Y 3d$^{3/2}$ (157.7 eV) and Y 3d$^{5/2}$ (155.6 eV) with a shift of 2 eV compared with the XPS spectra before heat treatment (Fig. 15.4 (b)), but the contribution

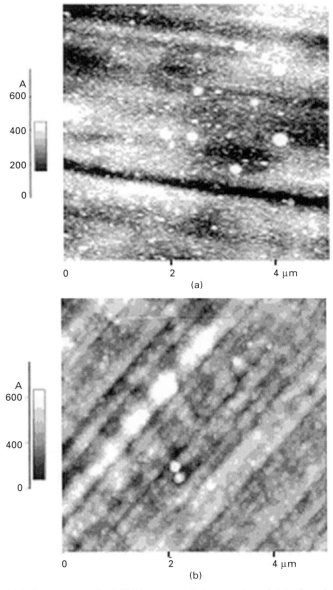

15.5 Contact mode AFM images of the coating: (a) before heat treatment, (b) after heat treatment.

of two other peaks (shoulders) has been identified for binding energies of 158.8 and 156.9 eV. The binding energies of the two peaks (Y $3d^{3/2}$ and Y $3d^{5/2}$) are consistent with the Y-O binding and confirm the transformation of the hydroxide into cubic yttrium oxide previously observed after heat treatment

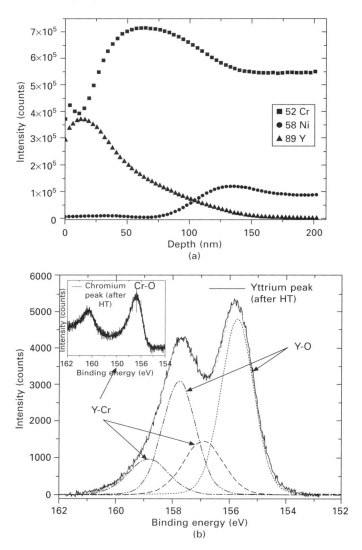

15.6 Surface analysis performed on Ni20Cr coated by yttrium hydroxynitrate hydrate ($|j| = 0.2$ mA.cm^{-2}, $t = 300$ s), after heat treatment at 950°C: (a) SIMS depth profiles and (b) XPS spectra.

[18]. In addition, Cr 2p$^{3/2}$ peak is detected after heat treatment, whereas it was missing before heat treatment. This observation is consistent with the increase of Cr content towards the gas/coating interface and can probably explain the existence of the two shoulders observed on 3d spectra. The Cr 2p$^{3/2}$ feature has a binding energy value of 576.5 eV which coincides with that of Cr-O binding. So, it is assumed that counterdiffusion of chromium

and oxygen could promote formation of Cr_2O_3, in agreement with the small weight gain recorded during heat treatment. This internal oxide could react with Y_2O_3 to form the $YCrO_3$ mixed oxide, leading to the wider yttrium signal observed on SIMS profile after heat treatment. Consequently, the two shoulders detected are likely to correspond to the contribution of Y-Cr binding.

These results confirm that heat treatment promotes dehydration and crystallisation of yttrium oxide but also that complex diffusion phenomena occur. Adhesion of the deposit to the substrate is thus expected to increase.

The isothermal oxidation curves obtained on Ni-20Cr-1.5Si substrate with and without Y-rich films (Fig. 15.7) indicate that the applied surface treatment strongly reduces the mass gain per surface unit, particularly during the first hours of oxidation. The total weight gain after 90 h at 850°C is around 10 times lower for the treated specimen. After 90 h of oxidation, the oxide scales formed on both samples are mainly composed of Cr_2O_3. No trace of NiO was detected but some traces of $NiCr_2O_4$ are present on the uncoated sample. In the case of the coated sample, NiO and $NiCr_2O_4$ are not observed but mixed chromium-yttrium oxide was detected indicating a solid state reaction between Y_2O_3 and Cr_2O_3. SEM observations (Fig. 15.8) showed that small addition of yttrium (here as Y_2O_3) significantly changes the morphology of the scale. The scale formed on the uncoated substrate contains large Cr-rich grains (around 2 µm) whereas in the case of the coated sample a continuous Y-rich scale, formed of small grains of about 0.3 µm, is surmounted by Cr-rich grains finer than the ones observed on the uncoated sample.

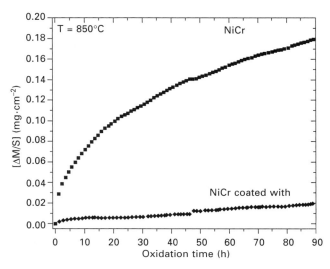

15.7 Weight gain curves of uncoated and yttria-coated samples oxidised at 850°C for 90 h.

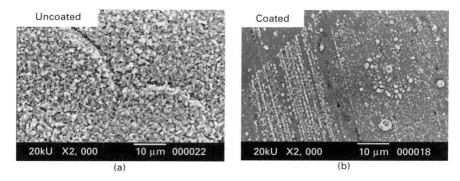

15.8 Oxide scale morphologies of (a) uncoated and (b) Y_2O_3-coated samples after 90 h of oxidation at 850°C.

The present results are consistent with the numerous studies on the effect of a reactive element on high temperature oxidation of chromia-forming alloys [22, 23]. Hou and Stringer [24] indicated that the surface deposition of Y_2O_3 reduced the base metal oxide (NiO) formation during the initial transient stage on Ni-25%Cr. The reactive elements present on the surface could act as nucleation sites for the initial formation of the chromia nuclei. The oxide scale formed is then constituted of smaller grains and the time required to obtain continuous protective scale is reduced. In addition, different studies have demonstrated the inversion of the scale growth mechanism in the presence of reactive element and attributed this inversion of the predominant diffusing species to the segregation of the reactive ions or of mixed chromium-reactive element oxide along the scale grain boundaries [24, 25]. In this study, the presence of mixed chromium-reactive element oxide ($YCrO_3$) has also been identified but no investigation has yet been carried out to localise this phase.

15.4   Conclusion

This study illustrates that electrolytic deposition is a promising process to obtain thin, uniform and adherent films containing a reactive element such as yttrium on metallic substrate for high temperature oxidation application. During the heat treatment, the yttrium hydroxynitrate film is then transformed into the corresponding oxide, Y_2O_3. Counterdiffusion promotes adhesion of the deposit to the substrate and formation of mixed chromium-reactive element oxide occurs. Oxidation tests indicate the beneficial effect of yttria coating on weight gain, improving selective oxidation of chromium and the formation of a continuous protective chromia scale. Longer oxidation tests as well as cyclic ones are currently in progress to evaluate the long-term effectiveness of the Y_2O_3 coating.

15.5   Acknowledgments

The authors are grateful to F. Jomard, from CNRS Meudon, and to C. Séverac from ICMMO Orsay, for their contribution to SIMS and XPS analysis, respectively. The authors also want to express their gratitude to the team of the Centre Commun d'Analyse (CCA) of the Université de La Rochelle for their technical assistance.

15.6   References

1. S.C. Tsai, A.M. Huntz and C. Dolin, *Mat. Sci. Eng.*, **A212** (1996) 6.
2. F. Czerwinski, J.A. Szpunar and W.W. Smeltzer, *Metall. Mat. Trans.*, **27A** (1996) 3649.
3. M.F. Stroosnijder, *Surf. Coat. Technol.*, **100–101** (1998) 196.
4. J.M. Brossard, J. Balmain, F. Sanchette and G. Bonnet, *Oxid. Met.* **64** (1/2) (2005) 43–61.
5. A. Weber, H. Suhr, H. Schumann and R.D. Köhn, *Appl. Phys.*, **A 51** (1990) 520.
6. S. Chevalier, G. Bonnet, J.P. Larpin and J.C. Colson, *Corros. Sci.*, **45** (2003) 1661.
7. Y. Akiyama, T. Sato and N. Imaishi, *J. Cryst. Growth*, **147** (1995) 130.
8. M.J. Capitan, S. Lefebvre, J.P. Dallas and J.L. Pastol, *Surf. Coat. Technol.*, **100–101** (1998) 202.
9. J. Kipkemoi and D. Tsipas, *J. Mat. Sci.*, **31** (1996) 6247.
10. X. Peng, T. Li and W.P. Pan, *Scripta Mater.*, **44** (2001) 1033.
11. C. Houngninou, S. Chevalier and J.P. Larpin, *Mat. Sci. Forum*, **461–464** (2004) 273.
12. F.J. Perez, F. Pedraza, M.P. Hierro, J. Balmain and G. Bonnet, *Oxid. Met.*, **58**(5/6) (2002) 563.
13. F. Czerwinski and J.A. Szpunar, *Thin Solid Films*, **289** (1996) 213.
14. F. Riffard, H. Buscail, E. Caudron, R. Cueff, C. Issartel and S. Perrier, *Corros. Sci.*, **45** (2003) 2867.
15. J.M. Brossard, J. Balmain, J. Creus and G. Bonnet, *Surf. Coat. Technol.*, **185** (2004) 275.
16. A Switzer, *Am. Ceram. Soc. Bull.*, **66**(10) (1987) 1521.
17. L. Gal-Or, I. Silberman and R. Chaim, *J. Electrochem. Soc.*, **138**(7) (1991)1939.
18. R. Siab, G. Bonnet, J. M. Brossard and J. F. Dinut, *Appl. Surf. Sci.*, **236** (2004) 50.
19. X. Lu, R. Zhu and Y. He, *Surf. Coat. Technol.*, **79** (1996) 19.
20. R. Chaim, G. Stark and L.Gal-Or, *J. Mat. Sci.*, **29** (1994) 6241.
21. S.K. Yeng, *Mat. Chem. Phys.*, **63** (2000) 256.
22. K. Przybylski and G.J. Yurek, *Mat. Sci. Forum*, **43** (1989) 1.
23. S. Chevalier, G. Bonnet, K. Przybylski, J.C. Colson and J.P. Larpin, *Oxid. Met.*, **54**(5/6) (2000) 527.
24. P. Hou and J. Stringer, *J. Electrochem. Soc.*, **134**(7) (1987) 1836.
25. B. A. Pint, *Oxid. Met.*, **45**(1/2) (1996) 1.

Part III

Test methods and service conditions

16

Reliable assessment of high temperature oxidation resistance by the development of a comprehensive code of practice for thermocycling oxidation testing (COTEST)

M SCHÜTZE and M MALESSA,
Karl-Winnacker-Institut der DECHEMA e.V., Germany

16.1   Introduction

In modern high temperature technology, materials play a key role with respect to performance, reliability, safety, economic profit and ecological compatibility. The advances in the development of energy conversion systems (low CO_2 emission fossil fuel-fired power stations, solid oxide fuel cells, waste and bio-mass combustion or gasification, coal conversion, etc.) and in engines for transportation (car engines, catalytic converters, advanced jet engines, etc.) are to the largest extent based on reliable long-term performance of high temperature materials. During operation of such high temperature technologies these materials are subjected to a complex interaction of temperature changes, oxidative and corrosive high temperature attack and mechanical stresses. This interaction determines whether components exhibit premature failure or show reliable and safe long-term performance and it also limits the upper service temperature, which decides the degree of efficiency and hence the economical and ecological performance of such a plant. A key role within this interaction is played by high temperature oxidation and/or corrosion and it is somewhat surprising that, although issues of high temperature corrosion have been dealt with for almost a century, in science and in industry no widely used standards or guidelines exist with regard to reliable testing under such conditions. The international situation with regard to standardisation of test methods characterising high temperature corrosion resistance (including issues of protective coatings and the measurement of their properties at room temperature) is summarised in Table 16.1, as far as this information has become available to the authors. This table shown that there is quite a significant number of standards for testing protective coatings on metallic substrates and for quantifying their mechanical properties (fracture, adhesion, etc.) but there is only a very limited number of test codes for the chemical aspects of high temperature corrosion. A former ASTM standard practice for single static oxidation testing has been discontinued in 2002 for reasons which are not known to the authors. For the testing of non-oxide advanced ceramics a

Table 16.1 Existing standards with relevance to high temperature corrosion

Work item title	Standard
Ceramic coating thickness by probe profilometer	EN 1071-1
Ceramic coating thickness by crater grinding	EN 1071-2
Adhesion of ceramic coatings by a scratch test	ENV 1071-3
Chemical composition of ceramic coatings (EPMA)	ENV 1071-4
Porosity of ceramic coatings (metallography)	ENV 1071-5
Abrasion resistance of ceramic coatings by a micro-abrasion wear test	ENV 1071-6
Measurement of metal and oxide thickness by microscopical examination of cross section	ASTM B487-85
Combined creep and temperature cycling of heating wires (FeCrAl-Alloys)	ASTM B78-90
Combined creep and temperature cycling of heating wires (NiCr and NiCrFe-Alloys)	ASTM B76-90
Standard Practice for simple static oxidation testing (discontinued 2002)	ASTM G54-84
Adhesion (Rockwell indentation test)	CEN/TC 184 EN(V) WI 131
Hardness and modulus by depth sensing indentation	CEN/TC 184 EN(V) WI 152
Hot salt corrosion resistance (VAMAS project)	
Fracture strain by 4-point bending	CEN/TC 184 EN(V) WI 158
Coating thickness by cross section	CEN/TC 184 EN(V) WI 159
Interfacial shear strength tensile testing	CEN/TC 184 EN(V) NWI
Adhesion of coatings by interfacial indentation	CEN/TC 184 EN(V) NWI
Adhesion of multi-pass fatigue testing	CEN/TC 184 EN(V) NWI
Adhesion evaluation by progressive indentation testing	CEN/TC 184 EN(V) NWI
Adhesion (Peel test)	ISO/TC206 WD 20503
Standard guide for high temperature	ASTM WD V3
Oxidation testing of nonoxide advanced ceramics at atmospheric pressures and low gas velocities	
Test method for continuous oxidation test at elevated temperatures for metallic materials	JIS Z 2281-1993
Method of cyclic oxidation testing at elevated temperatures for metallic materials	JIS 2282-1996

working draft exists within the ASTM committee work. For the high temperature corrosion testing of conventional materials, however, only two Japanese standards exist with draft English versions. A recently finished project which was funded by the European Commission under the acronym TESTCORR has provided a draft set of guidelines for discontinuous corrosion testing in high temperature gaseous atmospheres. This set of guidelines has, however, not yet been submitted to any standardisation initiative. As a consequence of this situation an ISO work group 'High Temperature Corrosion Testing' has been founded based on an initiative of Japanese industry in 2002. This work group no. 13 is part of the ISO technical committee 156 'Corrosion Testing'. As a first step of their work, the work group has defined five work items:

1. Thermogravimetric testing
   (work not yet started)
   Definition: *in-situ* mass measurements at elevated temperatures on a single specimen without intermediate cooling.
2. Continuous isothermal exposure testing (Fig. 16.1a)
   (ISO document ISO/TC 156CD21608)
   Definition: single post-exposure mass measurement on a series of specimens without intermediate cooling.
3. Discontinuous isothermal exposure testing (Fig. 16.1b)
   (document ISO/TC 156N1122 will be used as a basis for a combined document of continuous and discontinuous isothermal exposure testing)
   Definition: series of mass measurements on a single specimen with intermediate cooling at predetermined times not necessarily regular.
4. Cyclic oxidation testing (Fig. 16.1c)
   (ISO document ISO/TC 156 NWIP: Corrosion of Metals and Alloys-Test Method for Thermal Cycling Exposure Testing under High Temperature Corrosion Condition for Metalic Materials; this document is based on the results of the COTEST project)
   Definition: series of mass measurements on a single specimen with repeated, regular and controlled temperature cycles.

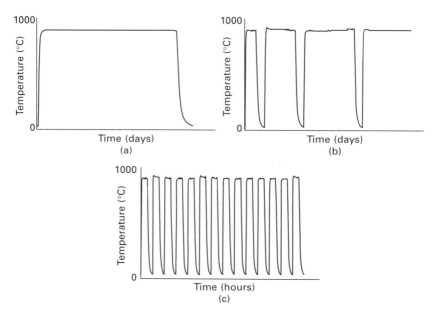

16.1 Temperature profiles in continuous isothermal exposure testing (a), discontinuous isothermal exposure testing (b) and thermal cycling oxidation testing (c).

5. General guidelines for post-exposure examination
    (a draft of a working document is presently under preparation in the ISO
    work group 13).

Among all these tests the cyclic oxidation test has become the most widely used in industry with regard to the number of specimens tested. However, each company and each research institute use their own modification of this type of test with different test parameters so that in the end no cross-comparison of the results from different laboratories is possible (Fig. 16.2). A set of standards or a code of practice which could be used by laboratories does not yet exist. A recent European workshop revealed that industry, in particular, has a very strong interest in the development of a standard for cyclic oxidation testing in order to get reliable and intercomparable data from such tests for design as well as for new alloy development programmes [1]. Owing to the number of parameters influencing materials behaviour under these conditions, the large number of users of this test and the different variants of tests used presently, it would have been impossible for a small group to work on a solution to this problem. The expertise in this field is scattered in industrial and scientific research laboratories all over the world. This was the reason why following the above-mentioned workshop a cyclic testing initiative group was formed whose aim was to work towards the establishment of a suitable standard. It was, however, realised that due to the situation at that time preliminary research was necessary in order to provide a basis for such a set of standards. Being aware of this situation the European Commission

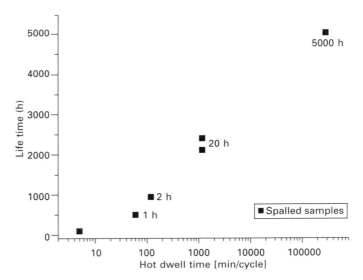

16.2 Life times of thermal barrier coatings as a function of hot dwell time as a test parameter (Data by Dr. Renusch, DECHEMA).

issued a dedicated call 'Measurement and testing – methodologies to support standardisation in community policies' within the Framework V GROWTH programme addressing this specific problem [2]. Following this call the European project 'Cyclic oxidation testing – Development of a code of practice for the characterisation of high temperature materials performance (COTEST)' was started with partners from 11 European countries. The contractors and the subcontractors of the project are listed in Table 16.2. The main technical and scientific objectives of the project which lasted from January 2002 to December 2004 were:

- to quantify the role of the test parameters that lead to scatter between the results of different laboratories
- to develop a reliable and meaningful test procedure for the cyclic oxidation

Table 16.2 Contractors and subcontractors of the COTEST project

Contractors		
DECHEMA (coordinator)	Research Institute	D
CESI SpA	Research and Engineering Company	I
MTU Aero Engines	Aero Engine Manufacturer	D
ARCELOR	Steel producer	F
Ugine-Savoie Imphy	Steel producer	F
ThyssenKrupp VDM GmbH	Manufacturer of Ni-base alloys and high-performance steels	D
Corrosion and Metal Research Institute	Research Institute	S
Technical Research Centre of Finland	Research Institute	FIN
Forschungszentrum Jülich GmbH	Research Institute	D
EU Joint Research Centre	Research Institute	NL
Cranfield University	University	UK
Universidad Complutense de Madrid	University	ES
NPL Mangement Ltd.	National Standards Laboratory	UK
University of Newcastle	University	UK
Alstom Power Ltd	Provider of Power generation solutions	CH
Subcontractors		
University of Liverpool	University	UK
Technical University Clausthal	University	D
University of Birmingham	University	UK
University of Siegen	University	D
DLR	Aerospace Research Centre and Space Agency	D
University of Krakow	University	PL
CEA	Commissariat à l´Énergie Atomique	F
Dunaferr	Steel producer	HU

test with three variants, to account for the most common technical applications

• to draft a code of practice based on the results of the project for submission to ISO/TC 156 as part of the work item 'Cyclic oxidation testing' of the work group 13.

The present paper describes the general concept of COTEST and serves as an introduction to the chapters that follow, which report in more detail about the results.

16.2   The concept of COTEST

The objective of the project was to create the technical basis that will permit the comprehensive definition of all the test-related aspects of the cyclic oxidation test. In order to achieve these objectives the following work packages and tasks were undertaken in two phases:

• Phase I: Evaluation of the effect of cyclic parameters on corrosion behaviour by combining information from the literature and the project contractors with results from an extensive experimental programme.
• Phase II: Development and experimental validation of a draft code of practice derived from the results of phase I and formulation of the final version of the code of practice.

Phase I of the project consisted of the following steps:

1. Evaluation of the presently used test procedures and experimental facilities for cyclic oxidation testing.
2. Evaluation of cyclic oxidation data available in the literature and from the participating laboratories.
3. Development of a set of test procedures, each adjusted to a wide range of industrial applications.
4. Materials procurement and modification of test rigs as required.
5. Testing of a series of reference materials under the test conditions defined.

In phase II the following tasks were accomplished:

6. Development of a draft code of practice for three types of cyclic oxidation testing based on the outcome of phase I.
7. Experimental validation of the code of practice.
8. Formulation of the final version of the code of practice and submission to ISO committee 156.

Each of these steps corresponded to a specific work package as described below.

16.3   Work package 1: Evaluation of the current test procedures and experimental facilities for cyclic oxidation testing

The aim of this work package was to obtain an overview of the test parameters in current and historical test procedures in different labs. This review included published literature as well as the test facilities of the contractors. The parameters addressed included:

- dwell time at oxidation temperature
- dwell temperatures
- cooling time
- temperature accuracy
- heating and cooling rate
- if used, methods used for forced cooling
- specimen handling and fixation
- methods of measuring corrosion rate
- environmental conditions.

The outcome of the work package was used for the definition of the experimental programme in work package 3 which offered the main basis for the final definition of the code of practice.

16.4   Work package 2: Evaluation of existing cyclic oxidation data available from the literature and those supplied by the various project contractors

The work package aimed to summarise the existing data for various groups of metallic materials with special emphasis on finding correlations between the measured corrosion behaviour of the different materials and the cycling parameters used. The output of this work package was needed for the development of the set of test parameters in work package 3 again. The main aim of this evaluation was to

- establish a large database as the starting point for the subsequent test programme
- derive cycling conditions which are relevant for a large variety of industrial applications
- estimate the response of different characteristic groups of metallic materials to variations in cyclic oxidation parameters
- summarise the available scientific background, which permits the description of the effect of changes in test parameters on materials behaviour.

The results of these two work packages are described in detail in Chapter 17.

16.5   Work package 3: Development of a set of test procedures suitable for standardisation

In industrial applications, metallic materials are subjected to different types of thermocycling. Therefore, it would be counter-productive to develop one single standard for the cyclic oxidation testing of all types of materials in all industrial applications. This work package, therefore, aimed at developing three different sets of test procedures, each of which produces data applicable to a range of service conditions, typical of a large number of industrial applications. The three different test types were selected in such a way that together they cover practically the entire range of service conditions in which high temperature materials are subjected to thermocycling. As a specific feature of the tests, standard statistical analysis methods were used to define confidence limits for sets of corrosion data. Statistical considerations were used to make the testing procedures as efficient as possible to produce maximum useful output. Details about the statistical design of the test programme are given in Chapter 18.

Basically three types of tests were addressed:

1. *Long dwell times*. This type of testing aims to simulate conditions in large-scale industrial facilities encountered in applications such as, e.g., in power generation plants, the chemical industry, waste incineration plants and the process industry. In these applications the metallic components are designed for extremely long-term operation, e.g. for typically 100,000 h. Thermocycling of materials occurs due to planned shutdowns, e.g. for regular maintenance or due to unplanned shutdowns as a result of offset conditions. Therefore, the time intervals between various thermocyclic cycles are relatively long and the number of cycles is quite small, e.g. typically around 50, considering the long operation time of the components.

2. *Short dwell times*. This type of thermocycling is typically experienced in applications such as industrial gas turbines, aero engines, heat treatment facilities, furnaces, etc. The intervals between start and shutdown of the facilities are generally much shorter than in applications described in 1. Also the design life and/or the time until complete overhaul/repair (typically 10,000 – 30,000 h) are much shorter and, depending on the specific practical application, the number of cycles is much higher than in the cases described in 1.

3. *Ultra short dwell times*. This type of testing mainly addresses applications of high temperature alloys as heating elements in the form of wires or

foils. Another typical application in which such short cycles prevail would be catalyst foil carriers, e.g. in cars. In such applications the number of cycles is, related to the overall design life (typically several hundred to a few thousand hours), extremely high and the time intervals between heating and cooling can be as low as minutes or even seconds. Such conditions are also commonly encountered in a number of other industrial applications such as, e.g., burners and hot gas filters but also in a large variety of domestic applications where metallic heating elements are used, e.g. in cooking plates, toasters, boilers, dryers, fryers, etc. For this test a special new test rig was developed and constructed by Cranfield University, based on currently existing test equipment in industry which was, however, not regarded as ideal.

The final design of the test matrix using statistical considerations is given in Tables 16.3(a) and (b). This test matrix consisted of five characteristic materials (see Section 16.6), three levels of upper dwell time, two levels of lower dwell time, and a comparison of dry and wet conditions. All in all, 36 possible combinations for each material followed from this test matrix including the selection of trials based on statistical concepts. In addition to the standard tests described so far, issues of mixed gas corrosion testing, deposits corrosion testing, low velocity burner rig testing, and high velocity burner rig testing were also addressed by COTEST.

16.6   Work package 4: Reference materials in the test programme

The number of primary reference materials tested in the COTEST programme was limited to five. This limited number allowed the establishment of a test programme concentrating on an extensive investigation of the test parameters, which was necessary to ultimately to be able to define test standards which are able to produce results relevant to industry. The five primary reference materials were selected such that each one is representative of a specific class of high temperature materials.

1. Ferritic 9–12% Cr steels which are commonly used as construction materials in power plants and in the chemical and process industries

Typical application temperatures range from about 500 to 750°C, and the estimated lifetime of components fabricated out of these materials are in most applications in the range of 100,000 h. As representative material the high strength 9% Cr (+ Mo, W) steel P91 was used which was also a reference material in the TESTCORR programme.

Table 16.3(a) Test matrix for long and short dwell testing;

Material	Upper dwell temperature	Environ-ment	Long dwell testing Upper/lower dwell time (h)		Short dwell testing Upper/lower dwell time (h)	
P 91	650°C	dry	4	4	1	0.25
P 91	650°C	wet	20	4	2	0.25
P 91	650°C	wet	4	4	1	0.25
P 91	650°C	dry	20	4	2	0.25
AISI 441	800°C	dry	4	2	0.5	1
AISI 441	850°C	dry	4	4	0.5	0.25
AISI 441	900°C	wet	4	4	0.5	0.25
AISI 441	800°C	wet	8	4	1	0.25
AISI 441	850°C	dry	8	4	1	1
AISI 441	900°C	dry	8	2	1	0.25
AISI 441	800°C	dry	20	4	2	0.25
AISI 441	850°C	wet	20	2	2	1
AISI 441	900°C	dry	20	4	2	0.25
Alloy 800	950°C	dry	4	2	0.5	1
Alloy 800	1000°C	dry	4	4	0.5	0.25
Alloy 800	1050°C	wet	4	4	0.5	0.25
Alloy 800	950°C	wet	8	4	1	0.25
Alloy 800	1000°C	dry	8	4	1	0.25
Alloy 800	1050°C	dry	8	2	1	1
Alloy 800	950°C	dry	20	4	2	0.25
Alloy 800	1000°C	wet	20	2	2	1
Alloy 800	1050°C	dry	20	4	2	0.25
CM 247	1100°C	dry	4	4	1	0.25
CM 247	1150°C	dry	20	4	2	0.25
CM 247	1150°C	dry	4	4	1	0.25
CM 247	1100°C	dry	20	4	2	0.25
Kanthal A1	1200°C	dry	4	4	1	0.25
Kanthal A1	1250°C	dry	20	4	2	0.25
Kanthal A1	1250°C	dry	4	4	1	0.25
Kanthal A1	1200°C	dry	20	4	2	0.25

Table 16.3(b) Test matrix for ultra short dwell testing

Material	Upper dwell temperature	Upper dwell time (min)	Environment[a]	Geometry
Kanthal A1	1200°C	2		0.7 mm ϕ wire
Kanthal A1	1200°C	5		0.4 mm ϕ wire
Kanthal A1	1200°C	10		70 μm foil
Kanthal A1	1250°C	2		0.4 mm ϕ wire
Kanthal A1	1250°C	5		70 μm foil
Kanthal A1	1250°C	10		0.7 mm ϕ wire
Kanthal A1	1300°C	2		70 μm foil

Table 16.3(b) Continued

Material	Upper dwell temperature	Upper dwell time (min)	Environment[a]	Geometry
Kanthal A1	1300°C	5		0.7 mm φ wire
Kanthal A1	1300°C	10		0.4 mm φ wire
Kanthal A1	1200°C	2	dry	
Kanthal A1	1200°C	5	humidity level 1	
Kanthal A1	1200°C	10	humidity level 2	
Kanthal A1	1250°C	2	humidity level 1	
Kanthal A1	1250°C	5	humidity level 2	
Kanthal A1	1250°C	10	dry	
Kanthal A1	1300°C	2	humidity level 2	
Kanthal A1	1300°C	5	dry	
Kanthal A1	1300°C	10	humidity level 1	
Alloy 800H	950°C	2	dry	
Alloy 800H	950°C	5	humidity level 1	
Alloy 800H	950°C	10	humidity level 2	
Alloy 800H	1000°C	2	humidity level 1	
Alloy 800H	1000°C	5	humidity level 2	
Alloy 800H	1000°C	10	dry	
Alloy 800H	1050°C	2	humidity level 2	
Alloy 800H	1050°C	5	dry	
Alloy 800H	1050°C	10	humidity level 1	

[a]humidity level 1: 4% specific humidity; humidity level 2: 10% specific humidity

2. Ferritic 16–18% Cr stabilised steels which are commonly used in the hottest parts of the automotive exhaust lines such as manifolds or in burners.

Their oxidation resistance relies on the formation of a protective chromia layer, on the good scale adherence and on the creep properties owing to the addition of stabilisers (Ti, Nb). Typical service temperatures are 700–950°C. As testing material in the COTEST programme the 18% Cr containing Ti and Nb stabilised AISI441 grade was used.

3. Austenitic, chromia forming (Fe, Ni)- and Ni-based materials which are commonly used as construction materials, e.g. in the chemical and process industry, gas turbines and aero engines

For oxidation resistance the alloys rely on the formation of protective chromia-based surface scales; however, mostly additional oxide phases, e.g. of the spinel type, are present in the surface scales. Depending on the exact alloy composition, e.g. depending on creep and oxidation resistance of the specific material, these alloys are commonly used in the temperature range 700–

1100°C. In the COTEST project the 32% Ni, 20% Cr containing austenitic steel Alloy 800 was used which had already been a reference material in the TESTCORR programme.

4. Nickel-based, γ'-strengthened super alloys primarily designed for components that have to withstand very high mechanical loads

The main alloying elements in these Ni- or Ni(Co)-base alloys are Cr, Al and Ti. Depending on the specific applications, the alloys contain additional γ'-stabilising elements such as Ta, Nb, and/or solid solution strengtheners such as W, Mo, and Re. The alloys are mainly used as blade and vane materials in stationary gas turbines and aero engines. Typical service temperatures are 850–1100°C. As a reference material in COTEST the material CM247 was tested.

5. FeCrAl-based heating element alloys

FeCrAl-based alloys possess far poorer high temperature strength than the Fe(Ni)- and Ni(Co)-based alloys discussed in (3) and (4). However, the FeCrAl-based materials are frequently used up to ultrahigh service temperatures (1000–1400°C) in cases where mechanical strength is not an important design issue. The excellent high temperature capability of these materials is related to protective alumina scales which form on the surfaces of the materials during high temperature service. Typical examples of applications of FeCrAl alloys are heating elements for industrial and domestic applications, catalyst carriers, burners, filters, and heat exchangers. In the COTEST project Kanthal A1 was tested as a reference material.

The compositions of all these reference materials in the present test programme are given in Table 16.4.

16.7   Work package 5: Experimental investigation of selected materials under cyclic oxidation conditions and evaluation of oxidation behaviour

The main aim of the extensive test programme in this work package was to assess the effect of the cyclic parameters on corrosion behaviour of the five primary reference materials investigated. This work package represented the most extensive part of the COTEST project and many of the results obtained in this work package are reported in other chapters in this book. In the present chapter only a few key aspects of the results will be addressed.

The first is the definition of a thermocycle in thermocyclic oxidation testing. This was based on a scientific approach in [3] where the beginning

Table 16.4 Composition of the reference materials

	P91	AISI 441	Alloy 800H	CM 247	Kanthal A1
Al	<0.005	0.005	0.16	5.81	5.6
C	0.087	0.015	0.063	–	0.026
Co				9.94	
Cr	9.33	17.9	20.98	8.43	21.2
Cu			0.08	0.0102	
Fe	88.03	80.46	45.66	0.0634	72.914
Hf				1.49	
Mn	0.5	0.22	0.63		0.08
Mo	0.93			0.732	
N	0.054	0.02			
Nb	0.061	0.45		0.0104	
Ni	0.38	0.2	30.92	60.45	
P	0.017	0.018	0.016	0.0046	
S	<0.005	0.0027	<0.005		
Si	0.39	0.57	0.35		0.18
Ta				2.78	
Ti		0.14	0.26	0.937	
Tl				0.0066	
V	0.22				
W				9.23	
Zr				0.0404	

of the hot dwell period was defined when the temperature reaches 97% of the final upper dwell temperature. This definition is based on kinetics calculations which show that already at a temperature below the final dwell temperature significant oxidation can take place and that the total amount of initial mass increase during the heating-up period can be represented by assuming that the hot dwell time starts already at 97% of the upper dwell temperature (see [3] for details). The end of the hot dwell period is reached as soon as temperature decreases (usually by removing the furnace from the specimen or the other way round) and the cooling period ends once a temperature of 50°C has been reached. A plot of a temperature profile is given in Fig. 16.3. The latter value has been set somewhat arbitrarily but was based on the assumption that in no laboratory in the world do ambient temperatures exceed 50°C. Another important observation in the tests was that test durations of at least 300 h should be used in order to evidence differences in the spalling or oxidation behaviour of the materials. An example is shown in Fig. 16.4. In this figure the influence of the environment and of the duration of the hot dwell period becomes evident after an exposure time of about 100 h. In the following 200 hours these differences become more marked. In the frame of this work package, the evaluation procedure of the test results was also developed. This procedure which aims at quantifying the oxidation rate constant k_n, the time of protective oxidation behaviour (to

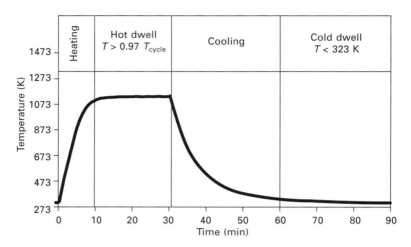

16.3 Temperature profile of a thermal cycle showing the four phases heating, hot dwell time, cooling and cold dwell time; the hot dwell temperature is 1123 K.

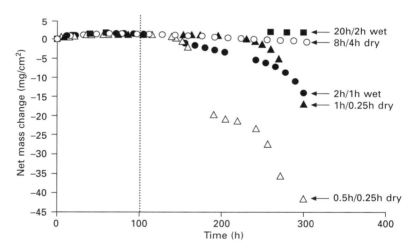

16.4 Example showing the need for a minimum test duration of 300 h accumulated hot dwell time; data show net mass change of Alloy 800H exposed in synthetic air at 1000°C.

the beginning of spalling or breakaway oxidation) $t_{protective}$ and a mass change value determined after test completion describing the oxidation behaviour after the end of the protective period is described in detail in [4]. As the evaluation shows, the k_n values (i.e. the oxidation behaviour in the protective period) do not depend on the cycling parameters (See Chapter 19). This is not a surprising result since in the protective time range oxidation is governed solely by diffusion of the oxide forming species in the scales. This situation changes once the protective time range is exceeded and in particular the values for the mass change at the end of the test very much depend on the cycling parameters as does the length of the protective period $t_{protective}$.

The significance of the influence of the different test parameters on these oxidation values has been evaluated by statistical methods and is described in detail in Chapter 19. It turns out that the cyclic oxidation behaviour described by these parameters depends very much on the type of material used, i.e. whether Fe-Cr-spinels, chromia or alumina scales are formed. This can be regarded as an indication that the test procedure is powerful enough to distinguish between oxide scales of different protective potential.

16.8   Work package 6: Development of a draft code of practice

Based on the outcome of the experimental studies in work package 5, in combination with the evaluation from work package 2, a draft code of practice for thermocyclic oxidation testing was developed, which can be viewed on the internet on the website of the project [4]. This code of practice was formulated along the guidelines used for putting up ISO test standards. Therefore, it addresses the following items:

- scope of the code of practice
- normative references
- definitions
- test apparatus (design, temperature monitoring, gas supply)
- test pieces (size and shape, characterisation prior to testing)
- test method (definition of a thermocycle, types and dwell times of thermocycles, test duration, supporting of test pieces, test environment in air oxidation testing, test parameters in complex corrosive environments, determination of mass change by oxidation)
- post-test evaluation of test pieces (macroscopic evaluation, metallographic cross section)
- reporting.

16.9   Work package 7: Experimental validation of the draft code of practice

All principal contractors and a number of subcontractors were involved in the experimental validation of the draft code of practice defined in work package 6. The tests were organised in such a way that each test was carried out in several laboratories. After testing, the contractors were requested to supply the outcome of the materials behaviour by:

- exact record of all test parameters during the course of the test
- gravimetric data for specimens as function of time
- gravimetric data for spalled oxide as function of time
- metallographic examination of corrosion damage
- characterisation of corrosion products by electron microscopy on selected specimens.

Since all tests had to follow the guidelines developed in the COTEST programme it was expected that no large scatter should occur between the results from the different laboratories. In fact, it turned out at the end of this work package that indeed in most cases there was only very little scatter between results from different laboratories and if so this scatter could be explained mainly by the different cooling rates inherent to the different furnaces. Examples of the good reproducibility of the data in particular in the protective time range are shown in Figs 16.5 and 16.6. However, even the duration of the protective time range and the behaviour in the non-protective time range in many cases showed a reasonable degree of agreement between the different laboratories. It was therefore concluded at the end of work package 7 that the guidelines developed so far are sufficiently powerful for industrial use in order to lead to reliable data with a good degree of intercomparison between different laboratories.

16.10   Work package 8: Formulation and fine-tuning of the final version of the code of practice

At the time of writing, fine-tuning of the code of practice was still continuing. This was mainly directed towards additional details in the test procedure

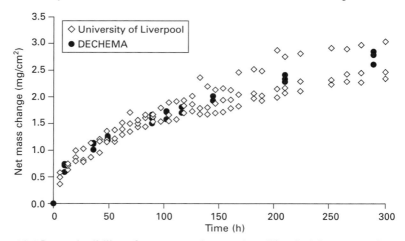

16.5 Reproducibility of net mass change data (Kanthal A1 exposed at 1250°C) collected at different facilities following the code of practice.

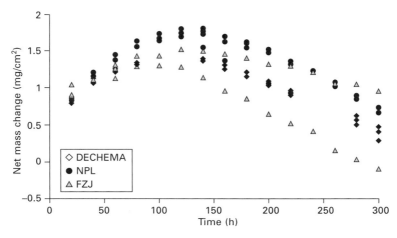

16.6 Reproducibility of net mass change data (Alloy 800H exposed at 1000°C) collected at different facilities following the code of practice.

which were detected as significant in work package 7. All in all, however, there was no need for a significant revision of the draft code of practice developed in work package 6. Work group 13 of ISO technical committee 156 has accepted the final report of the COTEST project as basis for a new work item proposal entitled "Corrosion of Metals and Alloys-Test Method for Thermal Cycling Exposure Testing under High Temperature Corrosion for Metallic Materials". The present draft of this document under work is available under: http://cotest.dechema.de/Draft+ISO+Standards.html/.

16.11   Concluding remarks

In the joint European COTEST project a solid basis has been developed for a reliable and meaningful standard for high temperature cyclic oxidation testing. This basis combines scientific approaches with industrial needs and for the first time allows an intercomparison of data from different laboratories. The draft standard developed in the frame of COTEST can be found on the internet [4]. In the following chapters and in [5], details of the work within the COTEST project and the results are reported.

16.12   Acknowledgement

The COTEST project was funded by the European Commission in the Measurement and Testing Activities of the Framework V programme under the project no. G6RD-CT-2001-00639, which is gratefully acknowledged by all project participants.

16.13   References

1. M. Schütze, W.J. Quadakkers (Eds.), *Cyclic Oxidation of High Temperature Materials,* EFC monograph No. 27, Institute of Materials, London, 1999.
2. http://europa.eu.int/comm/research/growth/
3. G. Strehl, G. Borchardt, *Materials at High Temperature*, in press.
4. The current version of the code of practice for thermal cyclic oxidation testing can be found on the internet under http://cotest.dechema.de/Code_of_Practice.html.
5. M. Schütze, M. Malessa (Eds.), *Standardisation of thermal cycling exposure testing*, EFC monograph No. 53, Woodhead Publishing, Cambridge, 2007.

17

Variation in cyclic oxidation testing practice and data: The European situation before COTEST

S O S G E R B Y*, Alstom Power, UK; Formerly National Physical Laboratory, UK and R P E T T E R S S O N, Swedish Institute for Metals Research, Sweden

17.1   Introduction

Cyclic oxidation testing is a key method to aid material selection and to predict service lifetime of components. However, it is a complex procedure that has many possible variables. Hence it is often difficult to compare data from different laboratories unless either (i) the procedures in each laboratory are similar, or (ii) the influence of any differences in procedure on the resultant data is known.

The extent of the problem is illustrated in Fig. 17.1. This shows the only pre-existing data to be submitted to the COTEST survey for the three iron-base alloys AISI 304, P91 and Incoloy 800H. The use of widely different temperatures, cycle lengths and other experimental parameters means that no direct comparison is possible between data from the two sources, but the data does provide a useful background for parameter selection.

The first task within the COTEST project was thus to survey existing facilities and results in order to establish current practice.

17.2   Methodology

A schematic representation of the test facilities required is shown in Fig. 17.2. The basic requirements of the facility are a zone for heating the specimens, a zone for cooling the specimens (this may be separate from the heating zone but is not necessarily so) and a method to transport the specimens between these two zones. Initial discussions within the COTEST consortium resulted in an agreed list of those parameters that were likely to differ between laboratories/facilities. These were classified into six categories:

* type of layout
* temperature control

---

*This chapter was written whilst Dr Osgerby was employed at the National Physical Laboratory, Teddington, UK.

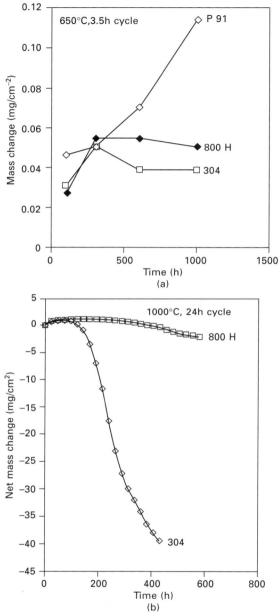

17.1 Pre-existing data reported in the COTEST survey for the iron-base alloys: AISI 304 (18Cr 9Ni), Incoloy 800H (20Cr 31Ni Al,Ti) and P91 (9Cr Mo,V) in dry synthetic air at (a) 650°C and (b) 1000°C. Sources: (a) J P T Vossen, P Gawenda, K Rahts, M Rohig, M Schorr, M Schutze, *Materials at High Temperature* 14 (1997) 387–401. Data reported to COTEST survey by M. Malessa, KWI, Dechema. (b) R Pettersson, L. Liu, S. Hänström, Influence of nickel, silicon and cold work on the oxidation resistance of stainless steels. SIMR report IM-2003-524. Data reported to COTEST survey by R. Pettersson, SIMR.

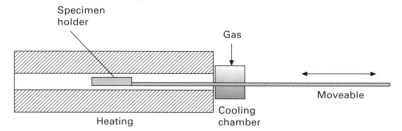

17.2 Schematic representation of cyclic oxidation test facility in horizontal layout.

- heating/cooling practice
- atmosphere
- specimens
- measurement techniques.

At this stage it was also decided to classify cyclic oxidation testing into three classes based upon the time at temperature during each cycle. The definitions were agreed to be as follows:

- long cycle – $t > 3.5$ h
- short cycle – 10 min $< t \leq 3.5$ h
- ultra-short cycles – $t \leq 10$ min

Existing data were requested for the five categories of structural materials to be covered by the COTEST experimental programme:

- ferritic Cr steels (9–12%Cr)
- ferritic Cr steels (16–18%Cr)
- austenitic Ni-Cr steels
- FeCrAl (RE)
- Ni-base alloys

A questionnaire in the form of an Excel spreadsheet was distributed to partners. This requested details of the test procedure and examples of test data. The form of the questionnaire allowed easy collation of returned data into an appropriate database. The database was then analysed to identify differences in test facilities, testing practice and the resultant data.

17.3   Differences in testing practice

Descriptions of 21 test facilities and operating procedures were received. All the required information was not available on every facility but sufficient information was forthcoming to allow conclusions to be drawn. The data supplied were analysed for differences in main parameters listed above, and each category was analysed according to whether the predominant use of the facility was for long, short or ultra-short cycle times.

17.3.1 Summary of layout

The first aspect to consider is the overall layout of the cyclic oxidation facility. The results are summarised in Table 17.1. There is an approximately even split between vertical and horizontal arrangements of furnace and cooling chamber. The 'other' category included a burner rig, a muffle furnace and a direct resistance heating facility for ultra-short cycle testing.

The choice of open or closed system is summarised in Table 17.2. Most facilities are 'closed': an arrangement that prevents any influence of local environment, such as chloride contamination at coastal sites or variations in laboratory humidity, affecting the test data. An 'open' system is one in which the laboratory atmosphere is also the atmosphere inside the furnace.

17.3.2 Temperature control

In most of the facilities, temperature was controlled and monitored by thermocouple. One piece of equipment, a direct heating system for ultra-short cycle testing, was controlled by pyrometer. Type R (Pt:Pt13%Rh) and S (Pt:Pt10%Rh) thermocouples were the most commonly used, although types K (NiCr:NiAl) and B (Pt6%Rh:Pt30%Rh) were also used. This information is summarised in Table 17.3.

The positioning of thermocouples is summarised in Table 17.4. One facility attached a thermocouple to a dummy specimen in the assembly, which is the most accurate way of measuring the surface temperature of specimens and it is perhaps surprising that more laboratories do not employ this practice. The

Table 17.1 Summary of overall layout of test facilities

Orientation	Long cycle	Short cycle	Ultra-short cycle	Total
Vertical	3	7	–	10
Horizontal	2	5	–	7
Other	1	1	1	3
Not specified	1	–	–	1

Table 17.2 Summary of open/closed choice for test facility

	Long cycle	Short cycle	Ultra-short cycle	Total
Open	1	4	–	5
Closed	5	8	–	13
Not specified	1	1	1	3

Table 17.3 Summary of thermocouple type

Thermocouple	Long cycle	Short cycle	Ultra-short cycle	Total
B	1	0	0	1
K	0	3	0	3
R	1	3	0	4
S	3	5	0	8
No thermocouple	0	0	1	1
Unspecified	2	2	0	4

Table 17.4 Summary of thermocouple positioning

Position	Long cycle	Short cycle	Ultra-short cycle	Total
Attached to dummy specimen	0	1	0	1
Adjacent to specimens	4	8	0	12
Furnace controller	1	2	0	3
Unspecified	2	2	0	4

Table 17.5 Thermocouple calibration periods

Calibration period	Long cycle	Short cycle	Ultra-short cycle	Total
Before each test	1	1	–	2
500 h useage	–	1	–	1
6 months	1	1	–	2
1 year	–	6	–	6
2 years	1	–	–	1
Not specified	4	4	–	8

majority of the facilities used thermocouples that were placed near to the specimen assembly whilst three laboratories relied upon the furnace controller. This latter practice is potentially misleading as the furnace tube and specimens will not be in thermal equilibrium for some time, if at all, after the start of any thermal cycle – a calibration exercise is strongly recommended if this practice is used.

The calibration period for thermocouples, Table 17.5, showed a wide spread in practice. The frequency should be governed by the operating temperature and type of thermocouple; therefore it is not surprising that there is a wide spread in the reported calibration periods. Best practice is, of course, to calibrate before each test but experience may be sufficient for laboratories to relax this procedure as data on thermal drift are recorded and analysed. It is somewhat worrying that several laboratories did not specify any calibration period for their thermocouples, all thermocouples require recalibration at intervals and this issue should be addressed in any future standard or code of practice.

The temperature stability that is required is a function of the upper temperature. The guidelines generated during the TESTCORR project [1] on discontinuous corrosion testing are sufficient to define the requirements of cyclic oxidation testing. The data generated during this survey are summarised in Table 17.6 and in general are adequate for this type of testing.

17.3.3 Heating/cooling practice

The majority of the facilities move the specimens from the hot to the cool zone of the test facility. Other methods employed include moving the furnace; opening the furnace door; and programming, or simply switching off, the furnace controller. The information supplied is summarised in Table 17.7.

The methods used to accelerate cooling are summarised in Table 17.8. There is an approximately even split between natural cooling and using a gas

Table 17.6 Temperature stability during hold period

Temperature stability ±°C	Long cycle	Short cycle	Ultra-short cycle	Total
<1	–	1	–	1
1–2	–	5	–	5
2–3	1	–	–	2
3–5	2	5	–	7
>5	1	1	–	2
Not specified	3	1	–	4

Table 17.7 Summary of cycling methods used

Cycling method	Long cycle	Short cycle	Ultra-short cycle	Total
Specimen moved	3	10	–	13
Furnace moved	–	1	–	1
Furnace opened	–	1	–	1
Furnace programmed	1	–	–	1
Furnace switched off	1	–	–	1
Not specified	2	–	1	3

Table 17.8 Summary of cooling methods

Cooling method	Long cycle	Short cycle	Ultra-short cycle	Total
Natural	3	5	1	9
Gas blast	1	6	–	7
Other	3	1	–	4
Not specified	1	1	–	1

blast. Two facilities cycled by means of programming the furnace controller. The type of cooling has a direct influence on the cooling rate that can be achieved and this is the next test parameter to be addressed.

Only ~50% of the responses included information regarding cooling rates and within these data there is a wide spread of values reported (Table 17.9). The slowest cooling rates were generated using programmed furnace controllers; however, it should be noted that in these tests the slow cooling rates were intended.

17.3.4 Gas flow

Where closed systems are used, the laboratories utilised bottled gas mixtures, and the reported gas flow rates in such cases are summarised in Table 17.10. As with temperature stability, the recommendations developed in the TESTCORR project should be adopted directly for cyclic oxidation testing, i.e. a linear flow rate of 1–10 mm s^{-1}. Where the tests are carried out in laboratory air in open furnace systems, the atmosphere is static apart from convection effects.

17.3.5 Specimens – geometry, preparation and handling

The majority of laboratories used rectangular specimens. Cylindrical specimens were used in short cycle tests, particularly for coatings where edge effects were considered to be important. One laboratory tested small components, as well as specimens in their facility. The data are summarised in Tabe 17.11.

Table 17.9 Instantaneous cooling rates

Cooling rate (K s^{-1})	Long cycle	Short cycle	Ultra-short cycle	Total
0.01–0.1	1	–	–	1
0.1–1	–	1	–	1
1–10	1	1	–	2
>10	–	3	–	3
Not specified	5	7	1	13

Table 17.10 Summary of gas flow rates

Flow rate (m s^{-1})	Long cycle	Short cycle	Ultra-short cycle	Total
$10^{-5} - 10^{-4}$	–	1	–	1
$10^{-4} - 10^{-3}$	–	1	–	1
$10^{-3} - 10^{-2}$	2	2	–	4
Not specified	4	5	1	10
Static/open systems	1	4	–	5

Table 17.11 Specimen geometry

Geometry	Long cycle	Short cycle	Ultra-short cycle	Total
Rectangular	5	8	–	13
Cylinder	–	4	–	4
Disk	2	–	–	2
Other	–	1	1	2
Not specified	–	–	–	–

Table 17.12 Range of specimen surface areas used

Surface area (mm^2)	Long cycle (7 facilities)	Short cycle (13 facilities)	Ultra-short cycle (1 facility)
Maximum	680	3117	189
Minimum	225	280	189

Table 17.13 Specimen surface finish

Final finish	Long cycle	Short cycle	Ultra-short cycle	Total
100/120 grit	–	3	–	3
180 grit	–	1	–	1
600 grit	2	3	–	5
1200 grit	3	1	–	4
2500 grit	1	–	–	1
1 micron polish	1	1	–	2
As-received	–	4	1	5

Specimen size/surface area is influential in controlling the cooling rate during thermal cycling. A wide range of specimen sizes are used, and the range of surface areas are given in Table 17.12. This test parameter is likely to affect the resultant data significantly and must be addressed in any code of practice for cyclic oxidation testing.

Specimen preparation is recognised as an important test variable. The surface finishes used by the different laboratories are summarised in Table 17.13. Coated specimens and components are tested in the as-received condition, whilst uncoated specimens normally undergo some preparation, the final finish being dependent upon material and application.

Ultrasonic cleaning using an organic solvent such as iso-propanol is the most popular method of degreasing, although several laboratories did not specify any method. One laboratory used a method that had been developed in the EU-funded LEAFA project [2], which it is claimed produces a very reproducible surface. The questionnaire returns are summarised in Table 17.14.

Table 17.14 Specimen cleaning procedure

Method	Long cycle	Short cycle	Ultra-short cycle	Total
Ultrasonic with organic solvent	4	5	–	9
Organic solvent without ultrasonic agitation	–	2	1	3
Other	1 (LEAFA)	–	–	1
Not specified	1	6	–	7

Table 17.15 Specimen support

Method	Long cycle	Short cycle	Ultra-short cycle	Total
Hook	3	1	–	4
Crucible	3	6	–	9
Cage	–	3	–	3
Other	–	1	1	2
Not specified	1	2	–	3

A wide range of methods to support the specimens were used, summarised in Table 17.15. The materials used to construct these supports were typically quartz, platinum or glass fibre for hooks alumina or quartz for crucibles; and metallic alloys (ODS or FeCrAl) for cages. The method used to support the specimens affects the thermal mass of the assembly and thus will influence the cooling rate during thermal cycling. During high temperature exposures it is also possible that the presence of silica [3] or platinum [4] may affect the oxidation kinetics of the material under test.

17.4   Cyclic oxidation data

17.4.1 Testing and evaluation parameters

A total of 131 datasets were reported for materials falling within the categories of structural materials covered by COTEST. The distribution of cycle duration at temperature for all the datasets is shown in Fig. 17.3. This falls well into line with the defined categories of ultra-short ($\leq 10\,min$), short ($10\,min < t \leq 8\,h$) and long cycle ($>8\,h$).

The number of cycles employed is shown in Table 17.16. The median values of the time at temperature were 636 h for long cycle testing, 930 h for short cycle testing and 500 h for ultra-short cycle testing. This gives an indication of the test duration that the participants have found necessary to assess the cyclic high temperature performance of materials, and as such provides a basis for future recommendations.

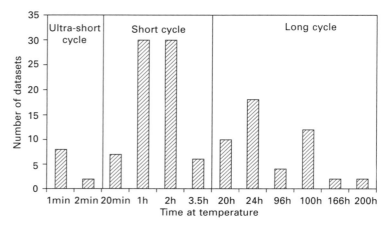

17.3 Cycle duration for all 131 reported data series and division into categories.

Table 17.16 Number of cycles employed for submitted data sets

Number of cycles	Long cycle	Short cycle	Ultra-short cycle	Total
1–9	8	–	–	8
10–99	31	11	–	42
100–499	3	22	–	25
500–999	1	16	–	17
> 1000	5	24	10	39
Total	48	73	10	131

Table 17.17 Measurement techniques used for evaluation of submitted datasets. A number of sets include several evaluation methods

Measurement technique	Long cycle	Short cycle	Ultra-short cycle	Total
Mass change (net)	34	54		88
Mass change (gross)	22	28		50
Mass change (after descaling)	–	7		7
Metallographic section	15	34	8	57
Rate constant only	2			2
End point only	8		10	18

The techniques that were employed for evaluation of the cyclic high temperature oxidation data are shown in Table 17.17. By far the most common method was recording of the net mass change as a function of time by intermittent weighings. Over half of the datasets also involved collection of the spalled scale to permit separate recording of the gross mass change. There were no instances in which mass change was recorded for ultra-short cycle testing; the usual parameter in this case was the number of cycles to failure of thin specimens. Metallographic evaluation of cross-sections was employed in an appreciable number of cases for all three types of testing. Post-test evaluation often also included determination of the scale composition in the SEM.

17.4.2 Materials and environments

The submitted data was fairly well distributed between materials categories, (Table 17.18). Because of the relative paucity of data for the ferritic steels, the two original categories, 9–12%Cr and 16–18%Cr, were combined. The largest group, almost half of the submitted datasets, was for nickel-base alloys.

More than half of the reported tests were carried out in laboratory air without facilities for controlling humidity (Table 17.19). This means that large variations in water vapour level are to be expected depending on the geographic location of the site, the time of year and the use of air conditioning/heating in the laboratory. These factors make this type of testing unsuitable for standardisation. Some laboratories used bottled synthetic air, which was used either dry or after controlled humidification. Only two corrosive gas mixtures were reported: Ar+5%HCl and Ar-5%H$_2$-1%H$_2$S. Each of these was used for a number of tests in the same two laboratories.

Table 17.18 Summary of reported testing data, by material type and cycle length

	Austenitic	Ferritic	FeCrAl (RE)	Ni-base	Total
Ultra-short			2	8	10
Short	15	9	3	46	73
Long	8	6	25	9	48
Total	23	15	30	63	131

Table 19.19 Summary of reported data, by material type and environment

	Austenitic	Ferritic	FeCrAl (RE)	Ni-base	Total
Dry air	13	4	8		25
Lab air	6	5	12	49	72
Moist air			6	4	10
Corrosive	4	6	2	10	22
Unspecified			2		2
Total	23	15	30	63	131

17.4.3 Variability of results

An understanding of the normal variability between replicate samples forms a vital background against which the effect of experimental variables should be assessed. One multiple data set submitted to the COTEST survey is shown in Fig. 17.4. This indicates a relatively small variation in the oxidation rate, as seen at 850°C. However, the time to spallation, seen in the data at 950°C, shows much larger specimen-to specimen variability.

17.4.4 Influence of experimental variables

Intercomparison between data from different sources was in most cases not possible because of the simultaneous variation in a number of test parameters. However, data from a single source was amenable to analysis since several closely allied datasets were frequently reported. The primary variables investigated were alloy composition and temperature, but systematic studies were also reported on the effect of the following variables:

- Specimen thickness for FeCrAl(RE) alloys. This is illustrated in Fig. 17.5. and is due to more rapid exhaustion of the matrix aluminium for thinner specimens.

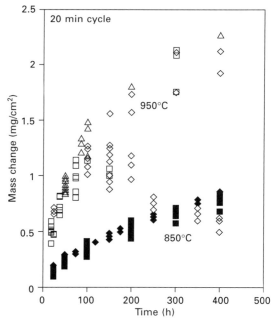

17.4 Results of multiple experiments on a ferritic 441 steel in laboratory air at two temperatures, showing extent of specimen-to-specimen variability.

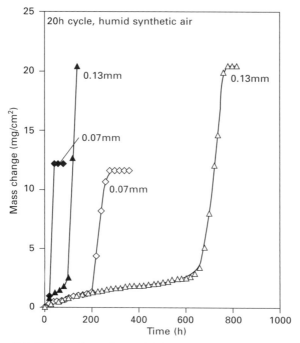

17.5 Effect of specimen thickness on time to breakaway for FeCrAl(RE) alloys at 1300°C (solid points) and 1200°C (open points).

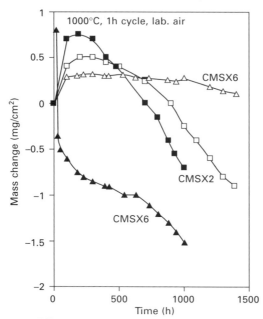

17.6 Effect of natural cooling (open points) or gas blast cooling (solid points) on cyclic oxidation of two nickel-base alloys.

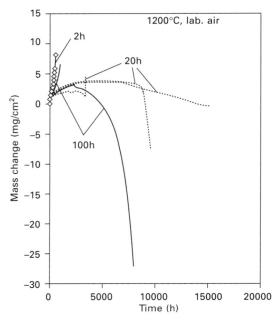

17.7 Effect of cycle duration on cyclic oxidation of PM2000.

- The use of natural or enforced cooling. The data for two nickel-base alloys in Fig. 17.6 indicates that spallation is promoted by rapid cooling, which is expected to create larger thermal stresses in the oxide.
- The effect of cycle duration for FeCrAl(RE) alloys, (Fig. 17.7). The effect cannot be unequivocally identified on the basis of the submitted data because of large specimen-to-specimen variation.

17.5   Conclusions

The results of these surveys underline the necessity of the COTEST project because little of the reported data from different sources is amenable to intercomparison. The role of testing variables on the outcome of cyclic oxidation testing at the time was not sufficiently clear, thus demonstrating the need for standardisation of testing procedures.

17.6   Acknowledgements

This work was part of the EU-funded project 'COTEST' contract number G6RD-CT-2001-00639. The contributions of the COTEST partners, who supplied information on their test facilities and resultant data; and Mr Tony Fry at NPL, who developed the questionnaire spreadsheet and database, are gratefully acknowledged.

17.7   References

1. A B Tomkings, J R Nicholls and D G Robertson, Discontinuous Corrosion Testing in High Temperature Gaseous Atmospheres, TESTCORR, Final Report EUR/19479, European Commission, 2001.
2. J R Nicholls and M J Bennett, in *Lifetime Modelling of High Temperature Corrosion Processes*, European Federation of Corrosion Publication No. 34 (2001), 3–14.
3. J E Rhoades-Brown and S R J Saunders, *Corrosion Science* 20 (1980) 457–460.
4. J R Nicholls, Discontinuous Measurements of High Temperature Corrosion, in *Guidelines for Methods of Testing and Research in High Temperature Corrosion* (eds H J Grabke and D B Meadowcroft), Institute of Materials, London, 1995, 11–36.

Designing experiments for maximum
information from cyclic oxidation tests and
their statistical analysis using half
normal plots (COTEST)

S Y C O L E M A N, University of Newcastle upon Tyne, UK
and J R N I C H O L L S, Cranfield University, UK

18.1　Introduction

Cyclic oxidation testing at elevated temperatures requires careful experimental
design and the adoption of standard procedures if reliable data are to result.
This is a major aim of the 'COTEST' research programme. Further, as such
tests are both time consuming and costly, in terms of human effort, to take
measurements over a large number of cycles, it is important to gain maximum
information from a minimum number of tests (trials). This search for
standardisation of cyclic oxidation conditions leads to a series of tests to
determine the relative effects of cyclic parameters on the oxidation process.
Standard oxidation conditions are required to develop a code of practice for
the characterisation of high temperature materials performance.

Applications for materials at high temperatures are diverse. They can
include materials for boilers, superheaters, heat exchangers, steam turbines,
industrial turbines and chemical process plant that operate hot for long service
intervals and are only shut down for maintenance or plant outages. This class
of service is characterised by long dwell times at high temperature. Short
dwell times are typical of aerogas turbine operation on short haul flights,
while ultra-short dwell times characterise heating elements in furnaces,
industrial plants, domestic heaters, toasters and other consumer products. A
representative set of alloys for such diverse applications was identified.

The test conditions could be summarised by a small number of factors
that would appear at a limited number of levels. To allow maximum information
to be obtained from the tests, it was decided to use statistically balanced test
matrices (orthogonal designs). For comparison purposes each design included
tests at 'standard settings'. The statistical designs ensured high levels of
replication for all factors leading to precise estimates of their effects. The
designs were balanced so that as far as possible each factor was tested across
the range of other factors, diminishing the effects of bias.

Initial data analysis provided some idea of the sample sizes required to
show significant effects of factors. However, in most cases these sample

sizes were larger than could sensibly be accommodated within the COTEST test matrix. Thus, some compromise on the size of the test matrix had to be made. A general test set of nine trials each with three replicate specimens was chosen for most cases of short and long dwell experiments. Due to additional factors being of interest (e.g. shape of specimen, wire or ribbon), 18 trials were required for ultra-short testing.

18.2   Experimental method and response summaries

One of the issues to arise from the data analysis was how to summarise the data so that results from different tests could be meaningfully compared. Mass change per unit area at several time points is a commonly used method for presenting high temperature oxidation/corrosion data and was provided by most partners as the output from their tests. In the COTEST test matrix, the number of time points varied from 2 to 4096 depending on the particular material and test condition. The use of the NASA 'winCOSP' programme (Smialek and Auping, 2002) was reviewed as a method of analysing the cyclic oxidation in the COTEST programme, but was not adopted in the end. Basically the program allows the user to enter their data and then fit a cyclic oxidation weight change curve by trial and error based on some elementary inputs and assumptions regarding model scale growth and spalling behaviour. Because there may be a number of competing curves that approximately fit the data, the method does not give a definitive set of responses. This would be problematic for a consortium to use and so 'winCOSP' was not adopted. The methodology for summarising mass change curves was agreed by all COTEST partners and is described in detail in Chapter 19 and outlined below.

It is assumed that the oxidation process can be approximated by a power law (parabolic if $n = 2$). The function $\Delta m = k_n * t^{(1/n)}$ was fitted to each dataset of net weight gain values to get the k_n value. (Note: $k_n = k_p$ (the parabolic rate constant) if $n = 2$: in fitting this data, the assumption is that these points correspond to oxidation without spallation.) This is illustrated in Fig. 18.1.

The 'time to spall' is defined as the time when the net and gross weight gains start to be 'different'. (For all specimens oxidised at 1050°C, there was no doubt about the definition of this time, as the differences in weight gains were clearly visible.) This is illustrated in Fig. 18.2.

The 'diff. Δm' value is defined as the difference between net and predicted gross weight gain (fitted through data less than the measured time to spall) for the last weight measurements for each specimen. This is illustrated in Fig. 18.2.

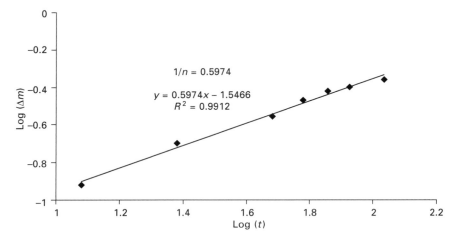

18.1 Evaluation of n value based on fit of the straight line to first seven points in Log (Δm) vs Log (t) plot.

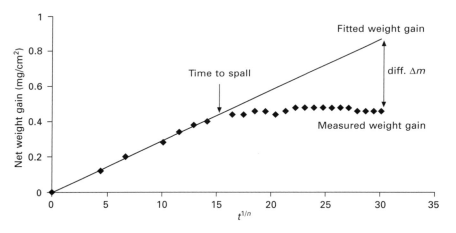

18.2 Evaluation of time for start of spallation and total weight of spalled scale (diff. Δm).

The key response values are therefore:

- the mass change curve
- gradient, n
- rate constant, k_n
- time to spall, t_s
- difference between fitted and measured weight gain, Δm.

18.3   Influential parameters

Following a review of the available literature, databases and the experience of partners to the COTEST project, the most influential parameters, upper dwell temperature (oxidation temperature) and time (hot time), lower dwell time (cold time) and environment, were investigated in partners' laboratories. It was decided to test upper dwell temperature at three levels, and equidistant from a reference temperature; to test upper dwell time at a reference, a higher and a lower time; to test lower dwell time at a reference and a higher time; and to test in wet and dry environments.

18.4   Design of experiments, results and operational observations

It was decided that four parameters or factors would cover the range of service conditions. Upper dwell time and temperatures were tested at three different values, lower dwell time at two different values and with two different environments. The number of combinations of conditions is therefore 36. A quarter fractional nine trial design necessarily has some 'confounding' and loss of information compared to a full factorial design, but retains the analysis of main effects and some synergies or interactions between factors.

An experiment, consisting of nine trials, was designed according to statistical criteria. The aim was to ensure that the results of the trials could be analysed statistically, to test the main linear and quadratic effects of upper dwell temperature and hot time and the main effects of lower dwell time (cold time) and environment.

There are many experimental designs and they are described in statistical texts such as Montgomery (2005), Wu and Hamada (2000) and Grove and Davis (1997). A balanced design can be constructed to test the main effects of four 'three-level factors' in nine experimental trials. It is possible to assign two-level factors (linear) to a three-level array using a method referred to as column collapsing (Grove and Davis, 1997, p. 177). Thus a four-factor mixed level design is fitted into nine trials. The modification that two factors have three levels and two factors have only two levels upsets the balance slightly but allows for a better estimate of experimental error.

The design can be written in two ways: as a Taguchi style experiment (see, for example, Breyfogle, 2000) as shown, in Table 18.1 and as a modified Graeco Latin square (see, for example, Cochran and Cox, 1957) as shown in Table 18.2. Notice that each row and column in Table 18.2 has equal numbers of 2 h and 4 h lower dwell time and wet and dry environments. Other designs (see, for example Metcalfe, 1994, and Wu and Hamada, 2000), such as central composites and star designs exist but would require more trials to obtain the same balance (see Appendix III).

Table 18.1 Experimental design expressed as a Taguchi design

Trial	Temperature(°C)	Upper dwell (h)	Lower dwell (h)	Environment
1	950	4	2	Dry
2	1000	4	4	Dry
3	1050	4	4	Wet
4	950	8	4	Wet
5	1000	8	4	Dry
6	1050	8	2	Dry
7	950	20	4	Dry
8	1000	20	2	Wet
9	1050	20	4	Dry

Table 18.2 Experimental design expressed as a Graeco Latin square design

Upper dwell/Temp(°C)	950	1000	1050
4	2 h, dry	4 h, dry	4 h, wet
8	4 h, wet	4 h, dry	2 h, dry
20	4 h, dry	2 h, wet	4 h, dry

Table 18.3 Example random orders to run the trials of the experiment

Order	Order	Order	Order	Order	Order	Order	Order	Order
9	7	5	8	1	1	5	3	2
3	8	9	2	5	8	3	7	1
4	3	7	4	8	4	7	9	5
6	5	4	5	3	6	2	4	3
2	9	6	6	4	5	4	2	7
7	1	8	9	7	3	9	5	6
5	2	2	7	9	2	1	6	8
8	6	3	3	2	9	6	8	4
1	4	1	1	6	7	8	1	9

The experimental procedure requires attention to a number of important aspects.

- A pilot trial should be run to see if all the parameters can be controlled.
- The order of the trials, should be randomised.
- The position of each specimen in the furnace should be randomised and its position recorded.

There are $9 \times 8 \times 7 \times 6 \times 5 \times 4 \times 3 \times 2 \times 1 = 362,880$ different random orders of nine trials. Most will avoid time trends biasing the experiment. A different order can be used for each experiment in each site. Some suitable random orders, which have been checked to avoid bias, are given in Table 18.3.

18.5   Analysis of results for Alloy 800

There are a number of ways of analysing the experimental data. The first step is to visualise the results by examining the mass change curves (see Chapter 19). Values for the four output summaries (n, k_n, t_s, and Δm) are calculated for each specimen in each trial. The data summaries are given in Appendix I.

18.5.1 Main effects plots

The result summaries can be examined by main effects plots which show the mean value of each output at each level of each factor. When there are replications, as here, it is also possible to investigate the effect of the parameters on the variability of the responses. The standard deviation of the responses for the three replications represents the variability of the output. The log transformation of the standard deviations is taken to make sure that the assumptions of approximate normality and constant variability are obeyed. Although these assumptions are not necessary for main effects plots, they are necessary for the statistical analysis described below.

For brevity, only the main effects plots for mean time to spall (t_s), mean diff (Δm) and sd diff. (sd(Δm)) are shown in Figs 18.3–18.5.

In Fig. 18.3 the effect of higher upper dwell temperature is to decrease the time to spall. The effect of lower cycle time at upper dwell temperature, i.e. increasing the number of cycles, is to decrease the time to spall. This is in line with the more stressful effect of a shorter oxidation cycle. There is no apparent effect of lower dwell time or environment on the time to spall.

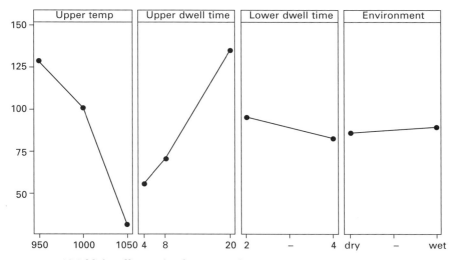

18.3 Main effects plot for mean time to spall, t_s.

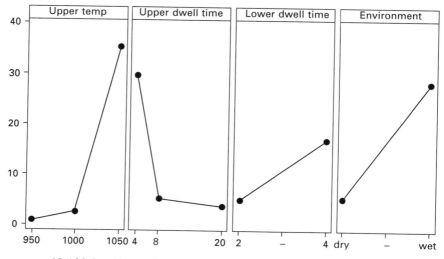

18.4 Main effects plot for mean difference, Δm.

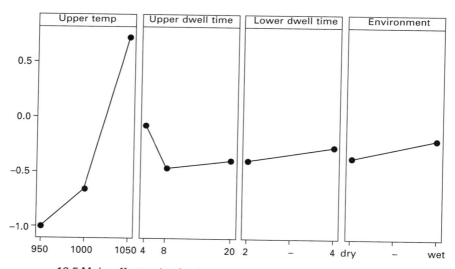

18.5 Main effects plot for log (standard deviation) of Δm.

It can be seen in Fig. 18.4 that higher temperature increases Δm as does the lowest upper dwell time (i.e the largest number of cycles for a given exposure). The Δm value is increased for a 4 h lower dwell time and for a wet environment, that is to say that the propensity of Alloy 800 to spall increases with increase in oxidation temperature, increased number of cycles, the longest cold dwell time and when water vapour is present in the test environment.

The main effects plot for the log (standard deviation) of Δm is given in Fig. 18.5. It can be seen in Fig. 18.5 that higher temperature increases the variation in Δm. Thus there is more scatter between the quantities of spall from Alloy 800 as the temperature increases. There is little or no apparent effect from the other factors.

18.5.2 Analysing for statistical significance using ANOVA and regression

The main effects plots show how the parameters affect the response variables. They do not, however, confirm which effects are statistically significant. In any plot, one factor will always have the greatest effect, but if it is not statistically significant then the apparent effect may not be real and therefore may not be repeatable.

Analysis of variance is the usual technique for analysing the results of designed experiments (see, Nicholls and Hancock, 1983, for examples of analysis of variance applied to oxidation and corrosion data). However, where interest is focussed on many different aspects of the data, then there may not be sufficient degrees of freedom left to enable an ANOVA. Regression is an alternative approach but this also requires sufficient degrees of freedom.

An alternative approach, which can be used when there are no or only one or two degrees of freedom, is to use the graphical method of 'half normal plotting' to show which effects are statistically significant. The results for Alloy 800 are analysed by half normal plots instead of analysis of variance or regression as there are only two degrees of freedom for the experimental error variance. Similar to ANOVA and regression, analysis by half normal plots still requires that the data are independent, approximately normal and with constant variance.

18.5.3 Analysing for statistical significance using half normal plots

Half normal plots give a visual indication of which factors are statistically significant. The technique is useful when there are few or no degrees of freedom available for a residual mean square in ANOVA or regression. Half normal plots are therefore useful when the experimental design is saturated and all effects are of interest. The half normal plot is a type of probability plot where a numerical value for the factor effects is plotted on the vertical axis against the expected normal order statistics on the horizontal axis (see Grove and Davis, 1997).

Factor effects are basically averages and differences of experimental results. If there are no statistically significant effects and the underlying experimental error is normally distributed, the factor effects will all be normally distributed with zero mean. If any factors are statistically significant then their factor effects will not be zero and their factor effects will show up as particularly large values on the half normal plot.

As the standard test matrix has been followed, the effects diagrams in Figs 18.3 to 18.5 above will give the same message as half normal plots. The advantage of half normal plots is that they show which effects are statistically significant. The numerical values of the factor effects for the half normal plot need to be found from orthogonal contrasts.

18.6   Orthogonal contrasts

In this experiment, there are nine trials; therefore up to eight independent comparisons can be made, for example the comparison of results for higher, medium and lower temperature. A comparison can also be thought of as testing the effect of a factor. The variation in the nine sets of results can be broken down into its constituent parts using sets of orthogonal contrasts. Each orthogonal contrast is a column of nine values summing to zero. An orthogonal array is a set of orthogonal contrasts with the property that the vector product of any two contrasts is zero. The vector product of an orthogonal contrast with the results column is called the contrast value. If the orthogonal contrast corresponds to the changing levels of a factor in the nine trials of the experiment, then the standardised[1] contrast value gives a numerical value for the factor effect (Cochran and Cox, 1957).

The orthogonal contrasts have the property that they completely partition the total variation in the results. The sum of squared factor effects is equal to the total sum of squares (see, for example, Cochran and Cox, 1957).

If a set of non-orthogonal contrasts is applied to the results, then the constituent parts will not sum to the total variation and they will not be independent of each other. The experimental design used here can be analysed in a number of ways corresponding to different sets of orthogonal contrasts. This is explored further in Appendix II.

Allowing for the full effect of the three level factors, a useful set of orthogonal contrasts leads to an assessment of

- four linear main effects
- two non-linear main effects
- two representations of experimental variation.

Because these contrasts are orthogonal, each effect can be evaluated independently of any other. Table 18.4 represents the different levels of the factors. Temp and Qtemp stand for the linear and quadratic effects of upper

---

[1]Standardisation is achieved by dividing by the square root of the sum of squared contrast cells. For example, if the contrast column is $(-1,0,1,-1,0,1,-1,0,1)^T$ then divide by $\sqrt{6}$.

dwell temperature, respectively, Hot and Qdwell stand for the linear and quadratic effects of upper dwell time, respectively, Cold stands for the effect of lower dwell time being either 2 h (2) or 4 h (-1) and Env stands for the effect of dry (-1) or wet (2) environment. Error 1 and Error 2 do not correspond to any systematic changes in cycling parameters and therefore represent random error. The set of contrasts is discussed further in Section 18.9 and Appendix II.

18.7   Synergies

Synergies between factors are when the effect of one factor depends on the level of the other factor. For example, it may be that the effect of environment is greater at higher temperatures than at lower temperatures. In statistical experimental design, synergies are referred to as interactions between factors.

The number of possible comparisons that can be made using the results of the nine trials is large, consisting of

- effects of factors, e.g. environment
- linear and quadratic effects of three-level factors, e.g. upper dwell time
- interactions between factors, e.g. environment and lower dwell time
- interactions between linear and quadratic effects of factors.

There are only nine trials in this experiment and interactions can only be examined if it is assumed that some of the other effects are non-significant. If the three-level factors can be assumed to have a linear effect only, then the interaction between some factors can be examined although the contrasts are not independent. There are only four degrees of freedom available for interactions even if the three-level factors are assumed to be linear. Alternative sets of comparisons available for the experimental design are:

- linear and quadratic effects of all four factors
- linear and quadratic effects and all interactions for the three-level factors only
- linear effects and some interactions for all four factors

Some interactions can be chosen *a priori* and tested. For example, we may be interested in the interaction between upper dwell time and environment and upper dwell time and lower dwell time. In this example, for time to spall, t_s, the quadratic terms are non-significant and so it is reasonable to test some interactions.

The experimental design used here is very concentrated in that it allows a lot of information to be extracted from a small number of trials. As explained, the disadvantage is that not all interactions between factors may be examined and there may be interactions between factors that bias the results.

18.8   Method of half normal plotting

The sets of contrast values have to be standardised so that they have the same standard deviation before they can be plotted in a half normal plot. Once the standardised contrasts have been found, they are placed in ascending order *without regard to their sign*. They are then plotted on the vertical axis against the expected half normal scores on the horizontal axis. Half normal scores are available in Minitab Statistical software or elsewhere, for example Grove and Davis (1997).

For eight contrasts, the half normal scores are:

0.077, 0.232, 0.394, 0.567, 0.760, 0.986, 1.281, 1.772

The set of orthogonal contrasts in Table 18.4 applied to the results for mean time to spall (see Appendix I for data) were used to produce the half normal plot in Fig. 18.6. For example, the standardised contrast value for 'hot' is 119. This is the vector product of the contrast for 'hot' and the results column divided by $\sqrt{6}$.[2]

The plot is interpreted by drawing a straight line through the origin and the lower few points. All points above this line correspond to effects that are probably significant. The linear effect of temperature and hot dwell time are significant as the points appear off the line.[3] A half normal plot can be constructed for any of the summary outputs. The plot in Fig. 18.7 is for Δm.

Table 18.4 Set of orthogonal contrasts for the experiment

Trial	Temp	Hot	Cold	Env	Qtemp	Qdwell	Error1	Error2
1	−1	−1	2	−1	1	1	0	−1
2	0	−1	−1	−1	−2	1	−1	1
3	1	−1	−1	2	1	1	1	0
4	−1	0	−1	2	1	−2	−1	0
5	0	0	−1	−1	−2	−2	1	−1
6	1	0	2	−1	1	−2	0	1
7	−1	1	−1	−1	1	1	1	1
8	0	1	2	2	−2	1	0	0
9	1	1	−1	−1	1	1	−1	−1

---

[2]As a check, the sum of the eight squared standardised contrast values should equal the corrected sum of squares of the nine results (i.e. standard deviation^2 × 8), provided all eight contrasts are orthogonal.

[3]Note that the orthogonal contrast for 'Qdwell' assumes that the levels are equally spaced whereas the factor levels are actually 4, 8 and 20 h which are not equally spaced. This will tend to bias the contrast value so that Qdwell is larger than it should be. The half normal plot should be looked at in conjunction with the main effects plot. The main effects plot for time to spall shows that the effect of upper dwell time is approximately linear which supports the lack of significance in the half normal plot in Fig. 18.6.

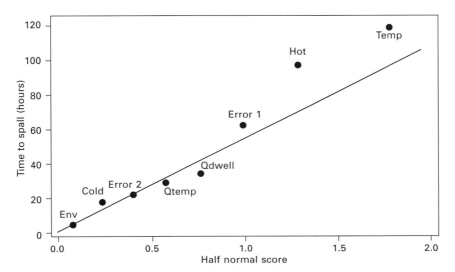

18.6 Half normal plot for mean value of time to spall.

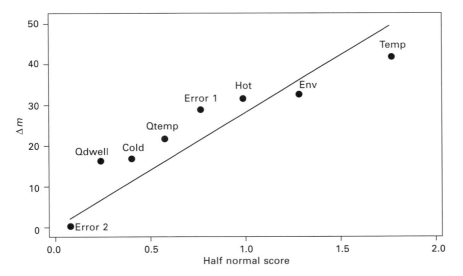

18.7 Half normal plot for mean value of Δm.

In the plot of time to spall, the effects of hot dwell time and temperature are clearly significant as they are off the line. However, in the plot for Δm none of the effects are off the line. This implies that none of the effects are significant as regards Δm.

Half normal plots can also be used to look for any effects of the factors on the variation in results. Figure 18.8 is the half normal plot for the logged standard deviation of Δm.

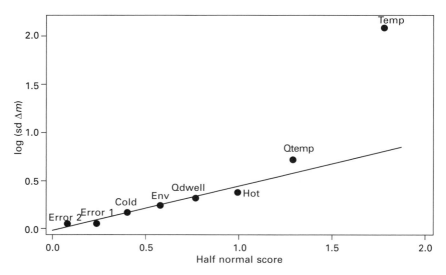

18.8 Half normal plot for log (standard deviation) of Δm.

In the measured standard deviations of the differences between measured net mass gain and that predicted, the factors upper dwell temperature and non-linear effect of upper dwell temperature are significant, demonstrating that the observed increase in scatter in the spall data is related to temperature. This suggests that a log transformation of the Δm should be used to stabilise the variance.

Figure 18.9 shows the half normal plot for the mean values of log Δm. It can be seen that the effects of temperature and hot dwell time are significant on the log of Δm whereas there were no significant factors for the Δm values.

18.9   Prediction of oxidation trends

It is useful to obtain a predictive equation for the responses so that responses for intermediate factor levels can be predicted with a confidence interval. This can be done using ANOVA[4] or regression if some of the factors can be assumed to be non-significant so that an error term can be estimated. The danger is that by choosing the smallest sums of squares the error estimate is biased to the small side. The half normal plot can be used to decide which factors should be included in the regression model.

---

[4]Note that if ANOVA is used to generate a predictive model, then all factor levels are treated equally and so there must be interpolation between the factor levels for hot dwell time because the factor levels are 4, 8 and 20 h which are not equally spaced.

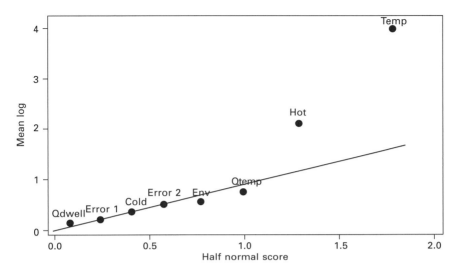

18.9 Half normal plot for mean log Δ*m*.

In this experiment, assuming that only main effects are important, the analysis can be carried out via multiple regression. The model for time to spall, t_s is given by

$$\text{mean } t_s = 1007 - 0.973 \text{ temp} + 5.06 \text{ upper dwell time}$$

The model explains 82% of the variation in the data and gives a reasonable fit as shown by the residual diagnostics in Fig. 18.10.

There are only nine trials and three terms in the model so the residuals are not independent. The Normal plot of residuals is close enough to a straight line, the I chart shows no particular pattern, the histogram of residuals is not markedly asymmetric and the residuals vs fits diagram shows no strong pattern. The model is therefore reasonable.

For an upper dwell temperature of 1020°C and upper dwell time of 16 h, the predicted time to spall is 95.1 h with a 95% confidence interval of 61 to 129 h. The experimentally determined values for Alloy 800 at 1000°C and 1050°C were 96 and 32 for 8 h upper dwell time and 160 and 40 for 20 h upper dwell time for comparison. Figure 18.11 shows the observed times to spall plotted against the times predicted by the model. Figure 18.11 shows a good fit between the observed values and the model because the points fall approximately on a straight line. The points for trials 4 and 9 are slightly off the line but there is no obvious connection between trials 4 and 9 to suggest a trend.

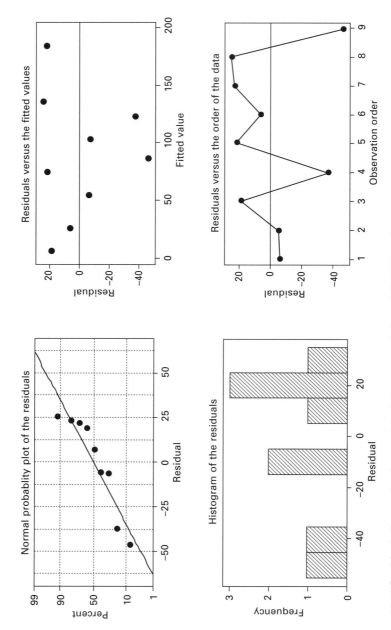

18.10 Residual diagnostics for model of time to spall, from MINITAB software.

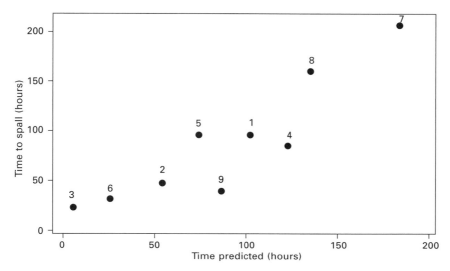

18.11 Plot of observed times to spall against times predicted by a model including linear effect of upper dwell temperature and time only. The trial numbers are shown.

18.10 Summary

The purpose of these experiments is to simulate service conditions. Rather than attempt to consider all service conditions, nine different scenarios were considered. From these results, the outcomes from other scenarios can be predicted. The precision of the prediction depends on the variability of the data and the fit of the regression model.

- Factors that are not significant will not affect the predictions and so some scenarios will have the same predicted values.
- Informative plots can be constructed to summarise the results provided a relevant summary variable is available. In the above analysis time to spall, t_s and difference between fitted and measured weight gain, Δm have been examined.
- The experimental design maximises the information that can be obtained from the experimental trials. It is not possible to balance for all interactions in only nine trials.
- The half normal, main effects and interaction plots can be obtained for all results using an Excel spreadsheet.
- Predictive equations can be constructed using regression. Care must be taken to include all important effects.

Statistically designed experiments help obtain a large amount of information from a small number of trials. Analysis via half normal plots is particularly useful when many factors are of interest, but the number of trials that can be carried out is limited due to cost, time or other consideration.

18.11 Appendices

Appendix I – Data

Table 18.5 Values for the gradient, n with mean for the three specimens in each trial

Trial	Environment	Temp	Upper dwell	Lower dwell	n1	n2	n3	Mean n
1	Dry	950	4	2	2.62	2.61	2.60	2.61
2	Dry	1000	4	4	2.22	2.09	2.11	2.14
3	Wet	1050	4	4	2.00	2.00	2.00	2.00
4	Wet	950	8	4	1.78	1.80	1.81	1.80
5	Dry	1000	8	4	2.76	2.95	3.15	2.95
6	Dry	1050	8	2	2.00	2.00	2.00	2.00
7	Dry	950	20	4	2.40	2.45	2.36	2.40
8	Wet	1000	20	2	2.56	2.69	2.76	2.67
9	Dry	1050	20	4	2.00	2.00	2.00	2.00

Table 18.6 Values for the time to spall, t, with mean for the three specimens in each trial

Trial	Environment	Temp	Upper dwell	Lower dwell	t1	t2	t3	Mean t
1	Dry	950	4	2	96.00	96.00	96.00	96.00
2	Dry	1000	4	4	48.00	48.00	48.00	48.00
3	Wet	1050	4	4	24.00	24.00	24.00	24.00
4	Wet	950	8	4	96.00	80.00	80.00	85.33
5	Dry	1000	8	4	96.00	96.00	96.00	96.00
6	Dry	1050	8	2	32.00	32.00	32.00	32.00
7	Dry	950	20	4	220.00	200.00	200.00	206.67
8	Wet	1000	20	2	160.00	160.00	160.00	160.00
9	Dry	1050	20	4	40.00	40.00	40.00	40.00

Table 18.7 Values for the difference between fitted and measured weight gain, d (fitted through data less than the measured time to spall) for the last weight measurement for each specimen, with mean and log standard deviation for the three specimens in each trial

Trial	Environ-ment	Temp	Upper dwell	Lower dwell	d1	d2	d3	Mean d	Log sd d
1	Dry	950	4	2	1.72	1.69	1.91	1.77	−0.92
2	Dry	1000	4	4	3.93	4.68	4.01	4.21	−0.39
3	Wet	1050	4	4	70.81	82.34	96.22	83.12	1.10
4	Wet	950	8	4	1.02	1.10	0.89	1.00	−0.97
5	Dry	1000	8	4	2.80	3.04	2.81	2.88	−0.87
6	Dry	1050	8	2	8.82	13.21	14.18	12.07	0.46
7	Dry	950	20	4	0.24	0.39	0.27	0.30	−1.10
8	Wet	1000	20	2	1.00	1.38	1.09	1.16	−0.70
9	Dry	1050	20	4	7.84	15.38	8.19	10.47	0.63

Table 18.8 Values for the rate constant, k, with mean and log standard deviation for the three specimens in each trial

Trial	Environ-ment	Temp	Upper dwell	Lower dwell	k1	k2	k3	Mean k	Log sd k
1	Dry	950	4	2	0.01	0.01	0.01	0.01	−3.18
2	Dry	1000	4	4	0.04	0.05	0.40	0.16	−0.69
3	Wet	1050	4	4	0.23	0.27	0.27	0.26	−1.65
4	Wet	950	8	4	0.02	0.02	0.02	0.02	−2.92
5	Dry	1000	8	4	0.04	0.04	0.04	0.04	−3.07
6	Dry	1050	8	2	0.26	0.25	0.26	0.25	−2.39
7	Dry	950	20	4	0.01	0.02	0.01	0.01	−2.96
8	Wet	1000	20	2	0.04	0.05	0.05	0.05	−3.12
9	Dry	1050	20	4	0.18	0.33	0.21	0.24	−1.09

Appendix II – Further technical details

1 Balance

The experimental design allows a balanced comparison of the levels of each factor. The design is a quarter fraction of the full 36 combinations of factor levels, however, and the balance is not perfect. For example, there is one result for wet environment at each temperature but two results for dry environment at each temperature; of the two dry results at 950°C, one is at 4 h hot dwell and one is at 20 h hot dwell; the 4 h hot dwell has 2 h cold dwell and the 20 h hot dwell has 4 h cold dwell; of the two dry results at 1000°C, one is at 4 h hot dwell and one at 8 h hot dwell; both have 4 h cold dwell time; of the two dry results at 1050°C, one is at 8 h hot dwell and one at 20 h hot dwell; the 8 h hot dwell has 2 h cold dwell and the 20 h hot dwell has 4 h cold dwell. This lack of balance arises because of the condensed nature of the design. It should not affect the relevance of the analysis but should be borne in mind in the interpretation.

2 Full set of effects and interactions

The full set of 35 effects and interactions (synergies) which can occur between two three-level factors, A and B, and two two-level factors, C and D, is as follows:

- four main linear effects – A_1, B_1, C, D
- two main non-linear effects – A_q, B_q
- four two three-level factor interactions – $A_1 \times$, B_1, $A_1 \times B_q$, $A_q \times B_1$, $A_q \times B_q$
- one two-level factor interaction – $C \times D$
- eight two-factor mixed level interactions
- twelve three-factor mixed level interactions
- four four-factor mixed level interactions.

It is only possible to analyse all of these effects and interactions if a full factorial experiment is carried out with 36 trials. The experimental design used in this analysis is a fractional factorial design in which a subset of nine trials have been carried out due to the cost and time required for each trial.

The variation in the nine sets of results can be broken up into constituent parts. These can correspond to the factors and interactions above. The constituent parts are obtained by calculating orthogonal contrasts of the results. There are several alternative sets of orthogonal contrasts. Not all sets correspond to the design, for example the following set is not meaningful for the experiment carried out here: (8, -1, -1, -1, -1, -1, -1, -1, -1), (0, 7, -1, -1, -1, -1, -1, -1, -1), and so on to (0, 0, 0, 0, 0, 0, 0, 1, -1). The analysis in this chapter is for main effects of all factors only, i.e. the first six effects listed above. The contrasts used are orthogonal to each other.

3 Quadratic effects and interactions

There is an alternative analysis where interactions between the three level factors are investigated but the two-level factors are ignored. The contrasts are shown in Table 18.9. Although these contrasts are orthogonal to each other, the interactions are confounded by the two two-level factors which do not appear explicitly in the contrasts but are confounded within them.

The matrix is orthogonal. Using the mean time to spall data gives the half normal plot in Fig. 18.12. In this plot, upper dwell temperature and time are significant as before. The interactions between temperature and time are not significant.

Table 18.9 Alternative set of orthogonal contrasts for the experiment

Trial	Temp	Hot	Qtemp	Qdwell	Tempx hot	Tempx Qdwell	Qtempx hot	Qtempx Qdwell
1	−1	−1	1	1	1	−1	−1	1
2	0	−1	−2	1	0	0	2	−2
3	1	−1	1	1	−1	1	−1	1
4	−1	0	1	−2	0	2	0	−2
5	0	0	−2	−2	0	0	0	4
6	1	0	1	−2	0	−2	0	−2
7	−1	1	1	1	−1	−1	1	1
8	0	1	−2	1	0	0	−2	−2
9	1	1	1	1	1	1	1	1

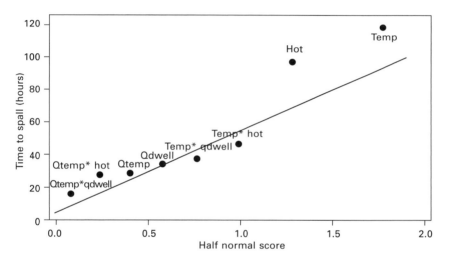

18.12 Half normal plot for time to spall with alternative contrasts.

4 Interactions between two- and three-level factors

There is another alternative set of contrasts which involves the four factors and some two-way interactions between them. In Table 18.10 the quadratic terms have been omitted. These contrasts, however, are not orthogonal. Half normal plots can be constructed for these sets of contrasts but their interpretation should be treated with caution because of the lack of orthogonality.

Overall, there are a number of alternative ways of sub-dividing the variation in the nine trial results but the set of orthogonal contrasts used in the analysis in this chapter are the most reliable. This is because they correspond to the experimental design and provide the most nearly balanced assessment of the effects of the cycling parameters.

Table 18.10 Alternative set of orthogonal contrasts including mixed level interactions

Trial	Temp	Hot	Cold	Env	Tempx hot	Cold* env	Error1	Error2
1	−1	−1	2	−1	1	−2	0	−1
2	0	−1	−1	−1	0	1	−1	1
3	1	−1	−1	2	−1	−2	1	0
4	−1	0	−1	2	0	−2	−1	0
5	0	0	−1	−1	0	1	1	−1
6	1	0	2	−1	0	−2	0	1
7	−1	1	−1	−1	−1	1	1	1
8	0	1	2	2	0	4	0	0
9	1	1	−1	−1	1	1	−1	−1

Appendix III – Alternative design

If interest does not lie in seeing the effects of all three-level factors at all levels of the other factors, then an alternative design can be used. This is a nine run design of four two-level factors with a centre point at a mid-way setting. The experimental design is shown in Table 18.11, followed by a set of eight orthogonal contrasts in Table 18.12. Note that there are now four

Table 18.11 Alternative set of orthogonal contrasts including mixed level interactions

Trial	Temperature(°C)	Upper dwell (h)	Lower dwell (h)	Environment
1	950	4	2	Dry
2	1050	4	2	Wet
3	950	20	2	Wet
4	1050	20	2	Dry
5	950	4	4	Wet
6	1050	4	4	Dry
7	950	20	4	Dry
8	1050	20	4	Wet
9	1000	12	3	Medium

Table 18.12 Set of orthogonal contrasts with centre point

Trial	Temp	Hot	Cold	Env	Tempx Hot	Hotx Env	Tempx Env	Curvature
1	−1	−1	−1	−1	1	1	1	−1
2	1	−1	−1	1	−1	−1	1	−1
3	−1	1	−1	1	−1	1	−1	−1
4	1	1	−1	−1	1	−1	−1	−1
5	−1	−1	1	1	1	−1	−1	−1
6	1	−1	1	−1	−1	1	−1	−1
7	−1	1	1	−1	−1	−1	1	−1
8	1	1	1	1	1	1	1	−1
9	0	0	0	0	0	0	0	8

trials at the longer cold dwell time and wet environment instead of three as in the original design used in this experiment. As in the three level designs above, the interactions are confounded with each other to a certain extent. The non-linearity in upper dwell temperature and time is evaluated via a single curvature contrast comparing the centre with the other points.

18.12 References

Breyfogle F.W. (2000) *Implementing Six Sigma*, Wiley, Chicheste.

Cochran, W.G. and D.R. Cox (1957) *Experimental Designs*, Wiley, Chichester.

Grove, D. and T. Davis (1997) *Engineering Quality and Experimental Design,* Longman, Harlow.

Metcalfe, A.V. (1994) *Statistics in Engineering – a Practical Approach*, Chapman and Hall, London.

Montgomery, D.C. (2005) *Design and Analysis of Experiments*, 6th edn, Wiley, Chichester.

Nicholls, J.R. and P. Hancock (1983) 'The analysis of oxidation and hot corrosion data – a statistical approach', *High Temperature Corrosion*, NACE-6, 198–210.

Smialek, J.L and J.V. Auping (2002) 'COSP for Windows: strategies for rapid analyses of cyclic oxidation behaviour', *Oxidation of Metals*, 57(5–6), June, 559–581.

Wu, C.F.J. and M. Hamada (2000) *Experiments – Planning, Analysis, and Parameter Design Optimization*, Wiley, Chichester.

19

Influence of cycling parameter variation on thermal cyclic oxidation testing of high temperature materials (COTEST)

M SCHÜTZE and M MALESSA,
Karl-Winnacker-Institut der DECHEMA e.V., Germany;
S Y COLEMAN, University of Newcastle upon Tyne, UK
and L NIEWOLAK and W J QUADAKKERS,
Forschungszentrum Jülich Gmbh, Germany

19.1   Introduction

During operation at high temperatures, materials are subjected to complex interactions of temperature changes, oxidative and corrosive environmental attacks and mechanical stresses. These interactions determine whether components exhibit premature failure or show reliable and safe, long-term performance, and also limit the upper service temperature, which determines the degree of efficiency and, hence, the economical and ecological performance of materials.

Thermal cycling oxidation tests are frequently used to simulate material damage occurring due to oxidation attack in combination with thermally induced stresses. Such laboratory investigations can, in most cases, simulate the industrial operation conditions only to a limited extent because of the complex loading conditions in practice and the much shorter testing times than the actual operating periods. Laboratory tests offer the invaluable potential of ranking materials behaviour. However, the large variety of test parameters used may lead to a situation where results from different laboratories cannot be directly compared. Furthermore, difficulties may arise for a meaningful assessment of their significance for service performance [1, 2].

In order to reach a quantitative understanding of the value of the different test parameters, the aim of the present study was to investigate the influence of cycling parameter variation (e.g. maximum oxidation temperature, dwell time at high and at low temperature, environmental humidity) on the oxidation and spallation kinetics of selected high temperature alloys. For this purpose, four commercial alloys were selected; these represent three common groups of metallic high temperature materials: i.e. ferritic steels, austenitic steels and Ni-based superalloys. The experimental times obtained, until the onset of spallation, and the relative amount of spalled oxide were statistically analysed using the ANOVA method [3]. The statistical analysis allowed determination of the significance of each varied parameter on the oxidation/ spallation kinetics from a limited number of experimental data.

19.2   Experimental methods

The following materials were used in the test programme: P91, AISI441, Alloy 800H and CM247. For the oxidation tests, samples of the selected materials, with nominal dimensions $12 \times 17 \times 1.5$ mm, were machined and subsequently ground down to 1200-grit surface finish. Additionally, half of the ground P91 specimens were heat-treated for 3 h at 1000°C in an Ar-5%H_2 atmosphere. The aim of this procedure was to enhance the oxidation rate by decreasing chromium diffusion in the surface deformation zone [4].

All tests in this study were performed according to the draft COTEST guidelines for cyclic oxidation testing [1]. The oxidation experiments were carried out in dry synthetic air (N_2 + 20vol.% O_2 with less than 10 ppm of water vapour) or wet synthetic air with 2% specific humidity. The oxidation experiments were conducted in a specially designed horizontal furnace facility in the case of long dwell time testing and a vertically moving furnace in the case of the short dwell time testing (Fig. 19.1). Due to expected spallation of the oxide scales, the specimens were placed in alumina crucibles so that the

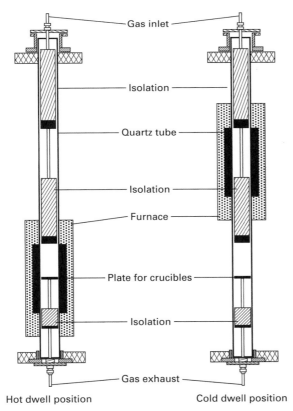

Gas inlet

Isolation

Quartz tube

Isolation

Furnace

Plate for crucibles

Isolation

Gas exhaust

Hot dwell position   Cold dwell position

19.1 Furnace design for short dwell thermal cycling oxidation testing.

amount of spalled oxide could be measured in addition to the weight change of the specimen (Fig. 19.2). For each experiment, three replicates were used to obtain statistically meaningful weight change data. The accumulated hot dwell time at high temperature for all experiments was 300 h using the definition for the hot dwell time derived by Strehl and Borchardt [5].

The variable parameters within the experiments were upper dwell time (UDT – maximum number of variations: three), lower dwell time (LDT – maximum two levels), environment (dry air or wet air) and temperature (maximum three levels). To cover a large number of situations frequently encountered in technical applications, two types of tests were performed, i.e. testing with short dwell times (UDT 0.5, 1, 2 h) and long dwell times (4, 8, 20 h).

After exposure, the surface scales formed were analysed by optical microscopy (OM – Axiplan 2 Zeiss) and scanning electron microscopy with energy dispersive X-ray spectroscopy (SEM/EDX – Leo 440/Oxford Inca), as well as X-ray diffraction (XRD – Siemens D5000). The bulk chemical compositions of the alloys were analysed by inductive coupled plasma-optical emission spectroscopy (ICP-OES – ARL 34000).

19.3   Results

The chemical compositions of the alloys determined by ICP-OES and the microstructures characterised by optical metallography are shown in Table 19.1

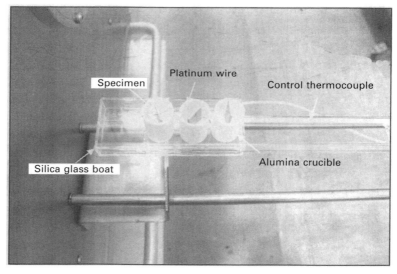

19.2 Specimen arrangement for long dwell thermal cycling oxidation testing.

Table 19.1 Chemical composition of the alloys investigated

	Fe	Cr	Ni	Mn	Co	Ta	W	Mo	Nb	Al	Si	Ti	Hf	P	S	N	O	C
							Mass %									ppm		
P91	89.2	9.2	0.3	0.7	–	–	–	0.7	–	–	0.3	–	–	130	20	500	40	700
AISI441	80.5	18.4	0.2	0.2	–	–	–	–	0.4	0.1	0.6	0.1	–	130	10	120	50	170
Alloy 800H	48.2	21.1	30.9	0.7	–	–	–	0.1	–	0.2	0.4	0.3	–	–	<10	45	30	650
CM247	–	8.7	61.2	–	10.5	3.8	11.9	0.7	–	5.7	–	1	1.5	–	<10	–	16	1470

and Fig. 19.3, respectively. Two of the four materials investigated (P91 and AISI441) were ferritic steels. The 9%Cr-steel P91 possesses a martensitic microstructure with carbide precipitates (Fig. 19.3(a)). The microstructure of this steel after heat treatment at 1000°C for 3 h in Ar-5%H_2 atmosphere exhibits, as desired for the present tests for making the material more susceptible for breakaway oxidation, substantial grain coarsening. The microstructure of the ferritic 18%Cr-steel, AISI441, consists of equiaxed ferrite grains with precipitates of Ti-rich carbo-nitrides (Fig. 19.3(b)). The austenitic steel, Alloy 800H possesses a typical recrystallised microstructure with annealing twins and carbide stringers ($M_{23}C_6$) in the rolling direction (Fig. 19.3(c)). The Ni-based superalloy, CM247, shows a typical γ' microstructure with precipitation of mixed carbides, rich in Hf, Ti, W and Ta which are located mainly at alloy grain boundaries (Fig. 19.3(d)).

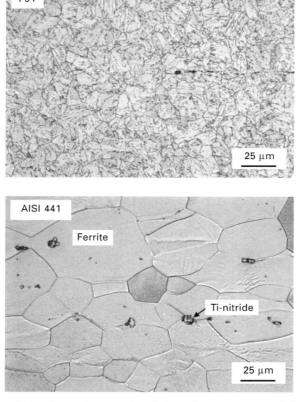

19.3 Optical micrographs of the microstructures of the alloys investigated in the 'as-received' state: (a) P91, (b) AISI441, (c) Alloy 800H, (d) CM247.

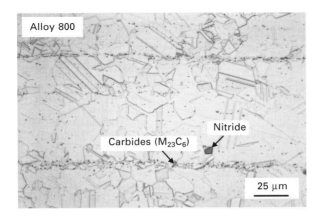

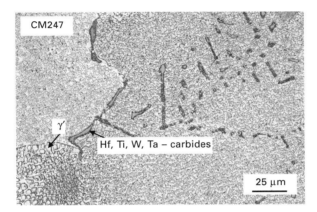

19.3 Continued

19.3.1 Oxide scale morphologies and growth rates

For the ferritic steel, P91, significant net weight changes were observed only in the case of the heat-treated specimens in humidified air (Fig. 19.4). The non heat-treated P91 specimens during oxidation in dry and wet air (not shown here) as well as the heat-treated specimens during oxidation in dry air showed hardly any weight change (Fig. 19.4). These negligible weight changes result from the formation of an extremely thin, protective oxide scale during the cyclic testing (Fig. 19.5(a)). The XRD and SEM/EDX measurements revealed that the oxide scale consisted of $(Fe,Cr)_2O_3$ and $(Fe,Cr,Mn)_3O_4$ spinel. In contrast, enhanced oxidation as the result of a breakaway process was observed for the heat-treated material in wet air (Fig. 19.5(b)). The non-protective scale consisted of an outer layer of hematite, inner layers of Fe_3O_4 (magnetite) and $(Fe,Cr)_3O_4$ spinel and a narrow internal oxidation zone with

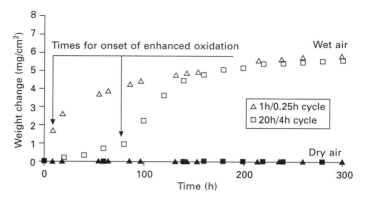

19.4 Net weight change as a function of time during 300 h cyclic oxidation at 650°C for heat-treated specimens of P91.

chromia stringers (Fig. 19.5(b)). Moreover, substantial in-scale void formation was observed. The time for the onset of enhanced oxidation seems to be dependent on the upper dwell time (or number of cycles). With decreasing upper dwell time or, in other words, increasing number of cycles in the same test duration, the time for the onset of enhanced oxidation was strongly decreased (Fig. 19.6).

The net weight change curves for AISI441 exposed at 800, 850 and 900°C are given in Fig. 19.7. Increasing the temperature resulted in an increase in the oxidation rate. In the early stages of exposure, the material behaviour was determined by near-parabolic oxide growth kinetics and only a minor difference between long and short dwell time testing was found (Fig. 19.8). The onset of spallation for short dwell time testing occurred earlier than in the case of long dwell time testing. This effect was more pronounced at higher oxidation temperatures. The AISI441 steel, oxidised at temperatures in the range of 800–900°C, formed on its surface an oxide scale that consisted of an outer $(Mn,Cr)_3O_4$ layer, and an inner layer of Cr_2O_3. The internal oxidation zone contained small precipitates of Al_2O_3 and TiO_2 (Fig. 19.9(a) and (b)). SEM/EDX analysis of the specimen surfaces after oxidation showed the formation of 'nodules' containing TiO_2 (Fig. 19.10). The number and average size of the 'nodules' increased with increasing oxidation temperature.

The net weight change curves for Alloy 800H during exposure at 950, 1000 and 1050°C are given in Fig. 19.11. The early stages of exposure were determined by a near-parabolic time dependence of the scale growth rate, and hardly any difference was found between long and short dwell time testing (Figs 19.12–19.15). The time for the onset of spallation varied strongly as a function of temperature and upper dwell time and, in general, decreased with decreasing UDT and increasing test temperature. Also, the extent of spallation was enhanced by shortening the UDT and/or increasing the oxidation

Specimen oxidised in dry air

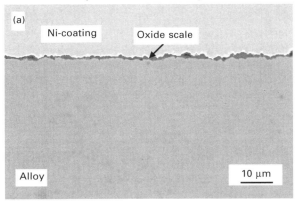

Specimen oxidised in wet air

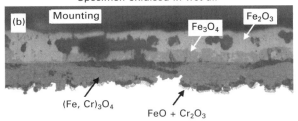

19.5 Microstructures of the oxide scales on P91 after 300 h cyclic oxidation at 650°C: (a) SEM/BSE micrograph of the untreated specimen – 20 h UDT/4h LDT, dry air; (b) optical micrograph of the heat-treated specimen – 4 h UDT/4h LDT, wet air.

temperature. Figure 19.16 shows cross-sections of the oxide scales formed on Alloy 800H after cyclic oxidation at 1000°C and 1050°C. During oxidation at 1000°C, the alloy formed on its surface an outer layer of $(Mn,Cr)_3O_4$ spinel and an inner layer of Cr_2O_3, as well as an internal oxidation zone. The substantial thermal expansion coefficient mismatch between the metal matrix (austenite) and the oxide [6, 7] leads, however, to spallation of the oxide scale. Similar results were found after oxidation at 950°C. Increasing the oxidation temperature to 1050°C resulted in enhanced spallation caused by significant changes in oxide scale morphology and phase composition (Fig. 19.16(b)). The oxide scale formed at 1050°C was heterogeneous and consisted of Fe,Cr,Ni-spinel and NiO-type oxide and a chromia layer formed near the

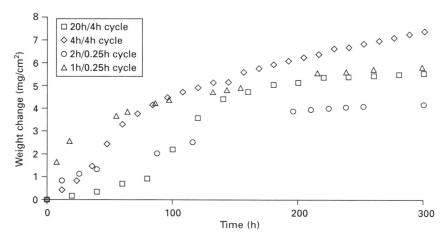

19.6 Net weight change as a function of time during 300 h cyclic oxidation at 650°C for heat-treated specimens of P91 in wet air.

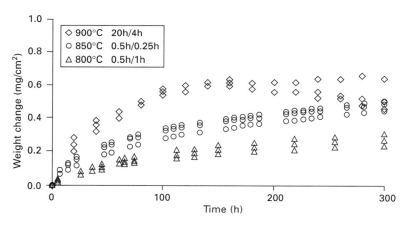

19.7 Net weight change as a function of time during 300 h cyclic oxidation for specimens of AISI441 in dry air.

metal/oxide interface (Fig. 19.16(b)). Moreover, substantial porosity and void formation were observed in the internal oxidation zone.

Typical weight change curves for the Ni-based superalloy, CM247, are given in Figs 19.17 and 19.18. In contrast to the aforementioned materials, the variable parameters showed only limited (oxidation temperature) or virtually no (dwell time) influence on the oxidation/spallation kinetics of the alloy (Fig. 19.18). The outer part of the oxide scales formed on the alloy surface spalled off during the first few cooling cycles. Increasing the oxidation temperature from 1100°C to 1150°C resulted in enhanced spallation during

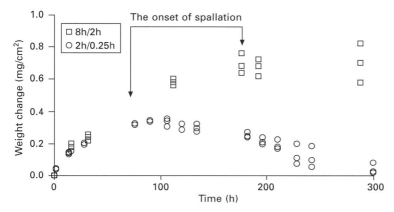

19.8 Net weight change as a function of time during 300 h cyclic oxidation at 900°C for specimens of AISI441 in dry air.

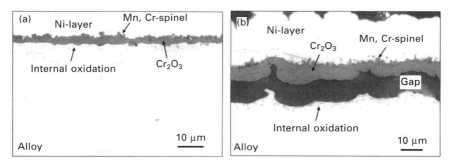

19.9 SEM/BSE micrographs of the oxide scales on AISI441 after 300 h cyclic oxidation: (a) 4 h/4 h at 850°C in dry air; (b) 4 h/4 h at 900°C in wet air.

the first 24 h of oxidation. Figure 19.19 shows a cross-section of the oxide scale after cyclic oxidation for 300h at 1100°C. It consisted of an outer layer of $NiAl_2O_4$ and an inner layer of Al_2O_3. In the internal oxidation zone, precipitates of HfO_2 could be observed [8]. The internal oxides were found at the alloy grain boundaries, i.e. they were formed out of the initially prevailing carbides (cf. Figs 19.3(d) and 19.19). The oxide spallation occurred mainly at the interface between the inner (Al_2O_3) and outer ($NiAl_2O_4$) layers. After spallation of large parts of the outer scale, the scale growth is determined by transport processes in the inner, dense alumina scale and the growth rate is substantially smaller than that during or just before the occurrence of spalling.

19.10 SEM micrograph showing oxide surface morphology on AISI441 after 300 h cyclic oxidation (4 h/4 h) at 850°C in dry air.

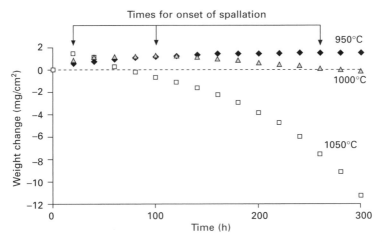

19.11 Net weight change as a function of time during 300 h cyclic oxidation (20 h/4 h cycles) for specimens of Alloy 800H in dry air at various temperatures.

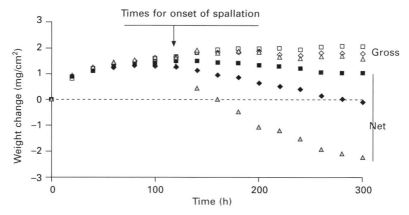

19.12 Net weight change as a function of time during 300 h cyclic oxidation (20 h/4 h cycles) for specimens of Alloy 800H in dry air at 1000°C. The different symbols represent different tests at different laboratories.

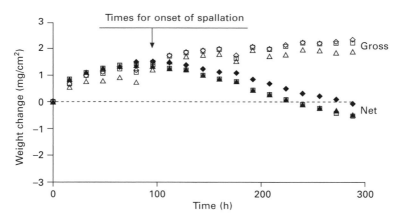

19.13 Net weight change as a function of time during 300 h cyclic oxidation (8 h/4 h cycles) for specimens of Alloy 800H in dry air 1000°C. The different symbols represent different tests at different laboratories.

19.3.2 Analysis of the net weight change curves

For evaluation of the effect of the cycling parameters on the oxidation behaviour, the net weight changes for the test pieces were plotted versus time, as shown in Fig. 19.20(a). The net weight change prior to the onset of spallation can be described by:

$$\left(\frac{\Delta m_{\text{net}}}{A}\right)^n = k_n t \rightarrow \frac{\Delta m_{net}}{A} = (k_n \cdot t)^{1/n} \qquad 19.1$$

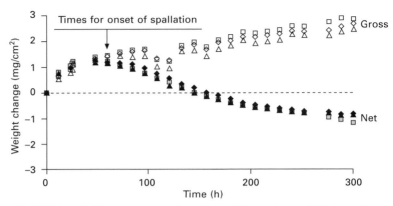

19.14 Net weight change as a function of time during 300 h cyclic oxidation (4 h/4 h cycles) for specimens of Alloy 800H in dry air at 1000°C. The different symbols represent different tests at different laboratories.

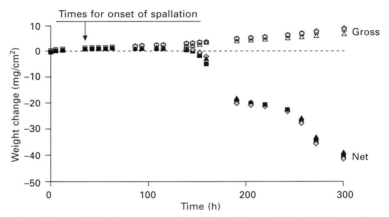

19.15 Net weight change as a function of time during 300 h cyclic oxidation (0.5 h/0.25 h cycles) for specimens of Alloy 800H in dry air at 1000°C. The different symbols represent different tests at different laboratories.

where Δm_{net} is the difference of weight of specimen between exposure time t and start of exposure.

A double logarithmic plot, given in Fig. 19.20(b), allows determination of the oxidation parameters k_n and n. A change in the oxidation mechanism (e.g. spallation or breakaway oxidation) becomes apparent by a change in the slope in the double logarithmic plot. Analysis of the linear part of the curve by linear regression using simple spreadsheet calculations yields the slope, $b = 1/n$, and the y-axis intercept, as shown in Fig. 19.20(b). The values for the oxidation parameter k_n and the exponent n as well as the number of

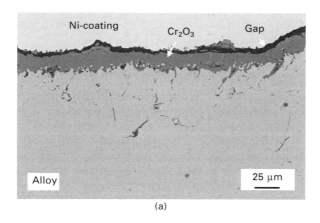

(a)

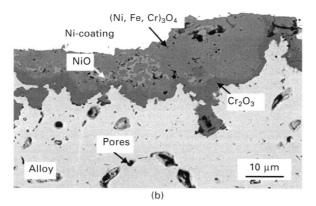

(b)

19.16 SEM/BSE micrographs of the oxide scales on Alloy 800H after 300 h cyclic oxidation in wet air: (a) 20 h/4 h at 1000°C, (b) 4 h/4 h at 1050°C.

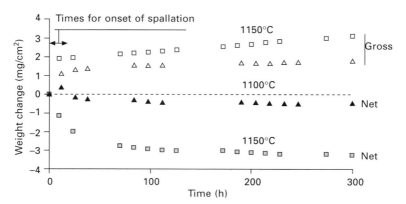

19.17 Net and gross weight changes as a function of time during 300 h cyclic oxidation (2 h/0.25 h cycles) for specimens of CM247 in dry air.

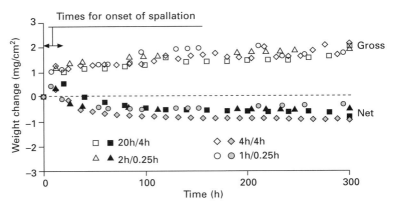

19.18 Net and gross weight changes as a function of time during 300 h cyclic oxidation for specimens of CM247 in dry air at 1100°C.

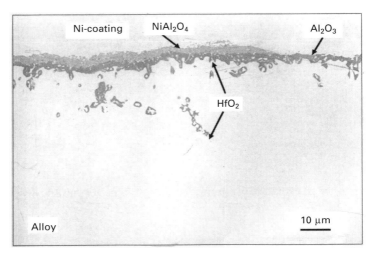

19.19 SEM/BSE micrograph of the oxide scale on CM247 after 300 h cyclic oxidation in dry air at 1100°C (4 h/4 h cycles).

cycles $N_{protective}$ or the accumulated hot dwell time $t_{protective}$ during which the material shows protective behaviour without spallation or breakaway are given in Tables 19.2 and 19.3 for AISI441 and Alloy 800H, respectively. The net weight change data for P91 and CM247 cannot be evaluated using this type of analysis because the data cannot be properly fitted by Eq. 19.1.

Additionally, the net weight change difference $\Delta m_{net}(t_{300h}-t_{protective})$ as defined in Eq. 19.2, was determined for all materials tested (Tables 19.2–19.4):

$$\Delta m_{net}(t_{300\ h}-t_{protective}) = \Delta m_{net}(t_{300\ h}) - \Delta m_{net}(t_{protective}) \qquad 19.2$$

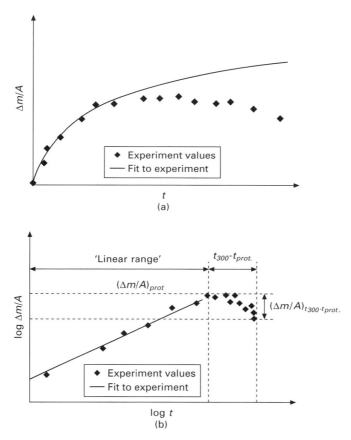

19.20 (a) Net mass change as a function of time; (b) double logarithmic plot of net mass change vs time.

where $\Delta m_{net}(t_{300\ h})$ is the net weight change of the test piece in (mg/cm2) after 300 h and $\Delta m_{net}(t_{protective})$ is the net weight change of the test piece at the time $t_{protective}$ which is represented by the last data point in the linear range of the double logarithmic plot (Fig. 19.20(b)).

19.3.3   Statistical evaluation using the ANOVA method

The test matrix design comprises four factors (temperature, hot dwell time, cold dwell time and humidity), which should be tested as three, three, two and two variations respectively in order to reveal the influence of each parameter. This results in a test matrix of 36 possible combinations. To limit the number of experiments, a subset of nine experiments was chosen, following statistical considerations for a balanced design. The balanced design approach

Table 19.2 Evaluation of the experimental net weight change data for AISI441.

T (°C)	Cycling conditions Dwell time (h) Hot*	Cold	Environment	n	$k_n \cdot 10^{-4}$	$t_{protective}$	$\Delta m_{net}(t_{300\,h} - t_x)$ (mg)
800	0.5	1	dry	1.82 ± 0.09	3.76 ± 0.33	>300	
800	1	0.25	wet	2.21 ± 0.12	0.84 ± 0.23	>300	
800	2	0.25	dry	1.98 ± 0.20	2.01 ± 0.51	>300	
800	4	2	dry	1.88 ± 0.06	1.57 ± 0.15	>300	
800	8	4	wet	1.36 ± 0.05	3.83 ± 0.26	>300	
800	20	4	dry	1.44 ± 0.20	4.20 ± 1.10	>300	
850	0.5	0.25	dry	2.06 ± 0.15	8.11 ± 0.28	>300	
850	1	1	dry	1.42 ± 0.36	8.28 ± 2.06	>300	−0.043 ± 0.017
850	2	1	wet	2.20 ± 0.12	3.01 ± 0.47	162	
850	4	4	dry	2.23 ± 0.17	9.30 ± 0.92	>300	
850	8	4	dry	1.8 ± 0.27	11.67 ± 0.50	>300	
850	20	2	wet	1.36 ± 0.21	18.03 ± 8.80	80–260	−0.23 ± 0.15
900	0.5	0.25	wet	2.19 ± 0.11	9.57 ± 1.77	109,5	−0.047 ± 0.014
900	1	0.25	dry	1.84 ± 0.05	11.67 ± 1.37	174	−0.184 ± 0.043
900	2	0.25	dry	1.91 ± 0.11	14.77 ± 2.22	120	−0.270 ± 0.024
900	4	4	wet	1.63 ± 0.06	33.33 ± 5.30	100	−0.28 ± 0.1
900	8	2	dry	1.67 ± 0.05	35.67 ± 0.41	166	−0.48 ± 0.15
900	20	4	dry	1.96 ± 0.31	29.33 ± 4.71	107	−0.47 ± 0.05

*Hot dwell times of 0.5 to 2 h are referred to as short dwell testing, hot dwell times of 4 to 20 h are referred to as long dwell testing.

Table 19.3 Evaluation of the experimental net weight change data for Alloy 800H

| | Cycling conditions | | | | | | |
| | Dwell time | | | | | | |
T (°C)	Hot*	Cold	Environment	n	$k_n \cdot 10^{-4}$	$t_{protective}$ (h)	$\Delta m_{net}(t_{300\,h}-t_x)$ (mg)
950	0.5	1	dry	2.54 ± 0.09	98.09 ± 3.83	215.5	-5.18 ± 1.80
950	1	0.25	wet	2.67 ± 0.02	133.22 ± 4.76	133	-1.43 ± 0.05
950	2	0.25	dry	2.60 ± 0.02	93.81 ± 7.61	196	-0.23 ± 0.05
950	4	2	dry	2.61 ± 0.01	135.00 ± 5.35	96	-1.22 ± 0.09
950	8	4	wet	1.80 ± 0.01	191.00 ± 9.80	85	1.00 ± 0.09
950	20	4	dry	2.40 ± 0.04	146.33 ± 8.96	207	0.30 ± 0.06
1000	0.5	0.25	dry	2.83 ± 0.05	263.02 ± 17.25	86.5	-41.92 ± 0.73
1000	1	0.25	dry	2.71 ± 0.19	258.13 ± 12.87	86	-16.27 ± 2.23
1000	2	1	wet	2.65 ± 0.06	369.06 ± 15.86	82	-9.90 ± 4.19
1000	4	4	dry	2.14 ± 0.06	430.33 ± 32.27	48	-2.11 ± 0.34
1000	8	4	dry	2.95 ± 0.16	378.00 ± 6.98	96	-2.00 ± 0.10
1000	20	2	wet	2.67 ± 0.08#	455.00 ± 6.16#	160#	-1.16 ± 0.16#
1050	0.5	0.25	wet	2.70 ± 0.08	637.90 ± 10.18	46.5	-40.25 ± 0.42
1050	1	1	dry	2.50 ± 0.05	661.68 ± 19.18	42	-52.91 ± 0.46
1050	2	0.25	dry	4.06 ± 0.57	301.32 ± 73.39	76	-39.03 ± 0.71
1050	4	4	wet	2.00*	2558.3 ± 182.95	24	-83.12 ± 10.39
1050	8	2	dry	2.00*	2529.0 ± 33.30	32	-12.07 ± 2.33
1050	20	4	dry	2.00*	2407.7 ± 666.47	40	-7.47 ± 3.51

*Hot dwell times of 0.5 to 2 h are referred to as short dwell testing, hot dwell times of 4 to 20 h are referred to as long dwell testing.
#Only few data points are available due to a 'non-24 h' operational mode and the evaluation is less accurate.

Table 19.4 Evaluation of the experimental net weight change data for heat treated P91 and CM247

Material	T (°C)	Cycling conditions Dwell time Hot	Cold	Environment	$\Delta m_{net}(t_{300\,h})$ (mg)
P91	650	1	0.25	dry	~ 0
P91	650	1	0.25	wet	5.62 ± 0.21
P91	650	2	0.25	dry	~ 0
P91	650	2	0.25	wet	4.55 ± 0.34
P91	650	4	4	dry	~ 0
P91	650	4	4	wet	6.96 ± 0.46
P91	650	20	4	dry	~ 0
P91	650	20	4	wet	5.37 ± 0.34
CM247	1100	1	0.25	dry	−0.25 ± 0.09
CM247	1100	2	0.25	dry	−0.32 ± 0.16
CM247	1150	1	0.25	dry	−3.86 ± 0.15
CM247	1150	2	0.25	dry	−3.08 ± 0.10
CM247	1100	4	4	dry	0.96 ± 0.05
CM247	1100	20	4	dry	0.98 ± 0.12
CM247	1150	4	4	dry	3.98 ± 0.34
CM247	1150	20	4	dry	4.17 ± 0.45

*Hot dwell times of 0.5 to 2 h are referred to as short dwell testing, hot dwell times of 4 to 20 h are referred to as long dwell testing.

allows investigations of the dependence of oxidation behaviour on these parameters from a smaller number of experiments [3].

There are four outcomes that have been considered in detail. Data are available for the exponent n, and the oxidation parameter k_n, for all of the 36 trials (9 trials for AISI441 and Alloy 800H each, including long and short dwell times). Data on the protective oxide growth time $t_{protective}$ is available for all of the Alloy 800H trials, but $t_{protective}$ exceeds the experimental maximum time of 300 h for most of the lower temperature AISI441 trials. The differences of net weight change between t_{300h} and $t_{protective}$ are available for all of the Alloy 800H trials but only for the higher temperature AISI441 trials.

The short time and long time experiments were conducted in various laboratories and there are expected to be differences between results because, for example, the cooling rates used were not identical.

The various outcomes are continuous variables and an analysis of variance (ANOVA) can be used to determine which parameters have a significant effect on the outcomes, e.g. experimental results. ANOVA assumes that the experimental error is approximately normal, that results are independent of each other and that the variance is constant. A log transformation is applied to the k_n parameter to stabilise the variance in the experimental error.

The experimental design was chosen such that each parameter was assessed over the same combination of other parameters, as far as possible. The design gives a balanced comparison of the effects of each parameter. Two parameters were required to be at three levels. This indicated the use of a nine trial design so that each level could be tested three times. The other two parameters therefore occur unequally often for each level. The design was chosen to make the easier level occur more often; therefore, there are three wet environments and six dry environments.

For the long dwell trials, there are three 2/h cold dwell times and six 4/h cold dwell times. For the short dwell trials, there are three 1/h cold dwell times and six 15/min cold dwell times as this fitted in with the laboratory procedures. A reference trial was included in each design. An example of a balanced design test matrix is shown for the AISI441 long dwell time experiments in Table 19.5.

Statistical evaluation for Alloy 800H

Figures 19.21 and 19.22 show the main effects for the protective oxide growth time ($t_{protective}$) in the case of Alloy 800H. It can be seen that increasing temperature has the effect of decreasing $t_{protective}$. Decreasing upper dwell time has the effect of decreasing $t_{protective}$. Environment appears to have virtually no effect. Only the effect of temperature ($p = 0.00$) is statistically significant for short dwell times. The effects of temperature (significance $p = 0.03$) and upper dwell time ($p = 0.05$) are statistically significant for long dwell times. Short and long dwell experiments are analysed separately because upper and lower dwell times are correlated across the two experiments and their separate effects cannot be evaluated reliably.

Figures 19.23 and 19.24 show the main effects for the log of the oxidation rate constant k_n for Alloy 800H. It can be seen that increasing temperature has the effect of increasing log (mean k_n). Environment appears to have virtually no effect. Only the effect of temperature ($p = 0.00$) is statistically significant for both short and long dwell experiments. The effect of dwell

Table 19.5 Test matrix for long dwell time testing of AISI441

Trial	UDT (h) (upper dwell time)	LDT (h) (lower dwell time)	Temperature (°C)	Environment
1	4	2	800	Dry
2	4	4	850	Dry
3	4	4	900	Wet
4	8	4	800	Wet
5	8	4	850	Dry
6	8	2	900	Dry
7	20	4	800	Dry
8	20	2	850	Wet
9	20	4	900	Dry

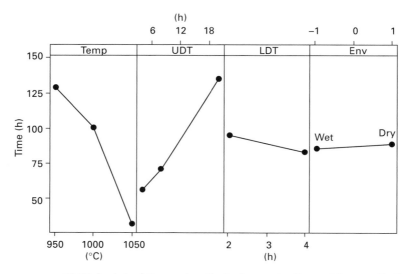

19.21 A plot of the main effects for protective oxide growth time ($t_{protective}$) times for Alloy 800H (long dwell times).

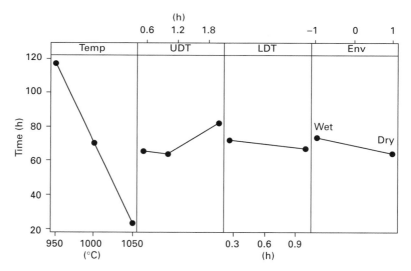

19.22 A plot of the main effects for protective oxide growth time ($t_{protective}$) for Alloy 800H (short dwell times).

time is not large enough to be statistically significant, probably because the nine-trial experiment is not sufficiently powerful.

Figures 19.25 and 19.26 show the main effects for the difference in net weight change between t_{300h} and $t_{protective}$ [$\Delta m_{net}(t_{300h}\text{-}t_{protective})$] for Alloy

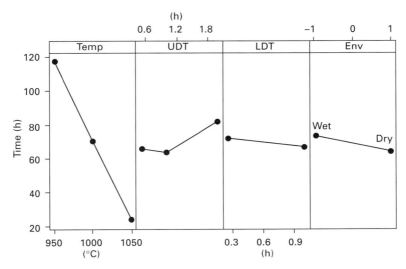

19.23 A plot of the main effects for log of the oxidation rate constant k_n for Alloy 800H (long dwell times).

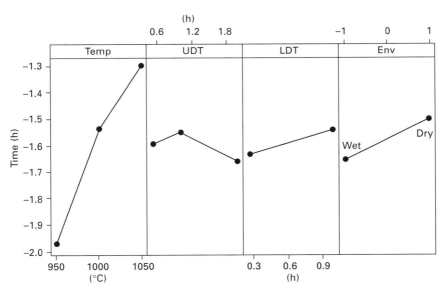

19.24 A plot of the main effects for log of the oxidation rate constant k_n for Alloy 800H (short dwell times).

800H. It can be seen that increasing temperature has the effect of increasing $\Delta m_{net}(t_{300h}\text{-}t_{protective})$ (in absolute terms). Environment appears to have virtually no effect. Decreasing upper dwell time has the effect of increasing $\Delta m_{net}(t_{300h}\text{-}t_{protective})$ (in absolute terms). Only the effects of temperature and

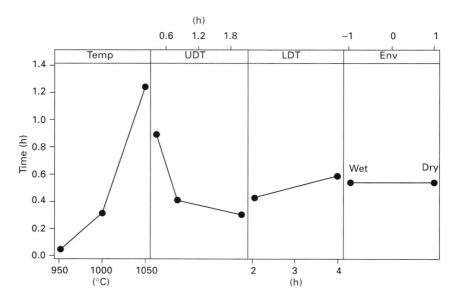

19.25 A plot of the main effects for the difference in net weight change between t_{300h} and $t_{protective}$ ($\Delta m_{net}(t_{300h}\text{-}t_{protective})$) for Alloy 800H (long dwell times).

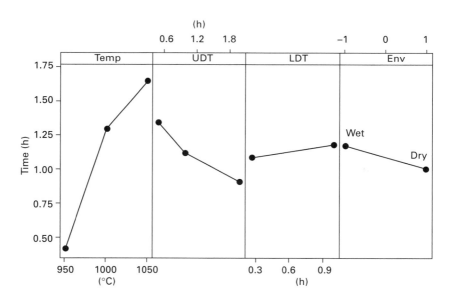

19.26 A plot of the main effects for the difference of net mass change between t_{300h} and $t_{protective}$ ($\Delta m_{net}(t_{300h}\text{-}t_{protective})$) for Alloy 800H (short dwell times).

upper dwell time are statistically significant for both short and long dwell experiments. The effect of lower dwell times is not large enough to be statistically significant.

There are no significant effects for the exponent of the growth law, n.

Statistical evaluation for AISI441

There were insufficient values for a statistical analysis of the net weight change; therefore, the analysis was carried out for the oxidation rate constant, k_n and the protective oxide growth time $t_{protective}$ only.

The long dwell time data show significant effects of upper dwell temperature. In particular, for the log of the rate constant, temperature ($p = 0.00$) has a significant effect, with higher temperature increasing the rate constant (Fig. 19.27). The short dwell time data show significant effects of temperature and environment. In particular, for the log of the rate constant, the temperature ($p = 0.00$) and environment ($p = 0.01$) have significant effects, with higher temperature increasing the oxidation rate constant and a wet environment decreasing the oxidation rate constant (Fig. 19.28).

For the protective oxide growth time ($t_{protective}$), upper dwell temperature ($p = 0.01$) has a significant effect, with higher temperature decreasing the time $t_{protective}$ (Figs: 19.29 and 19.30). For analysis, times to spall have been set at 300 h for those trials with $t_{protective}$ greater than 300 h.

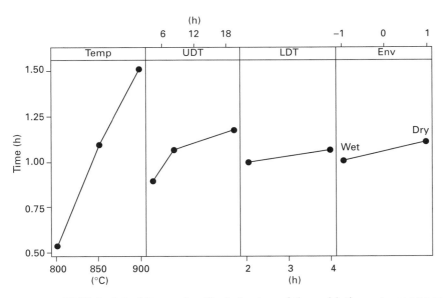

19.27 A plot of the main effects for log of the oxidation rate constant k_n for AISI441 (long dwell times).

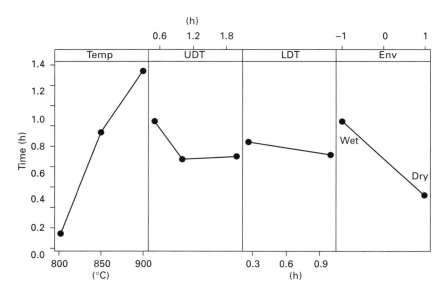

19.28 A plot of the main effects for log of the oxidation rate constant k_n for AISI441 (short dwell times).

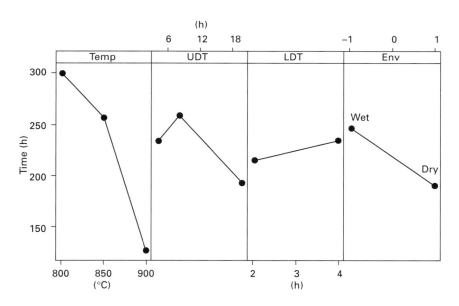

19.29 A plot of the main effects for the protective oxide growth time $t_{protective}$ for AISI441 (long dwell times).

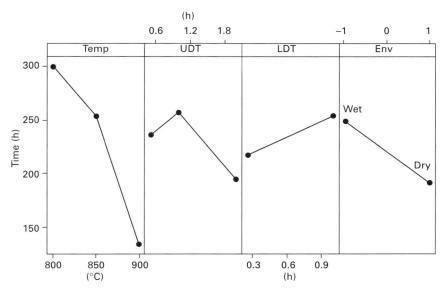

19.30 A plot of the main effects for the protective oxide growth time $t_{protective}$ for alloy AISI441 (short dwell times).

19.4   Discussion

The goal of this work was to gain a better understanding and to systematise the role of different test parameters on the oxidation/spallation behaviour of four selected commercial alloys. It should be underlined that reasonably good reproducibility was generally observed for net and gross weight gain kinetics in comparable experiments until the onset of spallation or until enhanced, breakaway type oxidation occurred. This is not surprising, because, until the onset of spallation/enhanced oxidation, the growth rates of the oxide scale can frequently be approximated by those for isothermal oxidation conditions [9, 10]. After the onset of spallation/enhanced oxidation, the kinetics are strongly affected by variations in the parameters used. Each of the four investigated materials exhibits different oxidation/spallation behaviour so that the effect of the test parameter variables differs from material to material. As a conclusion, some general statements can be made.

Increasing the oxidation temperature in the case of AISI441, Alloy 800H and CM247 results in an increase in the oxide scale growth rate. As long as no scale spalling or breakaway occurs, the oxidation kinetics can be treated as 'quasi-isothermal oxidation' [9, 10]. After the onset of spallation, the oxidation kinetics is strongly affected by the type of oxidised material and the oxidation parameters. For instance, both AISI441 and Alloy 800H exhibit a strong temperature dependence of the oxide scale growth, in contrast to

CM247, which after a short transient oxidation period (*ca.* 24 h), exhibits virtually no temperature effect on the scale growth rate at 1100°C and 1150°C. An increase in the oxidation temperature can strongly decrease the time for the onset of spallation and increase the total amount of spalled oxide. The effect was especially pronounced in the case of Alloy 800H and CM247. In the case of AISI441, the last two effects were less pronounced.

Based on the results presented, the upper dwell time seems to have a much more pronounced influence on the oxidation/spallation kinetics than the lower dwell time. First of all, decreasing the upper dwell time decreases the time for the onset of spallation/enhanced oxidation. This effect was especially evident in the case of Alloy 800H (cf. Figs 19.12–19.15); however, some indications of this effect were also found in the case of CM247 (Fig. 19.18). Moreover, decreasing the upper dwell time increases the total amount of spalled oxide in the case of the chromia-forming alloys. An interesting finding was that the relative scatter in the weight change data between specimens from one trial decreases when the upper dwell time was decreased. In other words, decreasing the upper dwell time improved the reproducibility of the weight change measurements even in the case when strong scale spallation occurred (cf. Figs 19.12 and 19.15). A substantial influence of the length of the lower dwell time on the oxidation/spallation kinetics could not unequivocally be derived.

The influence of environmental humidity at the used level of 2% specific humidity differed from alloy to alloy. However, it is well known that humidity can influence oxidation/spallation kinetics and also evaporation of volatile products formed in the presence of humidity, especially in the case of chromia-forming alloys [11–14]. The strongest effect was observed in the case of the heat-treated specimens of P91 where increasing the environmental specific humidity up to 2% led to a breakaway type oxidation [15, 16]. Enhanced oxidation in this case does not lead to spallation of the oxide scale. For untreated P91 specimens, as well as for AISI441, Alloy 800H and CM247, virtually no effect of humidity on the oxidation/spallation kinetics and amount of spalled oxide could be found.

For completeness, it should be mentioned that cooling rates also could have an influence on spallation behaviour [13, 17, 18]. This parameter has, however, not been investigated in the present work, due to the limited number of trials possible in the timescale of this project.

The statistical evaluation of the experimental data for Alloy 800H and AISI441 with the ANOVA method confirms the significance of these two out of four parameters investigated. According to the analysis presented in Figs 19.21–19.30, the 'statistically significant' parameters are temperature and upper dwell time. The lower dwell time and humidity are 'statistically insignificant'. The 'statistical insignificance' of the parameters does not necessarily mean that the parameters have no effect at all, but that the effect was not sufficiently strong to be detected in a small series of experiments consisting of only nine trials.

19.5   Conclusions

Receiving reliable oxidation/spallation kinetics requires that both the net and the gross weight gain are measured. For each experiment, specimens should be replicated (e.g. triplicated) in order to obtain information about the natural scatter in the weight change data within the specimens. The minimum accumulated upper dwell time should be 300 h or higher (depending on the alloy and oxidation conditions). Generally the most pronounced effects can be summarised as follows.

- Increasing the oxidation temperature results in:
  - increase in the oxidation rate
  - shorter periods of protective oxide growth and, therefore, earlier onset of spallation/enhanced oxidation
  - increase in the total amount of spalled scale (if spallation occurs).
- Decreasing the upper dwell time results in:
  - shorter periods of protective oxide growth and, therefore, earlier onset of spallation/enhanced oxidation
  - increase in the total amount of spalled scale (if spallation occurs)
  - improved reproducibility in weight change data after the onset of spallation.

Increasing the environmental humidity can lead to breakaway processes or to an increasing tendency to spallation of the chromia scales. The lower dwell time at the tested levels seems to have virtually no influence on oxidation/spallation kinetics of the investigated alloys.

Application of a balanced design test matrix combined with evaluation of the experimental data by the ANOVA method seems to be a meaningful method for assessment of oxidation/spallation kinetics data. This method allows the characterisation of the significance of the investigated parameters from a limited number of experimental data.

19.6   Acknowledgement

The work of the present paper was supported financially by the European Commission under the FP5 Measurement and Testing Program (contract number G6RD-CT-2001-00639) in the framework of the COTEST project. The findings of this work will form the basis for the development of an ISO standard for cyclic oxidation testing.

19.7   References

1. COTEST – Cyclic oxidation testing – development of a code of practice for the characterisation of high temperature materials performance, Contract G6RD-CT-2001-00639, European Commission, 2001.
2. J. R. Nicholls and M. J. Bennett, in *Cyclic Oxidation of High Temperature Materials* (Eds. M. Schütze, W.J. Quadakkers), EFC monograph No. 27, Institute of Materials, London, **1999**, 437–470.
3. J. R. Nicholls and P. Hancock, High Temperature Corrosion, *NACE-6*, **1983**, 198–210.
4. S. Leistikow, I. Wolf and H. J. Grabke, *Werkstoffe und Korrosion* **1987**, *38*, 556–562.
5. G. Strehl, G. Borchardt, in *Standardization of Thermal Cycling Exposure Testing* (Eds. M. Schütze, M. Malessa), EFC monograph No. 53, Woodhead Publishing, Cambridge, **2007**, 49–67.
6. L. Antoni and J. M. Herbelin, in *Cyclic Oxidation of High Temperature Materials* (Eds. M. Schütze, W.J. Quadakkers), EFC monograph No. 27, Institute of Materials, London, **1999**, 187–197.
7. M. A. Harper and B. Gleeson, in *Cyclic Oxidation of High Temperature Materials* (Eds. M. Schütze, W.J. Quadakkers), EFC monograph No. 27, Institute of Materials, London, **1999**, 273–286.
8. M. C. Stasik, F. S. Pettit, G. H. Meier, A. Ashary and J. L. Smialek, *Scripta Met. et Mater.* **1994**, *13*, 1645–1650.
9. B. A. Pint, P. F. Tortorelli and I. G. Wright, in *Cyclic Oxidation of High Temperature Materials* (Eds. M. Schütze, W.J. Quadakkers), EFC monograph No. 27, Institute of Materials, London, **1999**, 111–132.
10. P. Vangeli, in *Cyclic Oxidation of High Temperature Materials,* (Eds. M. Schütze, W.J. Quadakkers), EFC monograph No. 27, Institute of Materials, London, **1999**, 198–208.
11. H. Asteman, K. Segerdahl, J.-E. Svensson and L.-G. Johansson, *Materials Science Forum* **2001**, *369–372*, 277–286.
12. H. Asteman, J.-E. Svensson and L.-G. Johansson, *Oxidation of Metals* **2002**, *57*, 193–216.
13. J. L. Smialek, J. A. Nesbitt, C. A. Barrett and C. E. Lowell, in *Cyclic Oxidation of High Temperature Materials* (Eds. M. Schütze, W.J. Quadakkers), EFC monograph No. 27, Institute of Materials, London, **1999**, 148–168.
14. I. Armit and D. P. Holmes, The spalling of steam grown oxide from superheater and reheater tube steels, Technical planning study, pp.76–655, Final Report, Feb. 1978 EPRI, 3414 Hillview Ave., P.O. Box 10412, Palo Alto, CA 94303, USA, 1978.
15. Y. Ikeda and K. Nii, *Transaction of the Japanese Institute of Metals* **1984**, *26*, 52.
16. D. Naumenko, L. Singheiser and W. J. Quadakkers, in *Cyclic Oxidation of High Temperature Materials* (Eds. M. Schütze, W.J. Quadakkers), EFC monograph No. 27, Institute of Materials, London, **1999**, 288–306.
17. H. Echsler, E. Alija Martinez, L. Singheiser and W. J. Quadakkers, *Materials Science and Engineering A* **2004**, *384*, 1–11.
18. S. Rajendran Pillai, N. Sivai Barasi, H. S. Khatak, and J. B. Gnanamoorthy, *Oxidation of Metals* **1998**, *49*, 509–530.

20

Rapid cyclic oxidation tests, using joule heating of wire and foil materials (COTEST)

J R NICHOLLS and T ROSE, Cranfield University, UK
and R HOJDA, Thyssen Krupp VDM GmbH, Germany

20.1   Introduction

A major aim of the COTEST research programme is the development of suitable guidelines and the adoption of a standard set of procedures to ensure reliable data from cyclic oxidation testing at elevated temperatures. Different test procedures have been adopted, depending on the duration of the hot dwell time and the cooling rates that are required. Within the COTEST programme, three regimes have been identified:

1. 'long dwell' – applicable in boilers, superheaters, heat exchangers, steam turbines, industrial gas turbines and chemical process plants that operate for long service intervals and are only shut down for maintenance and plant outages;
2. 'short dwell' – typical of aero-gas turbine operations on short-haul flights and high temperature process plant requiring regular period shutdown;
3. 'ultra-short dwell' – which characterises many applications that involve heating elements, including industrial furnaces, domestic heaters, toasters and other consumer products, plus components such as automotive catalyst carriers, burners and hot gas filter elements.

This chapter addresses the latter – the development of a set of test procedures for the standardisation of rapid thermal cyclic oxidation testing. The duty cycle for such applications may involve temperatures up to, and occasionally in excess of, 1300°C, with heating periods (hot dwell time) between 1 and 10 min typically and cool periods (cold dwell time) typically of similar order, but usually 1 or 2 min.

Thus, the specific objectives of work package 3.3 – 'ultra-short dwell testing' – within the COTEST programme included:

- definition of suitable test conditions
- evaluation of alternative ultra-short cycle test methods
- down selection of the preferred test method

- design of a suitable test facility
- manufacture of a new prototype, rapid cyclic oxidation test facility
- evaluation of the test capability using reference materials
- definition of the ultra-short dwell test procedure.

This chapter reviews the activities undertaken within each of these objectives, with the aim of developing a rapid cyclic oxidation test procedure.

20.2   Background

To achieve the rapid heating rates (50–200°C/s) necessary to ensure rapid hot cycles with durations of 1–10 min, three possible test procedures may be adopted: 'joule heating', 'induction heating' or 'focused light heating'. In each case, the mass per sample surface area must be small, ensuring high power densities coupled with low sample heat capacities. Of these, joule heating and focused light heating show most promise.

20.2.1 Joule heating

Direct heating of resistance wires has been widely used by industry since the 1930s [1–6] for the characterisation of electrical resistance alloys (NiCr, FeNiCr and FeCrAl alloys). The test has been standardised by the ASTM for both austenitic [7] and ferritic [8] materials and has been adopted extensively by manufacturers of resistance materials as one of the tools for quality control during manufacture [4–6]. Life testing of wire samples by joule heating, as specified in these ASTM standards, follows methods originally proposed by Bash and Harsch in 1929 [1]. Over the years, the methodology has been progressively updated [2–6] to take into account new control methods and/or specific application needs. ASTM B76 [7] and ASTM B78 [8] are the latest standards in this areas, as formulated in 2001.

Currently, joule effect heating of resistance wires is used mainly as a comparative test for the control of product quality and uniformity in the resistance heating element industry. Companies such as Thyssen Krupp [5] and Kanthal [6] have developed their own variants of these ASTM standards, and thus one possibility is to develop a methodology for 'ultra-short cycle' oxidation testing based on this established industrial practice and the existing two ASTM standards (ASTM B76 and ASTM B78). Such an approach would readily be accepted by the relevant industries, as it is currently used 'in-house' as a comparative test of product quality. However, its applications would be limited to electrically conducting materials that can be fabricated in wire or ribbon form – essentially ductile metals – and the temperature cycling rate that can be achieved will depend on the material under test (its heat capacity and emissivity) and its geometric form. Temperature control – the main control parameter in the originally proposed test – is recognised as difficult, being a balance between the joule heating effect and heat dissipation largely by radiation, with the temperature of the wire or foil specimen measured using pyrometry, preferably two-colour pyrometry. Changes to the material dimensions and resistivity modifies joule heating, while changes to the surface emissivity due to oxidation will influence the degree of radiative heat loss. For these reasons, many industries have adopted a practice of controlling the test using constant power (representative of many commercial furnace applications) or constant voltage (representative of many domestic thermal loads). The constant voltage approach is considered to be conservative, giving an 'unfair' advantage in lifetime testing to materials that increase their resistance with time [6] since such behaviour will reduce the current drawn with time, resulting in a power loss and lower test temperature. Constant power tests, would lead to temperature variations throughout the test, as changes to the material can lead first to a drop in resistance due to metallurgical changes, then a resistance increase due to oxidation. However, overall the industry has adopted a 'constant power' approach over 'constant voltage' for these quality control applications. However, for a 'rapid thermal cyclic oxidation test', control methods have to be adopted to achieve constant temperature throughout the test. Compared to classical thermogravimetry [9], which is widely adopted for monitoring isothermal oxidation, joule heating studies on wires cannot directly measure oxygen uptake, the onset of spallation and spalling rates. Instead changes in 'hot' and 'cold' resistance are measured and related to wire/ribbon lifetime.

20.2.2 Induction heating

Like joule heating, induction heating is also limited to electrically conducting samples if rapid heating and cooling cycles are to be achieved. For non-conducting materials or samples of complex shape, a secondary susceptor can be used, but its heat capacity can limit heating and cooling rates that can be achieved. Induction heating provides more freedom in the design of samples, but still requires that temperature cycling rate depends on the sample heat capacity and again control and measurement of sample temperatures is difficult. Thermo-couples will suscept in the RF heating field and thus temperature measurement and control depends on pyrometry, similarly to the joule heating approach.

Overall, induction heating methods offer no advantage over joule heating, save the ability to use rectangular, disk or cylindrical specimens, rather than wire, ribbon or foil samples.

20.2.3 Focused light heating

Originally proposed within COTEST as an alternative to joule heating, the use of focused light as a heat source permits traditional thermogravimetric methods for monitoring oxidation, coupled with rapid heating and cooling cycles. Table 20.1 summarises the potential benefits of such a focused light heating approach, compared to joule heating. Two focused light test geometries were considered, one based on a focused light heated micro-thermobalance, with a balanced test specimen design aimed at mininising buoyancy effects, and the second aimed at using a multiple element focused light furnace to test multiple samples (up to six) in a single ultra-short dwell test.

Clearly, the use of focused light heating offers potential for providing an ultra-fast cycle test facility. There are no material limitations and few geometric constraints (only those classically associated with thermogravimetry). Again, as with the joule heating alternative, the temperature cycling rate is dependent on the sample heat capacity and emissivity. Unfortunately, at the time of preparing this standard, few facilities exist to evaluate material performance, therefore, there is no underlying established test methodology and as a result there is little industrial interest in this approach. Academically, it offers a new experimental approach to study rapid cycling behaviour and since the commencement of the COTEST project, one such facility has been constructed by D. Monceau and his group [10, 11]; both a single light furnace microbalance [10] and a multiple specimen capability are available. Figure 20.1 illustrates schematically the focused light furnace from the single microbalance, cyclic oxidation test facility at INPT in Toulouse [10].

Table 20.1 Performance comparison between focused light heating and joule heating for ultra-short dwell cyclic oxidation test procedures

Information requested	Joule	Focused light thermobalance	Multiple samples focused light
Resistance as $f(t)$			
Cold	++	0	++
Hot	++	0	++
End of test (lifetime)	++	++	++
Mass gain as $f(t)$	– (at test completion)	++	++ (discontinuous measurement)
Mass spall at $f(t)$	0 (?)	0	?
Breakaway point	?	++	?
Cooling rate	?	0(?)	?
Temperature absolute	Pyrometer	Pyrometer	Pyrometer thermocouple?
Cold			
Hot			
Time of cycle hot/cold	++	++	++
Industry interest	++	?	?
Approved system	++	?	?
Type of material	0	++	++

++ benefit   0 not possible
– limited   ? unknown whether measurable

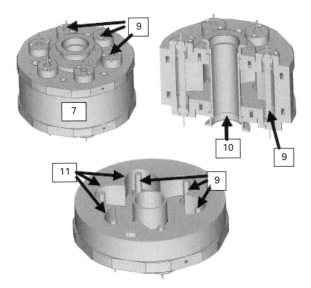

20.1 Schematic of a focused light furnace for a microbalance: (7) lamp furnace housing; (9) tungsten-halogen lamps; (10) specimen test cell; (11) Parabolic mirrors] [10].

Following this review of possible rapid heating/cooling methods, one must conclude that although a focused light approach offers promise, the only viable technology for an industrially accepted test methodology is joule heating.

20.3   Design of a joule heated, ultra-short dwell cycle test facility capable of materials testing in controlled environments

From a cyclic oxidation point of view, a major limitation of the current ASTM test standards is that tests are conducted in laboratory air, with no facility to control the oxidation environment. It was therefore decided that the COTEST programme should address 'ultra-short dwell' (rapid cycle) testing in controlled atmospheres. A four test station, joule heating, rapid thermal cycle test facility has been constructed at Cranfield University using individual 'bell jars' to provide controlled gas feeds and vacuum extraction of exhaust gases. Figure 20.2(a)–(c) illustrates the prototype rapid thermal cycling rig design.

This new facility can test to the ASTM standards with the test cycle controlled by either power, voltage or current. As explained earlier, power is preferred (see [6] for a more in-depth explanation). However, it is recognised that throughout the test the specimen temperature will change, because of

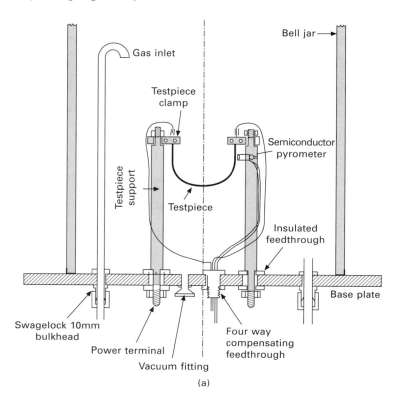

Gas inlet

Bell jar

Testpiece clamp

Semiconductor pyrometer

Testpiece support

Testpiece

Insulated feedthrough

Base plate

Swagelock 10mm bulkhead

Power terminal

Four way compensating feedthrough

Vacuum fitting

(a)

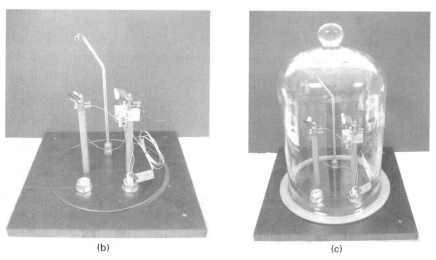

(b)   (c)

20.2 Prototype, environment controlled, joule heating, rapid thermal cycling rig: (a) schematic of rig design (b) prototype rig bell jar removed (c) prototype rig, within bell jar.

changes to the material's properties, microstructure and surface condition – the latter due to reaction with the controlled test environment resulting in the formation of corrosion scales and loss in material cross-section.

Controlling specimen temperature is exceedingly difficult, especially in a closed loop mode, because of the small sample dimensions, the need to use pyrometry, and the necessary incorporation of an environmental cell (the bell jar) around each sample. Thus in this study temperature is controlled by power to the wire/foil sample, pre-calibrated using a pyrometer. Intermittently through the test, manual temperature measurements using the pyrometer allows updating of the power/temperature calibration curve, so ensuring temperature control within ±10°C. Figure 20.3 illustrates these temperatures/power density calibration curves for the three sample geometries of Kanthal A1 (0.4 mm dia., 0.7 mm dia. and 1.25 mm × 70 μm ribbon) evaluated in this study.

Table 20.2 summarises the initial operating conditions for Kanthal A1 wire samples, to achieve a temperature of 1200°C. Wire diameters of 0.4 mm and 0.7 mm are tested with a length of 224 mm. Under these conditions, a higher surface loading (peak surface load of 29.5 W/cm^2) is required for the 0.4 mm dia. wire, than for the 0.7 mm dia. wire (peak surface load of 24.8 W/cm^2) to achieve the same operating temperature, 1200°C. However, due to the larger surface area of the large diameter wire, the 0.7 mm dia. wire requires more power to achieve 1200°C, 122 W for the 0.7 mm dia. wire compared to 83 W for the 0.4 mm dia. wire.

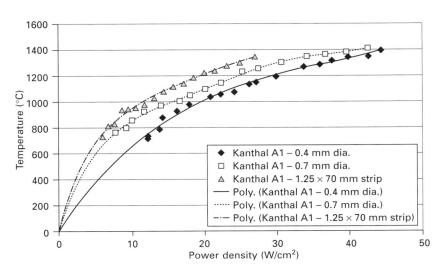

20.3 Specimen temperature/power density calibration curves for three specimen geometries, at start of test.

Table 20.2 Initial operating conditions for 1200°C peak temperature

0.4 mm dia. Kanthal A1		
Length = 224.0mm	Cross-section area = 0.126 mm^2	Start resistance = 13.062 ohms/m
Diam. = 0.40 mm	Surface area = 281.5 mm^2	Surface loading = 29.5 W/cm^2 (peak) 14.8 W/cm^2 (rms)
0.7 mm dia. Kanthal A1		
Length = 224.0mm	Cross-section area = 0.385 mm^2	Start resistance = 4.160 ohms/m
Diam. = 0.70 mm	Surface area = 492.6 mm^2	Surface loading = 24.8 W/cm^2 (peak) 12.4 W/cm^2 (rms)

By way of comparison, Jönsson *et al.* [6] used Kanthal AE and Kanthal AF as reference materials in their recent paper on 'Cyclic oxidation testing by resistance heating'. The wire samples were U-shaped, 224 mm long and 0.7 mm dia. and therefore similar to the larger diameter wires used in this study. The samples were tested in constant power mode using a 2 min on/2 min off cycle at a power of 155 W (a peak surface load of 31.5 W/cm^2). The temperature achieved at the hottest part of the sample was 1265°C. Such results agree well with the trends observed in this study, confirming our belief that this new rapid cycle, environmental test facility can test to the ASTM standard.

Under constant power conditions (32 W/cm^2 peak power density) and using a 2 min on/2 min off cycle – similar test conditions to those used by Jönsson *et al.* [6], but testing Kanthal Al rather than Kanthal AE or AF – the resistance was observed to initially increase rapidly, then more slowly before finally rapidly rising at failure (Fig. 20.4). The shape of curve was similar for both 0.4 mm dia. and 0.7 mm dia. wires, but the percentage change in resistance was higher for higher initial operating temperatures and the lifetimes were shorter for the higher initial temperature. Both results are consistent with more rapid oxidation and therefore increased rates of section loss at the higher operating temperature. Again comparing data with that published by Jönsson *et al.* [6], initial peak temperatures of the 0.4 mm dia. and 0.7 mm dia. wires were respectively 1235°C and 1328°C, compared to a reported value of 1265°C by Jönsson *et al.* [6]. One should note that although all alloys under test were FeCrAl alloys, the alloys were of different compositions between the current study and that of Jönsson *et al.*

One may conclude this section by stating that the new environmental controlled, rapid cyclic oxidation test facility can test to the ASTM standard for cycles controlled by power to the wire, or ribbon samples and that, within available data, the new equipment produces comparable results to those published previously in the literature.

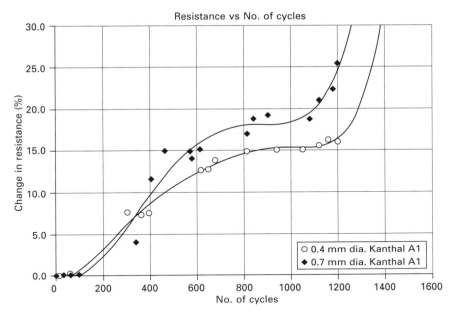

20.4 Constant power tests of Kanthal A1 using a 2 min on/2 min off cycle at a peak power density of 32 W/cm².

20.4 Statistical design of a test matrix to investigate critical parameters controlling ultra-short dwell cyclic oxidation tests

Using the above facility, a statistically designed test matrix based on a balanced 'Latin square' approach (see Chapter 18) has been used to investigate the role of:

- test temperature
- number of cycles
- hot dwell time
- sample geometry

on component (wire/foil samples) lifetime, as per COTEST work package 5. In the study reported in this chapter, the environment is kept constant (laboratory air) as is the lower temperature dwell time (2 min). Equally for this study only one reference condition is examined, namely:

- Kanthal A1
- oxidation in laboratory air at 1250°C
- 5 min hot dwell/2 min cool cycle (cold dwell)
- 0.4 mm dia. wire × 224 mm long.

In the complete study, both material and environmental effects will be addressed. A second reference material, Alloy 800, tested in laboratory air at 1000°C will be included, as well as the influence of two levels of water vapour on the ultra-short dwell cyclic oxidation behaviour. These results will be reported in subsequent papers from this European collaborative research study (COTEST).

The 'Taguchi designed' test matrix to be investigated in this study is summarised in Table 20.3, and consists of a $3 \times 3 \times 3$ matrix from which nine trials are taken in a balanced 'Latin square' configuration (see Chapter 18), with triplicate repeat tests under each trial condition. In total 27 ultra-short dwell cyclic oxidation tests were completed, either to sample failure, or 1200 cycles if samples survived the particular set of test conditions – in excess of 4000 h of rapid cyclic oxidation testing, for this one test condition.

Figures 20.5–20.8 illustrate typical results and Table 20.4 summarises all of the outcomes of the tests. Across all tests temperature control was better than $\pm 15°C$ (2 standard deviations). At 1200 and 1250°C the wire samples (either 0.4 mm dia. or 0.7 mm dia.) usually survived 1200 cycles of testing, whether the test cycle was 2 min, 5 min or 10 min hot dwell time; Figs 20.5 and 20.6 illustrate this behaviour at 1200°C. Of the 12 wire samples tested under these conditions, only two failed in under 1200 cycles. In contrast, the 70 µm foil samples all failed early (see Fig. 20.7, for example, where at 1250°C, the lifetime was 600 cycles × 5 mins hot time, i.e. 50 hot hours). The equivalent test for the 0.4 mm dia. wire, Fig. 20.7, lasted 1200 cycles, i.e. 100 hot hours.

At 1300°C, all but one of the samples failed in under 1200 cycles, and this one sample failed at 1204 cycles, irrespective of the hot dwell time. Figure 20.8 illustrates this high temperature behaviour for the 0.7 mm dia. wire tested using a 5 min on 2 min off power cycle. The illustrated sample had a lifetime of 115 cycles. This difference in behaviour between the 1300°C

Table 20.3 The test matrix for Kanthal A1 – influence of specimen geometry, displayed in a Taguchi design format

Trial	Upper dwell temperature	Upper dwell time	Specimen geometry	Number of repeat tests completed
1	1200°C	2 min	0.7 mm ϕ wire	✓✓✓
2	1200°C	5 min	0.4 mm ϕ wire	✓✓✓
3	1200°C	10 min	70 µm foil	✓✓✓
4	1250°C	2 min	0.4 mm ϕ wire	✓✓✓
5	1250°C	5 min	70 µm foil	✓✓✓
6	1250°C	10 min	0.7 mm ϕ wire	✓✓✓
7	1300°C	2 min	70 µm foil	✓✓✓
8	1300°C	5 min	0.7 mm ϕ wire	✓✓✓
9	1300°C	10 min	0.4 mm ϕ wire	✓✓

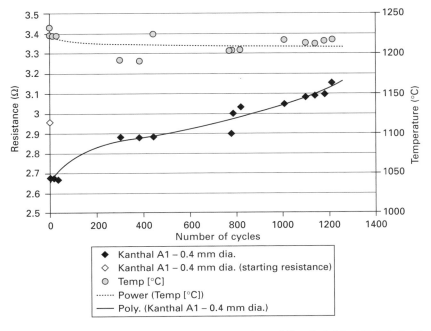

20.5 Ultra-short dwell cyclic oxidation of Kanthal A1: 1200°C, 2 min hot dwell/2 min cool in air, 0.4 mm dia. wire.

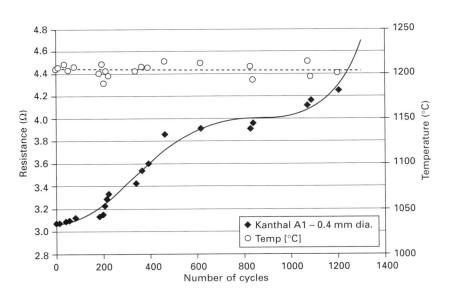

20.6 Ultra-short dwell cyclic oxidation of Kanthal A1: 1200°C, 5 min hot dwell/2 min cool in air, 0.4 mm dia. wire.

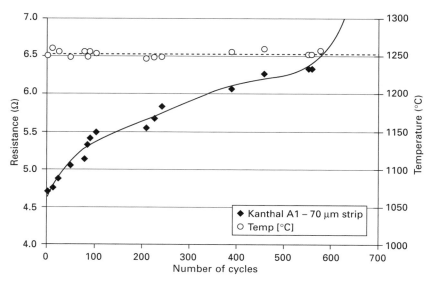

20.7 Ultra-short dwell cyclic oxidation of Kanthal A1: 1250°C, 5 min hot dwell/2 min cool in air, 70 µm thick ribbon specimen.

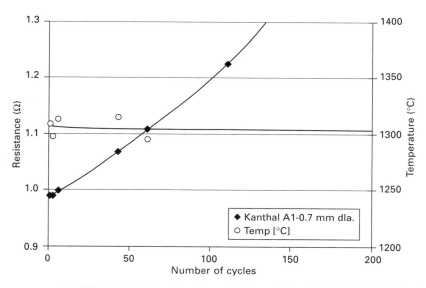

20.8 Ultra-short dwell cyclic oxidation of Kanthal A1: 1300°C, 5 min hot dwell/2 min cool in air, 0.7 mm diameter wire.

tests and those at lower temperatures for the wire samples is clearly evident in the mode of oxide scale growth, cracking and oxide spallation (see Fig. 20.9 which compares two samples one tested at 1300°C and one at 1200°C but after similar test durations –1200 cycles at 1200°C and 1204 cycles at

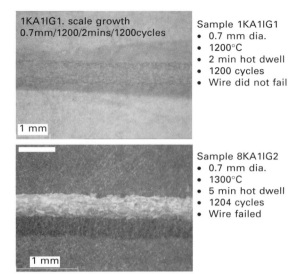

20.9 Comparison of Kanthal A1 wires after 1200 × 5 min hot dwell cycles at 1200 and 1300°C.

1300°C). At 1300°C, the oxide scale is severely cracked, with remaining oxide taking on an oyster shell appearance interspersed between regions of spall oxide, while at 1200°C the oxide that spalls is much finer and the 'oyster shell' remnant oxide is not observed. Clearly, at 1300°C, the more rapid creep within the substrate leads to specimen elongation. This, together with the more rapid oxide growth, contributes to this characteristic mode of oxide failure on high temperature, rapidly cycled FeCrAl based wire material.

In contrast, the foil material failed in a different way. First, the foil was observed to curl, forming a tube shape (Fig. 20.10). When the tubular shape had completely formed, then shortly afterwards the foil failed. Figure 20.10 illustrates such behaviour for two samples: one failed after 521 × 2 min hot cycles at 1300°C the second is pictured after 620 cycles but finally fails in a similar manner to the first after 673 cycles.

20.5   Statistical analysis of the rapid thermal cycle tests on Kanthal A1 wire/foil samples tested in laboratory air

The balanced 'Taguchi designed' test matrix (see Table 20.3), because of its 'Latin square' arrangement is ideally suited for statistical analysis of the rapid thermal cycle test behaviour. The main effects of test temperature, upper dwell time and specimen geometry can be investigated. The three

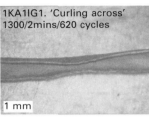

1KA1IG1. 'Curling across'
1300/2mins/620 cycles

1 mm

Sample 7KA1IG3
- 70 µm foil
- 1300°C
- 2 min hot dwell
- 620 cycles
- Foil failed at 673 cycles

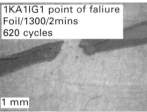

1KA1IG1 point of faliure
Foil/1300/2mins
620 cycles

1 mm

Sample 7KA1IG1
- 70 µm foil
- 1300°C
- 2 min hot dwell
- 521 cycles
- Foil failed

20.10 Comparison of two Kanthal A1 foil specimens tested at 1300°C, using a 2 min hot dwell/2 min cool in air test cycle.

repeats of each test condition provides sufficient degrees of freedom that the statistical significance of these main effects can be investigated using analysis of variance (ANOVA) (see Chapter 12 and 18). If a complete $3 \times 3 \times 3$ test matrix had been undertaken, then interactions between these main effects could also have been investigated in depth. However, in this current study only one-third of the total possible tests were undertaken, but in a balanced manner, such that the main effects that have a significant influence on the ultra-short dwell (rapid thermal cycle) test could be investigated. There are sufficient degrees of freedom to determine whether interaction effects may be important but not to uniquely identify them. Table 20.4 summarises the experimental results of these 27 rapid thermal cycle trials. From this data, the 'main effects' table, Table 20.5, can be constructed (see reference 11 for examples taken from oxidation and corrosion studies).

By inspection of Table 20.5, it can be seen that the mean cycle life decreases with increase in temperature and also decreases with increase in hot dwell time. These two observations are better illustrated in Figs 20.11 and 20.12. Figure 20.11 illustrates the effect hot dwell temperature has in reducing the cyclic lifetime of Kanthal A1 wire/foil material, while Fig. 20.12 illustrates how increasing the hot dwell time at 1200°C increases the change in resistance. Even though both samples survived 1200 hot cycles, it is evident that increasing the hot dwell time increases the extent of damage, as evident by the increased change in resistance of the 0.4 mm dia. wire samples. Given these observed trends, it would appear that both temperature and hot dwell time are important factors, but are they statistically significant relative to experimental errors in the test procedure?

Table 20.4 Measured lifetime data from rapid thermal cycle tests undertaken according to the Taguchi designed test matrix

Test conditions				Reference number	Measured experimental data			Change in resistance (%)	Change in power (%)
Temperature (°C)	Hot dwell (min)	Geometry	Dia./Thick. (mm)		Temperature (°C)	Cycle life (cycles)	Lifetime (h)		
1200	2.0	wire	0.700	1KA1IG1	1203.6	1273	42.43	13.65	24.76
1200	2.0	wire	0.700	1KA1IG2	1202.6	1206	40.20	5.11	16.57
1200	2.0	wire	0.700	1KA1IG3	1201.7	952	31.73	4.55	20.76
1200	5.0	wire	0.400	2KA1IG1	1204.9	1202	100.17	38.65	33.73
1200	5.0	wire	0.400	2KA1IG2	1201.2	1208	100.67	38.54	31.53
1200	5.0	wire	0.400	2KA1IG3	1205.1	1206	100.50	25.13	5.62
1200	10.0	foil	0.070	3KA1IG1	1199.8	1004	167.33	49.41	19.26
1200	10.0	foil	0.070	3KA1IG2	1201.4	815	135.83	43.03	19.37
1200	10.0	foil	0.070	3KA1IG3	1203.2	805	134.17	33.06	17.12
1250	2.0	wire	0.400	4KA1IG1	1253.5	1206	40.20	37.12	27.83
1250	2.0	wire	0.400	4KA1IG2	1252.5	1220	40.67	17.13	6.99
1250	2.0	wire	0.400	4KA1IG3	1250.2	1206	40.20	23.73	18.84
1250	5.0	foil	0.070	5KA1IG1	1252.3	611	50.92	39.52	24.17
1250	5.0	foil	0.070	5KA1IG2	1251.9	638	53.17	39.63	24.51
1250	5.0	foil	0.070	5KA1IG3	1253.5	90	7.50	3.86	0.31
1250	10.0	wire	0.700	6KA1IG1	1251.5	1200	200.00	21.83	25.75
1250	10.0	wire	0.700	6KA1IG2	1252.8	1207	201.17	22.02	24.15
1250	10.0	wire	0.700	6KA1IG3	1252.3	39	6.50	0.00	1.65
1300	2.0	foil	0.070	7KA1IG1	1301.3	521	17.37	43.22	26.16
1300	2.0	foil	0.070	7KA1IG2	1302.2	1153	38.43	34.53	12.75
1300	2.0	foil	0.070	7KA1IG3	1300.4	673	22.43	36.52	9.74
1300	5.0	wire	0.700	8KA1IG1	1303.5	115	9.58	19.10	16.42
1300	5.0	wire	0.700	8KA1IG2	1300.6	1204	100.33	47.44	40.87
1300	5.0	wire	0.700	8KA1IG3	1301.0	1112	92.67	37.04	32.95
1300	10.0	wire	0.400	9KA1IG1	1302.0	612	102.00	32.14	15.26
1300	10.0	wire	0.400	9KA1IG2	1298.3	641	106.83	34.36	16.16
1300	10.0	wire	0.400	9KA1IG3	1300.3	677	112.83	41.47	18.16

Table 20.5 Main effects table from rapid thermal cycle tests

Main effects	Mean temperature (°C)	Mean cycle life (cycles)	Mean lifetime (h)	Mean change in resistance (%)	Mean change in power (%)
	1202.6	1075	94.8	27.9	21.0
Temperature	1252.3	824	71.1	22.8	17.1
	1301.1	745	66.9	36.2	20.9
Main effects	Hot dwell (min.)	Mean cycle life (cycles)	Mean lifetime (h)	Mean change in resistance (%)	Mean change in power (%)
	2.0	1046	34.9	24.0	18.3
Hot Dwell	5.0	821	68.4	32.1	23.3
	10.0	778	129.6	30.8	17.4
Main effects	Geometry	Mean cycle life (cycles)	Mean lifetime (h)	Mean change in resistance (%)	Mean change in power (%)
	wire = 0.7 mm	923	80.5	19.0	22.7
Geometry	wire = 0.4 mm	1020	82.7	32.0	19.3
	foil = 0.07mm	701	69.7	35.9	17.0

In contrast, the overall lifetime would appear to increase with increase in hot dwell time, suggesting that under rapid thermal cycle test conditions failure is dominated by the number of cycles and not by the total hot time accumulated. Such behaviour agrees with lifetime trends expected within the industry [4, 6] where minimum life is associated with short duration cycles. The 2 min on/2 min off cycle is often used in industrial tests as the standard cycle test condition [6]. Change in resistance and change in power, necessary to maintain the test temperature, do not appear to vary systematically with test temperature or hot dwell time and therefore are unlikely to be significant factors. However, both of these parameters vary in a consistent manner with sample geometry. The percentage change in resistance increases, and the percentage change in power decreases, as the cross-sectional area of the test sample decreases, or the geometrical shape of the specimen changes. This latter effect is most notable for the foil samples. These had shorter cycle lives and lifetimes than the wire samples. Whether this is a significant factor must be determined based on the ANOVA tests.

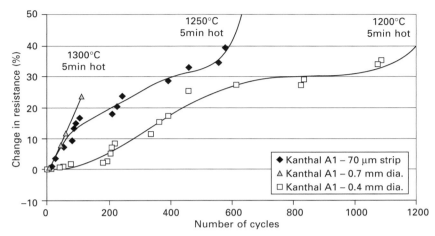

20.11 Influence of temperature on the ultra-short cycle cyclic behaviour of Kanthal A1, using a 5 min hot dwell/2 min cool in air test cycle.

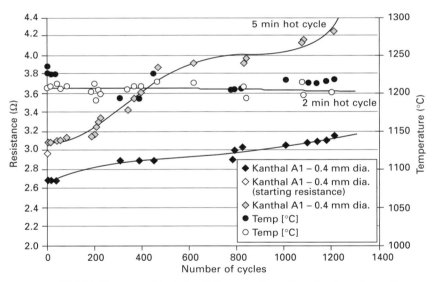

20.12 Influence of hot dwell time on the ultra-short cycle cyclic oxidation behaviour of Kanthal A1 at 1200°C.

20.5.1 ANOVA for rapid thermal cycle tests on Kanthal A1, over the temperature range 1200–1300°C

The analysis of variance (ANOVA) for these ultra-short dwell cyclic oxidation tests on Kanthal A1 wire and foil specimens is reproduced in Tables 20.6 and

Table 20.6 Summary statistics for ultra-short dwell oxidation tests

(a) Sum of squared deviations

Main effects	d.f.	Cycle life (cycles)	Lifetime (h)	Change in resistance (%)	Change in power (%)
Temperature	2	531946.9	4053.9	828.1	87.6
Hot dwell	2	372358.2	41574.0	345.7	184.5
Geometry	2	480530.7	872.1	1411.7	143.1
Interaction	2	257264.2	3838.2	478.2	90.6
Residual	18	2126756.0	32566.9	2186.9	1960.9
Total	26	3768856.0	82905.0	5250.6	2466.7

(b) Mean squared deviations

Main effects	d.f.	Cycle life (cycles)	Lifetime (h)	Change in resistance (%)	Change in power (%)
Temperature	2	265973.4	2026.9	414.0	43.8
Hot dwell	2	186179.1	20787.0	172.8	92.3
Geometry	2	240265.3	436.1	705.9	71.6
Interaction	2	128632.1	1919.1	239.1	45.3
Residual	18	14292.5	1809.3	121.5	108.9
Total	26	144956.0	3188.7	201.9	94.9

Table 20.7 Analysis of variance (ANOVA) for ultra-short dwell oxidation tests

(a) *F*-values

Main effects	d.f.	Cycle life (cycles)	Lifetime (h)	Change in resistance (%)	Change in power (%)
Temperature	2	18.61	1.12	3.41	0.40
Hot dwell	2	13.03	11.49	1.42	0.85
Geometry	2	16.81	0.24	5.81	0.66
Interaction	2	9.00	1.06	1.97	0.42
Residual	18	1.00	1.00	1.00	1.00
Total	26				

(b) Fisher *F* statistics

$F\,(0.050;2,18)$	3.55
$F\,(0.010;2,18)$	6.01
$F\,(0.005;2,18)$	7.21
$F\,(0.001;2,18)$	10.40

20.7. Table 20.6 reproduces the 'sum of squares' data for cyclic life, endurance (lifetime calculated from hot dwell time × number of cycles), change in resistance and change in power required to maintain constant temperature.

Also included are the main effects: 'temperature', 'hot dwell time' and 'geometry'; the associated degrees of freedom and estimates of the 'sum of squares' due to residual error and any interaction effects. For examples on how to undertake an ANOVA analysis, with case studies taken from oxidation and corrosion studies, see [11].

By dividing each 'sum of squares' by its associated 'degree of freedom', one calculates the 'mean squared error' (see also Table 20.6) for each main effect, any interaction effects and the residual error. To determine whether a given factor is significant, one compares the ratio between that factors' 'mean squared error' and the 'residual error'. These ratios are known as 'F-values' and can be compared with the Fisher-F statistic, used to compare variances – hence ANOVA – for a given probability of being in error and for two degrees of freedom that define this statistic. Table 20.7(a) presents the calculated F-values and Table 20.7(b) the Fisher-F statistics, for probability levels of 0.05, 0.1, 0.005 and 0.001 at 2, 18 degrees of freedom.

If the F-values are greater than the corresponding Fisher-F statistic, then the result is statistically significant at the given probability level, where the probability level defines the chance of being in error (i.e. $\alpha = 0.05$ corresponds to a 5% chance of being error, while $\alpha = 0.001$ corresponds to a 0.1% chance of error). If the F-value is less than the Fisher-F statistic, then this value is not statistically significant.

From Table 20.7, is can be deduced that, as far as cycle life is concerned, all the main effects are highly significant (0.1% change or error), also there is a significant interaction effect, although there is not sufficient tests/degrees of freedom to uniquely identify what this interaction may be. Inspection of Tables 20.4 and 20.5 suggest two possibilities: these are the already noted failure of the 70 μm foil/ribbon samples at less than 1200 cycles, irrespective of the test temperature or hot dwell time, and the observed early failures of all specimen geometries at 1300°C.

By calculating the wire/foil endurance, the hot lifetime – a product of hot dwell time and number of cycles – one finds that now only 'hot dwell time' is a significant life-limiting factor over the temperature range 1200–1300°C for Kanthal Al, irrespective of sample geometry. With this normalisation, both temperature effects and geometry effects are within the residual error of the measurements and therefore not statistically significant. This dependence of electrically heating wire lifetime on the hot dwell time was first identified in 1968 by Zawadzka [4]. He referred to the 'hot dwell time' as 'cycle time' and showed that relative to a 'non-cycled' test the lifetime rapidly dropped for short 'hot dwell times', before increasing again with increased 'hot dwell time' to values close to the 'non-cycled' life when few thermal cycles are involved. It must be for this reason that Jönsson *et al.* [6] noted that a 2 min on/2 min off cycle was the most exacting of cycles, often used in industry for quality control testing of heating element resistance wires [6].

20.6   Conclusions

- A test procedure for testing wire and foil samples under different environmental conditions, using rapid thermal cycling – ultra-short dwell – has been developed. The test procedure meets and can test to ASTM 76B and 78B, whilst extending these test standards by permitting wires/foils to be evaluated in controlled gaseous atmospheres at high temperature.
- The test procedure has been evaluated by using a Taguchi design, balanced subset of the 27 tests in a $3 \times 3 \times 3$ test matrix. Nine test conditions were evaluated with three repeat samples at each test condition.
- An analysis of variance (ANOVA) of this ultra-short dwell, rapid cyclic oxidation dataset confirmed that 'temperature', 'hot dwell time' and 'sample geometry' were significant factors controlling the cycle life of Kanthal A1 wire and foil samples when rapidly thermal cycled using joule heating. A notable interaction between these main effects was observed, with foil samples failing earlier than wire samples. For many wire samples of 0.4 mm dia. and 0.7 mm diameter no failures were observed up to test termination (1200 cycles) at 1200 and 1250°C.
- For tests on Kanthal A1 wire and foil samples, cycle life decreased with increase in temperature and also decreased with increase in 'hot dwell time'. The significant geometry effect is though to be due to the difference in behaviour observed between wire and foil specimens.
- When the hot endurance (wire/foil hot lifetime) is calculated by multiplying the 'hot dwell time' by the number of cycles only one statistically significant parameter results, that is the 'hot dwell time'. For short duration 'hot dwells' (2 min on/2 min off) the wire hot lifetime is significantly reduced, down to a quarter or a fifth of that observed for the longest cycle condition (10 min on/2 min off). This observation confirms the historical understanding within the heating element wire industry that the 2 min on/2 min off is a most exacting test condition for evaluating heating element wire quality control.

20.7   Acknowledgements

The authors wish to thank the European Commission for funding this research as part of contract G6RD-CT-2001-00639 Cyclic oxidation testing – development of a code of practice for the characterisation of high temperature materials performance (COTEST). The authors further thanks Prof. M. Schütze of DECHEMA for leading this project and Dr S. Coleman for her invaluable advice over the statistical analysis undertaken in support of this paper.

20.8   References

1. F. E. Bash and J. W. Harsch, 'Life Testing on Metallic Resistor Materials for Electrical Heating', *Proc. Amer. Soc. Test. Mater.* **29**, 506–522 (1929).
2. W. Fischer, *Int. Z. F. Elektrowarme* **10**, 59–64 (1940).
3. H. Frinken, *Int. Z. F. Elektrowarme* **24**, 217 (1966).
4. I. Zawadzka, Int. Z. *Electrowarme International* **26**, 123–132 (1968).
5. L. Rademacher and H. Stein, *Thyssen Eddst. Techn. Berr.* **9**, 52–60 (1983).
6. B. Jönsson, A. Westerlund and G. Landor, in *Cyclic Oxidation of High Temperature Materials* (Eds. M. Schütze and W. J. Quadakkers), EFC Publication No. 27, IoM Communications, London, 324–335 (1999).
7. ASTM B76 (2001) 'Accelerated Life Test of Nickel-Chromium and Nickel-Chromium-Iron Alloys for Electrical Heating'.
8. ASTM B78 (2001) 'Accelerated Life Test of Iron-Chromium-Aluminium Alloys for Electrical Heating'.
9. H. J. Grabke, Thermogravimetry in *Guidelines for Methods of Testing and Research in High Temperature Corrosion*, EFC-14, Institute of Materials, London 62–84 (1995).
10. J-C. Salabura and D. Monceau, *Material Science Forum* **461–464**, 689–696 (2004).
11. J. R. Nicholls and P. Hancock, *The Analysis of Oxidation and Hot Corrosion Data – A Statistical Approach, High Temperature Corrosion*, NACE-6, 198–210 (1983).

21

Thermo-mechanical fatigue – the route to standardisation (TMF-STANDARD)

T B E C K, Forschungszentrum Jülich GmbH, Germany;
P H Ä H N E R, Joint Research Centre, The Netherlands;
H - J K Ü H N, BAM, Labor V 21, Germany; C R A E,
University of Cambridge, UK; E E A F F E L D T, MTU
AeroEngines GmbH & Co. KG, Germany; H A N D E R S S O N,
SIMR, Sweden; A K Ö S T E R, ENMSP/ARMINES, France and
M M A R C H I O N N I, CNR-IENI – TEMPE, Italy

21.1   Introduction

During recent decades thermo-mechanical fatigue (TMF) testing has become an increasingly important method in design, materials performance and reliability assessment, and the residual life analysis of materials subjected to simultaneous thermal and mechanical loading, such as turbine blades, combustion chambers, pistons and cylinder heads of internal combustion engines. Detailed descriptions of current TMF testing techniques and results of TMF experiments on Ni-base superalloys, steels, intermetallics and aluminum alloys performed by several working groups worldwide can be found in [1–3].

Figure 21.1 shows the principle of thermally induced loading and its experimental simulation in a TMF test in terms of the example of a cooled turbine blade. Owing to the transient temperatures in the start-up and shut-down phases of a service cycle and the temperature gradients arising during stationary service, the thermal strain at the 'hot' outer side of the blade is partially constrained by the colder volume elements inside the blade. This leads to compressive loadings with increasing temperature at the 'hot' side ① and, as a consequence of the equilibrium of stresses, to tensile loadings on the 'cold' side ② of the component.

These thermally induced loadings are simulated by strain controlled TMF tests in such a way that the temperature–time path occurring, e.g. at location ① or ② of the component, is imposed on the gauge section of a laboratory specimen. At the same time the mechanical loading which results from the constraint of the local thermal strain is imposed on the specimen by controlling the total strain in an adequate way:

$$\varepsilon_t = \varepsilon_{me} + \varepsilon_{th} \qquad\qquad 21.1$$

where ε_t is the total strain at the specimen, ε_{th} is the thermal strain in the

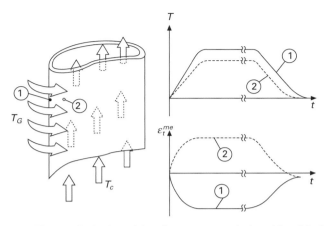

21.1 Thermally induced loadings at a cooled turbine blade and their simulation by TMF testing.

absence of external forces and ε_{me} is the mechanical strain, which exhibits a phase shift of 180° with respect to the temperature and thermal strain vs time path in the case of volume element ① of the considered component, or 0° phase shift in the case of volume element ②, respectively (so-called 'out-of-phase' and 'in-phase' TMF cycle, respectively).

From these basic considerations about TMF testing, it is concluded that there are various challenges in conducting such experiments, which go far beyond the practicalities of isothermal low-cycle fatigue (LCF) experiments, such as:

- the dynamic measurement and control of the temperature have to be assured within reliable tolerances
- the phase angle between temperature and mechanical strain has to be controlled as exactly as possible
- the thermal strain must be measured as a function of temperature and/or time to determine the control variable ε_t from the desired mechanical strain vs time path according to Eq. 21.1
- appropriate start-stop procedures to maintain the dynamic temperature equilibrium of the test setup from the first TMF cycle have to be defined.

These are among the topics that have to be addressed in the standardisation of TMF testing. The R&D work in the TMF-STANDARD project encompasses dedicated pre-normative research on the above-mentioned topics that has resulted in a preliminary (i.e. not yet validated) code of practice (CoP) for TMF testing. On this basis, validation testing is performed and subjected to thorough statistical analysis of results, in the light of which the CoP will be reviewed. The finally resulting validated CoP for TMF testing will be proposed to standardisation bodies, in particular, ISO. Dissemination of the results will be facilitated by a dedicated workshop which addresses all those who aim at improving the quality of TMF testing procedures and the intercomparability of the results. The present study will focus on the recent results of the pre-normative R&D work and will give an outlook on the further activities within the TMF-STANDARD project.

21.2   Test material

All experiments of the pre-normative research (except from the tests on dynamic temperature measurements which were conducted on IN 617) as well as the validation testing are performed on the Ni-base superalloy Nimonic 90. The metal was bought from two batches of a single cast produced by SpecialMetals and was delivered as 15 rods with a diameter of 26 mm and a length of about 4 m each. The heat treatment consisted of solution annealing at 1080°C for 8 h, quenching in water and an ageing heat treatment at 850°C for 5 h resulting in an over-aged state which was chosen to minimise the influence of different thermal pre-cycling procedures used within the TMF-STANDARD consortium. With this heat treatment, the hardness was kept within 311 ± 6 HV.

21.3   Reference TMF cycle

As a reference TMF cycle for all the tests in the pre-normative research as well as for half of the validation testing exercise, an out-of-phase TMF cycle (180° phase shift between temperature and mechanical strain) was chosen. The temperature cycle was triangular in shape with a maximum temperature $T_{max} = 850°C$ and a minimum temperature $T_{min} = 400°C$ and a heating and cooling rate of 5°C/s resulting in a cycle period of 180 s. The mechanical strain cycle was also triangular shaped with a strain ratio of $R_\varepsilon = -1$ and a strain range of $\Delta\varepsilon_{me} = 0.8\%$. Figure 21.2 shows the temperature- and strain–time paths of this cycle. In order to establish a benchmark against which the effects of various (deliberate) deviations in TMF test control can be assessed, five laboratories involved in the TMF-STANDARD project have each performed a set of three tests with those reference cycle parameters, resulting in a number of cycles to failure of 800 up to 1400 cycles.

21.4   Pre-normative research work

21.4.1 Dynamic temperature measurement and control

Comparative measurements using thermocouples (TCs) of several chemical compositions (Pt-PtRh, Ni-CrNi, Ni-NiAl, etc.), various configurations of attachment (ribbon type, spot welded, coaxial embedded) and different wire

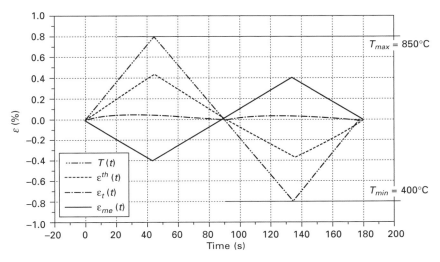

21.2 Reference TMF cycle.

diameters were performed at solid cylindrical and hollow cylindrical specimens with heating and cooling rates ranging from 2°C/s to 50°C/s and the same peak temperatures as in the reference TMF cycle. The test pieces were heated by induction. Cooling was achieved by heat conduction to the water cooled specimen grips and, if needed, by additional forced air cooling. The design of the induction coil was optimised in such a way that the longitudinal and the tangential temperature deviations within the gauge length were kept below ± 5°C.

Figure 21.3 shows the setup of the ribbon type (RT) and the spot-welded (SW) thermocouples, both with a wire diameter of 0.25 mm, Type R (Pt-PtRh). The ribbon type TC was wrapping along 180° around the surface of the gauge section resulting in a good thermal contact to the test piece, if roughening and/or oxidation of the surfaces is avoided. The spot-welded TCs were used in two configurations: one with the wires contacting the sample's surface along 90° starting from the welding point, the other with the wires tangentially pulled away from the surface of the test piece at the measuring point.

The temperature–time path measured with these TCs in the region of the maximum temperature is given in Fig. 21.4. The ribbon type TC as well as the spot-welded TC with the larger contacting length show nearly identical readouts whereas the spot-welded TC with the smaller contacting length results in distinctly lower temperature values due to the fact that a 'cold spot' is formed by heat conduction into the TC wires. The deviations of the maximum and the minimum temperature from spot welded TCs with different contacting lengths and of coaxial TCs placed in holes with different depths with respect

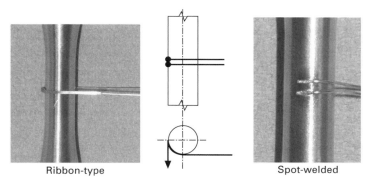

Ribbon-type   Spot-welded

21.3 Ribbon-type and spot-welded thermocouples.

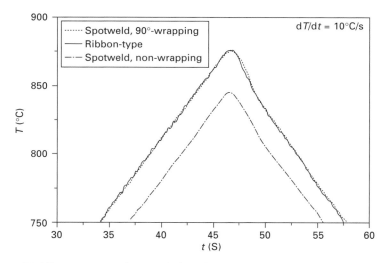

21.4 Temperature–time path for ribbon-type and spot-welded thermocouples.

to the values measured by a ribbon type TC which is taken as a reference are given in Table 21.1.

Obviously, the ribbon type, the spot-welded as well as the coaxial TCs are suitable for dynamic temperature measurement and control even at the highest temperature rates investigated if a sufficiently large contacting length is assured. This issue is more important for Pt-based TCs with their extremely high thermal conductivity than for Ni-based TC materials (e.g. Type N). The latter are more prone to oxidation meaning that especially at high maximum temperatures ribbon type TCs made of Ni-CrNi or Ni-NiAl wires may result in an underestimation of the sample temperature due to insufficient thermal contact.

Table 21.1 Deviations of T_{max} and T_{min} measured by SW and C TCs from the values determined by ribbon-type TCs

dT/dt (K/s)		SW (parallel)	SW (wrapping)	C (Fig. 21.6)	C (Fig. 21.8)
2	T_{max}	−14 K (1.6% of T_{max})	−3 K (0.4% of T_{max})	−36 K (4.2% of T_{max})	−2 K (0.2% of T_{max})
	T_{min}	−9 K (2.2% of T_{min})	−3 K (0.8% of T_{min})	−19 K (4.8% of T_{min})	−2 K (0.5% of T_{min})
5	T_{max}	−21 K (2.5% of T_{max})	/	−42 K (4.9% of T_{max})	−4 K (0.5% of T_{max})
	T_{min}	−7 K (1.8% of T_{min})	/	−18 K (4.5% of T_{min})	−2 K (0.5% of T_{min})
10	T_{max}	−24 K (2.8% of T_{max})	−1 K (0.1% of T_{max})	−55 K (6.6% of T_{max})	−5 K (0.6% of T_{min})
	T_{min}	−8 K (2.0% of T_{min})	−1 K (0.3% of T_{min})	−22 K (5.5% of T_{min})	−2 K (0.5% of T_{min})
50	T_{max}	−46 K (6.4% of T_{max})	−6 K (0.7% of T_{max})	/	/
	T_{min}	−11 K (2.8% of T_{min})	−1 K (0.3% of T_{min})	/	/

To investigate the possibility of temperature control without any TCs applied within the gauge length, a spot-welded TC was placed at the specimen shank. The setpoint program was adjusted in a way that temperature–time paths with T_{min} = 400°C, T_{max} = 850°C and constant heating and cooling rates of 5 and 10 K/s were achieved within the gauge length while the test was controlled by that thermocouple outside the gauge length. The temperature in the gauge length was measured by a ribbon-type TC. The results are given in Fig. 21.5. The maximum temperatures agree within ± 2 K (0.2% of T_{max}) with the desired value of 850°C. The largest deviation of the minimum temperature from the target value of 400°C was –6 K (1.5% of T_{min}). Obviously,

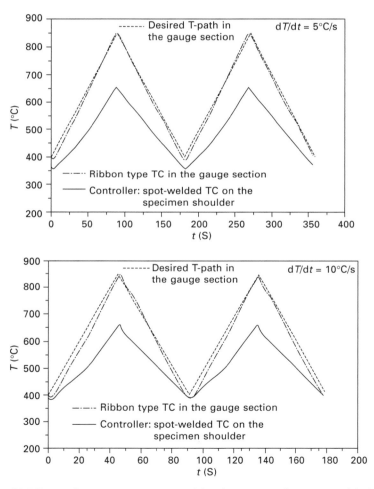

21.5 Dynamic temperature control by thermocouples spot welded at the specimen shoulder.

at the lower heating and cooling rate, the desired triangular temperature–time path is reproduced very well within the gauge length. At 10°C/s there are larger but still tolerable deviations in the shape of the *T-t* course.

Pyrometry measurements were performed by single-colour pyrometry at the Institut für Werkstoffkunde, Karslruhe and by two-colour pyrometry at the National Physical Laboratory, Teddington. Even after a pre-oxidation treatment of 2 h at the maximum temperature of the TMF cycle, the temperature paths measured by single-colour pyrometry are continuously shifting due to a varying thermal emissivity of the surface. The temperature deviation with respect to the first cycle reached nearly 100°C at $N = 50$ TMF cycles under reference conditions. The effect of a drift of the measured temperature can be avoided by the application of two-colour pyrometry. However, using this technique, intolerable temperature oscillations may occur because of the interference occurring at continuously growing thin oxide layers. Owing to these fundamental problems, temperature control by pyrometry requires thorough and time consuming feasibility investigations for each material and each temperature cycle, and it should be avoided in TMF testing whenever possible.

21.4.2 Influence of temperature tolerances on the TMF test results

Errors of the temperature path affect the cyclic deformation and lifetime behaviour in TMF and, therefore, lead to incorrect results, because plastic deformation is altered, oxidation processes are accelerated/decelerated and even undesired phase changes may be induced, if the temperature exceeds the target values. These influences, which are primarily dependent on the material as well as on the TMF temperature cycle, cannot be assessed by simple theoretical arguments, since they will interact, in general, in a complex way.

The majority of TMF tests are run in the time-based control mode. The command value for strain control is the sum of thermal and mechanical strain according to Eq. 21.1. Consequently, a temperature error leads to a thermal strain error which in turn leads automatically to a mechanical strain error. A positive temperature deviation $(+\Delta T)$ causes increasing thermal strain. If the command value is based on the nominal temperature the resulting additional thermal expansion is prevented by the closed loop strain control resulting in additional compressive loading. The other way round, at negative temperature deviations, undesired tensile loadings arise.

Figure 21.6 shows as an example the maximum and the minimum stresses and the peak temperatures vs the number of cycles of a TMF test in which temperature errors of ±20 K at both the minimum and the maximum temperature were applied and changed every 10 cycles. Obviously, an increase of the

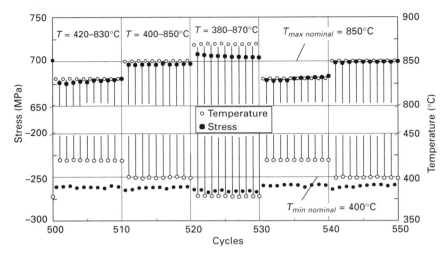

21.6 Maximum/minimum stress changes as a result of temperature errors.

temperature range increases the stress amplitude, and vice versa. Furthermore, due to the smaller amount of plastic deformation, a temperature error at the lower temperature influences the load much more than an error at the high temperature. Due to that, the minimum force is almost unaffected whereas the maximum force increases with decreasing minimum temperature. In the corresponding stress-mechanical strain loops the plastic deformation energy per loop differs by +28 or –24% with respect to the nominal cycle. Accordingly, the TMF tests with a smaller temperature range yield higher, the tests with a larger temperature range result in lower lifetimes. Up to 5°C deviation of T_{max} and T_{min} the influence of temperature errors on the number of cycle to failure cannot be clearly distinguished from the natural scatter of the material. For temperature errors of 10°C or higher, the lifetimes are clearly outside of values determined under the reference conditions. From these results the conclusion is drawn that the highest allowable temperature tolerance is 5°C, a value which can, according to the results presented above, be reached with appropriate experimental efforts.

21.4.3 Influence of temperature-strain phase deviations on the TMF test results

According to the definitions given in Fig. 21.7, the influence of phase angle deviations with respect to the reference out-of-phase cycle (180° T-ε_{me}-phasing) on the TMF cyclic deformation behaviour and lifetime was investigated.

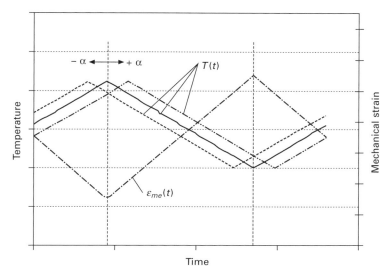

21.7 Definition of positive and negative phase errors.

As can be inferred from Fig. 21.8, showing cyclic deformation curves for several phase deviations, positive phase errors (strain earlier than desired) result in increasing plastic strain amplitudes and decreasing maximum stresses whereas the minimum stress is nearly unaffected. On the other hand, negative phase errors (strain later than desired) lead to smaller plastic strain amplitudes, while the maximum stress is decreasing in a similar extent as at positive phase shifts and the minimum stress again remains nearly unaffected.

The lifetime behaviour shows a weak reduction of the number of cycles to failure with respect to the reference cycle both at positive and negative phase errors up to 10°. At higher absolute values of the phase deviation, the lifetime increases and reaches at ±20° the same values as under reference conditions.

The results of temperature and phase angle deviations discussed so far consider the case of 'time based' compensation of the thermal strain $\varepsilon_{th}(t)$ in Eq. 21.1. Another possibility is to define $\varepsilon_{th}(T)$ as a function of the temperature which is referred to as 'temperature based' strain compensation. In this case, deviations of the peak temperatures and of the phase angle between temperature and mechanical strain affect the cyclic deformation and lifetime behaviour distincly less pronounced than for time-based compensation. However, a fit function has to be applied to feed $\varepsilon_{th}(T)$ in the control circuit of the testing machine which may result in considerable strain and stress errors especially in the case that one fit function is used for the heating as well as for the cooling part of the TMF cycle.

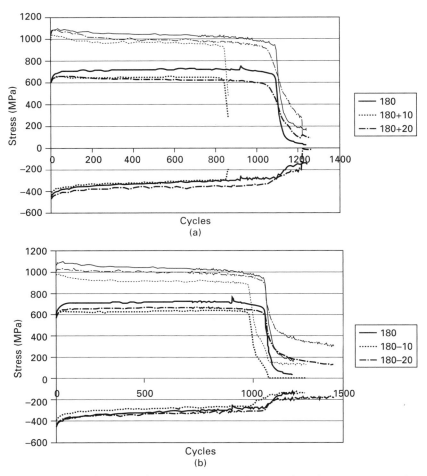

21.8 Influence of (a) positive and (b) negative phase errors on the cyclic deformation behaviour.

21.4.4 Influence of temperature gradients in different specimen geometries

The influence of temperature gradients on the TMF behaviour was investigated for test pieces with solid flat, solid cylindrical and hollow cylindrical gauge section.

In the first case (solid flat specimen), the temperature gradient was varied by different designs of the induction coil used for heating and by applying different heating and cooling rates. The temperatures inside the gauge section were measured by a specimen instrumented with seven coaxial thermocouples placed at different locations inside the specimen. Additionally, infrared thermography was performed to determine the surface temperature profiles.

In this way temperature differences within the gauge length ranging from 3°C up to 22°C were established. As expected, during heating the highest temperatures occurred near the surface because of the relatively low penetration depth of the inductive heating, while during cooling temperatures are lower at the surface as compared to the interior of the test sample. Only at the highest temperature gradient an influence on the lifetime was observed in terms of an increasing numbers of cycles to failure as compared to the tests performed with smaller T gradients.

The tests on specimens with solid cylindrical section (8 mm dia.) were also conducted with inductive heating. The temperature gradients were again measured by several thermocouples placed at positions inside the guage section and at the surface. Special attention was paid to the influence of the induction frequency (66 kHz and 400 kHz) on the radial T gradients at heating and cooling rates from 2°C/s up to 10°C/s. During heating with 2°C/s and 5°Cs at both induction frequencies, radial temperature differences up to 6°C and 12°C, respectively, were measured. At the highest heating rate of 10°C/s, an induction frequency of $f = 66$ kHz resulted in radial gradients up to 33°C, whereas at $f = 400$ kHz radial temperature differences up to 38°C were measured. However, with values from 10°C (at 2°C/s) up to 40°C (at 10°C/s) during cooling generally higher radial temperature differences arose than during heating, independent of the induction frequency. It was concluded that for solid round test pieces with gauge diameters up to 6 mm, temperature rates of more than 5°C/s should not be applied and that especially at high T rates the frequency of inductive heating should be kept as low as possible.

In the third set of tests performed using specimens with hollow cyclindrical gauge section (11 mm dia., 1 mm wall thickness) and heated by radiation in a bulb furnace, the influences of radial (achieved by forced air cooling from the inner side of the test piece), longitudinal (due to enforced water cooling of the specimen shoulders) and tangential (applied by an asymmetric configuration of the heating bulbs around the sample) temperature gradients were investigated. All types of temperature gradients affect the cyclic deformation behaviour in terms of the maximum and the minimum stresses observed. Whereas radial and longitudinal gradients did not affect the lifetime significantly, tangential gradients induced buckling in the gauge section leading to a significant decrease of the number of cycles to failure. As an example, the cyclic deformation curves of TMF tests with and without tangential gradients are shown in Fig. 21.9 together with typical fractures in the case of such gradients.

21.4.5 Best practice procedures for starting a TMF test

In numerous tests suitable procedures for determining the coefficient of thermal expansion, and the temperature dependent Young's modulus were

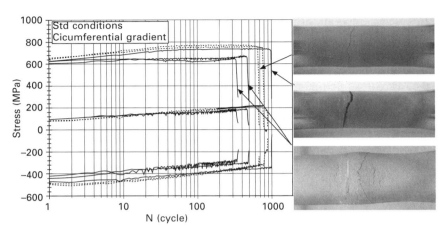

21.9 Influence of circumferential *T* gradients on the cyclic
deformation and fracture behaviour of hollow cylindrical test pieces.

evaluated. Before starting a TMF test, Young's modulus E has to be determined
as a function of the temperature. Two methods are applicable: the 'static'
method means that E is determined at stepwise constant temperatures by
appying cylic loadings in the elastic regime during every temperature step.
The 'pseudo-dynamic' method means that the same temperature cycle as
applied in the subsequent TMF test is superimposed on stress variations, and
E is calculated from the resulting strains. The static method results in clearly
defined E values at homogeneous temperatures within the gauge of the
specimen, whereas the pseudo-dynamic method does more closely reflect
the temperature gradients which arise during thermal cycling.

The test start procedure after having determined $R(T)$ is plotted schematically
in Fig. 21.10. A crucial prerequisite for the definition of strains in a TMF
cycle is the precise knowledge of the thermal strain ε_{th} as a function of time
or temperature in the time-based or temperature-based control mode,
respectively. Accordingly, after at least five force-free temperature cycles
which are applied to establish a sufficiently stable dynamic temperature
equilibrium in the test set-up, $\varepsilon_{th}(t)$ should be measured during five further
T cycles at zero force and afterwards an average of those thermal expansion
cycles be taken for thermal strain compensation.

In a next step, the machine is switched to total strain control and, in order
to validate the measured thermal strain path, $\varepsilon_{th}(t)$ is applied as the total
strain to the test piece for a couple of cycles. In the ideal case, the resulting
forces should stay at zero throughout these verification cycles ('zero-stress
test'). In real test practice, a stress range of up to 5% of the stress range in
the subsequent TMF cycles is considered to be tolerable. After the successful
verification cycles, the TMF test is started with the desired strain condition
according to Eq. (21.1).

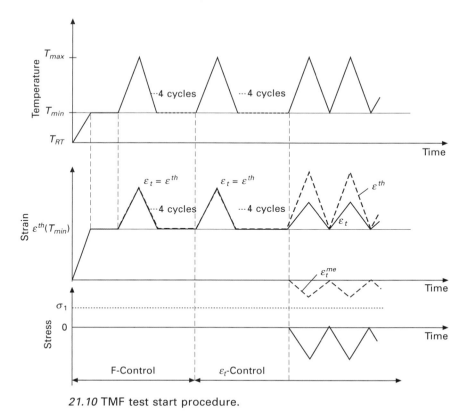

21.10 TMF test start procedure.

21.5   Conclusions and outlook

The pre-normative research within the TMF-STANDARD project has consisted of detailed investigations on quantifying the various effects of specific test procedures, of deviations from nominal values of critical test parameters, and of the specimen design on the TMF test results. This includes:

- Measuring the effect of the variability of the test procedure, concerning issues such as E modulus and thermal strain measurement, thermal strain compensation, temperature–strain path, temperature correction of the gauge length, etc. The goal consisted in establishing best practice procedures for the start/restart procedure of the TMF test.
- Establishing the effect of deviations from the nominal values of the test parameters that are specific to TMF testing. Methods for the measurement of temperatures that change with time, temperature and associated strain gradients over the volume of the specimen's gauge section, and phase angle shifts between thermal and mechanical loading cycles. For a thorough interpretation of the effect of the internal stresses caused by temperature

and strain gradients on the TMF test results, the experimental work has been accompanied by Finite Element Analysis. The goal of this part of pre-normative research consisted in the identification of appropriate measurement methods and allowable tolerance limits for TMF testing.

Based on the findings from the pre-normative testing a preliminary code of practice (CoP) for TMF testing has been drafted. Using this guideline, validation testing was performed in which the participants execute two series of TMF tests (reference out-of-phase cycle, and in-phase cycle) with equal nominal parameters using a common material, which was especially procured for the TMF-STANDARD project and microstructurally tested for its suitability. This validation testing was carried out in two stages: an 'inner circle' of nine participants performed TMF tests on specimens which have been centrally machined by the workshop at BAM, Berlin, Germany, to get surface conditions and gauge section geometries as uniform as possible. The 'outer circle' TMF testing was performed by ten associate contractors of the TMF-STANDARD project using specimens which have been individually manufactured at the respective laboratories according to the recommendations of the preliminary CoP.

All TMF-STANDARD partners reported their results according to a pre-defined format. Afterwards, the test results were compiled and submitted to an in-depth statistical analysis in order to establish how, and to what extent, the specifications of the draft CoP should be tightened or relaxed for enabling the definition of a commonly accepted standard TMF test procedure. After the statistical evaluation of the validation test results the CoP has been subjected to a final review taking into account the results of validation testing.

The project has been finalised in autumn 2005 with a dedicated workshop at which the final CoP and selected results of the research work of the project were presented to the public. This workshop addressed all the parties dealing with TMF testing with the aim of improving the quality (repeatability and reproducibility) of TMF testing. The proceedings of the workshop were published as a special issue of International Journal of Fatigue [5]. The validated final CoP for TMF testing [6] was presented to standardisation bodies, in particular, with a view to reviewing the corresponding non-validated draft ISO standard.

21.6   References

1. S. Kalluri, M. A. McGaw, J. Bressers and S. D. Peteves (eds), *Thermomechanical Fatigue Behavior of Materials: 4th Volume 4*, **ASTM STP 1428**, American Society for Testing and Materials, West Conshohocken, PA (2003).
2. M. J. Verilli, M. G. Castelli (eds), *Thermomechanical Fatigue Behavior of Materials: 3rd Volume 3*, **ASTM STP 1263**, American Society for Testing and Materials, West Conshohocken, PA (1996).

3. H. Sehitoglu (ed.), *Thermomechanical Fatigue Behavior of Materials: Volume 2,* **ASTM STP 1186,** American Society for Testing and Materials, West Conshohocken, PA (1993).

4. J. Bressers, L. Remy (eds), *Fatigue under Thermal and Mechanical Loading,* Proc. Int. Symp. 22–24 May 1995, Petten, NL, Kluwer Academic Publishers, Dordrecht/ Boston/London (1996).

5. P. Hähner, H. Klingelhöffer, T. Beck, M. S. Loveday and C. Rinaldi (guest eds), *Int. J. Fatigue,* Vol. 30 Issue 2 (2007).

6. P. Hähner, E. E. Affeldt, T. Beck, H. Klingelhöffer, M. Loveday and C. Rinaldi, Validated Code-of-Practice for Strain-Controlled Thermo-Mechanical Fatigue Testing, EC-Report EUR 22281 EN, 2006, ISBN 92-79-02216-6.

22

Oxidation behaviour of Fe-Cr-Al alloys during resistance and furnace heating

H ECHSLER, Forschungszentrum Jülich GmbH, Germany;
H HATTENDORF, Thyssen Krupp VDM GmbH, Germany
and L SINGHEISER and W J QUADAKKERS,
Forschungszentrum Jülich GmbH, Germany

22.1 Introduction

Because of their excellent oxidation resistance, Fe-Cr-Al alloys are commonly used construction materials for components that have to operate at high temperature. The oxidation resistance relies on the formation of slowly growing, well adherent alumina-based surface oxide scales, which form during high temperature exposure. Examples of the application of Fe-Cr-Al alloys are heating element strips or wires, fibre-based domestic and industrial burners, car catalyst carriers, furnace tubes, etc.

The oxidation limited lifetime of Fe-Cr-Al alloys is governed by the consumption of the bulk aluminium reservoir caused by scale growth and re-healing after scale cracking or spallation. As the Al reservoir for a given alloy decreases with component thickness, limitations in lifetime are of special concern in case of components fabricated of very thin foils in the thickness range of 20–50 μm. In the literature [1, 2] well-established models can be found that predict with reasonable accuracy the oxidation limited lifetime of such thin Fe-Cr-Al-strips based on Al-depletion. However, further modifications of such models are necessary to account for factors such as geometrical changes of the metal strips during high temperature service, the type of heating and hence the heating and cooling rates and the number of cycles for a given oxidation time.

In the present study, the oxidation behaviour of thin Fe-Cr-Al heating element strips during resistance and furnace heating are compared with respect to their oxidation limited lifetimes and geometrical changes prior to final failure. Isothermal and cyclic oxidation tests with varying total exposure time and cycle duration were performed. Scale characterisation was carried out by metallographic investigations on the respective cross-sections. Additionally, residual compressive stresses in the oxide scales were measured as a function of exposure time and cooling rate using the ruby fluorescence technique.

22.2   Experimental methods

The commercial Fe-Cr-Al alloys studied were the wrought alloys Aluchrom Y and Aluchrom YHf with the nominal composition of Fe-20Cr-5.5Al (wt.%). Both alloys contained minor additions in the range of a few hundred ppm of Y and Zr. Furthermore, alloy Aluchrom Y contained minor additions of Ti, whereas in Aluchrom YHf small amounts of Hf were present. The alloys were manufactured by conventional metallurgical methods (casting followed by hot and cold rolling). The width of the strips was 5 mm and the foil thickness varied between 30 μm and 100 μm, although most of the experiments were carried out using 50 μm thick foils.

Isothermal and cyclic oxidation tests were performed in an automatic, computer controlled, resistance heated test rig (SOMA GmbH, Lüdenscheid, Germany), in which four specimens could be tested simultaneously (Fig. 22.1). The specimens were fixed between two electroconductive clamps and heating was achieved via the joule effect by passing an electric current through the strips. The equipment allows the automatic recording and control of the electric current and voltage. The temperature was controlled by a 670 nm wavelength pyrometer. The hot and cold resistance of the samples were recorded throughout the test. During the resistance heating tests, the temperature was chosen as the controlling parameter and was set to 1050°C, 1100°C, 1150°C and 1200°C, respectively. Most of the tests were conducted using cyclic conditions with the cycle parameters 15 s 'on' and 5 s 'off' (hereafter designated as 15/5 s). However, further cycles with 120/15 s and 300/900 s as well as additional isothermal tests were carried out. Most tests were performed until occurrence of specimen failure. For mechanistic studies a

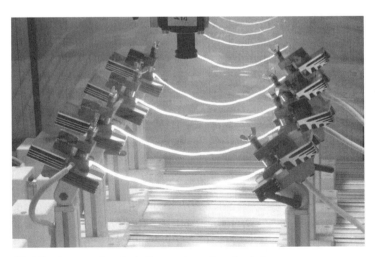

22.1 Resistance heated, thermal cycling test rig.

number of tests were carried out in which the test duration (cumulative time at oxidation temperature) was set to fractions of a defined, reference lifetime.

For comparison, isothermal and cyclic furnace oxidation tests were carried out using the same temperatures and durations as used in the resistance heating tests. These tests were either performed isothermally in a microbalance (SETERAM) in synthetic air or in a conventional resistance heated vertical tube furnace in laboratory air. In the latter tests each cycle consisted of 300 s heating and 900 s cooling. The test specimens used in the furnace tests were coupons of 20×5 mm in size which were cut from the 50 μm thick foils.

All specimens were tested in the as-received condition (cold-rolled). Scale characterisation was carried out by metallographic investigations on the respective cross-sections using conventional light and electron optical techniques. Additionally, residual compressive stresses in the oxide scales were determined using the ruby fluorescence technique. Details on the latter can be found in [3].

22.3 Results

Figure 22.2 shows the oxidation kinetics of Aluchrom YHf (batch HYH) measured in TGA tests at temperatures between 1050°C and 1200°C in

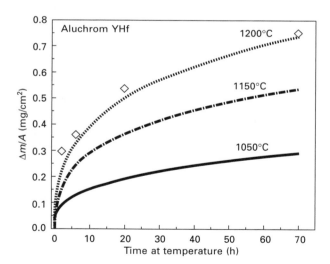

22.2 Isothermal oxidation kinetics of Aluchrom YHf obtained by TGA at different temperatures in synthetic air. The inserted open diamonds were derived from metallographic cross-sections on the respective resistance heated, isothermally oxidised specimens.

synthetic air. At all temperatures the oxidation kinetics could be described by a power law time dependence [4]

$$\Delta m = k \cdot t^n \qquad\qquad 22.1$$

In all cases near cubic oxidation kinetics were observed. The k-values (in mg/cm^2/h^n) were found to be 0.081, 0.149 and 0.205 for 1050°C, 1150°C and 1200°C, respectively, whereas the n-value was near to 0.3 at all temperatures. The inserted open diamonds in Fig. 22.2 refer to resistance heated specimens isothermally oxidised at 1200°C. The data were derived from metallographic cross-sections and calculated into weight changes, assuming the surface scales to consist solely of Al$_2$O$_3$. The results are in good agreement with the respective TGA curve illustrating the correctness of the temperature measurement of the resistance heated specimens by pyrometry. Apparently, the expected change in emissivity as a function of time did not lead to substantial errors in the temperature measurement.

Further TGA curves at 1200°C for Aluchrom Y (batch JUJ) and a further batch of Aluchrom YHf (JTT) are given in Fig. 22.3. Batch JTT shows slightly lower mass gains than the other two batches. Aluchrom Y exhibited breakaway oxidation, indicated by the steep increase of the mass gain, after approximately 50 h.

Figure 22.4 shows a comparison between experimentally obtained lifetimes during resistance heated thermal cycling and calculated oxidation limited lifetimes (t_B) using the Al depletion model for thin Fe-Cr-Al components proposed in [1]:

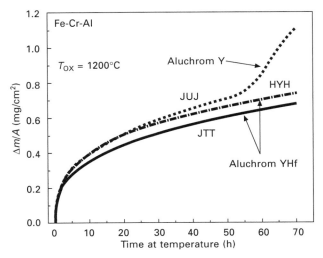

22.3 Isothermal oxidation kinetics of Aluchrom Y (batch JUJ) and different batches of Aluchrom YHf (batches HYH, JTT) obtained by TGA at 1200°C in synthetic air.

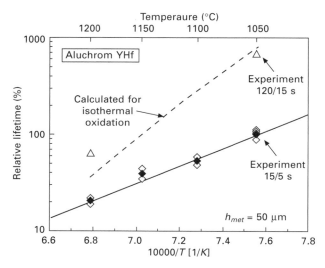

22.4 Experimentally obtained lifetimes during resistance heated thermal cycling at different temperatures compared with calculated lifetimes during isothermal oxidation showing the effect of temperature and cycle frequency. In the calculations the C_B value was assumed to be independent of temperature and/or specimen thickness and was set to 0.3 wt.%.

$$t_B = \left[4.4 \cdot 10^{-3} \cdot (C_0 - C_B) \cdot \frac{\rho \cdot h_{met}}{k} \right]^{1/n}.$$ 22.2

Here t_B denotes the time to breakaway (in h), C_0 the initial Al content in (wt.%), C_B the critical Al content at which breakaway occurs (in wt.%), ρ the alloy density (in g cm^3), h_{met} the initial foil or specimen thickness (in cm), k and n the kinetics constants defined in Eq. 22.1. In the calculations it was assumed that the critical Al content in Eq. 22.2 is independent of temperature and/or specimen thickness. The value of C_B was set to 0.3 wt.% as found for thin-foil Fe-Cr-Al alloys after oxidation at 1200°C e.g. in [5, 6]. It should be mentioned that, in former studies, indications were found that the C_B value depends on temperature. However, to the best of our knowledge, no extensive experimental data are available yet. In the diagram all the lifetimes are given in a normalised way, setting the obtained result at 1050°C for the short cycle (15/5 s) to 100%.

With increasing temperature the calculated lifetimes for isothermal exposure decrease due to the increase in the k value. The experimentally obtained lifetimes during resistance heated cycling show qualitatively the same decrease with increasing temperature, and in the case of the rapid cycle (15/5 s) the data points can as a first approximation be described by a straight line on the log t vs $1/T$ plot. The slope of the fitted curve is less steep than the calculated one for isothermal exposure. This leads for thermal cycling at 1200°C to lifetimes that are approximately 50% shorter than during isothermal exposures. This difference in lifetimes during isothermal and cyclic exposure becomes more pronounced with decreasing temperature. At 1050°C only around 15% of the lifetime calculated for isothermal oxidation was achieved during thermal cycling. The relatively small scatter of the test results indicated by the open diamonds illustrates the good reproducibility of the test data.

In the case of the longer cycle duration (120/15 s, indicated by the open triangles in Fig. 22.4), the observed lifetime at 1050°C is near to the value calculated for isothermal oxidation. At 1200°C the experiment yielded an even longer lifetime than expected for isothermal oxidation. Here, it has to be pointed out that Eq. 22.2 predicts the time to breakaway t_B, which is defined as the time after which Al_2O_3 formation can no longer be sustained. In contrast, final failure of the specimens during resistance heating testing is defined as fracture of the alloy strip. This fracture of the strip is caused mainly by local, rapid growth of iron oxides and is not necessarily identical to the t_B defined by Eq. 22.2. Especially for thin Fe-Cr-Al components, subscale chromia formation during isothermal oxidation has been observed for a substantial time period prior to final failure in e.g. [7, 8]. This subscale chromia formation, which results in a 'pseudo-protection' regime [7], was also observed in the present study in the case of the longer cycle duration, and thus the experimental lifetime is longer than that calculated assuming breakaway to occur if the value of C_B is reached.

Figure 22.5 shows the effect of the foil thickness on the normalised lifetimes. Again substantial differences between experimental cyclic data and the calculated lifetimes for isothermal oxidation prevail, and this difference is more pronounced in the case of thicker foils. Cycling reduces the lifetime for 30 μm thick foils by approximately a factor of two compared to a factor of around ten for the 70 μm thick foils.

In Figure 22.6 optical micrographs of resistance heated Aluchrom YHf specimens after various fractions of the total lifetime are depicted. The specimens were exposed at 1200°C with a cycle duration of 15/5 s. Up to an exposure of around 5% of the total lifetime the oxide/metal interface remains flat. After longer times a non-even interface starts to form and a slight alternation of contractions and expansions of the metal foil thickness is visible. This waviness increases in amplitude as well as in wavelength with increasing exposure time until at the end of the lifetime the thickness of the foil is locally reduced to virtually zero. This 'hour glass' waviness occurs not only in the vicinity of the failure area but over a large fraction of the specimen length. However, far away from the failure area this 'hour glass' waviness is less pronounced.

These observations, in combination with the results presented above, indicate that the lifetime of the rapidly cycled, resistance heated thin metal strips is

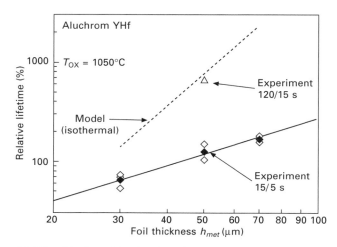

22.5 Double logarithmic plot of experimentally obtained lifetimes
during resistance heated thermal cycling at 1050°C compared with
calculated lifetimes during isothermal oxidation showing the effect of
foil thickness. In the calculations the C_B value was assumed to be
independent of temperature and/or specimen thickness and was set
to 0.3 wt.%.

strongly affected by the plastic deformation of the specimens. A similar type
of plastic deformation was recently found to occur during thermal cycling of
various NiAl + Pt and/or Hf alloys even for relatively thick specimens of
around 1 mm [9, 10]. The authors observed a strong effect of the cycle
duration on the amount of deformation, although it should be mentioned that
the cycle duration used was substantially longer than in the present study.

In order to verify whether the deformation in the present study is simply
caused by the large number of cycles or whether it is related to the resistance
heating procedure used, additional cyclic furnace tests and isothermal resistance
heated tests were carried out. Figure 22.7 shows optical micrographs of
thermo-cycled Aluchrom YHf specimens after 50 h exposure at 1200°C
using furnace heating (a) and resistance heating (b). The cycle duration in
both cases was chosen to be 300/900 s due to the slower heating and cooling
rates achievable during furnace heating. In both cases a deformation of the
specimens is visible, whereby in the case of the furnace heating the waviness
seems to be even more pronounced than for the resistance heated specimen.
However, the deformation is less pronounced than in case of the specimens
exposed to the short cycle during resistance heating testing shown in Fig.
22.6.

Figure 22.8 shows optical micrographs of resistance heated Aluchrom
YHf specimens after 70 h isothermal and cyclic exposure at 1200°C. Cyclic
exposure with 15/5 s (b) results in a similar extent of 'hour glass' waviness

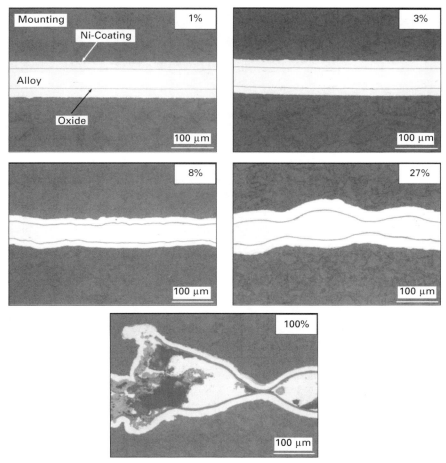

22.6 Optical micrographs of Aluchrom YHf specimens after various times of resistance heated thermal cycling at 1200°C (cycle: 15/5 s) showing the evolution of the 'hour glass' waviness during prolonged cycling. Test times are given as relative values of the maximum test times.

as for the specimens shown in Fig. 22.6 (27% and 100%). In contrast, the isothermally exposed, resistance heated specimen (a) shows hardly any waviness. These observations clearly indicate that the type of heating has only a minor effect on the plastic deformation and thus the lifetime of the specimen.

22.4   Discussion

Lifetime testing by resistance heating is used extensively by alloy manufacturers as a tool for quality control or as a screening test for alloy modification and

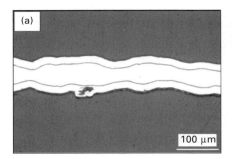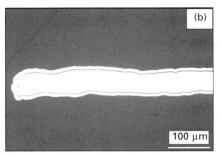

22.7 Optical micrographs of Aluchrom YHf specimens after 50 h of thermal cycling at 1200°C (cycle: 300/900 s) showing a comparison between furnace heating (A) and resistance heating (B).

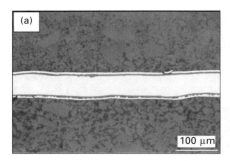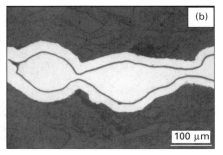

22.8 Optical micrographs of Aluchrom YHf specimens after 70 h of resistance heated exposure at 1200°C showing a comparison between isothermal (a) and cyclic exposure, 15/5 s (b).

improvement. In these tests electrical power is commonly used as the controlling parameter [11]. This approach is close to real applications and, additionally, control of the electric parameters is much easier to handle than temperature control by pyrometer because problems arising from the possible change of emissivity with time can be avoided. However, in the present study temperature was used as the controlling parameter because emphasis was put on the comparison of the results obtained from the resistance heating tests with those from furnace tests. Initial measurements of the emission coefficient of oxidised samples revealed a constant emission coefficient after a few hours of exposure. At the beginning of the oxidation process the emission coefficient was found to be lower than the constant value after longer exposure times and thus the actual temperature at the beginning of the resistance heating tests is expected to be slightly higher than set by the control system. This minor overshooting of the temperature can partly explain the reduction in lifetime during rapid cycling compared to the isothermal tests (Fig. 22.4). However, the comparison of the scale thickness measured after furnace and resistance heating showed that this difference is only marginal (Fig. 22.2).

The substantial reduction in lifetimes observed during rapid cycling compared to the calculated lifetimes for isothermal oxidation (Figs 22.4 and 22.5) are related mainly to the development of the 'hour glass' waviness caused by plastic deformation of the metal strip during prolonged cycling (Fig. 22.6). Final failure due to breakaway and thus rupture of the metal strip occurs in areas where the foil thickness and, in consequence, the Al reservoir is significantly reduced. Strictly speaking, at any given time the locally remaining Al reservoir can no longer be related to the initial foil thickness but rather to the actual thickness.

The test results clearly show that the time to failure is related not only to the time at temperature but strongly depends on the number of cycles for a given oxidation time. For instance, in the case of an oxidation temperature of 1050°C the lifetime for the 120/15 s cycle is approximately six times longer than for the 15/5 s cycle. Because the Fe-Cr-Al alloys studied exhibit only very poor creep strength at high temperature, it is likely that for the thin metal strips used, plastic deformation of the metal takes place during every cycle. The amount of this plastic deformation caused by creep of the metal depends on the oxide-to-metal thickness ratio and accumulates over the testing time.

Figure 22.9 shows the analytically calculated evolution of the in-plane stress in the metal during one cooling and re-heating cycle. In these calculations creep of the metal is taken into account. Details of the procedure and the thermo-mechanical properties of the metal and oxide can be found in [12]. In

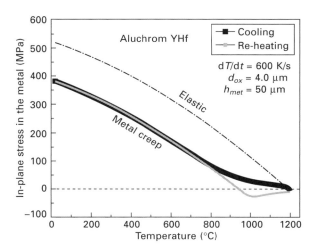

22.9 Analytically calculated in-plane stress evolution in the metal during cooling from and reheating to 1200°C assuming an ideal, flat oxide/metal interface. The calculations take into account stress relaxation by creep of the metal during heating and cooling.

the given calculations the oxide thickness is arbitrarily chosen as 4 μm and the cooling and re-heating rates are set to 600 K/s corresponding to the very fast heating and cooling rates prevailing during the resistance heating test. Further assumptions are: (i) flat specimen geometry without any interfacial roughness; (ii) same oxide scale thickness on both sides of the specimen and (iii) stress-free state at the oxidation temperature of 1200°C.

During cooling from oxidation temperature to approximately 800°C, most of the stresses are relaxed by metal creep and the difference compared to the elastic case amounts to around 120 MPa. Below 800°C no further stress relaxation takes place and at room temperature the stress in the metal reaches 380 MPa. Re-heating yields a congruent curve up to 800°C. At higher temperatures the curves diverge and a transition from tensile to compressive stresses in the metal occurs during re-heating reaching a maximum value at 1000°C. Further increase of the temperature leads to stress relaxation by creep under compression. Similar results were presented for chromia scales during oxidation of a Ni-30Cr alloy [13] and an austenitic steel Fe-25Ni-20Cr [14] at temperatures of 900°C and 1000°C. Also for alumina scales formed on Haynes 214 [15] and Kanthal Fe-Cr-Al alloys [16] during oxidation at temperatures of 1100°C and 1200°C, respectively, compressive stress formation in the metal during re-heating was illustrated.

Although these calculations are only valid for an ideal, flat oxide/metal interface it is likely that, once a certain waviness is present, the amplitude of the waviness increases with each cycle. Similar observations can be found during thermal cycling of electron beam physical vapour deposited thermal barrier coatings and the authors [e.g. 17] propose ratcheting as the responsible mechanism. In the respective literature three prerequisites are necessary for ratcheting to occur: (i) plastic deformation of the metal, especially during re-heating; in [17] this is expressed by yielding of the metallic bond coat; (ii) a generation of an oxide growth strain during the hot dwell time and (iii) an initial undulation of the metal/oxide interface. The former two factors can easily be assigned to the prevailing system, whereas the latter seems to be more difficult to explain.

Apparently a certain 'incubation time' exists where no significant waviness of the specimen or strictly speaking oxide/metal interface is visible. In the conditions leading to the micrographs shown in Fig. 22.6 this 'incubation time' is less than 10% of the total lifetime. Assuming that for a given temperature this incubation time is related to a certain minimum oxide thickness or oxide/metal thickness ratio, which is necessary to generate stresses in the metal that are sufficiently high for creep to occur, a decrease in temperature leads to an increase of the incubation time according to the slower oxidation kinetics.

During this stage of the lifetime, heterogeneities at the oxide/metal interface, such as, e.g., re-healed tensile cracks in the scale [8] or stress concentrations in the vicinity of yttrium-aluminium-garnets [8], are likely to serve as starting points for the initiation of the waviness. Once this initial waviness is established, 'ratcheting' can occur. This mechanism is related mainly to repeated plastic deformation and thus to the number of cycles and depends less substantially on the temperature and oxidation kinetics. These observations are confirmed by the time dependence of the electrical resistance of the metal strips during rapid cycling [18]: in particular, the cold resistance shows a steep increase by approximately 50% at the very end of the tests independent of the temperature.

If final failure of the specimens and the reduction of the lifetime during rapid cycling were related to the mechanisms mentioned above, the remaining Al content after final failure should be higher than the value of 0.3%, which was assumed in the calculations using the Al depletion model. This is not in agreement with the experimental findings at 1200°C. Final failure due to breakaway oxidation in the rapid cycling test is a local phenomenon, i.e. it might be caused by a local, 'total' Al consumption as a result of the hour-glass waviness. However, in failed specimens further spots of local chromia formation, indicating near total Al consumption, were often detected far away from the failure area. This is confirmed by EDX analysis showing the Al content after final failure at 1200°C far away from the failure location of the specimen given in Fig. 22.6 to be approximately 0.1%.

However, an increase in surface area due to the development of the waviness results in higher Al depletion rates than assumed in the Al depletion model. Assuming a sinusodial wave with an amplitude-to-wavelength-ratio $A/(2L)$ of 0.1 or 0.25 results in an increase of the total length of approximately 9% or 46%, respectively.

Assuming the initiation of the ratcheting effect depends on the oxide-to-metal thickness ratio, the dependence of lifetime on specimen thickness and the difference between lifetime during isothermal and cyclic exposure as shown in Fig. 22.5 can be explained. However, the increasing reduction in lifetime with increasing foil thickness or with decreasing temperature in Fig. 22.4 are presently not fully understood and further investigations are necessary to get a better insight into the prevailing failure mechanisms responsible for the lifetime reduction imparted by rapid thermal cycling. In particular, it should be elucidated whether the shortening of the lifetimes upon increasing specimen thickness and/or decreasing temperature is related solely to the rapid cycling or due to a change of one of the parameters in Eq. (22.2) (e.g. a temperature and/or specimen thickness dependence of C_B).

It should be mentioned that, in addition to the plastic deformation described as 'hour-glass' waviness (Fig. 22.6), further specimen deformations were frequently observed. Due to twisting/tilting or rolling of the specimen 'hot-tube' or 'corkscrew'-like deformation of the specimen occurs as illustrated in Fig. 22.10. In addition, a temperature gradient over the width of the

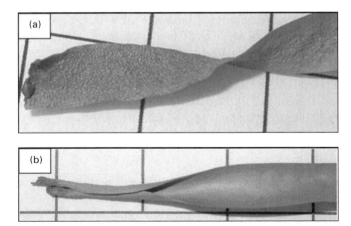

22.10 Macrographs of 50 μm thick Aluchrom YHf specimens after cyclic oxidation at 1050°C (15/5 s) showing 'corkscrew' (a) or 'hot tube'-type (b) plastic deformation.

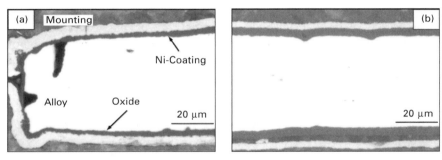

22.11 Optical micrographs of Aluchrom YHf specimens after 70 h of isothermal, resistance heated exposure at 1200°C showing a thinner oxide scale at the edge (a) than in the centre of the specimen (b).

specimens was observed during resistance heating testing. As a result of the temperature gradient, the oxide scale is remarkably thinner at the specimen edges than in the centre of the specimen, as illustrated in Fig. 22.11. This causes further non-uniform stress development and relaxation, which results in additional changes of the specimen shape. This non-uniform stress development and/or relaxation was verified by residual stress measurement in the oxide for isothermal resistance heating of the specimens and the results are shown in Fig. 22.12. In particular, after longer exposure times a remarkable difference in compressive stress values in the oxide between the centre and the near edges area, around 20 μm away from the edge, exists. This is caused by a combined effect of differences in oxide thickness and temperature [12]. However, a detailed treatment of this effect falls outside the scope of this chapter.

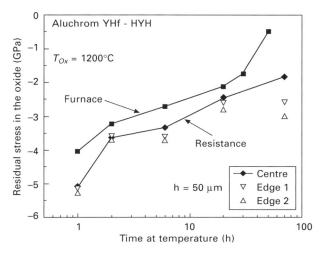

22.12 Residual stress in the oxide scales grown on 50 μm thick Aluchrom YHf specimens as function of exposure time at 1200°C showing the effect of heating type and the developing temperature gradient over the width of the resistance heated specimens.

22.5   Conclusions

The lifetime of thin heating element strips of Fe-Cr-Al alloys during rapid thermal cycling is shorter than during isothermal exposure. The lifetime increases with increasing cycle duration and hence decreasing number of cycles at a given time at temperature. This lifetime decrease is related to an 'hour glass' waviness of the specimens developing during prolonged cycling. A two-step mechanism is introduced combining an oxidation kinetics related time to the onset of significant waviness with an enhancement of this waviness similar to the ratcheting effect proposed in the literature. The latter depends strongly on the number of cycles and on the plastic deformation generated during each cycle rather than on the total time at temperature. Additionally, this development of an 'hour glass' waviness leads to an enhanced aluminium depletion due to an increase of the specimen surface area.

During the resistance heating tests, a temperature gradient develops over the specimen width due to the poor aspect ratio of the specimens, which causes additional deformation phenomena termed 'hot tube' or 'corkscrew' behaviour.

22.6   References

1. W.J. Quadakkers and K. Bongartz, *Werkstoffe und Korrosion*, **1994**, 45 232.
2. J. Nicholls, M. Bennett and R. Newton, *Materials at High Temperatures*, 20, (3), **2003**, 429.

3. D.M. Lipkin and D.R. Clarke, *Oxidation of Metals*, **1996**, 45, 267.
4. W.J. Quadakkers, D. Naumenko, E. Wessel, V. Kochubey and L. Singheiser, *Oxidation of Metals*, **2004**, 61, 17.
5. H. Al-Badairy, G. Tatlock, H. Evans, G. Strehl, G. Borchardt, R. Newton, J. Nicholls, D. Naumenko and W.J. Quadakkers, *Lifetime Modelling of High Temperature Corrosion Processes* (eds M. Schütze, W.J. Quadakkers), EFC monograph No. 34, Institute of Materials, London, **2001**, 50.
6. D. Naumenko, L. Singheiser and W.J. Quadakkers, *Cyclic Oxidation of High Temperature Materials* (eds M. Schütze, W.J. Quadakkers), EFC monograph number 27, Institute of Materials, London, **1999**, 287.
7. M. Bennett, R. Newton, J. Nicholls, H. Al-Badairy and G. Tatlock, *Materials Science Forum*, **2004**, 461–464, 463.
8. D. Naumenko and W.J. Quadakkers, Report Forschungszentrum Jülich, Jülich, Germany, Jül-3948, **2002**.
9. B.A. Pint, *Surface & Cooatings Technology*, **2004**, 188–189, 71.
10. B.A. Pint, J.A. Haynes, K.L. More and I.G. Wright, submitted to *Superalloys*.
11. B. Jönsson, A. Westerlund and G. Landor, *Cyclic Oxidation of High Temperature Materials* (eds M. Schütze, W.J. Quadakkers), EFC monograph No. 27 Institute of Materials, London, **1999**, 324.
12. H. Echsler, E. Alija Martinez, L. Singheiser and W.J. Quadakkers, *Materials Science and Engineering*, **2004**, A 384, 1.
13. J.J. Barnes, J.G. Goedjen and D.A. Shores, *Oxidation of Metals*, **1989**, 32, 449.
14. H.E. Evans, *Cyclic Oxidation of High Temperature Materials* (eds M. Schütze, W.J. Quadakkers), EFC monograph No. 27, Institute of Materials, London, **1999**, 3.
15. H.E. Evans, S. Osgerby and S. Saunders, *John Stringer Symposium on High Temperature Corrosion 2001* (eds P. Totorelli I. Wright, P. Hou), **2003**, 757–758, 122.
16. V.K. Tolpygo and D.R. Clarke, *Acta Materialia*, **1999**, 47, 3589.
17. M.Y. He, J.W. Hutchinson and A.G. Evans, *Acta Materialia*, **2002**, 50, 1063.
18. H. Hattendorf, ThyssenKrupp VDM GmbH, Werdohl, Germany, unpublished results.

23

Service conditions and their influence on the oxide scale formation on metallic high temperature alloys for the application in innovative combustion processes

G TENEVA-KOSSEVA, H KÖHNE and
H ACKERMANN, Oel-Wärme-Institut gGmbH, Germany
and M SPÄHN, S RICHTER and J MAYER,
Central Facility for Electron Microscopy, Germany

23.1   Introduction

In modern small-and medium-sized burners for industrial gas oil (IGO), the technique of recirculation of external exhaust gas is used to achieve a low level of NO_x emissions and soot formation. Part of the exhaust gases are recirculated into the combustion zone by means of a head-piece or flame tube. As high temperature alloys with good resistance to high temperature corrosion, sufficient mechanical strength at elevated temperatures, and good workability are available, most of the flame tubes are manufactured from sheets of high temperature alloys. During service the flame tubes are exposed to extreme thermal and atmospheric conditions. Future developments aim at reducing the dimensions of the heating systems, thereby, further raising the temperature experienced by the tube material.

As the combustion atmosphere contains a mixture of different compounds (CO, CO_2, N_2, NO_x, O_2, H_2O, SO_2, SO_3, carbon hydroxides and their radicals), it is unlikely that the corrosion resistance measured in air is applicable to the situation in burners. The scope of this study is to provide industry with data concerning the corrosion resistance of metallic high temperature alloys under conditions characteristic for oil combustion. Of particular value would be the temperature limit for the use of commercial alloys.

The present chapter describes the service conditions at the flame tube of an intermittent operating recirculation burner measured in a burner rig (maximum temperature experienced by the material 1000°C). Three different austenitic high temperature Ni-base alloys are being tested in the rig in an ongoing experiment. The effect of the parameters material temperature, duration of the air ventilation after burner shutdown and sulphur content of fuel on the structure and growth of the oxide scale on the alloys after 50 h exposure time have been investigated.

23.2   Experimental methods

23.2.1   Burner rig

The burner rig is a recirculation burner arranged in a standard water-cooled test boiler. The test boiler is cylindrical with 225 mm inner diameter and 540 mm length. The burner is equipped with a commercial mixing device consisting of a simplex pressure nozzle for fuel atomization and an air nozzle. The combustion air is supplied through the air nozzle with a swirling motion. Utilizing the injection effect, exhaust gas is recirculated through a circumferential slit in the flame tube. Here, air and recirculated exhaust gas are mixed with fuel and stabilization of the flame is achieved. The standard flame tube of the burner is manufactured from a 1 mm thick sheet of a high temperature alloy. Figure 23.1 shows the front plate of the burner rig with the mounted burner.

In all experiments a burner power of 16.7 kW and CO_2 content of 13.4 ± 0.2 vol.% in the exhaust gas (corresponding to an air ratio of 1.15) were utilized. The burner operated intermittently. Each cycle lasted 20 min and consisted of 15 min burner operation followed by a 5 min pause. According to the safety instructions, air ventilation of approximately 12 s before burner start, as well as 5 s after burner shutdown, was integrated generally. The liquid fuel used was IGO. According to the appropriate German standard, it contained up to 2000 mg of sulphur per kg.

For simultaneous testing of different materials the flame tube comprised six segments, manufactured from different alloys. The dimensions of the sample tube were identical to those of the standard tube of the burner. The tube and a single segment are shown in Fig. 23.2.

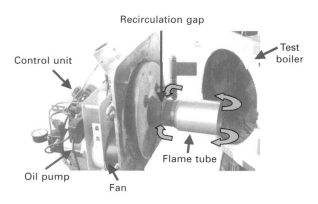

23.1 Burner rig.

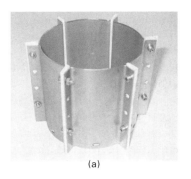 

(a)   (b)

23.2 (a) Flame tube, and (b) a single alloy segment. The sheet thickness is 1 mm.

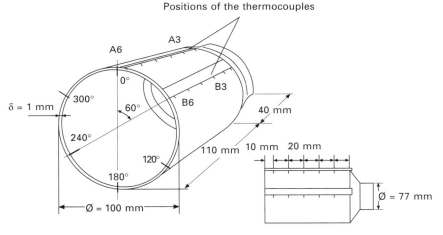

23.3 Positions of the thermocouples in the axial and the tangential directions for measurement of temperature of tube material during operation of the burner.

23.2.2 Measurement of temperature and gaseous environment

A maximum temperature of 1000°C experienced by the material was preset for the experiments. In order to achieve such a high temperature, the burner was fired into a metallic cylindrical chamber installed in the boiler (hot combustion chamber). The flame tube temperature in the axial and the tangential direction was measured using Ni/Cr-Ni thermocouples, positioned in special canals in the material and pressed against the metal surface by a metal plate as shown in Fig. 23.3. The distance between the thermocouples in the axial direction was 20 mm. Measurements of the flame tube temperature in six tangential positions (0°, 60°, 120°, 180°, 240°, 300°) were carried out by rotating the flame tube.

The composition of the gas environment near the wall along the flame tube was measured by using a suction probe made from stainless steel with an inner diameter of 2 mm. The concentrations of CO and CO_2 were measured by infrared spectrometry and that of O_2 with a magneto-mechanical method (minimal detectable concentration 0.1 vol.% O_2).

23.2.3 Alloys and exposure tests

Three austenitic Ni-based alloys were investigated and their chemical compositions are given in Table 23.1. The Cr content of the alloys ranged between 22 and 29 wt.%. Alloy 602 and 693 are aluminium doped, and Alloy 603 contains silicon instead of aluminium.

To test the influence of duration of air ventilation after burner shutdown, two experiments with 5 s and 0.1 s air ventilation were conducted. For examination of the effect of sulphur, two different qualities of IGO were used: standard quality with a sulphur content of 1700 mg/kg and IGO with reduced sulphur content of 20 mg/kg sulphur. The duration of each test was 50 h or 203 burner cycles.

23.2.4 Metallographic analysis

In order to avoid contamination of the surface, 1 cm × 2 cm coupons were cut from the sheet metal by a laser process. Deposition of removed material during the process was prevented by a steady gas flow. The cut pieces were prepared metallographically in the same way as specimens of steel. However, for near-surface analysis the samples were coated with a thin gold layer and then nickel-coated in an electrolytic bath before starting the cross-section preparation. Measurements were carried out in an electron microprobe (Camebax SX50) equipped with four wavelength-dispersive spectrometers. For EPMA analysis the following measuring conditions were employed: the electron gun was operated at 15 keV beam energy and 20 nA beam current, with an effective probe size of about 0.5 μm under these conditions. Elemental

Table 23.1 Chemical composition of the alloys tested

| Alloy | Chemical composition, wt.% | | | | |
	Ni	Cr	Al	Fe	Other
Alloy 602 CA 2.4633	62.6	25.1	2.29	9.25	balance
Alloy 603 XL	>73.2	21.97	<0.01	0.08	Si = 1.39 Mo = 3.06 balance
Alloy 693	61.6	29.1	3.36	4.56	balance

mappings were acquired, i.e. the electron beam was scanned over an area of 60 μm × 60 μm and simultaneously characteristic X-ray intensities of the corresponding elements were detected in each pixel. The total sum of pixels was 256 × 256 with a measuring time per pixel of 50 ms. Backscatter electron images (BSE) and secondary electron images were taken from each analysed area.

23.3   Results

23.3.1 Service conditions

Figure 23.4 shows the band of material temperatures measured for the tangential directions as a function of axial position. The temperature monotonically increases towards the outlet of the flame tube, where it approaches a saturation value. As requested, the highest temperature is 1000°C. The variation range of the temperature in the tangential direction is 60 K at most at the cold side of the tube and about 20 K at the outlet.

As shown in Fig. 23.5, the composition of the gas environment varies along the flame tube for the six tangential positions. It differs from the composition of the exhaust gas after complete combustion. The dry exhaust gas contains 13.4 vol.% CO_2, 3 vol.% O_2 and max. 5 ppm CO.

For axial positions between 0 mm and 40 mm at 60° and 240°, the concentration of oxygen drops below the detection limit of the analyser. The metallographic cross-sections nonetheless showed that an oxide scale had formed.

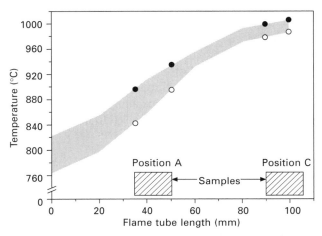

23.4 Material temperatures measured in axial and tangential direction. The solid symbols represent the highest and the open ones the lowest temperature experienced by the material at the position of sampling for the metallographic investigation.

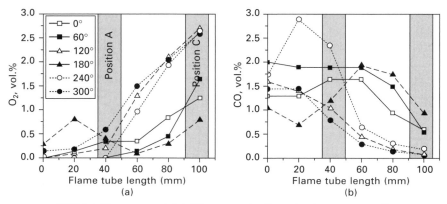

23.5 (a) Oxygen and (b) CO concentrations in dry gas near flame tube wall measured along the flame tube. The rectangles show the positions of the sampling for the metallographic investigation.

23.3.2 Characterization of the oxide scale

After 50 h exposure time the following layer structure on the alloy surface was detected: an outer layer of Cr_2O_3 and underneath precipitates of Al_2O_3 (Alloy 602 and 693) or of SiO_2 (Alloy 603).

The Cr_2O_3 layers on Alloys 602 and 603 are uniform, in contrast to Alloy 693, where regions with accumulated thick oxides and others with a thin Cr_2O_3 layer were observed. On Alloy 603 the Cr_2O_3 layer was thinner than those on Alloys 602 and 693 (see Fig. 23.8). For each alloy, particles with matrix composition are embedded in the Cr_2O_3 layer.

For Alloy 602, selective internal oxidation of aluminium occurred along the alloy grain boundaries (Fig. 23.6). On Alloy 693 Al_2O_3 formed a thin inner layer directly underneath the outer Cr_2O_3 layer. In other places it also precipitated to a depth of about 10 μm preferentially along grain boundaries parallel to the metal surface. On Alloy 603 SiO_2 grew as a non-continuous layer below the outer Cr_2O_3 scale. In addition, silicon was internally oxidized in some places (Fig. 23.8). The minor alloy constituents titanium and manganese are enriched in the oxide layer at the gas and oxide interface on Alloy 602 and Alloy 693 (Fig. 23.6).

23.3.3 Effect of the duration of the air ventilation after burner shutdown

The samples analysed for the investigation of ventilation effect were taken from position C corresponding to material temperatures between 980°C and 1010°C (see Fig. 23.4). The frequency of minimum scale thickness was obtained by subdividing the BSE images into sections of 2.2 μm widths and determination of the minimal scale thickness of the Cr_2O_3 layer for each

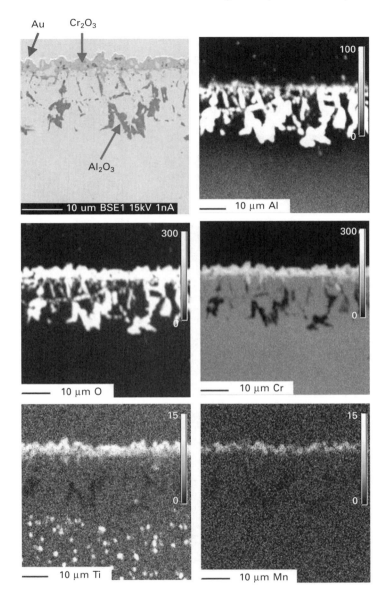

23.6 Layer structure of Alloy 602CA after 50 h at position C corresponding to material temperatures between 980°C and 1010°C (see Fig. 23.4): backscattered electron image and elemental maps. The bar in each elemental map image describes the colour scale between the minimum and maximum intensity of the corresponding X-ray line.

section. The distribution functions of frequency for the three alloys and the role of different ventilation times after shutdown (0.1 and 5 s) are shown in Fig. 23.7 (a–c). For the ventilation time of 0.1 s the minimal scale thickness is between 0.5 μm and 1.5 μm for Alloy 602, between 0.5 μm and 1 μm for

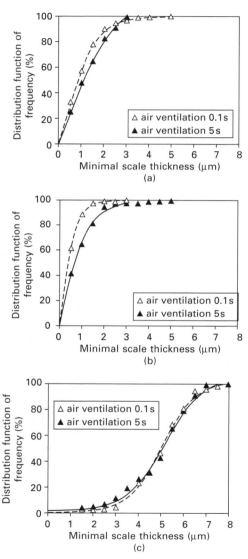

23.7 Distribution function of frequency for the minimal scale thickness: (a) Alloy 602, (b) Alloy 603 and (c) Alloy 693 after 50 h for air ventilations of 0.1 s and 5 s and at position C corresponding to material temperature between 980°C and 1010°C.

Alloy 603, and between 1.5 µm and 6 µm for Alloy 693 for 80% of the surface, respectively. An effect – albeit small – of air ventilation time is observed for Alloy 602 and 603. For these alloys the distribution moves towards slightly higher values (+0.5 µm). This trend is not observed on Alloy 693 (Fig. 23.7 (c)).

23.3.4 Effect of the sulphur content in the fuel

In order to investigate the effect of sulphur, samples were taken from position A corresponding to material temperatures between 840°C and 940°C (see Fig. 23.4). Figure 23.8 shows the layer structure formed on the alloys exposed to the combustion products of the different fuel qualities. The Cr_2O_3 scale formed when firing with IGO with a reduced sulphur content of 20 mg/kg is thinner than that formed when standard quality IGO was used.

The relative amount of chromium oxide was quantified by estimation of the area of its layer in two BSE images. The two images covered a total surface length of 150 µm for each alloy. The results shown in Fig. 23.9 confirm the visual impression of Fig. 23.8.

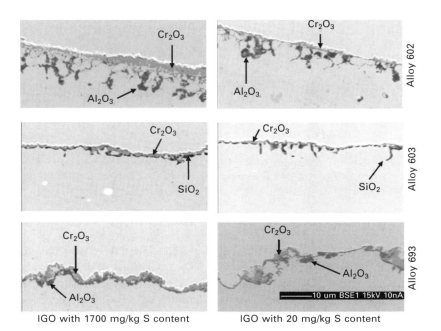

23.8 BSE images of Alloy 602, 603 and 693 after 50 h exposure at position A (see Fig. 23.4). The images on the left side show the layer structures formed by firing standard quality IGO (1700 mg/kg sulphur) and on the right side those by using IGO with 20 mg/kg sulphur content.

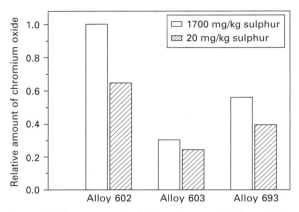

23.9 Relative amount of chromium oxide for the two different oil qualities tested.

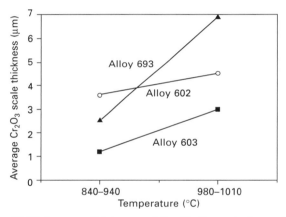

23.10 Average thickness of the Cr_2O_3 scale formed at temperatures between 840°C and 940°C and between 980°C and 1010°C after 50 h testing.

23.3.5 Effect of temperature

The thickness of the chromium oxide scale was determined from BSE images of the samples at two different temperatures (see positions A and C in Fig. 23.4). The average Cr_2O_3 scale thickness is shown in Fig. 23.10, which increased with the temperature. This temperature effect was greatest for Alloy 693 and less so for Alloy 603 and Alloy 602.

23.4   Discussion

A typical two-layer oxide scale structure was observed on the surface of the Ni-based alloys studied: an outer layer of Cr_2O_3 and below it Al_2O_3 (Alloy

602, 693) or SiO_2 percipitates (Alloy 603). The thickest chromium oxide scales were observed on Alloys 602 and 693. On Alloys 602 and 603 the Cr_2O_3 layer formed is more uniform than on Alloy 693. In addition, aluminium was internally oxidized, predominantly along the grain boundaries, in Alloy 602. This two-layered structure of the oxide scale on Alloy 602 was also found previously [1] at temperatures of 1000°C and 1100°C after respectively 18 one-day cycles and 40 one-day cycles in an oxidizing atmosphere. In Alloy 693 Al_2O_3 formed a thin inner layer. This layer was detected directly under the chromium oxide, or about 10 µm below the Cr_2O_3 matrix interface along grain boundaries parallel to the surface. On Alloy 603, SiO_2 grew as a non-continuous layer below the outer Cr_2O_3 layer and silicon was internally oxidized in some places.

The oxide scale characterization studies described in Section 23.3 focused on the main constituents, but in addition information was derived on the behaviour of two other minor alloy constituents, manganese and titanium. On Alloys 602 and 693, diffusion of manganese and titanium towards the surface was detected. For manganese it is assumed that it forms $MnCr_2O_4$ spinel at the gas oxide interface. This layer is less protective than Cr_2O_3 and tends to spallation [1, 2]. Enrichment of manganese at the chromium oxide gas interface has been reported [3] on the surface of SS304 in metal dusting exposures at 600°C after 300 h and on 310 stainless steel oxidized for 168 h at 600°C in O_2 + 40% H_2O [4]. In the present work, titanium was enriched in a thin layer on the oxide scale surface indicating that it diffused across the Cr_2O_3. In the bulk alloy it formed globular carbides. A similar distribution of titanium was found for 321 stainless steel in dry oxygen at 1000°C after 100 h [2].

Temperature measurements showed that after burner shutdown when air ventilation is continued the temperature drops from 1000°C to 700°C within 30 s. During this period the oxygen partial pressure is at a high level of 21 vol.% typical for air. The present experiments indicated a different rate of oxidation between cooling down with and without air ventilation in the case of Alloys 602 and 603. For these alloys higher values of the minimal thickness of the chromium oxide scale were found. On Alloy 693 this effect was absent.

As discussed previously, the sulphur level in the liquid fuel oil affects the growth of the chromium oxide scale. Evidence was found that the scales formed on all the alloys studied were thicker in the case of firing with IGO with sulphur content of 1700 mg/kg compared to those formed when the fuel oil contained only 20 mg/kg sulphur. In addition, the Cr_2O_3 scale observed on the alloy surface after combustion with a reduced fuel sulphur content is non-continuous. This was particularly noticeable for Alloy 693 as regions with thick oxide accumulations were detected beside places where the oxide scale was thin or absent. The effect of the sulphur content on the growth and structure of the Cr_2O_3 scale is in agreement with the results of other authors [5, 6], who reported that the presence of SO_2 in air or exhaust gases of combustion processes may accelerate the high temperature corrosion of several metals. In contrast, investigations on iron carried out by other authors describe an inhibitive effect of traces of SO_2 in oxygen for temperatures below 600°C and no effect of the sulphur at higher temperatures [7].

The oxidation rate in the current experiments increased with increasing temperature. In agreement with this, the thickness of the chromium oxide layer for temperatures between 840°C and 940°C was smaller than for temperatures between 980°C and 1010°C for the three alloys. The maximum oxide scale increase was detected on Alloy 693 followed by Alloy 603 and Alloy 602. Strikingly, Alloy 602 showed the lowest oxide scale increase at the upper temperature although at the lower temperature it formed the thickest oxide scale. This behaviour might be caused by chromium oxide evaporation (CrO_3 gas) at temperatures of about 1000°C [8] or by scale spallation. In the latter case at the high temperature the critical oxide thickness for spallation was reached earlier on Alloy 602 than on the other alloys.

23.5   Conclusion

After 50 h exposure time the following layer structure on the alloy surface was detected: an outer layer of Cr_2O_3 and underneath precipitates of Al_2O_3 (Alloys 602 and 693) or SiO_2 (Alloy 603). The Cr_2O_3 layers on Alloys 602 and 603 are more uniform than on Alloy 693. On Alloy 603, a thinner Cr_2O_3 layer formed compared with those on Alloys 602 and 693. For each alloy, particles with matrix composition were embedded in the Cr_2O_3 layer. Titanium and manganese migrated from the bulk towards the metal surface on Alloys 602 and 693.

Aluminium was internally oxidized predominantly along the grain boundaries in Alloy 602. On Alloy 693, Al_2O_3 formed a thin inner layer directly underneath the chromium oxide or at a depth of about 10 μm preferentially along grain boundaries parallel to the metal surface. On Alloy 603, SiO_2 grew as a non-continuous layer below the outer Cr_2O_3. In addition, silicon was internally oxidized in some places.

After a longer ventilation time at temperatures between 980°C and 1010°C, the minimal scale thickness of the chromium oxide scale on Alloys 602 and 603 slightly increased. This trend is not observed on Alloy 693.

For temperatures between 840°C and 940°C, the Cr_2O_3 scale formed when firing with IGO with reduced sulphur content of 20 mg/kg is thinner than that formed when standard quality IGO with sulphur content of 1700 mg/kg was used. Increase of temperature led to formation of thicker Cr_2O_3 scales, being most marked for Alloy 693 and less so for Alloys 603 and 602.

23.6   References

1. B. Li, Ph.D. Thesis, Ames, Iowa, **2003**.
2. F.H. Scott and F. I. Wei, *MST,* **1989**, 5, 1140.
3. E.M. Müller-Lorenz and H.J. Grabke, *Improvement of stainless steels for use at elevated temperatures in aggressive environments*, European Commission, Report EUR 20103 EN.
4. H. Asteman, J.-E. Svensson and L.-G. Johansson, *Corros. Sci.*, **2002**, 44, 2635.
5. P. Kofstad, *High Temperature Corrosion,* Elsever Applied Science, London, New York, 1988.
6. H. Xu, M. G. Hocking and P.S. Sidky, *Oxid. Met.*, **1993**, 39, 371.
7. A. Järdnäs, J.-E. Svensson and L.-G. Johansson, *Oxid. Met.*, **2003**, 55, 427.
8. R. Bürgel, *Handbuch Hochtemperaturwerkstofftechnik*, Vieweg, Braunschweig, 2001.

24

Reducing superheater corrosion in wood-fired boilers

P J HENDERSON, C ANDERSSON,
H KASSMAN and J HÖGBERG, Vattenfall Research
and Development, Sweden and P SZAKÁLOS and
R PETTERSSON, Corrosion and
Metals Research Institute, Sweden

24.1   Introduction

In the last few years, there has been a move away from burning fossil fuels through the co-utilisation of biomass and coal and finally to 100% biomass such as wood and waste wood products. Unfortunately, burning of biomass causes widespread fouling of superheater tubes and corrosion can occur rapidly under the sticky alkali chloride deposits. Even at today's maximum steam temperatures of 500 to 540°C there are some severe corrosion problems when burning 100% wood-based fuel [1]. It is also desirable to be able to burn other environmental fuels such as straw, demolition wood or other waste wood products, to reduce production costs and avoid dumping waste at landfill sites. This, however, makes the corrosion and fouling problems even more serious [2, 3].

A complete set of superheaters for a 100 MW combined heat and power boiler costs in excess of €1m. The durability of superheaters is thus an important factor in determining the long-term production costs. Unplanned outages due to leaking superheaters are also very expensive. As well as causing corrosion problems, the build-up of deposits reduces the heat uptake to the superheaters which leads to lower efficiency. Consequently, ways are being sought to reduce superheater corrosion.

Most biomass fuels have a high content of alkali metals and chlorine, but they contain very little sulphur compared to fossil fuels. Potassium chloride, KCl, is found in the gas phase, condenses on the superheater tubes and forms complex alkali salts with iron and other elements in the steels. These salts have low melting points and are very corrosive. Vattenfall has developed and patented an instrument for *in-situ* measurement of gaseous alkali chlorides which gives an indication of how corrosive the flue gases are [4]. This instrument is called an *in-situ* alkali chloride monitor (IACM). Vattenfall has also developed and patented a concept with a sulphate containing compound called 'ChlorOut' [5], which is sprayed into the flue gases after combustion is complete, but before the flue gases reach the superheaters, and effectively converts KCl into potassium sulphate, K_2SO_4. This compound is much less corrosive than KCl. In the experiments reported here, the sulphate used in ChlorOut was ammonium sulphate. This is also used for the reduction of NOx.

This study reports on measures taken to reduce superheater corrosion in two fluidised bed boilers burning wood-based fuels, using the ChlorOut additive to control the KCl levels and more corrosion-resistant steels.

24.2   Experimental methods

24.2.1   Descriptions of plants used for testing

105 MW$_{tot}$ boiler in Nyköping, Sweden

This plant is situated about 120 km south of Stockholm. The plant consists of a bubbling fluidised bed (BFB) steam boiler (Boiler 3) for combined heat and power (CHP) operation, two circulating fluidised bed (CFB) boilers for hot water production and a hot water accumulator. Tests were performed in Boiler 3, for CHP production.

The CHP unit produces 35 MW of electricity and 69 MW of heat and has been in operation since the end of 1994. The final steam temperature is 540°C and the pressure 140 bar. The fuel mix consists of wood chips (logging residues), demolition wood and coal.

The ChlorOut was sprayed into the flue gases upstream of superheater 2, the first superheater that the flue gases meet on their passage through the boiler. Deposit and corrosion probe testing were performed immediately before superheater 2 and flue gas measurements immediately after (see Fig. 24.1).

The signal from IACM was used to control the amount of ChlorOut sprayed into the boiler. A fuel such as demolition wood will increase the KCl level in the flue gas, which means that more ChlorOut has to be injected into the boiler to bring the KCl levels under control. The ChlorOut solution was diluted with water from the flue gas condensation unit. The ChlorOut patent covers a number of sulphates; in this case ammonium sulphate was used to aid the reduction of NOx.

98 MW$_{tot}$ boiler in Munksund, Sweden

This is a circulating fluidised bed (CFB) boiler situated at SCA's pulp and paper mill in Munksund, near Piteå, in the north of Sweden, about 75 km south of the Arctic Circle. The boiler was taken into operation in 2001 and can produce up to 25 MW of electricity as well as process steam. The final steam temperature is 480°C and pressure 60 bar. The boiler is fired mainly with bark (>80%), but also sawdust, woodchips and approximately 6% of

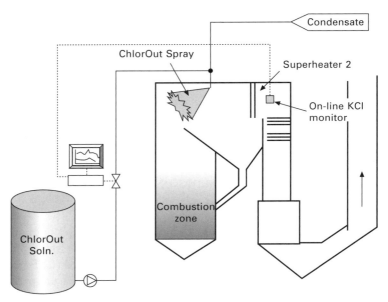

24.1 Schematic diagram of the ChlorOut system in BFB boiler, Nyköping.

waste. The waste is mostly plastic from tape and packaging, obtained from the reprocessing (recycling) of cardboard cartons.

The ChlorOut was sprayed into the flue gases at the entrance to the cyclone, and deposit and corrosion probe testing were performed immediately before superheater 2 where the flue gas temperature is 680°C (see Fig. 24.2).

Measurement techniques

Deposit, corrosion and flue gas measurements were performed under normal conditions (no ChlorOut) and with ChlorOut in Nyköping and Munksund.

The chemical composition of the flue gas at the superheaters/corrosion probes was determined by several measurement techniques, including *in-situ* alkali chloride monitoring (IACM). IACM is a new method, developed and patented by Vattenfall AB, for continuous and *in-situ* measuring of alkali chlorides in the gas phase in the area around the superheaters in power stations [4]. A schematic diagram of a set-up with IACM is shown in Fig. 24.3. IACM measures the overall concentration of NaCl and KCl, which during wood fuel firing can be expressed as the level of KCl and works on the principle of optical absorption. It works in the temperature range 600–1500°C. IACM was also used to measure SO_2 in these tests. Light in the wavelength range of 200–380 nm is used for the analysis of both alkali chlorides and sulphur dioxide. The instrument has a sampling time of 5–10 s and the

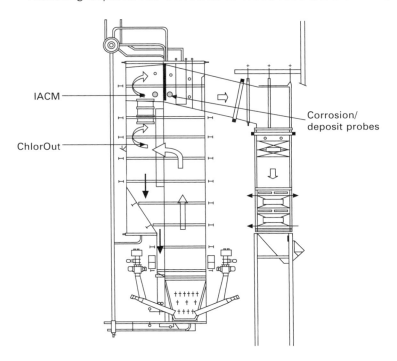

24.2 Schematic diagram of the ChlorOut system in the CFB boiler, Munksund.

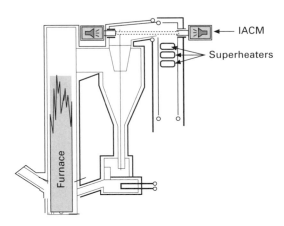

24.3 Schematic diagram showing how the *in-situ* alkali chloride monitor, IACM is placed in a furnace before the superheaters. The beam is transmitted across the flue gas passage.

detection limit at a 5 m measuring length (width of the flue gas channel) is 1 ppm for KCl and NaCl and 4 ppm for SO_2.

Deposits were collected on stainless steel rings on temperature controlled deposit probes for a period of 3 h and were analysed by energy dispersive x-ray analysis (EDX) in a scanning electron microscope (ASEM), which penetrates several microns. Some deposits were analysed by time-of-flight secondary ion mass spectrometry (TOF-SIMS). The secondary ion mass spectrum gives information about the elemental and chemical composition of the material. Since the secondary ions originate only from the 1–5 outermost atomic/molecular layers of the material, this technique is highly surface sensitive. Chemical profiles through the depth of the oxide, formed *under* the deposit on ring specimens of 13CrMo44, were obtained by glow discharge optical emission spectroscopy (GDOES). In this technique, a plasma bombards the surface of the specimen (the cathode) and sample erosion occurs by cathode sputtering. Since the sputtering rate is known for a specific material, the depth from which atoms are coming can be calculated. The atoms are excited in the plasma and subsequent photoemission from electron de-excitation is analysed by optical spectroscopy. Thus simultaneous material removal and multi-element acquisition is obtained. It was not possible to perform GDOES on the deposits.

Corrosion specimens were exposed as rings on internally cooled probes placed near the superheaters. Details of probe construction can be found in [6]. Each ring was measured at 15° intervals on two circumferences before and after testing to obtain values of metal loss caused by corrosion or oxidation. Internal corrosion (or selective corrosion such as grain boundary attack) was also measured. The total corrosion is the sum of metal loss and internal corrosion values. The microstructures of the exposed rings were examined by light optical and scanning electron microscopy and energy dispersive X-ray analysis. The chemical compositions of the steels used are shown in Table 24.1.

24.3   Results

24.3.1   Deposit and flue gas chemistry

Figure 24.4 shows how the amount of gaseous KCl varies with the fuel mix and presence of ChlorOut (ammonium sulphate) in the Munksund plant. The results are shown as three hourly measurements for bark, bark + waste, and bark + waste + ChlorOut. Deposit probe exposures lasting 3 h were taken at the same time as these measurements were being recorded. ChlorOut reduced the amount of KCl (g) in the flue gas from about 20 to 2 ppm when burning bark plus waste (the normal fuel mix for this plant). This led to a large decrease in the chloride content of the deposits and a reduction in deposit

Table 24.1 Chemical composition in wt.% of the steels tested. The balance is Fe

Steel	Cr	Ni	Mo	Mn	Other
15Mo3	–	–	0.30	0.52	C 0.16, Si 0.26
13CrMo44	0.86	0.07	0.48	0.46	C 0.12, Si 0.21
10CrMo910 (T22)	2.1	–	0.92	0.43	C 0.12, Si 0.22
X20CrMoV12 1	10.45	0.70	0.88	0.6	C 0.18, Si 0.22, V 0.26
T92	9.15	0.26	0.5	0.46	W 1.7, Si 0.22, Nb 0.6, N 0.05, V 0.2, B 0.003, C 0.11
Esshete 1250	14.9	9.65	0.94	6.25	C 0.084, Si 0.58, Nb 0.86, V 0.22, B 0.004
TP 347 H	17.6	10.7	–	1.84	Si 0.29, C 0.05, Nb 0.6

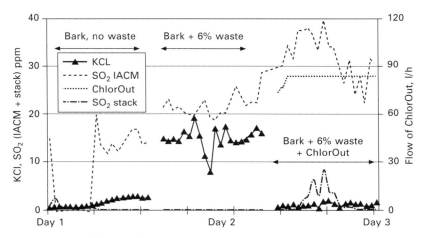

24.4 KCl and SO_2 levels at the superheaters measured during fuel tests at the Munksund plant. Measurements made for 3 h on three consecutive days. Day 1, bark no waste; day 2 bark and 6% waste; day 3 bark, 6% waste and ChlorOut. The combustion of waste increases the KCl levels, indicating a higher corrosivity. ChlorOut reduces the KCl levels.

growth rate of at least 50%, as shown in Table 24.2. The key elements from the deposit and flue gas analyses are also given in Table 24.2. The Cl level was below the limit of detection by ASEM when ChlorOut was used.

In Nyköping during the 3 h test period ChlorOut reduced the average gaseous alkali chloride levels in the flue gases from 30 to 17 ppm. Chemical analyses of deposits collected for 3 h at 600, 500 and 400°C showed that ChlorOut removed or greatly reduced the Cl content of the deposit by converting KCl into K_2SO_4. The key elements from the analyses are shown in Tables 24.3 and 24.4. During these tests the fuel burnt was 5% coal, 47.5% demolition wood and 47.5% logging residues.

Table 24.2 Results from Munksund. Deposit analyses in wt.% (by ASEM) of deposits collected on rings on cooled probes. Ring temperatures 400 and 500°C. Deposits were analysed by ASEM on the windward side (facing the flue gas) and the leeward side (away from flue gas). Total chloride levels were measured by dissolving the deposits in de-ionised water, analysing by ion chromatography and compensating for deposit mass. Weight gains are the increase in mass of the deposit rings during a 3 h period. Gaseous KCl and SO_2 measurements were made by IACM in the flue gas. Fuel is bark with 6% waste from cardboard recycling

Method	Element	With ChlorOut				No ChlorOut			
		400°C wind	400°C lee	500°C wind	500°C lee	400°C wind	400°C lee	500°C wind	500°C lee
ASEM	Na	2.4	1.3	2.5	1.6	2.4	1.7	2.4	1.7
	S	10.0	6.1	10.1	5.8	5.5	2.6	5.5	3.1
	Cl	0	0	0	0	13.8	4.2	9.6	4.3
	K	6.9	5.0	8.3	5.0	17.2	7.2	14.0	7.4
	Ca	12.3	24.7	13.0	24.5	13.8	21.4	15.4	21.6
Ion Chr.	Chloride	0.14		0.22		3.6			
	Weight gain	31 mg		24 mg		66 mg		50 mg	
IACM	SO_2	22–40 ppm				23–28 ppm			
	KCl (g)	0.5–1.9 ppm				14–19 ppm			

Table 24.3 Results from Nyköping. Deposit analyses of key elements, in wt.% (by ASEM) and corresponding KCl levels in the flue gas. Ring temperatures 500 and 600°C. Deposits were analysed by ASEM on the windward side (facing the flue gas) and the leeward side (away from flue gas). Fuel 5% coal, 47.5% demolition wood and 47.5% logging residues

	With ChlorOut KCl (g) 17 ppm				No ChlorOut KCl (g) 30 ppm			
Element	600°C wind	600°C lee	500°C wind	500°C lee	600°C wind	600°C lee	500°C wind	500°C lee
Na	1.5	1.4	1.1	1.8	0.3	1.0	0	0.7
S	11.4	9.7	10.3	8.0	5.8	13.0	5.7	6.8
Cl	0	0	0	0.1	1.3	0.3	2.9	2.2
K	17.4	11.9	16.5	9.9	8.4	21.4	8.0	11.2
Ca	6.4	6.7	6.3	4.2	19.5	12.3	19.3	17.2

Table 24.4 Key elements and compounds found on leeward side of deposits collected at 400°C, in Nyköping. Same conditions as in Table 24.3 Elements in wt.% by ASEM and compounds by TOF-SIMS. The results show that KCl has been converted into K_2SO_4 by ChlorOut

(ASEM) Element	No ChlorOut (wt.%)	With ChlorOut (wt.%)	(TOF-SIMS) Compound	No ChlorOut Relative conc. (total = 1)	With ChlorOut Relative conc. (total = 1)
Na	1.0	1.5	NaCl	0.044	0.055
S	5.2	8.9	Na_2CO_3	0.005	0.007
Cl	7.5	0	Na_2SO_4	0.022	0.063
K	20.7	6.7	KCl	0.628	0.067
Ca	10.4	2.2	K_2CO_3	0.016	0.021
			K_2SO_4	0.284	0.787

24.3.2 Corrosion results

The results of GDOES profiling through the oxide surface of 13CrMo44 rings exposed in Munksund showed that after 3 h exposure the corroded layer was approximately 2 µm thick in the case with no ChlorOut, but only 0.5 µm thick with ChlorOut (see Figs 24.5 and 24.6). Analytical scanning electron microscopy of the surface after exposure confirmed that there were large numbers of iron chloride particles present in the ring exposed without ChlorOut and that these were largely absent from the specimen exposed with ChlorOut.

The chemical compositions of all the steels tested are given in Table 24.1. The results of long-term corrosion probe testing of superheater steels in Nyköping are shown in Fig. 24.7. T92 (a development of the 9% Cr steel T91) or T91 itself, are sometimes chosen as replacements for X20; they have higher strengths and can therefore be used with thinner wall thicknesses. Figure 24.7 shows that it is not a good idea to use 9% Cr steels for biomass

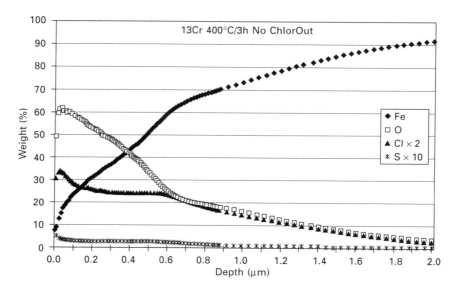

24.5 Composition profiles for Fe, O, Cl and S from the surface into the steel 13CrMo44 after exposure for 3 h in Munksund. Fuel is bark with 6% waste.

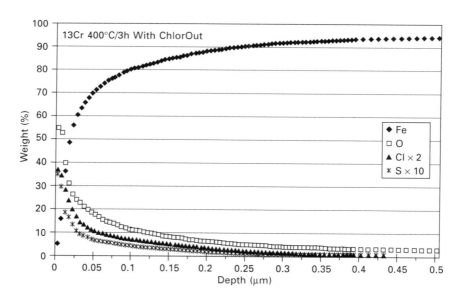

24.6 Composition profiles for Fe, O, Cl and S from the surface into the steel 13CrMo44 after exposure for 3 h with ChlorOut in Munksund. Fuel is bark with 6% waste.

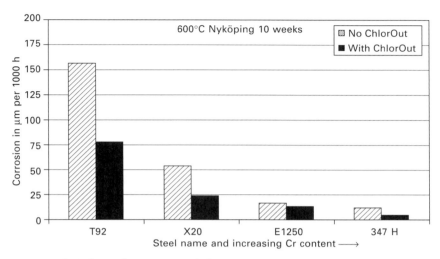

24.7 Corrosion per 1000 h for various steels tested in Nyköping. Exposure time was 10 weeks. Average metal temperature was 580°C without ChlorOut and 602°C with ChlorOut. In spite of a 20°C higher temperature the corrosion rates were lower with ChlorOut.

plants with 540°C steam. The ChlorOut additive reduced corrosion by 50% for the 9–12% Cr martensitic steels T92 and X20.

A closer examination of the X20 steel revealed a high $FeCl_2$ content in the metal/oxide interface in the test without ChlorOut (see Fig. 24.8). It may be noted here that iron chloride is deliquescent, a fact which can explain both the volume expansion of this phase after surface preparation and the high oxygen content in the EDS analysis. The oxide formed on this material had an outer iron oxide layer, probably hematite Fe_2O_3, and an inner mixed Fe-Cr oxide layer, which is seen in the top part of Fig. 24.8. The growth rate, structure and composition all indicate that this is a spinel-type oxide (M_3O_4). At the interface to the metal there was an oxide border with a higher chromium content, again most probably a mixed spinel. There were no significant amounts of alkali metals at the metal oxide interface in either the oxide or chloride phases. In the specimen exposed with ChlorOut, the oxide on X20 seems to be somewhat thinner and more adherent to the metal substrate (Fig. 24.9). Distinct iron sulphide layers were integrated within the oxide and the Cl content was much lower. This type of layer appears only in the case of ChlorOut additions.

The two austenitic stainless steels Esshete 1250 and TP347 H showed much lower corrosion rates, with or without ChlorOut. In the plant in Nyköping the original X20 superheaters were replaced with Esshete 1250 in 1998, because of corrosion problems and have since shown low corrosion rates. In Munksund the use of ChlorOut approximately halved the corrosion rate of X20 with a probe temperature of 530°C.

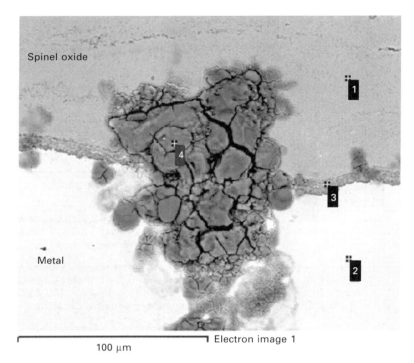

Spinel oxide

1

4

3

Metal

2

100 μm ⊢ Electron image 1

24.8 SEM-BEI, X20 (Fe-10Cr) without ChlorOut. Reaction zone with high FeCl$_2$ content in metal/oxide interface. (Iron chloride is deliquescent which explains the volume expansion and high oxygen content.) No detectable Ca and K content (± 0.02%), i.e the steady state corrosion in the reaction front is independent of alkali content.

No increased risk for low temperature corrosion at the back end of the boiler was detected when the ChlorOut additive was in use. The air pre-heaters in Nyköping are made of the austenitic stainless steel AISI316 (SS2343, WNr 1.4436) which contains 16–18% Cr, 11–14% Ni and 2.5–3% Mo. Corrosion probe tests near the air pre-heaters, with a temperature gradient of 60–130°C, showed that corrosion with and without ChlorOut was negligible in both cases [7].

Figure 24.10 shows that increasing the chromium content of steels from the 0% of 15Mo3 to 1% in 13Cr, 2% in 10Cr or even 10% in X20 did not lead to a reduction in corrosion rates. However the stainless steel Esshete 1250, containing 15% Cr, showed very low corrosion rates. Detailed metallographic analysis of the stainless steel has yet to be performed.

24.3.3 Thermo Calc modelling

Calculations with the Thermo Calc system using the SSUB database were performed in order to predict the stable species under conditions which may

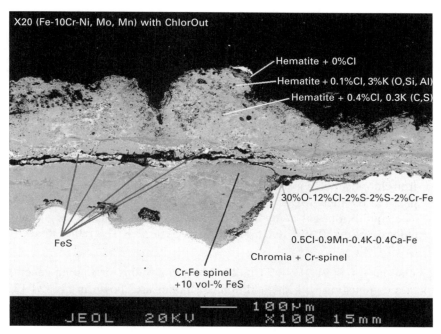

X20 (Fe-10Cr-Ni, Mo, Mn) with ChlorOut

Hematite + 0%Cl
Hematite + 0.1%Cl, 3%K (O,Si, Al)
Hematite + 0.4%Cl, 0.3K (C,S)

30%O-12%Cl-2%S-2%S-2%Cr-Fe

FeS

0.5Cl-0.9Mn-0.4K-0.4Ca-Fe

Chromia + Cr-spinel

Cr-Fe spinel
+10 vol-% FeS

JEOL   20KV   100μm   X100   15mm

24.9 SEM-BEI, X20 (Fe-10Cr) with ChlorOut. Significantly better adhesion in metal/oxide interface with lower Cl content. A few spots with high Cl content were detected, again with no connection to detectable alkali contents. Massive FeS-layers were detected. (A 'clean' signal from these layers typically gave 45%S–49%Fe in atomic%.).

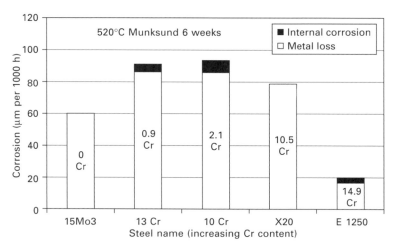

24.10 Corrosion rates per 1000 h for various steels tested in Munksund.

be encountered during biomass combustion. In order to simulate the situation occurring under a deposit a calculation was made for the iron-chlorine system in a gradient of S, O and K (see Fig. 24.11). Since low levels of these elements will occur at the metal-oxide interface and higher levels at the oxide/deposit or oxide/gas interface, the diagram can basically be seen as a cross-section through the oxide. The relative amounts of the phases in the calculation are not significant here, since it is dependent on the selection of the input values. Instead it is the regions in which the various phases occur that is important. There are two particularly significant features apparent from Fig. 24.11:

- Iron chloride, $FeCl_2$ is predicted to be the stable solid chloride-containing phase at the metal/oxide interface.
- The equilibrium gas composition at low p_{O_2} is dominated by gaseous chlorides – e.g. 0.3% HCl, 5.10^{-3}% $FeCl_2$, 10^{-5}% KCl – rather than Cl_2 (5.10^{-15}%).

Calculations for an 91%Fe–9%Cr alloy under conditions simulating general biomass combustion with the addition of sulphur are shown in Fig. 24.12.

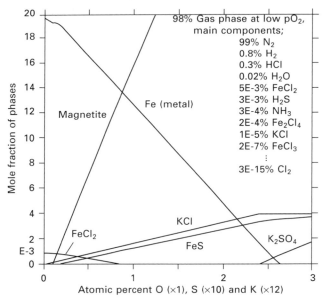

24.11 Results of Thermo Calc modelling for Fe-10at.%Cl at 500°C in equilibrium with a S, O and K gradient. The left-hand side corresponds to the metal/oxide interface with $p_{O_2} \leq 10$–25 bar and the right-hand side to the oxide-deposit interface with $p_{O_2} \geq 10$–18 bar. Note also that the equilibrium amount of hydrogen chloride and metal chlorides in the gas phase is several orders of magnitude higher than the Cl_2 content.

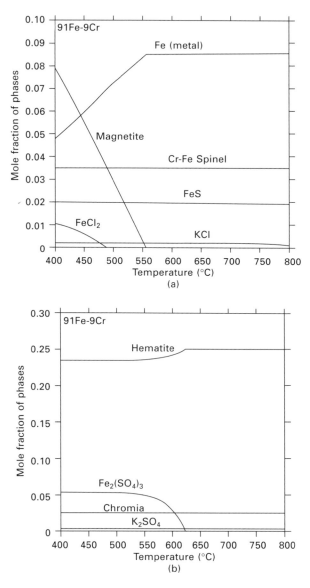

24.12 Results of Thermo Calc modelling. Temperature dependence of stable phases for Fe-9Cr exposed in Cl, S, O, K and H containing environment, simulating general biomass combustion with a sulphur containing additive. (Input, one mole: 0.1Fe, 0.01Cr, 0.1O or 0.4O, 0.01Cl, 0.01S, 0.001K and balance: H.): (a) at $p_{O_2} \leq 10{-}20$ atm, i.e a simulation of the conditions under a deposit or oxide scale, (b) at $p_{O_2} = 10{-}15$ atm or (much) higher, i.e a simulation of the oxide/gas or deposit/gas interface.

The situation at the metal interface, i.e. under a deposit or oxide scale where the oxygen partial pressure is low is shown in Fig. 24.12(a) and the situation at the gas interface, i.e. at the top of a deposit or oxide scale where the oxygen partial pressure is higher, is shown in Fig. 24.12(b). In both cases the levels of Cl, S, and K are held constant and the temperature varied. It is to be noted that these conditions will vary slightly from case to case. For example, an increase in the amount of plastic waste or demolition waste being co-combusted will increase the amount of Cl, while the amount of sulphur, sulphate or even coal or peat used in combustion will affect the S levels. Nevertheless, two significant features can be identified from these diagrams.

- With an increase in temperature $FeCl_2$ ceases to be a stable solid phase at the metal interface (in Fig. 24.12(a) at temperatures above 500°C) whereas FeS is stable over a wide temperature range at the metal-scale interface.
- Solid sulphate phases dominate at the gas interface, in agreement with analysed deposit compositions.

24.4   Discussion

Most biomass fuels have high contents of alkali metals and chlorine, but they contain very little sulphur compared to fossil fuels. The alkali metal of major concern in wood is potassium. The majority of potassium is released into the gas phase during combustion and is present mainly as potassium chloride, KCl, and potassium hydroxide, KOH. The alkali metals form compounds with low melting temperatures and can condense as chlorides causing widespread fouling of superheater tubes and other operational problems during combustion [3].

The use of 'ChlorOut' (ammonium sulphate) reduced the long-term corrosion rates of the steels in Nyköping. The overall gaseous reactions in the case of ammonium sulphate and KCl are as follows with KCl being converted into HCl and K_2SO_4. (The ammonia produced is used to reduce NOx.)

$$(NH_4)_2SO_4 \rightarrow 2\ NH_3 + SO_3 + H_2O$$

$$SO_3 + H_2O + 2KCl \rightarrow 2HCl + K_2SO_4$$

A deposit containing only alkali sulphates has a higher first melting point than a deposit containing alkali chlorides [8] and, as molten phases increase the corrosion rate, alkali sulphates are preferred to alkali chlorides in the deposits.

It is known that co-firing wood fuel with sulphur containing fuels like coal decreases the corrosion rate [9]. It has been shown that the addition of SO_2 to a He-O_2-HCl mixture led to a decrease in corrosion under deposits made from waste incinerator fly-ash and that no chlorides were found in the deposits after the experiments [10]. It has also been shown that the addition of SO_2 to pure oxygen reduced the oxidation rate of T22 in the range 500–600°C when no deposits were involved [11]. In this case the suppression of oxidation was suggested to be due to the formation of small amounts of iron sulphate which interfered with the oxidation process.

In the present work it has been shown that FeS layers form within the oxide when ChlorOut is added to the system. It is proposed here that these FeS layers are themselves beneficial in that they reduce both ion and gas transport through the mixed oxide/sulphide layer. As a consequence, it seems that any sulphur addition that may form iron sulphide is beneficial. In the case of ChlorOut, the reactive compound is ammonium sulphate, which easily decomposes. Other sulphates, which are more thermodynamically stable, like calcium sulphate (gypsum), may not have any beneficial effect at all.

Elemental sulphur, mixed directly with the wood fuel is the most obvious source of sulphur for use in 100% biomass firing, but it has been previously shown that sulphur leads to a great increase in the SO_2 level of the flue gases at the superheaters and an increase in the acidity and sulphate content of water from condensed flue gas at the cold end of the boiler [12]. This carries with it an increased risk for low temperature corrosion. In Nyköping ChlorOut showed only positive effects in terms of reduced corrosion and fouling and none of the negative effects such as increased low temperature corrosion. This is probably because ChlorOut is sprayed as a liquid into the flue gases after combustion, thus ensuring good mixing.

In Nyköping, the corrosion rate of steels decreased with increasing alloying content, principally Cr and Ni. In Munksund, no such trend could be seen. This lack of correlation between alloying content and corrosion behaviour in biomass or waste fired boilers has been seen before [e.g. 6, 13]. However the 15% Cr steel Esshete 1250, which also contains 6% Mn, performed well in both boilers.

The use of ammonium sulphate resulted in a 50% reduction in the corrosion rate of ferritic/martensitic steels. Because it also reduces NOx, the additional cost is offset by a reduction in the use of ammonia. Ammonium sulphate also reduces deposit growth, leading to a reduction in unplanned outages.

24.5   Conclusions

Superheater probe testing showed that ammonium sulphate reduced the deposit growth rate and halved the corrosion rate of ferritic/martensitic steels in a wood-fired boiler. Under normal conditions (without sulphate) iron chloride particles were present at the metal/oxide interface. With the addition of the sulphate, iron sulphides were formed within the oxide, which are believed to have hindered the corrosion process and iron chlorides were largely absent. The oxide was found to adhere better to the metal interface. The stainless steel Esshete 1250, which contains only 15% Cr also showed very low corrosion rates compared with other common superheater steels.

24.6   Acknowledgements

We thank Mattias Mattsson and Annika Stålenheim, Vattenfall Utveckling, for their invaluable work on this project. We also thank Thomas Björk, Institute for Metals Research, Stockholm, for the GD-OES measurements and Peter Sjövall, SP, Borås, for the TOF-SIMS results. This work was financed by Vattenfall's 'Renewable Fuels' programme, the EU project CORBI, no. ENK5-CT2001-00532 and the Swedish Energy Authority's programme 'Materials Technology for Thermal Energy Processes' project KME 135.

24.7   References

1. P. Henderson, A. Kjörk, P. Ljung and O. Nyström, Värmeforsk report 700. June 2000. Värmeforsk Service AB, Stockholm. (In Swedish.)
2. P. J. Henderson, A. Karlsson, C. Davis, P. Rademakers, J. Cizner, B. Formanek, K. Göransson and J. Oakey, *Proc. 7th Liège Conf. Materials for Power Engineering*, p. 785 (Eds J. Lecomte-Beckers, M. Carton, F. Schubert and P.J. Ennis). Forschungszentrum Jülich, Germany (2002).
3. H. P. Nielsen, F. J. Frandsen, K. Dam-Johansen, and L. L. Baxter, *Progress in Energy and Combustion Science*, **26**, 283 (2000).
4. IACM. European Patent EP1221036 'A method and device for measuring, by photospectrometry, the concentration of harmful gases in the fumes through a heat-producing plant' (2006).
5. ChlorOut. European Patent EP1354167 'A method for operating a heat-producing plant for burning chlorine-containing fuels' (2006).
6. P. J. Henderson, P. Ljung, S.-B. Westberg and B. Hildenwall, *Proc Conf. Advanced Materials for 21st Century Turbines and Power Plant*, p. 1094 (Eds A. Strang *et al.*) IOM, London (2000).
7. B. Hildenwall, Vattenfall Utveckling. Private communication (2002).
8. K. Iisa, Y. Lu, and K. Salmenoja, *Energy & Fuels*, **13**, 1184 (1999).
9. P. J. Henderson, Th. Eriksson, J. Tollin and T. Åbyhammar, *Proc Conf. Advanced Heat Resistant Steels for Power Generation* (Eds R. Viswanathan and J. Nutting), p 507. IOM Communications, London (1999).
10. M. Spiegel, *Materials at High Temperatures*, **14**, 221 (1997).
11. A. Järdnäs, J.-E. Svensson and L.-G. Johansson, *Materials Sci. Forum*, **369–372**, 173 (2001).
12. P. Henderson, H. Kassman and C. Andersson, *VGB PowerTech*, **84**, June 2004, 58.
13. P. Makkonen. *VGB PowerTech.*, **80**, 8 (2003).

Hot erosion wear and carburization in petrochemical furnaces

R L D E U I S, A M B R O W N and S P E T R O N E,
Quantiam Technologies Inc., Canada

25.1   Inroduction

Materials utilized in the construction of ethylene pyrolysis fittings and coils are subjected to severe operating conditions with regard to both environment and service temperature. During ethylene pyrolysis, a hydrocarbon feedstock (ethane, propane, butane, naphtha or gas oils) is mixed with steam, and flows through furnace coils of tubes and fittings that are heated externally between 1000 and 1150°C. Steam cracking of the hydrocarbons occurs and generally lower olefins (ethylene, propylene and butenes) are produced along with the deposition of coke onto the internal surfaces of the coils. In an effort to remove coke build-up, an ethylene furnace is periodically subjected to a decoking treatment where a steam-air mixture is introduced into the furnace at a temperature between 800 and 1000°C for a defined time period. Therefore, these furnace component alloys experience both chemical compositional change and microstructural degradation with service life. As a result of oxidation and carburization, there is a depletion of constituent alloying elements within the alloy's microstructure [1]. The turbulent fluid flow conditions established within the pyrolysis tube promotes erosion wear to occur on the internal surfaces by the impact of carbonaceous particles of various shapes and mechanical properties, and other debris. The velocity of these particles (erodent velocity) contacting the surface is estimated to range between 10 and 100 m s^{-1}.

Highly alloyed austenitic stainless steels are commonly used in pyrolysis furnace tube application. The common alloys utilized are Incoloy™ 800H (20Cr-30Ni-Fe), HK40 (25Cr-20Ni), HP-Modified (25Cr-35Ni-Fe), and 35Cr-45Ni-Fe [2–7]. Alloying elements to the steel matrix of nickel, chromium and silicon, and additives inclusive of molybdenum, tungsten, titanium, aluminum and niobium are reported to increase the carburization resistance of these high temperature alloys [2, 4]. As ethylene manufacturers increase process temperatures in an effort to improve furnace performance, there is a trend towards the development of stainless steels with higher nickel and chromium content for this application [2, 3]. At an operating temperature of 1000°C, it has been cited that to provide effective resistance against carburization, the chromium content should be at least 25 wt.% [4]. The material's properties such as processing history (static as-cast or centrifugal casting), carbon content and microstructural features (grain size, constituent segregation and precipitate distribution) influence the service performance and component life [3].

A commercial micro-alloyed austenitic stainless steel, 25Cr-35Ni-Fe was selected as the main reference steel within the present work. Carburization results for this alloy and that of similar materials have been reported previously [1–7]. To date there has been no published literature regarding the high temperature erosion wear properties of 25Cr-35Ni-Fe. However, findings related to wear degradation phenomena termed either erosion-corrosion or material wastage within fluid-bed combustion furnaces (FBCF) components (power generation) have been extensively published [8–11]. Common alloys utilized for this application are Incoloy™ 800H, Inconel™ 738 (16Cr-4Al-3.4Ti-Ni) and iron-chromium alloys [8]. Furnace operating conditions are not as severe with respect to pyrolysis furnaces. Temperatures within an FBCF are maintained at 500–700°C. Erodent velocities are also less extreme with values cited between 1 and 4.5 m s^{-1} [8, 10, 11]. It is hoped that the experimental results presented in this study will enhance the understanding of both carburization and erosion wear behavior of this commonly accepted pyrolysis tube alloy HP-Modified (25Cr-35Ni-Fe).

25.2   Experimental methods

A novel manufacturing technology was developed by Quantiam Technologies that enabled the deposition of an erosion wear and corrosion-resistant macro-coating (500–5000 μm) onto a variety of stainless steel substrates, namely 304, 316L and higher alloyed steels such as 800H, 25Cr-35Ni-Fe and 35Cr-45Ni-Fe. This technology involved the deposition of a coating containing a range of compositions onto the substrate and then the subsequent formation of a metallurgical coating during a heat treatment process. The wear-resistant coating was identified as Q1100 macro-coating materials system. Coupons of the 25Cr-35Ni-Fe used as the substrate steel were machined from a statically as-cast 180°-return bend fitting. Due to the widespread acceptance of this highly alloyed austenitic stainless steel in the fabrication of pyrolysis furnace components, its selection as the primary base alloy was considered to be a suitable choice. The hot erosion wear study investigated the properties of the Q1100 coating and two stainless steels, 25Cr-35Ni-Fe (statically cast) and a preliminary assessment of 35Cr-45Ni-Fe (centrifugally cast). The chemical compositions of these stainless steels are shown in Table 25.1. In this study, the majority of the work focused on the use of coupons $35 \times 18 \times 8$ mm^3 in

Table 25.1 Chemical composition (wt.%) of 25Cr-35Ni-Fe and 35Cr-45Ni-Fe steels

Steel	Ni	Cr	W	Nb	Si	Mn	Mo	C	Other	Fe
25Cr-35Ni-Fe	34–37	24–28	0.5–1.5	1.0–1.8	<2.0	<1.5		0.4–0.6		bal.
35Cr-45Ni-Fe	40–46	30–35		0.50–2.0	<2.0	<2.0	0.50	0.4–0.6	Ti, Al	bal.

size. Where appropriate, examples of coated industrial components are also illustrated.

25.2.1 Coating microstructure

Microstructures of the coating and reference steel were examined initially using optical microscopy (OM) and X-ray diffraction (XRD). OM instrument was a Zeiss model Axiovert 25 inverted microscope with differential interference contrast (DIC) capabilities. The XRD unit was a Bruker AXS model D8 Advance with an maximum operating voltage of 40 keV utilizing a CuK$_\alpha$ radiation source. The unit is fitted with a Göbel mirror for beam collimation and a glancing incidence stage for higher surface sensitivity. It also has a controlled atmosphere-high temperature stage to 1600°C to monitor phase stability and phase evolution as a function of temperature–time and atmosphere.

25.2.2 Performance characterization

Micro-mechanical properties

Micro-mechanical properties were determined for the Q1100 coating and the uncoated steel substrate using a Fischerscope H100C microhardness indenter. Measurements were taken on mounted polished cross-section coupons. Several regions within the coating were surveyed. A fracture toughness value was determined for the coating using a hardness indentation approach [12, 13]. The indentation test conditions and mechanical properties measured are shown in Table 25.2.

Thermal shock

Thermal shock resistance tests were completed on the coating in order to assess its compatibility to thermal cycling and to study the coatings' coefficient of thermal expansion (CTE) property. Five coated coupons were placed within a horizontal tube furnace and heated in an air atmosphere. A sample was removed from the furnace after being exposed to a specific temperature, namely, 400, 500, 600, 800 and 1000°C. The dwell time for each temperature was 30 min. The removed sample was quenched in water at room temperature.

Table 25.2 Micro-mechanical test conditions and properties determined for the coating

Machine	Fischerscope H100 Ultra-Microhardness Tester
Indenter geometry	Berkovich Diamond
Nominal load	300 mN
In/decreasing load time	20 s
Number of measurements per sample/region	5–15
Indentation depth at maximum test load, h_{max}	µm
Elastic component of material displacement work, η	%
Hardness at maximum test load, H	GPa
Vickers hardness, H_v	kg mm^{-2}
Plain strain modulus, E'	GPa
Young's modulus, E	GPa
Fracture toughness, K_C	MPa m$^{1/2}$

All quenched coated coupons were then sectioned, mounted and polished. These mounted samples were then examined by OM for any evidence of thermal shock damage such as cracking and spallation. It is recognized that water quenching from temperature is extremely severe relative to real-world conditions.

Oxidation

The Q1100 coating and reference 25Cr-35Ni-Fe were evaluated for oxidation resistance within a laboratory muffle furnace. Three coated coupons were exposed to a temperature of 1100°C, in air, for exposure times of 100, 500 and 1000 h, respectively. These test samples were thicker than the previously stated coupon dimensions, so that the oxidation behavior of the coating and reference steel could be studied on the same coupons, under the same exposure time (without, for example, corrosion interference from cut edges). At the completion of each test time, the coupon was sectioned, mounted and polished. The microstructure (transverse section) was then examined using OM, XRD and auger electron spectroscopy (AES) methods. The AES instrument was a Perkin Elmer model PHI 600 scanning Auger multiprobe with an imaging resolution of ≈25 nm and an effective Auger lateral resolution of ≈200 nm. The top and bottom regions of each coupon represented the oxidized behavior of the coating and 25Cr-35Ni-Fe, respectively.

Carburization

The coating and reference steel were studied for their resistance to carburization. Four coated coupons were exposed to test conditions of 1100°C and a carbon activity of approximately 1 ($a_c \approx 1$), for defined exposure times within a horizontal tube furnace. Even though the 25Cr-35Ni-Fe is reported to be suitable for long-term service at temperatures up to 1093°C [7], it was considered that a laboratory test temperature of 1100°C was necessary in order to simulate an accelerated (severe) carburizing environment for the applications targeted. The exposure times were 250, 500, 750 and 1000 h. The carbon activity was achieved by applying graphite paint (2-propanol based) to each coupon and then embedding the coupons into an alumina boat filled with graphite powder. The boat was then covered with an alumina lid. This carburization technique was selected to simply carburize a defined depth of the samples. A controlled atmosphere was used for the testing with initial evacuation to $\approx 5 \times 10^{-3}$ Torr, several argon purge cycles, and finally a low dynamic argon gas flow providing a steady-state pressure of 1-2 Torr. In general, these coupons were significantly thicker so the carburization behavior of the coating and reference steel could be studied on the same coupon (without, for example, cross-interference from cut edges). At the completion of each test, the carburized coupon was removed from the furnace, ultrasonically cleaned in acetone and then mounted and polished. The microstructure (transverse section) was then examined using OM, XRD, AES and field emission scanning electron microscopy-energy dispersive spectroscopy (FE-SEM/EDS) analytical methods. The FE-SEM unit was a JEOL model JSM-6300F with a lateral resolution of 1.5 nm and a maximum operating voltage of 30 keV. The top and bottom regions of each coupon represented the carburized behavior of the coating and reference steel, respectively.

25.2.3 Erosion wear

High temperature materials degradation facility

The erosion study was performed in the high temperature materials degradation facility at Quantiam Technologies, design arrangements of which are given in Figs 25.1 and 25.2. This erosion wear rig consists of three main components, namely 1 – the vibratory screw powder feeder, 2 – the furnace chamber and 3 – the spent erodent effluent containment unit. The unit has an ability to evaluate erosion wear behavior under a broad range of test parameters, namely: impingement angles 30°, 45°, 60° and 90°; sample temperature 25–1160°C; erodent particle loading 0.2–50 g/min; and atmospheres protective, oxidizing, carburizing and corrosive or a combination thereof.

A description of the operating details for a hot erosion wear test is as follows. The coupon sample to be characterized is affixed to the sample holder. The sample holder is then inserted into the lower region of the furnace chamber, 2. With the holder mechanism bolted into position, the distance between the sample's top surface and the exit opening of the nozzle tube is

25.1 Arrangement of the materials degradation facility, where '1' represents the vibratory screw powder feeder, '2' the furnace chamber, and '3' the spent erodent effluent containment unit.

always a constant value. The furnace atmosphere gas (inert, carburizing or oxidizing) is delivered via gas cylinders into a 6 mm O.D. 316 stainless steel coil positioned within a heat resistant alloy chamber tube that is situated in a 3-zone 1200°C vertically-hung furnace. The furnace gas is pre-heated and is transported out of the top of the furnace and into a smaller nozzle tube that then redirects the heated gas vertically down into the center of the furnace, as shown in Fig. 25.2. The exit opening of the nozzle tube is positioned in the lower center of the furnace heating zone. A vibratory screw feeder, 1, transports the erodent particulate (SiC powder) using argon carrier gas to a 'T' -joint connection to the nozzle tube above the furnace (Fig. 25.2). The erodent stream is aspirated into the nozzle tube at the 'T'-joint connection. Average erodent particle velocity is ≈ 25 ms^{-1}. The erodent stream is heated via mixing with the furnace gas and the mixture is directed towards the exit opening of the nozzle tube. The erosion test sample is positioned at a defined distance below the exit opening of the nozzle tube within a sample holder. The holder affixes the sample's test surface at a defined orientation with respect to the erodent trajectory (impingement angle). Two type-K

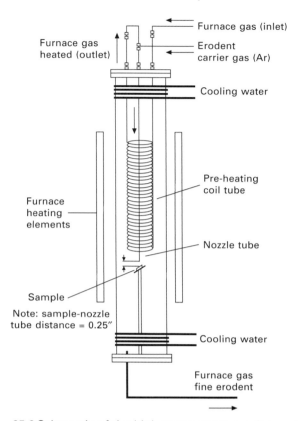

Furnace gas (inlet)

Furnace gas
heated (outlet)

Erodent
carrier gas (Ar)

Cooling water

Pre-heating
coil tube

Furnace
heating
elements

Nozzle tube

Sample

Note: sample-nozzle
tube distance = 0.25″

Cooling water

Furnace gas
fine erodent

25.2 Schematic of the high temperature erosion wear rig.

thermocouples welded onto the holder monitor furnace testing temperature. During an erosion test, the heated erodent stream impacts the sample's surface and the coarser size fraction is collected in the bottom of the sealed furnace tube. This size fraction is removed once testing has been completed. The finer size fraction of the erodent is transported out of the furnace chamber with exiting furnace gas through a stainless tube. The tube passes through a heat exchanger unit located directly below the furnace. The cooled fine erodent is transported into a cyclone, 3 and classified (Fig. 25.1). The undersize erodent size fraction collects into a bin positioned below the cyclone and the finer size fraction is collected in a 0.5 μm filter column.

The standard erosion sample size adopted had dimensions $35 \times 18 \times 10$ mm^3. The sample surface to nozzle-tube distance was fixed at 6 mm. Erosion coupons machined from the reference steel (uncoated substrate) were $35 \times 18 \times 8$ mm^3. A shim was used to compensate for differing sample thicknesses in order to maintain constant sample surface to nozzle distance during testing. All erosion samples had a ground flat surface finish of Ra ≈ 0.8 μm. Erosion wear rate was defined as the weight loss of the coupon divided by the erodent particle loading (erodent weight) and expressed in units of g/g.

Carburization and erosion wear

In ethylene pyrolysis, the internal surfaces of a 180°-return bend fitting are exposed to high temperature erosive wear, oxidation and carburization simultaneously during service. In an attempt to assess the performance of the Q1100 coating and reference steel under similar environmental conditions, pre-carburized coupons of these materials were subjected to erosion wear testing. Due to the severity of the erosion wear process and the time dependent nature of diffusion-based carburization, it is very difficult to emulate these two environmental degradation processes simultaneously. With reference to pages 448–9, it was decided that a carburizing condition of 250 h and 1100°C would provide coupons that exhibited an adequate and uniform carburized surface, suitable for the erosion wear study.

Prior to the start of this erosion study, an optimum test time was determined that represented erosion wear conditions localized only within the carbide denuded region (upper carburized surface layer, exclusive of the carbide precipitated and substrate regions). In order to study the influence of carburization on erosion wear, it was important that the wear scar for this investigation was contained within the upper carburized surface layer (denuded zone). The optimum test time to achieve this was determined to be 8 min (valid for temperatures, 25–900°C). The erosion behavior test parameters for the pre-carburized coating and 25Cr-35Ni-Fe samples are shown in Table 25.3.

Erosion wear behavior of 25Cr-35Ni-Fe, 35Cr-45Ni-Fe and Q1100 coating

The erosion wear behavior of as-received 25Cr-35Ni-Fe and 35Cr-45Ni-Fe, under an oxidizing atmosphere (air) as a function of temperature was

Table 25.3 Erosion wear conditions for the pre-carburized coating and 25Cr-35Ni-Fe

Temperature range	200–1040°C
Impingement angle	30°
Test time	8 min
Furnace gas flow-rate	20.0 SLPM*
Carrier gas flow-rate	22.3 SLPM*
Erodent particulate (SiC)	+220 grit ($\approx$70 μm)
Particle feed rate (loading)	1.4 g min^{-1}
Particle velocity	$\approx$25 ms^{-1}

*Reference gas, Ar

Table 25.4 Erosion wear parameters for 25Cr-35Ni-Fe
and 35Cr-45Ni-Fe (oxidizing and inert atmospheres)

Temperature range	25–1040°C
Impingement angle	30°
Test time	15 and 20 min
Furnace gas flow-rate	20.0 SLPM*
Carrier gas flow-rate	22.3 SLPM*
Erodent particulate (SiC)	+220 grit ($\approx$70 μm)
Particle feed rate (loading)	1.4, 13 g min^{-1}
Particle velocity	$\approx$25 ms^{-1}

* Reference gas, Ar

investigated. These results were then compared to the erosion behavior for both steel alloys under an inert atmosphere (Ar) for the same temperature range. The experimental parameters for this study are given in Table 25.4.

The influence of an oxidizing atmosphere (air) on the hot erosion wear properties of the Q1100 coating and 25Cr-35Ni-Fe was also studied. The experimental conditions for the work were identical to that reported in Table 25.4, except that the test time duration was 15 min.

25.3   Results

25.3.1 Coating characterization: structure and properties

A polished cross-section of the Q1100 coating system is shown in Fig. 25.3. Typically the macro-coating system revealed good coating-substrate bonding, evident by the absence of interfacial macro-porosity formations. Average coating thickness was determined to be $\approx$1200 μm. An arrow on the micrograph in Fig. 25.3 indicates the location of the interface. The formation of the coating was associated with relatively low dilution content ($\leq$ 5%).

Examination of the substrate beneath the coating revealed a primary chromium carbide network structure present at the grain boundaries and a secondary carbide precipitate phase within the grain structure. A similar carbide structure was present within the reference 25Cr-35Ni-Fe. This indicated that the coating process imposed little microstructural change on the steel substrate.

A micrograph of the coating-substrate interfacial region is provided in Fig. 25.4 and shows the differences in the micro-mechanical properties of the coating and the 25Cr-35Ni-Fe substrate. Data from the micro-indentation survey is shown in Table 25.5. The micro-indentation results revealed three distinct coating zones, namely, surface, middle and interfacial. The interfacial zone exhibited the highest hardness and the greatest ductility measurement for the survey.

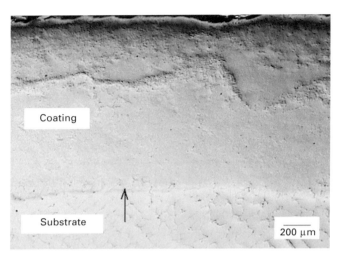

25.3 Optical micrograph (transverse section, DIC, unetched) of the Q1100 macro-coating deposited onto 25Cr-35Ni-Fe, the arrow indicating the coating/substrate interface.

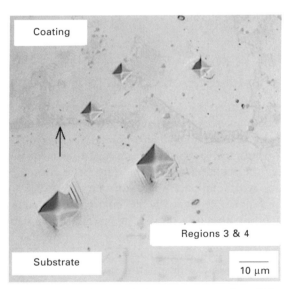

25.4 Optical micrograph (transverse section, DIC, unetched) showing micro-mechanical properties at the interface (regions 3, 4) of the Q1100 coating, where the arrow indicates the interface.

The ductility property was indicated by the W_{rc}/W_t ratio which was represented by η. The η term evaluates the proportion of recovered elastic deformation that occurred during plastic deformation as a result of indentation. W_{rc} indicates the capability of the material to absorb the deformation energy

Table 25.5 Micro-indentation survey of Q1100 coating and 25Cr-35Ni-Fe (Fischerscope H100C tester)

Region of indentation	H (N/mm^2)	H_v (kg/mm^2)	h_{max} (μm)	η (%)	E' (GPa)	E (GPa)	K_c (MPa m$^{1/2}$)
1 – Coating, surface	6965.0	813.4	1.175	33.86			
2 – Coating, middle	6465.9	716.7	1.219	30.81			
3 – Coating, interface	8168.6	1000.5	1.074	36.90			
4 – Substrate, interface	3215.8	296.4	1.767	14.26			
Coating average	7199.8	843.5	1.156	33.86	298.8	271.9	4.25
Substrate	3462.9	324.3	1.700	15.58			

during indentation without incurring any plastic deformation. W_t corresponds to the area under the load-displacement curve that was generated during the micro-indentation test and represents total deformation energy. It has been reported that materials exhibiting high W_{rc} and η values generally display superior resistance to impact and abrasive wear [14].

The middle zone exhibited the lowest hardness and ductility values. Even with the variation in coating micro-mechanical properties, the coating displayed approximately double the hardness and ductility properties of the 25Cr-35Ni-Fe substrate. The fracture toughness of the coating was determined to be ≈4.25 MPa√m. This indicated the coating exhibited a toughness value comparable to that of a wear resistant ceramic such as alumina. A typical hardness impression within the Q1100 coating showing the indentation surface cracking pattern utilized in the fracture toughness calculation is shown in Fig. 25.5.

With regard to CTE evaluation, the coated coupon quenched at 400°C was the only sample to show no macro-cracking within the coating. The other four samples (quenching temperatures of 500, 600, 800 and 1000°C) all suffered some level of cracking within the coating due to the severe thermal shock treatment. The propagation of the macro-cracks was in a perpendicular direction towards the coating/substrate interface. All cracks were arrested at the interface. The cracks became more noticeable as the quench temperature increased. Even though macro-cracks were evident within the samples tested at temperatures higher than 400°C, their presence did not

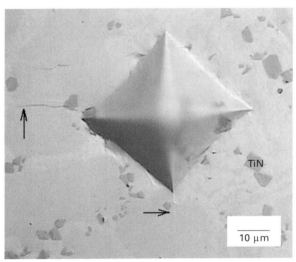

25.5 Optical micrograph (transverse section, DIC, force ≈9.81 N) of a hardness impression in Q1100 coating, arrows indicate crack growth from indentation corner.

result in any coating spallation, since no lateral cracks were observed at the interface of any of the thermally shocked coupons. In summary, the Q1100 coating exhibited good resistance to thermal cycling. The coating's CTE value appears to be compliant with the substrate's CTE value. As was noted earlier, the testing severity is extreme considering the targeted industrial applications. A microstructure of Q1100 coupon quenched in water from 1000°C is shown in Fig. 25.6.

25.3.2 Oxidation

The oxidation behavior of the coating is depicted in Fig. 25.7. Internal oxidation increased with time. In comparison to the 25Cr-35Ni-Fe, the coating displayed minimal scale formation for the exposure times investigated. The predominant surface oxide species were identified by XRD and AES techniques. An interfacial barrier was noted to form and its formation appeared to be time-dependent. An examination of the substrate near the coating interface (Fig. 25.7) compared with that of the as-received structure, Fig. 25.8, revealed the characteristic primary carbide network had been unaffected by the oxidizing environment.

The reference steel exhibited significant oxidation damage after exposure for only 100 h (Fig. 25.9). Significant metal loss was evidenced by surface scaling (external oxidation) and internal oxidation attack. Internal degradation manifested itself as an inter-dendritic attack that affected the primary chromium carbide network. The extent of the degradation due to the oxidizing environment became progressively worse as exposure time increased. XRD analysis

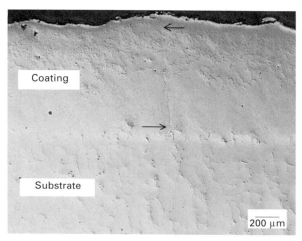

25.6 Optical micrograph (transverse section, DIC, unetched) of Q1100 coating thermally shocked at 1000°C, where arrows indicate the presence of a crack.

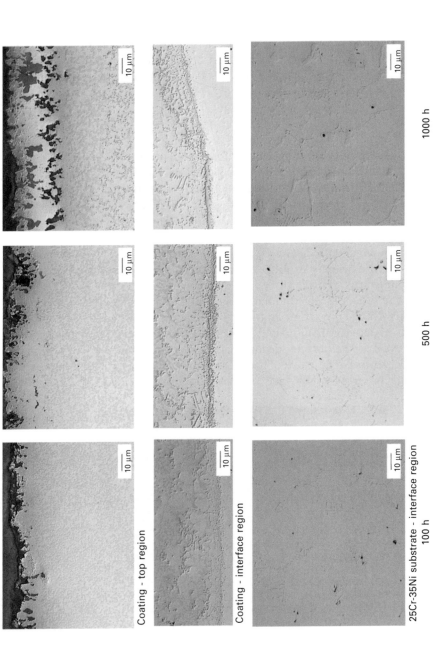

Coating - top region

Coating - interface region

25Cr-35Ni substrate - interface region

100 h   500 h   1000 h

25.7 Optical micrographs (transverse section, DIC, unetched) of the Q1100 coating (top and interfacial regions) and 25Cr-35Ni-Fe substrate (interfacial region) depicting oxidation behavior as a function of exposure time (1100°C, air).

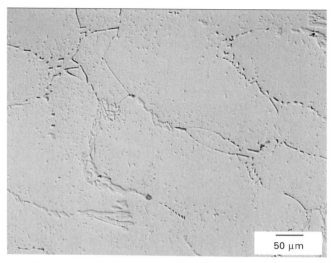

25.8 Optical micrograph (transverse section, DIC, unetched) of as-received 25Cr-35Ni-Fe.

indicated the predominant surface oxide species were chromia (Cr_2O_3), a Mn-Cr spinel structure, and magnetite (Fe_3O_4).

25.3.3 Influence of carburization on erosion wear

Carburization

The Q1100 coating appeared to exhibit minimal structural degradation due to the carburizing environment (Fig. 25.10 – coating top region). The coating offered significant protection to the 25Cr-35Ni-Fe substrate as indicated in Fig. 25.11 (interface region). Microstructural features associated with the as-received alloy (Fig. 25.8) such as the primary carbide structure present within the interdendritic regions and secondary carbides positioned within the grains were still evident in the coated 25Cr-35Ni-Fe substrate (interfacial region). No significant carburization attack was noted in the substrate near the interfacial region (Fig. 25.10).

SEM/EDS characterization of a cross-section of the Q1100 coating subjected to exposure for 750 h provided an insight into the carburization process with respect to microstructure. No significant macro-porosity was observed within the carburized Q1100 coatings.

Interfacial structure

Within the Q1100 coating, an interfacial structure formed with exposure to the carburizing atmosphere (Fig. 25.11). As with the oxidation study, the

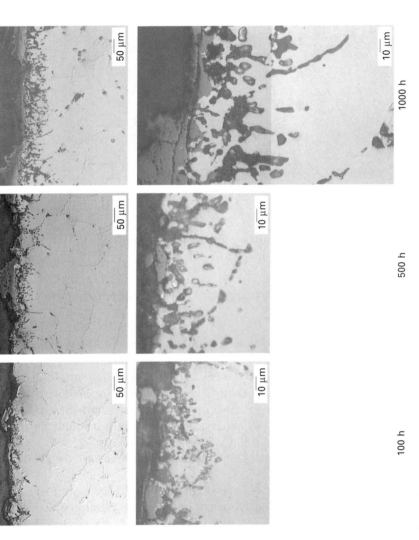

100 h   500 h   1000 h

25.9 Optical micrographs (transverse section, DIC, unetched) displaying the oxidation behavior of 25Cr-35Ni-Fe at the surface as a function of time (1100°C, air).

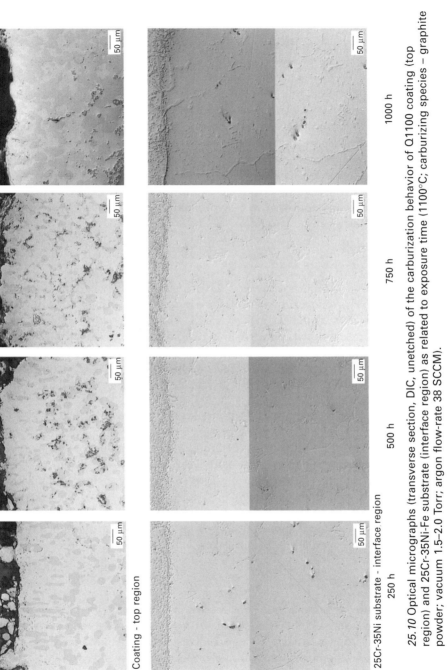

Coating - top region

25Cr-35Ni substrate - interface region

250 h   500 h   750 h   1000 h

25.10 Optical micrographs (transverse section, DIC, unetched) of the carburization behavior of Q1100 coating (top region) and 25Cr-35Ni-Fe substrate (interface region) as related to exposure time (1100°C; carburizing species – graphite powder; vacuum 1.5–2.0 Torr; argon flow-rate 38 SCCM).

development of the barrier at the steel interface also appeared to be time-dependent. SEM/EDS and AES analytical techniques indicated that the barrier was composed primarily of coating species. Optical micrographs of the interfacial barrier are shown in Fig. 25.11.

The 25Cr-35Ni-Fe alloy did not offer the same resistance to carburization as did the Q1100 coating (Fig. 25.12). The extent of the steel's structural degradation due to the carburizing atmosphere with time is clearly evident. As carburizing time increased, so did the volume density of chromium-based carbides in the near-surface region. In other words, as exposure time increased, a significant change was observed in the carbide structure at the grain boundaries and within the grains. This is exemplified by a chromium-depleted zone in the immediate near-surface (little or no carbide structure and depletion of overall chromium), followed by a significant increase in newly-precipitated carbides at the grain boundaries.

Carburization-erosion wear synergy

Pre-carburized Q1100 coating and 25Cr-35Ni-Fe reference coupons were subjected to erosion wear conditions for a range of temperatures. The erosion wear behavior is displayed in Fig. 25.13. At a temperature of 500°C, the Q1100 coating exhibited approximately four times the erosion wear resistance of the reference steel. The coating displayed *double* the erosion wear resistance of the 25Cr-35Ni-Fe at 900°C. The test temperature of 900°C is significant, since this value represents a typical service temperature for a key commercial application targeted.

For steel specimens evaluated within the temperature range of 25 – 900°C, the wear scar was confined to the upper carburized region's denuded zone, as indicated in Fig. 25.14. At temperatures higher than 900°C, it was found that under the test conditions used, the wear scar was not contained within the targeted denuded zone, requiring further study and testing optimization to probe this temperature regime.

In summary, under conditions of carburization, the erosive wear resistance of the Q1100 coating demonstrated superior erosion wear resistance to that of the 25Cr-35Ni-Fe of up to four times.

Commercial steel performance

The influence of atmosphere on the erosion wear behavior of both 25Cr-35Ni-Fe and 35Cr-45Ni-Fe is shown in Fig. 25.15. For each alloy, it is evident that the erosion wear rate increases with temperature for each atmosphere. At temperatures above 800°C, the erosion wear rate was significantly higher within the air atmosphere for both alloys. This is a very surprising result as it is commonly believed within the olefins manufacturing

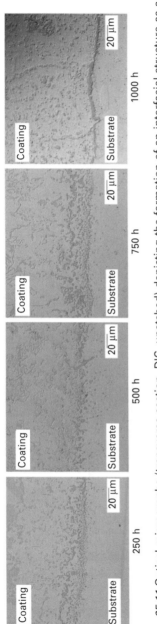

25.11 Optical micrographs (transverse section, DIC, unetched) depicting the formation of an interfacial structure as a function of carburization time within the Q1100 coating (1100°C; carburizing species – graphite powder; vacuum 1.5–2.0 Torr; argon flow-rate 38 SCCM).

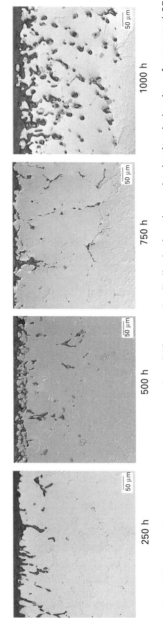

25.12 Optical micrographs (transverse section, DIC, unetched) displaying the carburization behavior of uncoated 25Cr-35Ni-Fe as a function of exposure time (1100°C; carburizing species – graphite powder; vacuum 1.5–2.0 Torr; argon flow-rate 38 SCCM).

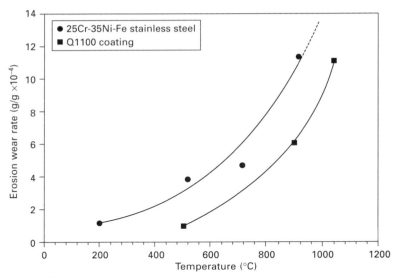

25.13 Influence of pre-carburization on the erosion wear behavior of Q1100 coating and 25Cr-35Ni-Fe (test parameters: impingement angle 30°, erodent SiC 70 μm, particle loading 1.4 g min^{-1}, furnace gas flow-rate (Ar) 20 SLPM, carrier gas flow-rate (Ar) 22 SLPM, test time 8 min).

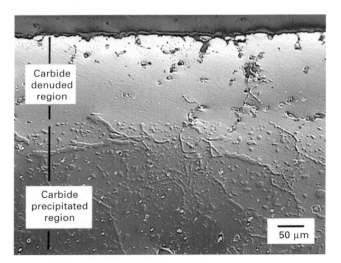

25.14 Cross-section of the middle of the wear scar showing that the extent of the erosion wear was confined to within the carbide denuded region, 25Cr-35Ni-Fe (pre-carburization: 1100°C, 250 h; carburizing species – graphite powder; vacuum 1.5–2.0 Torr; argon flow-rate 38 SCCM; erosion test: impingement angle 30°, erodent SiC 70 μm, particle loading 1.4 g min^{-1}, furnace gas flow-rate (Ar) 20 SLPM, carrier gas flow-rate (Ar) 22 SLPM, 500°C, test time 8 minutes).

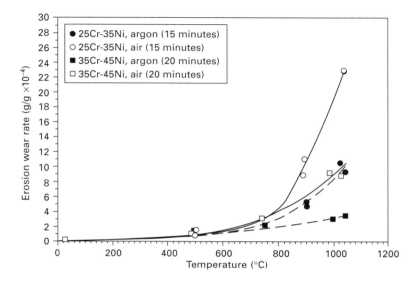

25.15 Influence of atmosphere (oxidizing, air and inert, argon) on the erosion wear behavior of 25Cr-35Ni-Fe and 35Cr-45Ni-Fe (test parameters: impingement angle 30°, erodent SiC 70 μm, particle loading 13 g min⁻¹, furnace gas flow-rate (Ar or Air) 20 SLPM, carrier gas flow-rate (Ar) 22 SLPM, test time 15 and 20 min).

industry that a temperature exceeding 1000°C is required prior to a significant increase in erosion wear occurring under oxidizing conditions. The result clearly provides evidence that this transition temperature may actually be 800°C or approximately 200°C lower that previously expected. It was also observed that 35Cr-45Ni-Fe exhibited a higher erosion wear resistance with respect to 25Cr-35Ni-Fe for both atmospheres studied.

The influence of an oxidizing atmosphere on erosion wear is shown in Fig. 25.16, for the Q1100 coating and 25Cr-35Ni-Fe. For the temperature range studied, the Q1100 coating offered a superior erosion wear resistance with respect to 25Cr-35Ni-Fe.

The results reveal the major influencing effect that atmosphere has on erosion wear behavior of highly alloyed austenitic stainless steels. Erosion wear within an oxidizing environment appears to be more severe when compared to a protective atmosphere for each alloy studied. There is probably some synergistic relationship between the erosive wear process and high temperature oxidation. The Q1100 coating demonstrated a higher resistance to erosion wear (oxidizing atmosphere) when compared to 25Cr-35Ni-Fe.

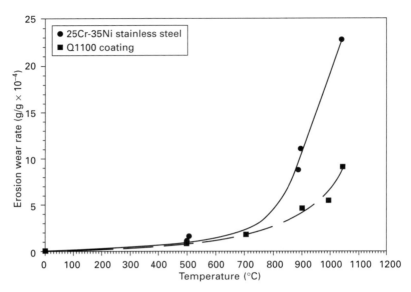

25.16 Influence of an oxidizing atmosphere (air) on the erosion wear behavior of 25Cr-35Ni-Fe and Q1100 coating (test parameters: impingement angle 30°, erodent SiC 70 μm, particle loading 13 g min^{-1}, furnace gas flow-rate (Air) 20 SLPM, carrier gas flow-rate (Ar) 22 SLPM, test time 15 min).

25.4   Discussion

25.4.1   Coating microstructure and micro-mechanical properties

A novel coating process developed by Quantiam allowed for the successful deposition of a metallic-based coating onto a highly alloyed austenitic stainless steel, 25Cr-35Ni-Fe. The coating system studied in the recent work, identified as Q1100, exhibited good metallurgical bonding with the substrate, a very low dilution content and minimal macro-porosity. Using this proprietary process, coating thicknesses greater than 1000 μm are possible with no significant detrimental change in the substrate's microstructure.

Compared to the 25Cr-35Ni-Fe substrate, the Q1100 coating revealed a higher hardness and ductility value η, approximately three times greater, suggesting that the coating should display superior resistance to both impact and abrasive wear mechanisms [14]. A fracture toughness value determined by an indentation method [12, 13] indicated that the coating exhibited a value comparable to that of a wear resistant ceramic. The coating is best described as a composite with a metal matrix and a fine distribution of hard particles. Generally, metal matrix composites exhibit superior wear resistance and high fracture toughness values compared to monolithic materials, especially if the hard reinforcement phase is an *in-situ* formed precipitate [15, 16].

Good thermal shock resistance results for the Q1100 coating indicated that the coefficient of thermal expansion of the coating and substrate were compatible and that a good coating-substrate bond was developed during the coating process.

25.4.2 Oxidation behavior

Both the uncoated 25Cr-35Ni-Fe and the Q1100 coating suffered microstructural degradation from the oxidation environment. The 25Cr-35Ni-Fe exhibited significant external scale formation and subsequent internal oxidation primarily within the interdendritic region. The extent of the oxidation attack increased with exposure time. The oxidation behavior for the 25Cr-35Ni-Fe typically followed normal trends previously published for this grade of stainless steel [17].

The Q1100 coating exhibited minimal scale formation due to the oxidizing conditions. Associated with the coating was the formation of an interfacial barrier that appeared to be time-dependent. The microstructure of the steel substrate revealed no obvious damage due to oxidation attack. The Q1100 coating system was effective in protecting the steel from the effects of the oxidizing environment.

The presence of the interfacial barrier most likely protected the steel substrate from any oxidation attack by acting as a diffusion barrier.

25.4.3 Carburization resistance

The Q1100 coating appeared to offer superior carburization resistance to the 25Cr-35Ni-Fe substrate. No obvious microstructural degradation was evident for the exposure times investigated. The carburized coating displayed no significant outer scale formation, no localized surface pitting or inter-granular corrosion attack. As mentioned previously in Section 25.3, the 25Cr-35Ni-Fe beneath the coating showed no microstructural changes as a result of the carburizing environment.

The uncoated 25Cr-35Ni-Fe displayed significant damage due to the carburizing conditions. The four carburization zones, i.e. (1) outer scale formation with localized pitting attack, (2) inter-granular corrosion, (3) carbide denuded region, and (4) region with carbide precipitates, published elsewhere for highly alloyed austenitic stainless steel [5, 18] were clearly identified within the present work. For exposure times of 250 and 1000 h, the combined inter-granular corrosion and carbide denuded depth was approximately 150 and 350 µm, respectively.

For both environmental degradation processes (high temperature oxidation and carburization) the existence of this diffusion barrier at the coating–steel interface appears to be a critical element to the Q1100 coating's ability to impart service protection to the 25Cr-35Ni-Fe substrate. In summary, the Q1100 coating protected the 25Cr-35Ni-Fe from the carburizing atmosphere under the accelerated laboratory test conditions provided. The uncoated reference steel exhibited relatively poor carburization resistance.

25.4.4 Erosion wear and carburization

The main failure mechanisms associated with the service performance of ethylene pyrolysis coils and fittings have been identified, in order of severity, as internal carburization, bulging, cracking and erosion wear [2, 3, 5, 19]. Fittings such as return bends have been observed to suffer from the effects of both carburization and erosive wear [19]. The as-cast return bend surfaces often contain surface cavities that collect coke particles during normal furnace operation. These localized surface sites are associated with abnormally high coke deposits. The high coke accumulation and the geometry of the fitting with respect to gas flow-rate both promote structural degradation of the fitting by the simultaneous interaction of erosive wear and carburization. Very little research has been published concerning the possibility of a synergistic relationship between carburization and erosion wear at elevated temperatures.

The erosion wear study of the pre-carburized Q1100 coating and 25Cr-35Ni-Fe showed the superior service performance of the coating towards these two degradation mechanisms. The carburized Q1100 coating offered better mechanical integrity when subjected to the erosive wear test conditions. The carburized 25Cr-35Ni-Fe exhibited significant microstructural damage and displayed poor erosive wear resistance. These findings suggest that there could be a synergistic relationship between carburization and erosion wear, especially concerning the surface (upper) carburized regions (inter-granular corrosion and carbide denuded regions) of the steel. It is reasonable to propose that during furnace operation, the highly eroding upper carburized region (denuded zone) could keep reforming (during the service life of the coil) and then be subjected to high erosion wear when the erosive load on the surface reached a critical level. Failed 180°-return bends autopsied at Quantiam exhibited heavily eroded and carburized macroscopic regions that were suggestive of this degradation sequence.

At temperatures exceeding 800°C, the erosion wear rate for both 25Cr-35Ni-Fe and 35Cr-45Ni-Fe steels was greater in air, when compared to the inert atmosphere. The erosive wear behavior of the two commercial alloys with respect to atmosphere indicated that oxidation, directly influenced the erosion wear resistance. No literature was found regarding the erosion wear properties of 25Cr-35Ni-Fe at elevated temperatures. Only data concerning the carburization resistance of 25Cr-35Ni-Fe at elevated temperatures (pack-cyclic tests at 850–1150°C) has been cited [7]. The Q1100 coating provided good erosion wear resistance within an oxidizing environment.

Research has been reported relating to the erosion wear properties of Incoloy™ 800H, Inconel™ 738 and Fe-Cr alloys in air at temperatures of 500 to 700°C, and an erodent velocity range of 1 to 4.5 ms^{-1} [8–11]. Even though the reported erodent velocities were lower than the value used in the present work, these researchers identified an erosion wear-corrosion map containing three distinct regions: namely erosion dominated; erosion-corrosion dominated; and corrosion dominated (Fig. 25.17). The term 'corrosion' used in this research referred to the oxidation process. At a constant temperature, an increase in erodent velocity resulted in an erosion-dominated wear mechanism. Maintaining a constant erodent velocity with an increasing temperature created the conditions for an oxidation-dominated wear process. In the present work, the erodent velocity was ≈25 ms^{-1} and the temperature increased from 25–1040°C. Under these severe conditions, as temperature increased, it is possible that an erosion-oxidation dominated wear mechanism was active. Therefore, a synergistic relationship between erosion wear and oxidation could have accounted for the higher wear rates in the air-tested stainless steel coupons.

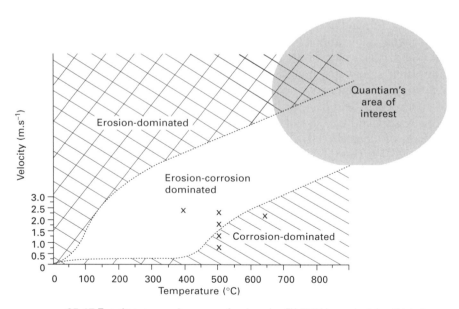

25.17 Erosion-corrosion map for Incoloy™ 800H eroded for 24 h in fluidized bed containing 200 μm silica [9], showing Quantiam's area of interest.

25.4.5 Commercial opportunities and future work

The Q1100 coating affords significant oxidation and carburization protection to the 25Cr-35Ni-Fe, under laboratory test parameters that can be realistically translated to commercial ethylene pyrolysis furnace conditions. The novel coating technology developed at Quantiam allows for the deposition of a coating using non-line-of-sight technology and coating thicknesses of 500 to 5000 μm. Internal curved surfaces such as return bends, tube sections and the external surfaces of thermo-wells have been successfully coated using this proprietary coating process.

Field trials to-date have included 300 mm diameter elbow fittings (Incoloy™ 800H) and 150 mm diameter, 180°-return bend fittings (25Cr-35Ni-Fe) that have been internally coated with the Q1100 coating and their service performance evaluated in commercial ethylene furnaces. Typical operating conditions within these furnaces were metal-skin temperature of 800–1125°C with both a hot erosion and a hot corrosion load. The expected fitting service life is 3–7 years.

In one field trial, the Q1100 coated elbow fitting with ≈2 mm of coating thickness was installed in the same furnace as an identical elbow fitting coated with a industry-standard overlay coating of 10–15 mm in thickness. After 12 months of service both elbow fittings were removed and inspected. The Quantiam coated elbow showed no measurable deterioration and was evaluated by the client's inspection group as acceptable for re-installation. The reference industry-standard overlay coating showed significant wear requiring re-application of a new coating prior to re-installation. In summary, the Q1100 coated elbow fitting (Fig. 25.18), afforded the component acceptable protection against the high temperature erosion wear, and carburizing and oxidizing environments.

In a second field trial, two Q1100 coated 180°-return bend fittings have been in service in excess of 24 months with no indication of wear failure. Pictures of a coated 180°-return bend prior to furnace installation is shown in Fig. 25.19. The Q1100 coating was applied only to the 'bottom' stepped region within the entry opening of each fitting where hot erosion is extensive.

The coating process developed by Quantiam has the potential to enhance the service life of components used within the chemical-petrochemical, oil-sands mining, and other industries. The Q1100 and other coating formulations can be deposited onto surfaces where protection is required from different degradation processes such as wear (abrasion and erosion), carburization or oxidation for a wide range of service temperatures (25–1050°C).

Regarding future work, Quantiam is investigating the hot erosion-corrosion properties of other furnace alloys such as 300 series stainless steels, Inconel and Incoloy alloys. Further research will also be conducted on the synergistic relationship between hot erosion wear and corrosion. This work will include

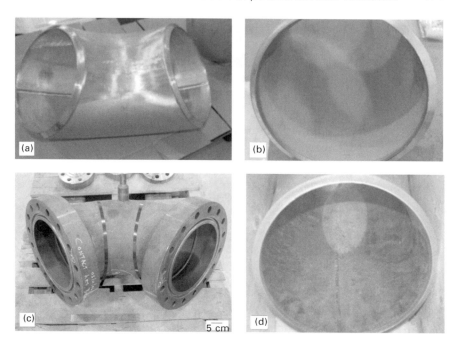

25.18 Pictures of 300 mm diameter elbow fitting (Incoloy™ 800H) showing various stages of processing: (a) as-received, (b) Q1100 coated interior, (c) after heat treatment, and (d) fabricated elbow fitting.

the following: hot erosion wear properties of oxidized materials, other regimes of erodent velocity, temperature and atmosphere (as indicated in Fig. 25.17), and other erodent media with different morphology and mechanical properties.

The ultimate goal of the research is the development and commercialization of new coatings that increase lifetime of critical-use industrial components. Further field trials are continuing for the Q1100 coating (1050°C maximum service temperature), targeting a two-to threefold performance improvement over conventional materials. For a service temperature up to 1000°C, a four- to fivefold enhancement in hot erosion wear resistance for the Q1100 coating is projected. A new coating formulation for use in operating temperatures up to 1125°C and targeting a two- to threefold enhancement in wear resistance is also under development. This 1125°C-coated product should soon be available for field trials. A pilot manufacturing facility is currently under construction to enable coating of a broad range of commercial-scale products.

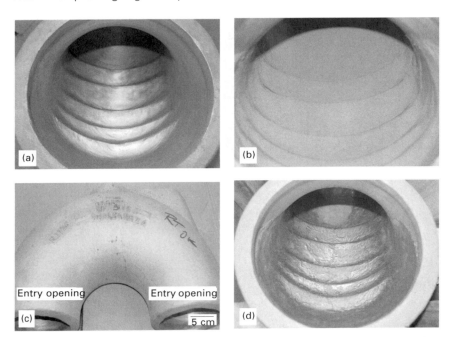

25.19 A 180°-return bend (25Cr-35Ni-Fe) showing various stages of processing: (a) fitting geometry, (b) pre-cleaned internal surface, (c) Q1100 coated interior, entry opening, and (d) after heat treatment.

25.5   Conclusions

Based upon the findings of the present work, the following conclusions can be made.

- Q1100 macro-coating offers protection to 25Cr-35Ni-Fe from oxidizing and carburizing environments.
- The oxidizing and carburizing resistant properties of the Q1100 coatings are partially related to the interfacial barrier that forms within the coating upon exposure to these severe environments. The interfacial barrier can be regarded as a highly element-selective diffusion barrier.
- A synergistic relationship exists between erosion wear and carburization for the 25Cr-35Ni-Fe, but not for the Q1100 coating.
- At temperatures exceeding 800°C, a synergistic relationship between erosion wear and oxidation for both 25Cr-35Ni-Fe and 35Cr-45Ni-Fe steels was found. This degradation process could be described as an erosion-oxidation dominated wear mechanism.
- A novel coating process has been developed that offers the deposition of a non-line-of-sight macro-coating with wear, oxidizing and carburizing resistance properties onto a range of stainless steel substrates with a thickness range of 1–5 mm.

25.6   Acknowledgements

The authors wish to thank the National Research Council of Canada – Industrial Research Assistance Program (NRC-IRAP Canada, Project No. 480189) for partial funding of the project, and Quantiam Technologies for permission to publish aspects of the work. Thanks are also expressed to Mr J. Chen for operation and assistance in the analysis and interpretation of FE-SEM/EDS and AES results, to Mr X. Zuge for the operation of the XRD instrument and assistance with data analysis, and Mr C. Doell for assistance with the design and construction of the hot erosion test facility.

25.7   References

1. Kaya, A.A., Krauklis, P. and Young, D.J., *Materials Characterization* **49** 11–21 (2002).
2. Yamazaki, D., Hirata, I. and Morimoto, T., *Mitsubishi Heavy Industries Technical Review* **19** (2) 149–155 (1982).
3. Petkovic-Luton, R., *Canadian Metallurgical Quarterly* **18** 165–170 (1979).
4. Nishiyama, Y., Otsuka, N. and Nishizawa, T., *Corrosion* **59** (8) 688–700 (2003).
5. Wu, X.Q., Yang, Y.S., Zhan, Q. and Hu, Z.Q., *J. Materials Engineering and Performance* **7** (5) 667–672 (1998).
6. Zhu, M., Xu, Q. and Zhang, J., *J. Materials Science and Technology* **19** (4) 327–330 (2003).
7. Kubota Alloy KHR35CW, *Alloy Digest* Dec. 2–3 (1999).
8. Stack, M.M., Stott, F.H. and Wood, G.C., *Materials at High Temperatures* **9** (3) 153–159 (1991).
9. Stack, M.M., Stott, F.H. and Wood, G.C., *Materials Science and Technology* **7** (12) 1128–1137 (1991).
10. Stack, M.M., Stott, F.H. and Wood, G.C., *Wear* **162–164** 706–712 (1993).
11. Stack, M.M., Chacon-Nava, J.G. and Stott, F.H., *Materials Science and Technology* **11** 1180–1186 (1995).
12. Evans, A.G. and Charles, E.A., *J. American Ceramic Society* **59** (7–8) 371–372 (1976).
13. Lawn, B.R. and Wilshaw, R., *J. Materials Science,* **10** 1049–1081 (1975).
14. Lui, R., Li, D.Y., Xie, Y.S., Llewellyn, R. and Hawthorne, H.M., *Scripta Materialia* **41** (7) 691–696 (1999).
15. Deuis, R.L., Subramanian, C. and Yellup, J.M., *Wear* **201** 132–144 (1996).
16. Deuis, R.L., Subramanian, C. and Yellup, J.M., *Composites Science and Technology* **57** 415–435 (1997).
17. Lai, G.Y., *High-Temperature Corrosion of Engineering Alloys*, 3rd edn, ASM International, Metals Park, Ohio, 15–46 (1997).
18. Bennett, M.J. and Price, J.B., *J. Materials Science* **16** 170–188 (1981).
19. Moller, G.E and Warren, C.W., *Proc. Conf. Corrosion '81*, Toronto, Canada, NACE, Paper No. 237 1–2 (1981).

26

High temperature corrosion of structural materials under gas-cooled reactor helium

C CABET, A TERLAIN, P LETT, L GUÉTAZ
and J-M GENTZBITTEL, CEA, France

26.1   Introduction

The Generation IV International Forum (GIF) has identified promising nuclear energy systems for further collaborative investigations and development. For these future concepts, the GIF board has defined high-level goals in the four broad areas of sustainability, economics, safety and reliability. Among the six selected candidates, two are gas-cooled reactors (GCR), namely the very high temperature reactor (VHTR) a thermal spectrum reactor, and the gas-cooled fast reactor (GFR). The most near-term system is the VHTR, likely to be available by about 2030. In this case, the fuel elements as well as the core structures are made of large amounts of graphite. The cooling helium heats up in contact with the hot graphite and transfers its calories to a secondary fluid within compact heat exchangers. The secondary coolant then generates electricity via large-scale gas turbines or is used for massive hydrogen production without emission of greenhouse gases.

In order to meet the demanding criteria of the Gen IV roadmap, especially regarding the high-efficiency energy conversion, the primary helium should supply a vessel output temperature that is as high as reasonably achievable, typically in the range of 900°–1000°C. Furthermore, the VHTR should be licensed for a 60-year lifespan. Not all the components could be designed for such a long duration, but time spans between two replacements should be long and lifetime has to be well assessed. The thermal, environmental, and service life conditions will thus make the selection and qualification of the structural alloys very challenging, particularly for the components at the highest temperatures. IHX materials will, for instance, develop large hot surfaces in contact with the cooling gases whereas its wall must remain thin.

Materials for application in GCRs were studied extensively in the 1970s and 1980s. The results of these earlier works identify some critical items that must be addressed to guarantee the integrity of the metallic structure over the reactor operation time, such as the microstructural stability, the creep strength and the compatibility with the cooling gas. This latter issue appeared to be especially problematic at high temperatures as some alloys could suffer major corrosion damages under GCR environment. The best alloys, namely Alloy 617, Hastelloy X and Nimonic 86, were nickel based, Cr rich and strengthened by molybdenum. Alloy 617 is the most creep resistant but, in GCR environment such as the Prototype Plant for Nuclear Process Heat (PNP) (Table 26.1), it corrodes actively above approximately 920°C.

In recent years, tungsten has tended to replace molybdenum as an additive element for HT reinforcement. The Ni-Cr-W alloys have been reported to have superior high temperature strength, hot workability, and structural stability under long time exposure at elevated temperature. The main goal of our program is to find an alternative structural material with high creep strength and improved corrosion resistance. Based on its promising creep properties, recently developed Haynes 230 (14.2wt.% W) was chosen for further investigation. Figure 26.1 compares the creep properties of Haynes 230 at 850°C to those of two Ni-Cr-Mo alloys. Haynes 230 performs better than Hastelloy X and its creep strength appears to be as high as that of reference Alloy 617. However, the hot corrosion behaviour of Haynes 230 in GCR helium has not yet been investigated.

The rich literature on the corrosion of Ni-Cr-Mo alloys in GCR helium is briefly reviewed. Then, we report our preliminary results on the exposure tests of Haynes 230 in GCR representative conditions and we compare resistance of this material to the corrosion features of some reference Ni-Cr-Mo alloys.

26.2   Overview of Ni-Cr-Mo alloy and corrosion in GCR helium

26.2.1 Characteristics of GCR helium

The helium coolant of the GCRs inevitably contains residual pollution. The sources of impurities are, for instance, the degassing of adsorbed species

Table 26.1 Impurity content in the coolant helium of experimental reactors and composition of some standard test gases

	H_2	H_2O	CO_2	CO $(10^{-6}$ bar)	CH_4	O_2	N_2
DRAGON	20	1	<0.4	12	3		3
AVR	300	30	100	100			
Peach Bottom	225	<12		12	15		12
PNP	500	1,5		15	20		5
HHT	50	5	5	50	5		5
HTGR-SC	200	10	<1	20	20		15
Jaeri-typeB	200	1	2	100	5		<5

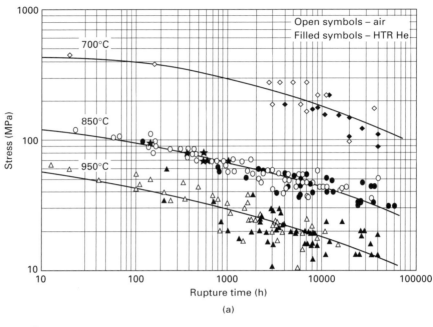

(a)

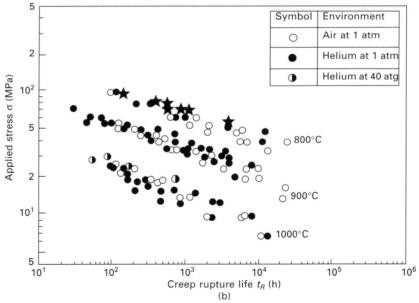

(b)

26.1 Creep properties of Haynes 230 (stars) at 850°C in air and in vacuum compared to the creep behaviour of (a) Alloy 617 (σ/rupture time curves after Ennis [19]) and (b) Hastelloy X (σ/creep rupture life curves after Hada *et al.* [20]).

from the massive graphite structures as well as some air ingress or in-leakage [1]. The major part of the contamination is removed within the purification unit but the remaining pollutants, transported in the high velocity gas flow, are ready to react with the core graphite and other hot materials. Table 26.1 gives the available data on the operation of three experimental reactors together with some standardised atmospheres used for lab testing. The GCR helium contains hydrogen, carbon monoxide (and dioxide), methane, and nitrogen ranging from a few to hundreds of microbars; the water vapour content is extremely low, in the microbar range. In this specific environment, the oxidising potential is low while the carbon activity, set by carbon bearing species, is possibly significant. Furthermore, due to the short dwell time as well as the low contamination level, GCR helium is not in thermodynamic equilibrium [2]. Instead, a dynamic balance between the pollution rates, the reactions in the core and the purification efficiency establish the impurity nature and concentrations [1]. It is therefore not possible to describe the gas phase by the classical thermodynamic potentials of carbon and oxygen.

26.2.2 Gas/metal surface reactions in GCR helium

Owing to the low impurity concentrations, the collision probability between gaseous molecules is very low and the gas/metal reactions can be regarded as independent [3]. The previous testing programmes on Ni-Cr-Mo alloys identified several possible reactions between helium impurities and metallic surfaces. They also highlighted the major role of chromium as alloying element. Brenner and Graham [4] considered a set of five reactions, illustrated below with Cr as metal and the chromium carbide $Cr_{23}C_6$:

$$3H_2O + 2Cr = Cr_2O_3 + 3H_2$$
Oxidation/reduction by water and H_2 26.1

$$6CH_4 + 23Cr \rightarrow Cr_{23}C_6 + 12H_2$$
Decomposition of methane   26.2

$$6CO + 27Cr \xrightarrow{T<T_A} 2Cr_2O_3 + Cr_{23}C_6$$
Carbon monoxide splitting   26.3

$$6H_2O + Cr_{23}C_6 \rightarrow 23Cr + 6CO + 6H_2$$
Attack of carbides by water   26.4

A further reaction, named the 'microclimate reaction' by Warren [5], occurs at the highest temperatures. Though it is not a solid/solid reaction, but rather a gas phase catalysed reaction, it formulates globally:

$$2Cr_2O_3 + Cr_{23}C_6 \xrightarrow{T<T_A} 6CO + 27Cr$$
26.5

For a given chromium activity and a given level of CO, Eq. 26.5 reverses at a critical temperature T_A that represents the changeover point for the relative stabilities of oxide and carbide [6]. At temperatures above T_A, Cr_2O_3 and $Cr_{23}C_6$ can no longer co-exist within the alloy and Eq. 26.5 continues up to completion until either all the carbide or all the oxide is removed. This precludes any passive corrosion behaviour, based on surface oxide development.

26.2.3 Kinetics of the surface reactions

As previously noted, the gas phase reaches a steady state rather than thermodynamic equilibrium. Based on the kinetic study of the gas/metal reactions, Quadakkers and Schuster [7] were able to describe the steady state carbon activity and oxygen partial pressure in helium at the metal surface. Through specially designed experiments, they studied the kinetics of the competing reactions Eq. 26.6 and Eq. 26.7 that locally control the carbon transport to and from the metal surface.

$$CH_4 = [C] + 2H_2 \qquad\qquad 26.6$$

$$[C] + H_2O = CO + H_2 \qquad\qquad 26.7$$

Considering that Eq. 26.7 is the sum of two partial reactions, Eq. 26.8 and Eq. 26.9:

$$[C] + \tfrac{1}{2}O_2 = CO \qquad\qquad 26.8$$

$$H_2O = H_2 + \tfrac{1}{2}O_2 \qquad\qquad 26.9$$

Eq. 26.8 is rapid compared to the other reactions (especially Eq. 26.6 and Eq. 26.9). Assuming that Eq. 26.8 is always near equilibrium, it is possible to express a_C^{ss} as:

$$a_C^{ss} = a_C^{eq.26.7} \cdot \left(1 - \frac{k_6 \cdot P_{CH_4}}{k_9 \cdot P_{H_2O}}\right) \qquad\qquad 26.10$$

in which $a_C^{eq.26.7}$ is the carbon activity determined by equation Eq. 26.7 and k_6, k_9 are the reaction constants respectively for the carbon transfer by methane and the oxygen transfer by water. The oxygen partial pressure $P_{O_2}^{ss}$ is reported to be:

$$(P_{O_2}^{ss})^{1/2} = (P_{O_2}^{eq.26.9})^{1/2} \cdot \left(1 - \frac{k_6 \cdot P_{CH_4}}{k_9 \cdot P_{H_2O}}\right) \qquad\qquad 26.11$$

Here, $P_{O_2}^{eq.26.9}$ stands for the equilibrium oxygen partial pressure of Eq. 26.9.

Quadakkers and Schuster estimated the reaction rates for the nickel base alloy Alloy 617 around 950°C [7]. The ratio k_6/k_9 is said to be on the order of 1/100. Thus, given that the ratio P_{CH_4}/P_{H_2O} is not greater than 100/1, a_C^{ss} and $P_{O_2}^{ss}$ are respectively given by:

$$a_C^{ss} = k_8 \cdot \frac{P_{CO}}{(P_{O_2}^{ss})^{1/2}} \qquad\qquad 26.12$$

$$P_{O_2}^{ss} = \left(k_9 \cdot \frac{P_{H_2O}}{P_{H_2}} \right)^2 \qquad\qquad 26.13$$

Where k_8 and k_9 are the reaction constants for equilibrium in Eq. 26.8 and Eq. 26.9 respectively.

26.2.4 Critical conditions for the surface reactions

Since the corrosion of metals in helium is highly sensitive to environmental parameters, the large number of experimental results has been fairly scattered. Schematic maps, namely Ternary Environmental Attack (TEA) diagram [3, 4, 8] and modified stability diagram [7, 9], were therefore introduced in order to correlate the surface reactivity of Ni-Cr-Mo alloys with the levels of the main gaseous impurities at given temperature and given chromium activity. Those maps delineated domains for the stable Cr compounds (metal, oxide and carbides) as a function of gas parameters (P_{CH_4}/P_{H_2}, P_{H_2O}/P_{H_2}, P_{CO}/P_{H_2} or $P_{O_2}^{ss}$, a_C^{ss}). From a practical point of view, oxidation (Eq. 26.1) must prevail over all other reactions Eq. 26.2–26.5 so that a protective film forms (see below). The GCR environment must thus promote the oxide domain that meets the following criteria:

- an oxidising gas toward Cr i.e. $P_{O_2}^{ss} > P_{O_2}^*$,
- a sufficient CO level to balance the 'microclimate reaction' Eq. 26.5, i.e. $P_{CO}^{ss} > P_{CO}^*$,
- a fairly low carbon activity that precludes carbide formation, i.e. $a_C^{ss} < a_C^*$,

where the asterisk refers to critical values related to the alloy. Because of various factors, the determination of these critical values (or diagrams) has to be partly experimental: the Cr activity in complex alloys is unknown, the alloying elements greatly affect the properties of the oxide layers as well as the relative stability of the metallic carbides, and the sub-surface alloy composition is time-dependent.

26.2.5 Corrosion damage of Ni-Cr-Mo alloys in GCR helium

The long-term resistance of structural alloys at high temperature relies on the formation of a compact, adherent and slow-growing surface oxide layer via the reaction Eq. 26.1. If the oxide fails to develop or if it does not offer sufficient protection, the GCR helium would exchange carbon with the metal through the reactions Eq. 26.2–26.5. This carbon transfer is not only likely to modify the surface corrosion products (oxides and/or carbides), but also to induce in-depth changes in the alloy microstructure, detrimental to the mechanical properties:

- *Carburisation*: Carbon deposition can induce the precipitation of coarse internal carbides that embrittle the Ni-Cr-Mo alloys at low temperature [10]. Moreover, the creep strain is strongly affected by carburisation [11, 12]. However, it is worth noting that the creep rupture lifetime of Hastelloy X under 'carburising' helium is not far below that in air [13].

- *Decarburisation*: Carbon removal causes the dissolution of carbides degrading the creep properties of carbide-hardened alloys. Occurring at high temperature, this damage is fast and deep. Shindo and Kondo [14] show that the creep rupture life of Hastelloy XR shortens greatly under 'decarburising' helium.

26.2.6 Oxidation of Ni-Cr-Mo alloys in GCR helium

Thus, any 'carburising' or 'decarburising' environment must definitely be avoided and oxidation by water vapour (Eq. 26.1) must prevail over anyother reactions. In this case, a protective surface film can form and then the alloy performance depends on the oxide properties. The corrosion may be slow but the rate constant can differ by several orders of magnitude. Several factors influence the oxidation kinetics: the nature of the oxide scale, chromia or alumina, the significant, and sometimes controversial, effects of the minor alloying elements such as Mn, Si, Ti or Al on the surface layer features, the evaporation of chromia based scales at high temperature [15], as well as the surface oxide spalling [16]. Hastelloy X, for instance, forms a protective Cr_2O_3 layer that has a marked tendency to spall and to evaporate above 1000°C. The up-graded Hastelloy XR was specially designed to improve the adherence and stability of the Cr_2O_3 film. Moreover, beneath the surface layer, internal oxidation of a strong oxide forming elements can occur especially in aluminium containing alloys (Cr_2O_3 or Al_2O_3 formers). A deep internal oxidation can initiate fatigue or creep cracks and can promote the desquamation of the scale.

26.3   Experimental methods

26.3.1 Materials

Table 26.2 gives the chemical composition of Haynes 230 and the two other alloys that were tested. Haynes International provided the as-received material in the shape of a 9.52 mm thick plate (heat treatment: solutioning at 1177–1246°C/quench). The grain size is in the range 30–100 μm. Microscopic observations show large carbides, lined up along the cold-work direction (see Fig. 26.2). TEM analysis reveals that these coarse carbides, presumably

Table 26.2 Chemical composition of the materials (wt%)

	Ni	Fe	Cr	Mo	W	Ti	Al	Co	Si	Others
Haynes 230	base	1.4	21.8	1.3	14.2	<0.1	0.3	0.18	0.39	Mn: 0.5
Alloy 617	base	1.1	22.3	8.7		0.35	1.26	11.7	0.11	Cu: 0.1
Hastelloy X	base	18.3	21.6	8.8	0.5			0.89	0.22	Mn: 0.49

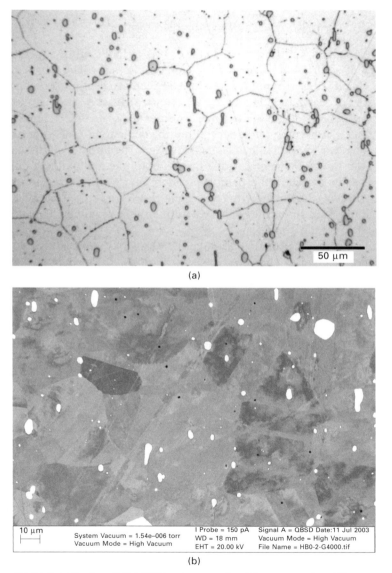

(a)

(b)

26.2 (a) optical and (b) SEM micrograph of the as-received Haynes 230.

with the M_6C structure, are rich in W and contain Ni and Cr. It also shows that fine carbides, of the $M_{23}C_6$ type rich in Cr, precipitate at grain boundaries. The other alloys were purchased at Krupp VDM.

Coupon specimens of 5 cm^2 were machined. The specimen surfaces were mechanically ground to 1200 grit emery paper, and then cleaned ultrasonically in an acetone/alcohol mixture.

26.3.2 Conditions of the tests in helium

Corrosion tests were performed in the CORALLINE facility, a specific device equipped with two test sections at ambient pressure. The exposures ran at 750°C and 950°C for up to about 800 h. Each test section is composed of a horizontal quartz tube heated by an electrical furnace and can house only four specimens at the same time, in order to avoid interference between coupons. Furthermore, in order to limit the depletion of the impurities, each specimen was placed in its own gas flow line. The impure helium is supplied from pre-mixed cylinders. At the test-section inlet and outlet, a mirror-type hygrometer monitors the helium moisture and a gas chromatograph (He-ionisation detector) analyses the permanent gases. We verified that no significant depletion of the impurities occurred during the tests. Table 26.3 summarises the experimental conditions.

It was assumed initially that Haynes 230 and Ni-Cr-Mo alloys would have common corrosion characteristics. So taking into account the model by Quadakkers and Schuster [7] and the composition of the standard atmospheres (Table 26.1), we designed test helium with a likely composition containing H_2, CO and CH_4. No water was purposely added but the residual amount of water vapour was about 2 ppm. The selected gas mixture should promote oxidation of the materials, provided that their Cr activity is higher than about 0.4 at 950°C. Table 26.3 gives the steady state potentials in the test atmosphere using the equations developed by Quadakkers and Schuster [7].

Table 26.3 Conditions of the tests in helium

Temp (°C)	Flow rate (L.h^{-1})	Time (h)	Log (Po$_2$)	Log (a_C)	Gas composition (ppm)					
					H$_2$	CH$_4$	CO	H$_2$O	N$_2$	O$_2$
750	10	216 576 888	−23.5	−2.9	195±5	20±1	56±2	~2	<1.2	<0.1
950	15	239 518 813	−19.8	−3.8	189±6	20.1±0.5	50.6±1.4	~2.1	<8	<0.1

26.3.3 Conditions of the tests in air

We also carried out isothermal tests at 750°C and 950°C under flowing air in a furnace and in a symmetric thermobalance.

26.3.4 Post-exposure observations

After exposure the specimens were weighed and metallographically evaluated. The surface scales were observed using optical microscope, and SEM equipped with EDX-system; selected specimens were analysed by XRD.

26.4   Results

26.4.1 Oxidation kinetics

TGA was performed at 750° and 950°C in air and the kinetics follow parabolic laws (see Fig. 26.3). The k_p local analysis after the method of Monceau and

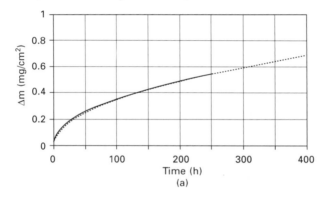

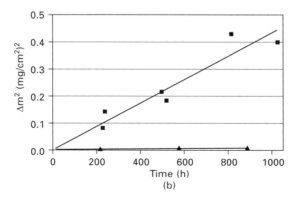

26.3 (a) TGA of Haynes 230 at 950°C in air (parabolic fit: k_p = 3.4 × 10^{-3} g² cm⁻⁴ s⁻¹] and (b) experimental mass gains after exposure to impure helium at 750° (▲) and 950°C (■).

Pieraggi [17] shows a 50 h transient stage in the early stage of oxidation, then the mass gains exhibit pure parabolic behaviour. Though lacking experimental data on the impure helium, we assumed that the parabolic law, valid in air, might also be applied to the oxidation under He. With these assumptions, tentative k_p have been estimated (see Table 26.4). It can be seen that k_p is independent of the atmosphere at 750°C. This is in agreement with the chromia scale growth mechanism depicted by Kofstad [18]. The oxide would indeed develop outward via interstitial Cr^{3+} diffusion and the parabolic rate constant would be independent of the Po_2. On the other hand at 950°C, the kinetics are slower under He than in air indicating an effect of the atmosphere.

26.4.2 Corrosion at 750°C

Haynes 230 oxidised and corroded a little at 750°C over 800 h. Figure 26.4 illustrates the surface of the specimen exposed in the test helium. A thin

Table 26.4 Tentative k_p (in g^2 cm^{-4} s^{-1}) for Haynes 230 oxidation at 750° and 950°C in air and in test helium

k_p ($g^2 \cdot cm^{-4}$ s^{-1})	Air	GCR helium
750°C	2.1×10^{-15}	2.2×10^{-15}
950°C	3.4×10^{-13}	1.1×10^{-13}

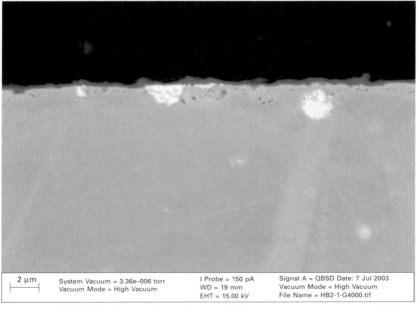

2 μm	System Vacuum = 3.36e–006 torr	I Probe = 150 pA	Signal A = QBSD Date: 7 Jul 2003
	Vacuum Mode = High Vacuum	WD = 19 mm	Vacuum Mode = High Vacuum
		EHT = 15.00 kV	File Name = HB2-1-G4000.tif

26.4 SEM picture of the Haynes 230 specimen exposed for 888 h under impure helium at 750°C

chromia layer has formed and there is some evidence of minor intergranular oxidation of Al.

26.4.3 Corrosion in helium at 950°C

After treatment at 950°C, on the other hand, Haynes 230 exhibits thicker surface layers. Figure 26.5 shows a cross-section of a 813 h coupon. The treatment has produced a duplex scale with a loose Cr-Mn spinel at the outside and a dense Cr-rich oxide inward. The alloy/oxide interface is rough and metallic islands remain included within the oxide scale. Beneath the surface layer, fine internal oxide precipitated, possibly rich in Cr. Oxidation of Al at the alloy grain boundaries is observed up to 40 μm deep.

Besides the surface corrosion products, the corrosion has also induced a change in the alloy microstructure. Thermal treatments during 1000–4000 h at 850°C produce the precipitation of secondary carbides at the alloy grain boundaries. These lamellar carbides of the $M_{23}C_6$ type are richer in Cr than the initial large M_6C carbides. Figure 26.5(b) shows that the secondary precipitation took place within the alloy bulk but did not occur in a sub-surface zone.

It is worth noticing that the corrosion morphology after a treatment in air looked quite similar (except for the loose outer layer) but every damage feature was more pronounced.

26.4.4 Comparison Haynes 230 to reference Ni-Cr-Mo alloys

In the previous section, we identified changes in the microstructure that describe the alloy corrosion. The set of parameters given in Table 26.5 is used to compare the oxidation resistance of different materials in GCR representative conditions at 950°C. Weight measurements indicated that Alloy 617 oxidised notably more than the two other casts. Figure 26.6 shows the surface of the three alloys after a 813 h exposure in impure helium at 950°C. Alloy 617 suffered a severe intergranular as well as internal oxidation of Al and an irregular, rather thick, chromia scale significantly doped in Ti and Mn has grown. As already mentioned, Haynes 230 shows some inner oxidation of Cr and Al limited to the grain boundary areas. Hastelloy X exhibited a dense thin surface layer and presented no sign of internal oxidation.

The depth of the carbideless zone seems to be in direct relation to the scale thickness, so that the carbide-free area is larger in Alloy 617 than in Hastelloy X or Haynes 230.

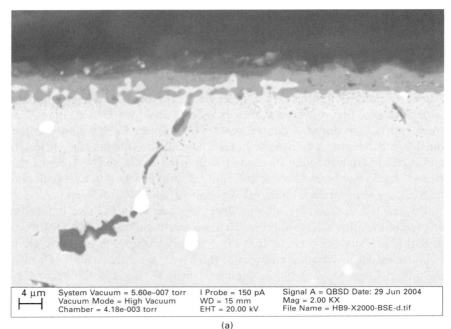

4 µm	System Vacuum = 5.60e–007 torr	I Probe = 150 pA	Signal A = QBSD Date: 29 Jun 2004
⊢─┤	Vacuum Mode = High Vacuum	WD = 15 mm	Mag = 2.00 KX
	Chamber = 4.18e-003 torr	EHT = 20.00 kV	File Name = HB9-X2000-BSE-d.tif

(a)

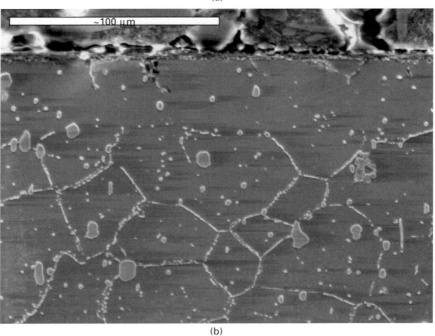

~100 µm

(b)

26.5 SEM pictures of the Haynes 230 specimen exposed for 813 h under impure helium at 950°C (a) without and (b) with electrochemical etching.

Table 26.5 Corrosion characteristics: mass gain, thickness of the dense oxide scale, depth of internal oxidation, depth of the carbide-free zone for Haynes 230, Alloy 617 and Hastelloy X specimens exposed to impure helium at 950°C

		Mass gain (mg cm^{-2})	Dense oxide scale (μm)	Internal oxidation (μm)	Carbide-free zone (μm)
Haynes 230	239 h	0.38	1.8	13.8	30
	518 h	0.43	1.4	11.2	30
	813 h	0.66	3.1	18.6	55
Alloy 617	239 h	0.45	1	9	35
	518 h	0.70	1.8	11	45
	813 h	1.19	4.6	16	65
Hastelloy X	239 h	0.27	1.2		9
	518 h	0.40	1.8		20
	813 h	0.39	1.8		30

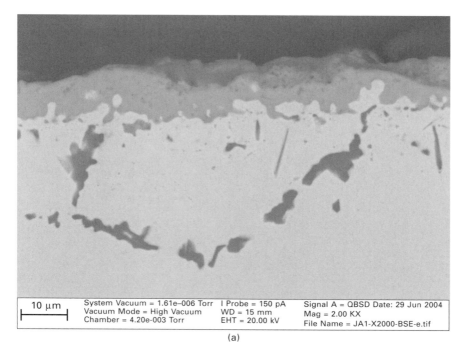

10 μm	System Vacuum = 1.61e–006 Torr	I Probe = 150 pA	Signal A = QBSD Date: 29 Jun 2004
	Vacuum Mode = High Vacuum	WD = 15 mm	Mag = 2.00 KX
	Chamber = 4.20e-003 Torr	EHT = 20.00 kV	File Name = JA1-X2000-BSE-e.tif

(a)

26.6 SEM pictures of (a) Alloy 617, (b) Hastelloy X, and (c) Haynes 230 after 813 h at 950°C in impure helium.

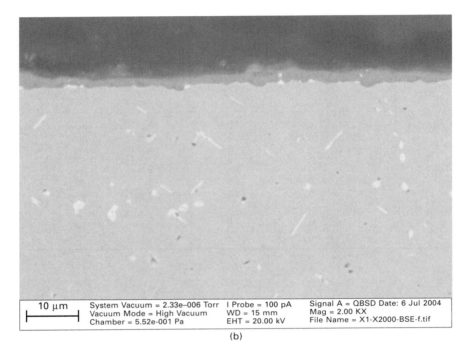

10 μm	System Vacuum = 2.33e–006 Torr	I Probe = 100 pA	Signal A = QBSD Date: 6 Jul 2004
	Vacuum Mode = High Vacuum	WD = 15 mm	Mag = 2.00 KX
	Chamber = 5.52e-001 Pa	EHT = 20.00 kV	File Name = X1-X2000-BSE-f.tif

(b)

10 μm	System vacuum = 6.75e–007 Torr	I Probe = 150 pA	Signal A = QBSD Date: 29 Jun 2004
	Vacuum mode = high vacuum	WD = 15 mm	Mag = 2.00 KX
	Chamber = 4.19e-003 Torr	EHT = 20.00 kV	File Name = HB10-X2000-BSE-a.tif

(c)

26.6 (Continued)

26.5   Summary and conclusion

Ni-Cr-W alloys are claimed to have higher creep strength than the Ni-Cr-Mo alloys together with superior corrosion properties. Thus, Haynes 230 could replace the reference Hastelloy X or Alloy 617 as a structural material for application in an advanced GCR that operates at temperatures higher than 920°C. However, if the corrosion of Ni-Cr-Mo alloys has been extensively studied in the past, data on the behaviour of Ni-Cr-W alloys in GCR representative conditions are required. First assessment of the creep properties at 850°C indicated that Haynes 230 is as creep strong as Alloy 617 and is more resistant than Hastelloy X. Furthermore, our preliminary corrosion results showed that Haynes 230 has a good compatibility with 'slightly oxidising' helium at 950°C. It develops a dense Cr-based oxide scale, although the interface metal/oxide is somewhat convoluted. Cr and Al oxidise internally at grain boundaries up to about 20 μm after 800 h. The tendency to internal oxidation in chromia-former alloys seems to be correlated with the Al, and possibly Ti, contents [14]. The use of a low Al cast of Haynes 230 might therefore improve the oxidation resistance of the alloy toward intergranular precipitation of alumina.

26.6   Acknowledgement

The authors wish to thank the Region Ile de France for its financial support to acquire a SEM.

26.7   References

1. L.W. Graham *et al.*, *Gas-cooled reactors with emphasis on advanced systems Vol. I*, International Atomic Energy Agency, Vienna, **1976**, 319.
2. K. Krompholtz, J. Ebberink and G. Menken, *Proc. of the 8th Int. Congress on Metallic Corrosion Vol. II*, Mainz, FGR, **1981**, 1613.
3. K.G.E. Brenner and *Gas-cooled Reactors Today*, BNES, London, **1982**, 191.
4. K.G.E. Brenner, L.W. Graham, *Nucl. Techol.*, **1984**, 66, 404.
5. M.R. Warren, *High Temperature Technology*, **1986**, 4, 119.
6. H.-J. Christ *et al.*, *Mat. Sci. Eng.*, **1987**, 87, 161.
7. W.J. Quadkakers, *Werkstoffe und Korrosion*, **1985**, 36, 141 and 335.
8. L.W. Graham, *High Temperature Technology*, **1985**, 3, 3.
9. W.J. Quaddakers and H. Schuster, *Nucl. Technol.*, **1984**, 66, 383.
10. P.J. Ennis, D.F. and Lupton D.F, *Proc. of the Petten International Conference on Behaviour of high temperature alloys in aggressive environments*, The Metals Society, London, **1980**, 979.
11. H.E. McCoy, J.P. Strizak and J.F. King, *Nucl. Technol.*, **1984**, 66, 161.
12. R.H. Cook, *Nucl. Technol.*, **1984**, 66, 283.
13. T. Nakanishi, H. Kawakami, *Nucl. Techol.*, **1984**, 66, 273.
14. M. Shindo, T. Kondo, *Nucl. Techol.*, **1984**, 66, 429.

15. Y. Kurata, Y. Ogawa, H. Nakajima and T. Kondo, *Proc. of the Workshop on structural design criteria for HTR*, Jülich, Germany, **1989**, 275.
16. T. Mutoh, Y. Nakasone, K. Hiraga K. and T. Tanabe, *J. Nucl. Mat.*, **1993**, 207, 212.
17. D. Monceau and B. Pieraggi, *Oxidation of Metals*, **1998**, 50, 477.
18. Kofstadt P., *High Temperature Corrosion*, Elsevier Applied Science, London, **1988**.
19. P.J. Ennis, *Proccedings of the 6th International Charles Parsons Turbine Conference*, A. Strang *et al.* (eds), Maney, **2003**, 1029.
20. K. Hada, M. Ohkubo and O. Baba, *Nuclear Engineering and Design*, **1991**, 121, 183.

27

Geometry effects on the oxide scale integrity during oxidation of the Ni-base superalloy CMSX-4 under isothermal and thermal cycling conditions

R O R O S Z and H - J C H R I S T, Universität Siegen, Germany
and U K R U P P, University of Applied Science, Germany

27.1   Introduction

Generally, the application of high-temperature alloys depends on the capability to form dense and slow-growing, protective oxide scales such as chromia and/or alumina with low defect densities [1]. Among these alloys, CMSX-4 is a single-crystalline nickel-base superalloy used for gas turbine blades due to its excellent mechanical properties combined with high-temperature oxidation resistance. This alloy is designed for long-term exposure, in the case of small gas turbines or aero engines under thermal cycling conditions. These conditions may lead to an increase in the susceptibility to oxide scale failure on CMSX-4, and reduce service life significantly. As a consequence of repeated scale rehealing accompanied by element depletion, formation of less protective oxide scales, transition to internal oxidation and nitridation may occur in combination with the dissolution of the strengthening γ'-phase, leading to a significant degradation of the properties of nickel base superalloys [2, 3].

Besides temperature changes there are, of course, several other factors determing the life of alumina-forming high-temperature alloys. Pint *et al.* [4] have considered a wide range of experimental data obtained on alumina-forming alloys, among those also Ni-base alloys, in order to illustrate the effect of cycle frequency and test procedures. In particular, the influence of the minor element sulfur was found to be detrimental on the scale integrity of different modifications of the single-crystalline Ni-base superalloy René N5, proven by thermogravimetrical measurements.

In particular thin-walled components with tapered edges suffer from an accelerated depletion of scale-forming elements, like Al and Cr, due to concentrations below their critical contents for protective scale formation and consequently lead to premature breakaway oxidation [5]. Since this kind of non-protective oxidation accelerates the kinetics of subsequent internal oxidation and nitridation processes, it is complicated to study the onset of changes in the corrosion processes by using rectangular-shaped specimens.

To overcome this problem, wedge-shaped specimens of Fe-20Cr-5Al were used by Al-Badairy and Tatlock [6] in order to study the chemical processes involved, with the focus on the degradation of the oxide scale just prior to the transition to non-protective breakaway oxidation.

In the present study the characterization of the effect of specimen geometry and the effect of exposure under thermal cycling conditions on the oxide scale failure mechanisms were the main objectives. Kinetics of corrosion product formation as well as changes in the chemical composition of both spalled oxide scale segments and internal corrosion products were determined to evaluate the material's resistance to high-temperature corrosion attack.

27.2   Experimental methods

The well-established single-crystalline Ni-base superalloy CMSX-4 test material was used for the present study. Its chemical composition is given in Table 27.1.

Isothermal and cyclic thermogravimetric studies were carried out on specimens having a conventional rectangular (Fig. 27.1(a)) and a wedge geometry (Fig. 27.1(b)). The dimensions of the specimens in Fig. 27.1(a) were $11 \times 7 \times 2.5mm^3$. The wedge-shaped specimens exhibit the same side lengths ($11 \times 7mm^2$) with a wedge angle of $12°$. A thermobalance system was used equipped with a computer-controlled lift to move the specimens, which were suspended by a quartz wire, periodically into and out of the hot zone of the furnace within a few seconds to allow rapid heating and cooling. Each temperature cycle of the cyclic oxidation experiments included a hot dwell period of 5 h at 1100°C followed by a $^1/_4$h cooling period at 40°C. The thermogravimetric tests should clarify the effect of different specimen geometries and the effect of temperature control (isothermal/thermal cycling)

Table 27.1 Chemical composition of CMSX-4 (in wt.%).

Ni	Cr	Al	Ti	Co	Ta	Mo	W	Re	Hf	S
Bal.	6.0	5.6	1.0	10.0	6.0	0.6	6.0	3.0	0.1	<12ppm

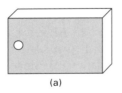

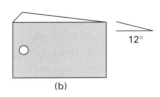

(a)   (b)

27.1 Shape of the specimens used for isothermal (a) and thermal cycling (a+b) thermogravimetric studies.

on the overall corrosion process (internal/external) and on the oxide spallation behaviour.

Rectangular specimens were used to quantify the chemical composition of the spalled oxide products and the fraction of the oxide scale remaining on the specimen surface after different stages of thermal cycling exposure. For this purpose the specimens were exposed to laboratory air at 1100°C for 1 h followed by a $^1/_4$h cooling period at room temperature. During cooling, the specimen surface was located close to an adhesive tape to collect the spalling corrosion products. In order to document the change in the chemical composition of these corrosion products, the specimens were removed after different numbers of thermal cycles.

All specimens were wet-ground using SiC abrasive paper down to 2500 grit and polished with 6 μm and 1 μm diamond suspension. Prior to the corrosion tests, the specimens were cleaned ultrasonically in ethanol.

Scanning electron microscopy (SEM) in combination with energy-dispersive X-ray spectroscopy (EDS) was used to examine the morphology as well as the chemical composition of the spalled and remaining oxide products.

27.3 Results

Isothermal thermogravimetric measurements showed a continuous mass gain during 100 h exposure at $T = 1100$°C, even on a wedge-shaped specimen. Contrary to this, thermal cycling thermogravimetric measurements showed already after two cycles an appreciable amount of weight loss for both the rectangular-shaped and the wedge-shaped specimen as shown in Fig. 27.2.

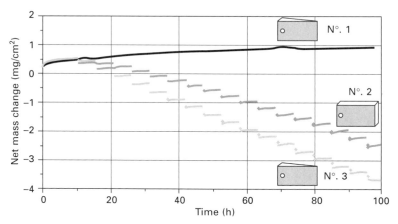

27.2 Net mass change vs exposure time for different specimen shapes. Isothermal exposure at $T = 1100$°C (No. 1) and thermal cycling exposure for 5 h hot dwell times at $T = 1100$°C and 0.25 h cooling down periods to $T = 50$°C (Nos 2, 3).

The cross-section of the isothermally exposed specimen does not show any internal precipitation as a consequence of chemical or mechanical oxide scale failure. Scale integrity was maintained even in the tip region of the wedge-shaped specimen (Fig. 27.3(a)). Contrary to this, the specimen exposed under thermal cycling conditions, developed a blunted tip region (Fig. 27.3(b)). It was observed that internal corrosion attack, proceeded along the longitudinal axis of the specimen, while the external scale in the longitudinal direction away from the tip remained protective. However, due to the internal corrosion attack, cracks were formed in the vicinity of the tip (arrow in Fig. 27.3).

The chemical composition of the tip area is given in Fig. 27.4 by EDS-element mappings giving evidence that the stage of breakaway oxidation has been exceeded. The blunted wedge exhibits mainly less-protective and fast-growing Ni and Co oxides. TiN (arrow in Fig. 27.4) was formed due to the loss of the ability of the alloy to maintain the formation of a protective

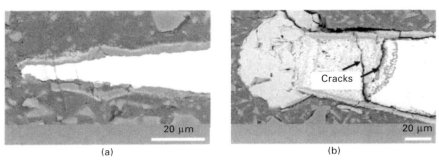

(a)   (b)

27.3 Cross section of the wedge-shaped specimens after (a) an isothermal test (No. 1) and (b) after thermal cycling exposure (No. 3 in Fig. 27.2) to laboratory air at $T = 1100°C$.

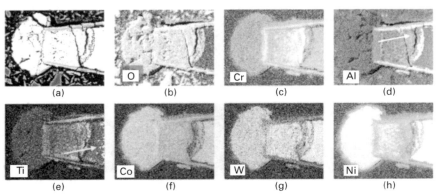

27.4 EDS-element mapping of the tip of the wedge-shaped specimen in Fig. 27.3(b) ($T = 1100°C$ thermal cycling exposure) after thermal cycling exposure.

alumina scale. An indication for aluminium depletion below a critical value is the transition from external to internal Al_2O_3 formation (arrow in Fig. 27.4(d)), a mechanism, which can generally be described by means of the theories of Wagner [7] and Maak and Wagner [8].

The effect of the specimen geometry on the corrosion kinetics is documented by the thermogravimetrical curves of a wedge-shaped specimen and a rectangular-shaped specimen in Fig. 27.2. Assuming that both net mass measurements were performed under reproducible conditions, the comparison of the two results reveals a mass loss difference, manifesting itself in a 35% higher loss in the case of the wedge-shaped specimen after 100 h. The surface area of the tip region, which is significantly affected by spalling of corrosion products, comprises just about 2% of the total specimen surface area. Thus, a substantial fraction of the overall spalling of corrosion products can be attributed to the wedge geometry.

In order to determine the microstructural changes in the vicinity of wedge-shaped edges just before the onset of 'breakaway oxidation', an additional test was performed. This test included pre-oxidation for a duration of 12 h at 1100°C. Afterwards, the specimen was submitted to a single thermal cycle (10 min heating at 1100°C followed by 15 min cooling down to 50°C). Figure 27.5 shows the cross-section of the sample and indicates that only one thermal cycle is sufficient to cause enhanced corrosion attack as well as deformation of the tip, probably as a consequence of thermal mismatch between the oxide and the underlying metal during cooling. Different to the specimen shown in Fig. 27.3(a), the tip depicted in Fig. 27.5 is decorated by a seam of less stable oxides after just 10 min of oxidation in the thermal

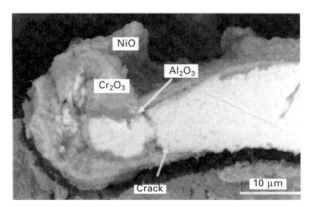

27.5 Initial stage of oxidation and creep deformation of the wedge-tip region of a wedge-shaped specimen after preoxidation for t = 12 h at T = 1100°C. High-temperature exposure was continued for 10 min after cooling down.

cycle. Aluminium depletion and scale cracking close to the tip has apparently accelerated this process.

In order to document the changes in the chemical composition of the spalled corrosion products during thermal cycling, exposure tests on rectangular-shaped specimens were carried out. Spalled oxide fragments were collected during the cooling period of the specimen and characterized by means of EDS analysis.

These analyses revealed that a high amount of corrosion products consists of less stable Cr-, Ni- and Co-containing oxides in the early stages (after 4 and 16 cycles) of exposure. Finally, after 60 cycles, a change in the element composition documenting the presence of Al_2O_3 became evident. An overview of the observed changes in corrosion product formation is given in Table 27.2.

An explanation for this evolution is the presence of selective oxidation, i.e., less noble oxide compounds are exposed to the oxidizing atmosphere above the alumina scale and are spalling primarily. Afterwards, the alumina scale underneath starts spalling off the metallic surface.

In order to quantify the extent of the spalled scale area, a representative section (2 mm × 2 mm) of the oxidized surface was chosen. By adjusting the contrast of the SEM image one could distinguish between covered and uncovered areas of the surface. By means of the image-analysis software ImageJ® the ratio between the remaining oxide scale and the uncovered metallic surface was determined after different numbers of thermal cycles. These results are shown in Fig. 27.6.

The remaining oxide scale decreased with increasing number of thermal cycles. The fraction of spalled oxides stayed below 5% until the 26th cycle and increased abruptly to values of 15–20% at about the 60th cycle. A complete spallation down to the metallic surface became apparent during the late stages of thermal cycling exposure. At this stage, the alloy loses its protection against internal corrosion attack as can be seen at local sites of breakaway oxides.

Table 27.2 Spallation behaviour during thermal cycling exposure of CMSX-4

Number of thermal cycles	Results concerning spalled corrosion products
4–16	Low amount of spalled oxide fragments; high content of Cr, Ni and Co in the oxide
26	Slight increase in the amount of spalled oxide fragments; slight decrease in the content of transient oxides like Ni and Co
60 and higher	The spalled oxide is almost pure alumina, leaving bare metal surface behind. Obvious increase in percentage and area of spalled oxide

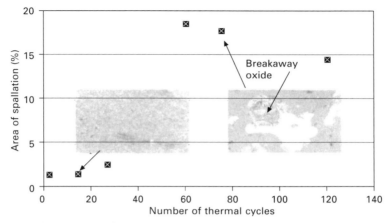

27.6 Spalled area fraction (%) on the surface of oxidized CMSX-4 specimens after different numbers of thermal cycles at *T* = 1100°C in air.

27.4   Discussion

The use of wedge-shaped samples to study the material degradation mechanism by high-temperature corrosion is an appropriate testing method to evaluate the influence of critical geometries, since they are applied to technical components, e.g. turbine blades or substrates for automotive catalytic converters. Furthermore, fundamental scientific reasons exist to characterize the influence of geometry. Since kinetics of scale growth, the state of stress between external scale and substrate, the adherence of the oxide layer and thermodynamics (in terms of their 'breakaway failure' [9]) can be affected strongly.

Under conditions of cyclic oxidation, the arrangement of the various corrosion products seems to be rather independent of specimen geometry (see Fig. 27.7 and compare to Fig. 27.3(b)). The corrosion products formed on the external scale consist of transient oxides in both retangular-shaped and wedge-shaped specimens (mainly containing Ni and Co). Underneath this unprotective oxide other corrosion products precipitate. There is some remaining chromium oxide on the surface (Fig. 27.7) and internal aluminium oxide and aluminium nitrides form a seam-like internal zone. In regions further away from the tip (Fig. 27.3(b)) or with increasing penetration depth (Fig. 27.7), respectively, needle-like titanium nitrides are formed.

The disadvantage of the use of rectangular specimens is the local and temporal uncertainty of the onset of breakaway oxidation. As an example, Fig. 27.7 shows that 76 thermal cycles were required to find sporadic sites where the transition to internal oxidation had occurred, while only one cycle was sufficient to reach this state during oxidation of wedge-shaped specimens,

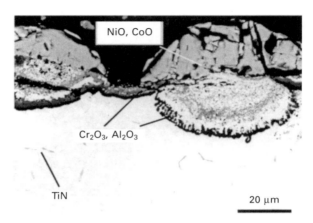

NiO, CoO

Cr$_2$O$_3$, Al$_2$O$_3$

TiN

20 μm

27.7 Corrosion product formation on a rectangular-shaped CMSX-4 specimen after 76 thermal cycles (1 h hot dwell at T = 1100°C/0.25 h cold dwell at 40°C).

as shown in Fig. 27.5. Furthermore, the Al depletion zones, especially in thin-walled specimens, may affect the complete cross-section of the substrate. After this stage, both, the mid-section and the interfacial Al concentration will reduce with time. An appropriate treatment of this mechanism was provided by Cowen and Webster [10] deriving a suitable solution of the Fickian diffusion differential equation. The mathematical concept was applied by Evans and Donaldson [11] to the depletion profile calculation of thin-walled austenitic steels.

Besides the oxidation-induced depletion effects, mechanical failure of the external alumina scale by spallation accelerates element depletion. This is the reason for a fast degradation process of the alloy during exposure under thermal cycling conditions. In this context, mechanical oxide scale failures can be attributed to mechanical stresses arising from differences in the coefficients of thermal expansion in the oxide-substrate system. Investigations on the effect of edges and corners [12] revealed the significance of the interaction between thermal stresses resulting from cooling and substrate plasticity. Besides this, oxide growth stresses contribute to the mechanical stresses [13, 14].

One important result of the thermal cycling experiments using wedge-shaped specimens is that previous observations are confirmed indicating that thermal mismatch stresses are much more significant for spallation than growth stresses [14]. Exposure of specimens of the same geometry under isothermal conditions revealed that the shape of the tip and the adhesion of the external alumina scale were sustained during the exposure time. Obviously, oxide growth stresses that arise do not have a great impact on the scale integrity. Even though the tip of a specimen is the most critical site under thermal cycling conditions, creep and plastic deformation is an important relaxation mechanism in the thin-walled area [11]. A slight deformation of the tip, as can be seen in Fig. 27.5, suggests, that stress relief by creep should not be neglected.

The change in the chemical composition of the spalled oxide fragments with increasing number of thermal cycles on rectangular shaped specimens is less pronounced compared to wedge-shaped specimens in the vicinity of the tip. Here, the corrosion process is governed mainly by an enhanced Al depletion and, therefore, a loss of the scale-rehealing capacity results after just a few cycles.

27.5   Conclusions

- The use of wedge-shaped samples is an appropriate testing method to study the influence of geometry effects on the material degradation mechanism by high-temperature corrosion.
- Comparison between thermogravimetric measurements under isothermal and thermal cycling conditions on wedge-shaped specimens revealed that thermal mismatch stresses seems to be much more detrimental for the integrity of the external oxide layers than growth stresses.
- Deformation of the tip by creep favours stress relief during cyclic oxidation.
- The corrosion process is governed mainly by an enhanced element depletion in the vicinity of the tip and, therefore, a loss in the scale-rehealing capacity results after just a few cycles.

27.6   Acknowledgements

The financial support of Deutsche Forschungsgemeinschaft (DFG) under grant No. KR 1999/2-1 and the material supply by Alstom Power, Baden, Switzerland are gratefully acknowledged.

27.7   References

1. P. Kofstad, Defects and transport properties of metal oxides, *Oxidation of Metals*, **1995**, 44, 3.
2. S. Chang, U. Krupp and H.-J. Christ, The influence of thermal cycling on internal oxidation and nitridation of nickel-base alloys, *Proc. Cyclic Oxidation of High Temperature Materials* European Federation of Corrosion, **1999**, 27, 63.
3. U. Krupp, S. Y. Chang and H.-J. Christ, Microstructural changes in the sub-surface area of Ni-base superalloys as a consequence of oxide scale failure, *Materials Science Forum*, **2001**, 369–372, 287.
4. B. A. Pint, P. F. Tortorelli and I. G. Wright, Effect of cycle frequency on high-temperature oxidation behavior of alumina-forming alloys, *Oxidation of Metals*, **2002**, 58, 73.

5. I. Gurrappa, S. Weinbruch, D. Naumenko, and W. J. Quaddakkers, Factors governing breakaway oxidation of Fe-Cr-Al-based alloys, *Materials and Corrosion*, **2000**, 51, 224.

6. H. Al-Badairy and G. J. Tatlock, The application of a wedge-shaped technique for the study of breakaway oxidation in Fe-20Cr-5Al base alloys *Oxidation of Metals*, **2000**, 53, 157.

7. C. Wagner, Reaktionstypen bei der Oxydation von Legierungen, *Zeitschrift für Elektrochemie*, **1959**, 63, 772.

8. F. Maak, and C. Wagner, Mindestgehalte von Legierungsbestandteilen für die Bildung von Oxidschichten hoher Schutzwirkung bei höheren Temperaturen, *Werkstoffe und Korrosion*, **1961**, 12, 273.

9. G. Strehl, D. Naumenko, H. Al-Badairy, L. M. Rodriguez Lobo, G. Borchardt, G. J. Tatlock and W. J. Quadakkers, The effect of aluminium depletion on the oxidation behaviour of FeCrAl foils, *Materials at High Temperatures*, **2000**, 17(1), 87.

10. H. C. Cowen and S. J. Webster, in *Corrosion of Steels in CO_2*, British Nuclear Society, London, **1974**, 349.

11. H. E. Evans and A. T. Donaldson, Silicon and chromium depletion during the long-term oxidation of thin-sectioned austenitic steel, *Oxidation of Metals*, **1998**, 50, 457.

12. D. Renusch, S. Muralidharan, S. Uran, M. Grimsditch, B. W. Veal, J. K. Wright and R. L. Williamson, Effect of edges and corners on stresses in thermally grown alumina scales, *Oxidation of Metals*, **2000**, 53, 171.

13. V. K. Tolpygo, and D. R. Clarke, Competition between stress generation and relaxation during oxidation of an Fe-Cr-Al-Y alloy, *Oxidation of Metals*, **1998**, 49, 187.

14. V.K. Tolpygo, J.R. Dryden and D. R. Clarke, Determination of the growth stress and strain in α-Al_2O_3 scales during the oxidation of Fe-22Cr-4.8Al-0.3Y alloy, *Acta Materialia*, **1998**, 46, 927.

28

What are the right test conditions for the simulation of high temperature alkali corrosion in biomass combustion?

T. B L O M B E R G, ASM Microchemistry Ltd, Finland

28.1   Introduction

High temperature corrosion of heat transfer surfaces in biomass fired boilers is the most important obstacle preventing the utilization of its energy potential to produce electricity. Corrosion is initiated with metal temperatures above 400°C and increases moderately up to 500°C. Above the metal temperature of 500°C the corrosion rate increases rapidly to an unacceptable rate and prevents the use of steam temperatures above 480°C without the risk of shortening the lifetime of superheater tubes to a few years. The highly corrosive nature of the biomass fuels was not anticipated when the first biomass boilers for electricity production were built, because chemical analyses did not reveal anything alarming. From the chemical analysis one can conclude that pure biomass is relatively clean fuel and the concentrations of harmful substances (S, Cl, Na, K, and heavy metals), with the exception of potassium, are normally lower than in coal. Yet serious corrosion was encountered and has slowed down the utilization of biomass energy in electricity production. Analyzing the deposits from biomass fired boilers has revealed KCl and K_2SO_4 as the major deposit-forming substances, KCl being the substance associated with high corrosion rates. Without KCl in the deposit, the corrosion rates are normally low and formation of K_2SO_4 seems to have a beneficial effect on the protection of tubes against corrosion. Because the saturation vapor pressure of K_2SO_4 is very low even at 1000°C, it has been assumed that it is not directly condensed on the tubes, but forms heterogeneously within the deposit by reactions between KCl and SO_2 or SO_3. There is some controversy whether this is the right assumption, because it has been shown that a molten phase is needed for the rapid sulfation of KCl, but the melting temperature of KCl is 774°C, a value over 200°C higher than the typical tube metal temperatures [1]. It has been proposed that KCl can form a molten phase by first forming a eutectic mixture with K_2SO_4 or some other substance in the deposit. On the other hand, short-term deposit tests support the deposit formation mechanism where initially deposit starts to grow by gas phase condensation and formation of sticky surface, after which other formation mechanisms like impaction of particles start to play a more significant role [2]. If KCl alone were the initial substance condensing on the surface, one would assume it to solidify rapidly on the tube surface, and the strength of adherence with the tube, the stickiness towards particles and reactivity towards sulfation should be low.

28.2   Biomass vs fossil fuels

Although the chemical analysis of biomass does not reveal directly any alarming concentrations of corrosive elements, there is one fundamental difference between fossil and biomass fuels that can be derived from it. This is the excess of alkali metals compared to the amount of sulfur plus chloride. Even if one assumes that all the sulfur in the fuel is reacted to alkali sulfates and all the chloride is reacted to alkali chlorides, there is still excess alkali left in biomass whereas with the fossil fuels this is not the case. The free alkali index has been used to predict the existence of alkali hydroxides in a biomass boiler [3]:

$$A_f = \frac{(Na_{sol} + K_{sol} - (2 \cdot S + Cl))}{LHV} \qquad 28.1$$

A_f = free alkali index, mol/MJ
Na_{sol} = acetic acid soluble sodium, mol/g
K_{sol} = acetic acid soluble potassium, mol/g
S = sulfur content of the fuel, mol/g
Cl = chlorine content of the fuel, mol/g
LHV = lower heat value of the fuel, MJ/g

Soluble fractions of the alkali metals are used to emphasize the fraction that is evaporated in the gas phase during combustion. However, sulfur is not totally released to the gas phase either, so equally well the total alkali can be used. Additionally, the amount of soluble alkali is very close to the amount of total alkali in biomass fuels and using the total alkali instead may give overestimated free alkali indexes mainly for fossil fuels. As an example, fuels analyzed by Kurkela [4] are sorted with the help of the free alkali index in Table 28.1.

28.2.1 Thermodynamic equilibrium calculations

HSC chemistry 4.1 was used to calculate the equilibrium concentrations of alkali species for the fuels in Table 28.1. The main gas phase components simulating the flue gases in a boiler were fixed in all calculations to be: 70 mol.% N_2, 14 mol.% CO_2, 12 mol.% H_2O and 3.5 mol.% O_2. The remaining

Table 28.1 Example fuels [4] assorted by free alkali index, A_f.

Fuel	K (ppm)	Na (ppm)	Cl (ppm)	S (%)	LHV (MJ/kg)	K (mol/kg)	Na (mol/kg)	Cl (mol/kg)	S (mol/kg)	A_f (mol/GJ)
Straw (wheat)	10910	230	2890	0.08	17.2	0.279	0.010	0.082	0.025	9.16
Forest residue	2110	110	240	0.04	19.7	0.054	0.005	0.007	0.012	1.37
Eucalyptus	890	670	940	0.02	18.5	0.023	0.029	0.027	0.006	0.70
Bark (pine)	1120	50	110	0.03	20.1	0.029	0.002	0.003	0.009	0.45
Sawdust (pine)	500	40	15	0.01	19.0	0.013	0.002	0.000	0.003	0.41
Wood chips (pine)	1060	30	75	0.03	18.9	0.027	0.001	0.002	0.009	0.40
Peat, Surface	690	380	180	0.10	18.5	0.018	0.017	0.005	0.031	−1.80
Peat, Carex	440	330	270	0.20	21.0	0.011	0.014	0.008	0.062	−5.08
Rhein brown coal	140	300	250	0.30	24.1	0.004	0.013	0.007	0.094	−7.37
Iowan Rawhide coal	570	1140	25	0.50	26.0	0.015	0.050	0.001	0.156	−9.56
Polish bituminous coal	1420	450	760	0.70	29.2	0.036	0.020	0.021	0.218	−13.77
Columbian bituminous coal	2600	440	130	1.00	28.4	0.066	0.019	0.004	0.312	−19.08
Illinois no 6, bituminous coal	3610	1420	1210	2.90	25.2	0.092	0.062	0.034	0.905	−67.03

0.5 mol.% was divided for S, Cl, Na and K according to their relative molar abundance in Table 28.1. The species included in the equilibrium calculations are presented below. Figures 28.1–28.2 show the results.

- *Gaseous species:*
  $N_2(g)$, $CO_2(g)$, $Cl(g)$, $Cl_2(g)$, $HCl(g)$, $H_2O(g)$, $H_2S(g)$, $H_2SO_4(g)$, $K(g)$, $K_2CO_3(g)$, $KCl(g)$, $KOH(g)$, $K_2S(g)$, $K_2SO_4(g)$, $Na(g)$, $NaCl(g)$, $NaOH(g)$, $Na_2SO_4(g)$, $O_2(g)$, $S(g)$, $SO_2(g)$, $SO_3(g)$
- *Condensed species:*
  KCl, $NaCl$, K_2CO_3, KOH, K_2SO_4, Na_2CO_3, $NaOH$, Na_2SO_4, K_2S, Na_2S

Figures 28.1–28.2 show that if the free alkali index is positive, thermodynamic equilibrium predicts the formation of alkali hydroxides, carbonates, chlorides and sulfates; if the index is negative, the only alkali species are chlorides and sulfates. It has to be noted that the free alkali index gives only a rough estimate of the alkali hydroxides in the real flue gases, because the reactions of alkali metals in the furnace with other substances, such as silica in the ash or bed material, are not counted for. On the other hand, $CaCO_3$ is often injected in the furnace to reduce sulfur emissions, thus reducing the available SO_2 to capture alkali hydroxides and leading to their possible existence even when the free alkali index calculated from the fuel would be negative. Similarly, formation of alkali sulfates requires oxygen in the flue gases. If the oxygen content is set to zero, sulfates are no longer stable and sulfur is not able to bind hydroxides. This would be the case in the flue gases from gasification plants or combustion in stoichiometric or reducing conditions (low NO_x combustion). Additionally, imperfect mixing of the flue gases before entering the convection section makes it possible for a non-equilibrium state to exist, which would enable the alkali hydroxides to exist in places where the equilibrium calculations would not predict it.

Nevertheless, it seems as though with a simple parameter derived from the chemical analysis of the fuel one can separate the biomass-based fuels from the fossil fuels. This difference may be very important for the high temperature corrosion behavior of these fuels as will be discussed further.

28.2.2 KOH diffusion through the boundary layer vs reaction with CO_2

If one assumes that potassium hydroxide is still present in the flue gases surrounding the superheater tubes in a gas temperature above 700°C, the obvious question arising is: is it also present on the tube surface? The tube surface temperature in the hottest superheaters is normally between 500 and 600°C. This temperature is low enough for K_2CO_3 to be stable and with the high vapor pressure of CO_2 in the flue gases, all KOH would quickly react to form K_2CO_3. The essential question now becomes: does KOH first condense on the tube surface and then react with CO_2 (HCl and SO_2) in the flue gases, or does KOH react with CO_2 in the boundary layer as it diffuses towards the tube surface and K_2CO_3 forms before KOH reaches the tube surface?

To answer these questions one would need to know the diffusion constant of KOH in the flue gases and the kinetics of the reaction:

$$2KOH(g) + CO_2(g) \Rightarrow K_2CO_3(s) + H_2O(g) \qquad 28.2$$

To the best of my knowledge, experimentally determined values are not available for either unknown. The estimation of the diffusion time of KOH(g) from the flue gases on the tube surface is presented next.

The diffusion constant of KOH can be considered to be similar to molecules with similar geometry and molecular weight. HCN has a similar molecular geometry to KOH and its diffusion constant in air has been experimentally determined [5]. Because only an order of magnitude estimation is needed, the following equation is used to estimate the diffusion constant of KOH in air at 0°C and 1 atm:

$$D_{KOH} = \frac{\sqrt{M_{HCN}}}{\sqrt{M_{KOH}}} \cdot D_{HCN} = \frac{\sqrt{27}}{\sqrt{56}} \cdot 0.17 \, \frac{cm^2}{s} = 0.12 \, \frac{cm^2}{s} \qquad 28.3$$

The temperature and pressure dependence of the diffusion constant is:

$$D = D_0 \left(\frac{T}{T_0} \right)^{\frac{3}{2}} \left(\frac{p_0}{p} \right) \qquad 28.4$$

At 700°C and 1 atm this leads to a diffusion constant of 0.81 cm²/s

The flow field around a superheater tube is very complex involving both laminar and turbulent boundary layers and the estimation of the local boundary layer thicknesses (velocity, diffusion and thermal boundary layers) around the tube requires computer simulations with computational fluid dynamic (CFD) software packages. However, for this rough analysis an average value of the thermal boundary layer thickness around the tube is enough and can be estimated if the average Nusselt number around the tube is known [6, p. 286]:

$$N_{Nu} = \frac{d}{x} \qquad 28.5$$

where d = tube diameter, and x = thickness of the thermal boundary layer

For typical convection pass designs, the gas flow velocities are 10–30 m/s. With 40 mm tube diameter and with air at 700°C this leads to Reynolds numbers of 3000–10 000. The corresponding average N_{Nu} numbers (determined experimentally) for air flowing normally to a single tube are 30–60 [6, p. 322].

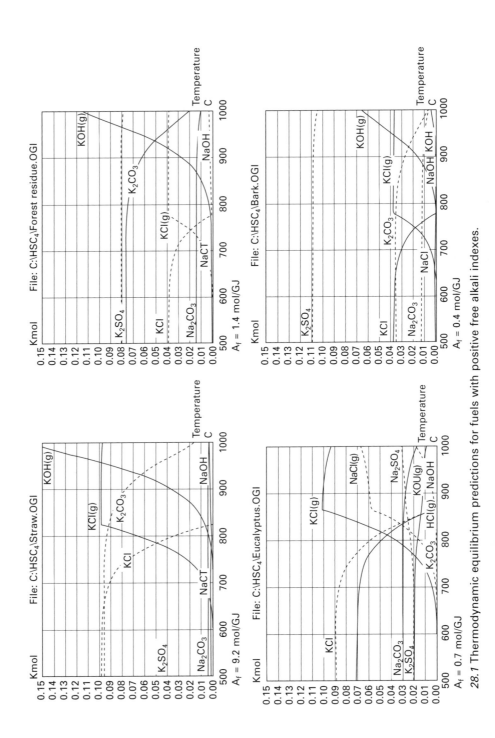

28.1 Thermodynamic equilibrium predictions for fuels with positive free alkali indexes.

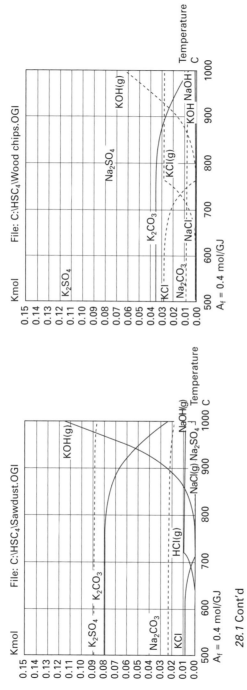

28.1 Cont'd

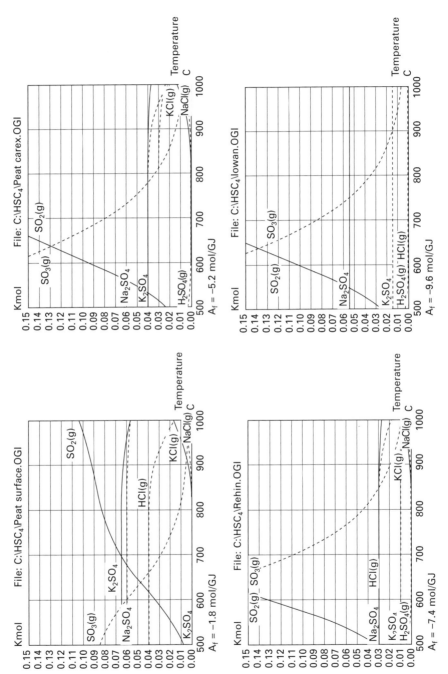

28.2 Thermodynamic equilibrium predictions for fuels with negative free alkali indexes.

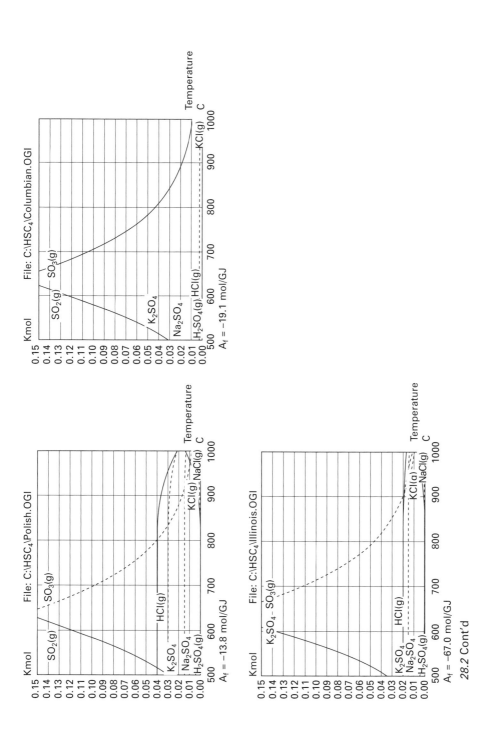

28.2 Cont'd

With a 40 mm tube diameter, Eq. 28.5 leads to thermal boundary layer thicknesses of 0.7–1.3 mm.

Finally, the characteristic time for diffusion through the boundary layer is:

$$t_D = \frac{x^2}{4 \cdot D_{KOH}}$$ 28.6

This leads to diffusion times of 1.5–5.2 ms.

The estimation of the reaction rate is substantially more complicated as one would need to take into account the nucleation kinetics of K_2CO_3. The theory of homogeneous nucleation is still disputed and the rates have to be determined experimentally. It is still worth noting that homogeneous nucleation always requires a high supersaturation of the condensing substance and heterogeneous nucleation on the surface would always be energetically more favorable. With K_2CO_3 the high supersaturation could, in fact, be impossible to achieve owing to its instability in gas phase.

28.3   Discussion

Knowing the form in which free alkali reaches the tube surface is essential for the high temperature corrosion estimation of biomass fuels as KOH can form a molten phase from metal temperatures of 400°C upwards [3], whereas K_2CO_3 melts at 891°C and even the system $(Na,K)Cl-(Na,K)_2CO_3-(Na,K)_2SO_4$ has a lowest possible melting point of 511°C [7].

If KOH would react in the gas phase to form K_2CO_3, then one could consider formation of KCl on the tube surface by direct condensation of KCl(g) or by the reaction:

$$K_2CO_3(s) + 2HCl(g) \Rightarrow 2KCl(s) + 2CO_2(g) + H_2O(g)$$ 28.7

And the formation of sulfate:

$$2KCl(s) + SO_2(g) + O_2(g) \Rightarrow K_2SO_4(s) + Cl_2(g)$$ 28.8

$$2KCl(s) + SO_2(g) + \tfrac{1}{2}O_2(g) + H_2O(g) \Rightarrow K_2SO_4(s) + 2HCl(g)$$

    28.9

$$2KCl(s) + SO_3(g) + \tfrac{1}{2}O_2(g) \Rightarrow K_2SO_4(s) + Cl_2(g)$$ 28.10

$$2KCl(s) + SO_3(g) + H_2O(g) \Rightarrow K_2SO_4(s) + 2HCl(g)$$ 28.11

$$K_2CO_3(s) + SO_2(g) + \tfrac{1}{2}O_2(g) \Rightarrow K_2SO_4(s) + CO_2(g)$$ 28.12

$$K_2CO_3(s) + SO_3(g) \Rightarrow K_2SO_4(s) + CO_2(g)$$ 28.13

No molten phases are involved in these reactions in the typical tube temperatures encountered in superheaters and if a molten phase would be formed, it would have to be formed by the formation of eutectic mixtures between KCl, K_2CO_3 and K_2SO_4. The other issue worth questioning with this mechanism would be the stickiness of solid K_2CO_3 towards the tube metal. Even if K_2CO_3 and KCl would both condense simultaneously on the tube, the mixture would possess a minimum eutectic temperature of 629°C [7]. This is still well above typical superheater tube surface temperatures.

If KOH first condenses on the tube surface and then reacts heterogeneously with the flue gases, one has to take into account the corrosion activity of molten KOH as well as the formation of KCl, K_2SO_4 and K_2CO_3 heterogeneously by reactions of HCl, SO_2, SO_3 and CO_2 with molten KOH on the tube surface:

$$KOH(l) + HCl(g) \Rightarrow KCl(s) + H_2O(g) \qquad 28.14$$

$$2KOH(l) + SO_2(g) + {}^1/_2O_2(g) \Rightarrow K_2SO_4(s) + H_2O(g) \qquad 28.15$$

$$2KOH(l) + SO_3(g) \Rightarrow K_2SO_4(s) + H_2O(g) \qquad 28.16$$

$$2KOH(l) + CO_2(g) \Rightarrow K_2CO_3(s) + H_2O(g) \qquad 28.17$$

The KCl(s) formed in Eq. 28.14 can continue reactions according to Eqs 28.8–28.11 and the K_2CO_3(s) formed in Eq. 28.17 can continue reactions according to Eqs 28.7, 28.12 and 28.13. The KOH(g) $\Rightarrow$ KOH(l) condensation from the flue gases would provide a continuous supply of KOH(l) consumed in reactions 28.14–28.17. This mechanism would explain the melt formation as low as 400°C regardless of deposit composition so no eutectic mixtures would be needed. This could also explain the high corrosion rate of KCl below its melting point (or below the KCl-K_2CO_3 eutectic) in cases where KCl is practically the only alkali species found in the deposit.

Recent findings support the latter mechanism. Blomberg *et al.* measured the melting temperature of the deposit *in situ* in a biomass boiler burning forest residue with a high temperature galvanic probe [3]. This showed a melting temperature around 400°C indicating the possible presence of KOH on the tube.

In the publications by Schofield [8, 9] the alkali condensation from flames with different Cl and S concentrations was shown to be irrelevant of the form of alkali in the flame, the only relevant parameter being the alkali concentration in the flame, and the same deposition rates were evident with metallic alkali, hydroxide or chloride. Alkali deposition resembled the titration of alkali from the flame. However, the substance formed on the deposition probe depended on the sulfur and chloride impurities in the flame. If the concentration of sulfur was enough to bind all alkali, the only substance formed was sulfate. If there was too little sulfur but some chloride, the result was a mixture of sulfate and chloride and if there was too little sulfur and chloride, the result was a mixture of sulfate, chloride and carbonate. Because alkali carbonate cannot exist in the gas phase, the author concluded that the deposition mechanism has to be heterogeneous. However, it remains unclear if the carbonate was formed in the boundary layer or by KOH condensation and subsequent reaction with CO_2 on the surface.

These recent findings suggest that condensation of KCl(g) is not the only, and perhaps not even the most important, deposit growth initiation mechanism in biomass fired boilers. Figure 28.3 presents the proposal for the test environment that could be used to better simulate the alkali corrosion in biomass boilers.

The benefit of using the test conditions with KOH(g) rather than KOH(l) is that no matter which mechanism is responsible for K_2CO_3 formation, it will be correctly simulated. If the more likely heterogeneous mechanism is assumed, KOH could also be introduced directly in the molten phase on the specimen surface, if that would result in a more simplified experimental setup.

28.4   Conclusions

The mechanism of deposit formation and alkali corrosion in biomass-fired boilers is currently not completely understood. It is the suggestion of the author that potassium hydroxide condensation and subsequent heterogeneous reactions with HCl, SO_2, SO_3 and CO_2 should be included in the experimental and theoretical simulations of the fouling and high temperature corrosion of heat transfer surfaces in biomass-fired boilers. Further experimental work is needed to solve the questions related to the K_2CO_3 formation on heat transfer surfaces. The diffusion constant of KOH in the flue gases and the kinetics of the reaction $2KOH(g) + CO_2(g) => K_2CO_3(s) + H_2O(g)$ should be determined in order to verify the applicability of the proposed mechanism.

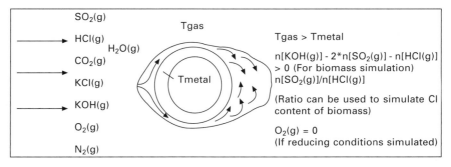

28.3 Proposed test environment for the simulation of high temperature alkali corrosion in biomass boilers.

28.5   References

1. Frandsen, F.J., Nielsen, H.P., Dam-Johansen, K., *Energy & Fuels*, 1999, 13(6), 1114–1121.
2. Frandsen, F.J., Nielsen, H.P., Baxter, L.L., Morey, C., Sclippab, G., Dam-Johansen, K., *Fuel*, 2000, 79, 131–139.
3. Blomberg, T., Makkonen, P., Hiltunen, M., *Materials Science Forum* (High Temperature Corrosion and Protection of Materials 6), 2004, 461–464, 883–890.
4. Kurkela, E., *Formation and Removal of Biomass Derived Contaminants in Fluidized-Bed Gasification Processes*, VTT Publications 287, 1996, Espoo.
5. Klots, I.M., Miller, D.K., *Journal of the American Chemical Society*, 1947, 69(10), 2557–2558.
6. McCabe, W.L., Smith, J.C., Harriot, P., *Unit Operations of Chemical Engineering*, 4th Edn, McGraw-Hill, 1985, 981.
7. Tran, H., Gonsko, M., Xiaosong, M., *Tappi Journal.*, 1999, 82(9), 93–100.
8. Schofield, K., Steinberg, M., *Combustion and Flame*, 2002, 129, 454–470.
9. Schofield, K., *Energy and Fuels*, 2003, 17, 191–203.

Part IV

Modelling

29

Optimisation of in-service performance of boiler steels by modelling high-temperature corrosion (OPTICORR)

L H E I K I N H E I M O,VTT Industrial Systems, Finland;
D B A X T E R, JRC Petten, The Netherlands; K H A C K,
GTT-Technologies, Germany; M S P I E G E L, Max-Plank-
Institut für Eisenfoschung, Germany; M H Ä M Ä L Ä I N E N,
Helsinki University of Technology, Finland; U K R U P P,
University of Applied Science, Osnabrück, Germany and
M A R P O N E N Rautaruukki Steel, Finland

29.1   Introduction

The present challenges for the power industry are dictated by global and European agreements, such as the Kyoto and EU waste directives. These, together with the fact that the utilisation of hydrocarbon fuels for conventional power generation will remain high (90%) in years to come (2020), are the basis for the development of high efficiency low emission boiler technologies. The main demands today are:

* decrease in CO_2 emissions
* increase in efficiency
* increase in steam temperature
* use of flexible fuels (co-combustion of coal and bio/waste fuels).

Raising the steam values in a power plant up to supercritical (SC) levels will improve the fuel-to-power and power-to-pollutants ratios substantially in the future. Also the combustion of less CO_2 producing fuels, e.g. bio/waste based, will play an important and economically large role in the future power industry. Thus, the use of pure bio-fuels (wood chips, etc.) or mixed coal or peat and fresh bio-fuels are very topical issues today.

A material and component life assessment concept, that is traditionally based on the theory of creep only, has to be established for the new fuel types (bio-based wood chips, straw, peat, recycled waste, etc.). The OPTICORR approach aims at development of modelling tools for high-temperature active oxidation and corrosion to predict metal loss as a function of the service conditions and gas compositions. The development is based on selected cases and corresponding experimental studies.

29.2   Thermodynamic data collection

Thermodynamic data collection is the first step for the selected approach and it is explained in detail in [1]. The first task of the project was to define the chemical system to be treated. The alloys under investigation contain many elements, the gas phase in a power plant/waste incinerator can contain a multitude of compounds, and the salt deposited on the heat exchangers may lead to dissolution of alloy components or even to solid-liquid equilibria among the salt phases. As a result, the whole data package has been split into metal, salt, oxide and sulphide subsystems, (Fig. 29.1).

In the metal (alloy) subsystem, the components are being treated with respect to their Gibbs energies in the FCC_A1 and BCC_A2 states in order to treat not only austenite and ferrite bulk materials, but also superalloys. In the salt subsystem, the combination Al, Ca, Fe(2+,3+), K, Mn, Na, Ni, Pb, Zn // Cl, SO_4, CrO_4, OH, MoO_4 has to be dealt with. The phases to be considered are the liquid solution and a large number of solid stoichiometric compounds. In the oxide subsystem, the component oxides are all solid, mostly stoichiometric compounds but there are also solubility effects to be considered, such as in wüstite (Fe^{2+}, Fe^{3+}, V, Ni, Mn)O, corundum (Al, Cr, Fe)$_2O_3$ or spinel FeO.(Al,Cr,Fe)$_2O_3$. Tailor-made thermochemical data files of these components are now available. In the sulphide subsystem, the components are FeS, FeS_2, CrS, Cr_2S_3, MnS, NiS, etc. Moreover, higher order stoichiometric sulphides, such as $FeCr_2S_4$ may appear.

The systems Fe-Cr-O_2, Fe-Si-Cr-O_2 and Fe-Si-Cr-O_2-Mn, etc., have been calculated using FactSage [2] (Fig. 29.2), and the phase sequences correlated with the experimental results [3–7]. Also, for the salt systems Fe-Na-K-Zn-Cl_2, stability diagrams (T vs log (p(Cl_2)) have been calculated to evaluate the presence of molten phases in the system. In the Fe-Na-K-Cl_2 salt system, there

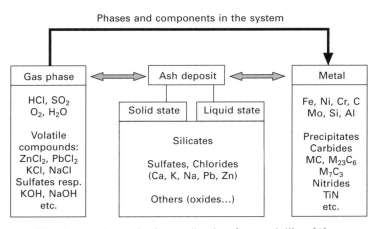

29.1 The thermodynamic data collection for modelling [1].

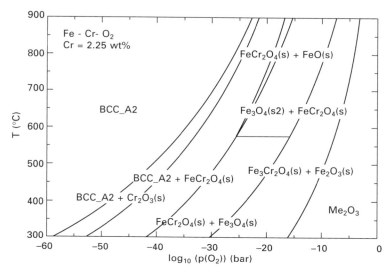

29.2 Fe-Cr-O stability diagram – $p(O_2)$ vs temperature – for 2.25% Cr alloy realised using FactSage [1].

exists a large stability field with salt melt coexisting with solid $FeCl_2$. As log $p(Cl_2)$ is increased to –2, there is a salt melt consisting of alkali chlorides and $FeCl_3$ which is important due to its high volatility (Fig. 29.3). Such diagrams will form the basis of a phase diagram 'atlas' within the project.

29.3   Corrosion and oxidation modelling

The development of modelling and simulation tools is split into two different approaches (Fig. 29.4). First, modelling of general oxidation and corrosion using a tool implemented in Excel and separate Fortran-based libraries to calculate the thermodynamic equilibria using ChemApp-library is carried out. This technique, called ChemSheet, is applied to model and simulate salt deposit-metal interactions for the KCl-$ZnCl_2$/steel case. Second, modelling of inward corrosion and internal corrosion with the InCorr tool [10] based on ChemApp (local thermodynamic-equilibrium calculations) and the finite-difference technique (FD model) has been developed for oxidation and internal corrosion. The tool enables simulations for prediction of internal corrosion depths and inward corrosion of the metal. For both of these, a self-consistent thermochemical database for high and low alloyed steels has been collected and corresponding comprehensive one- and two-dimensional equilibrium mappings to show the phase formations in corrosive systems have been created. The full modelling approach is presented in Table 29.1.

The diffusion data, in Fig. 29.4, have been defined using common statements in the development of the modelling tools. The adapted statements are as follows.

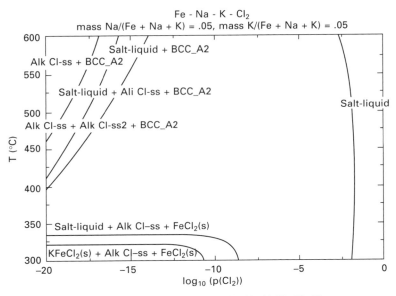

29.3 Phase diagram for the salt system Fe-Na-K-Cl$_2$ [8, 9].

- Thermodynamics give: what will be formed: possibility and order of phases.
- Kinetics give: how fast and how much.
- Typically, diffusion coefficients are determined using the diffusion couple method giving D_v (volume diffusion in parabolic growth) and diffusivities are determined from exposure or TG-analyses of a specific phase.
- Constraints for the use of classic analytical methods to solve the diffusion parameters in simulation of 'real cases' are:
  - there are always multiphase reaction products present
  - many of the products are not line/stoichiometric compounds
  - two phase reaction layers are possible.
- In this project, the definition of effective diffusion coefficient, D_{eff}, from the Wagner solution is adapted and bulk oxygen diffusivity data are taken from the open literature and experimental test results available. The grain boundary diffusivity values are approximated to yield 100 times the bulk values.

29.3.1 ChemSheet-based modelling tool for high temperature corrosion

The CorrApp model based on ChemSheet, developed at VTT and described in [11], can be used to simulate any initially stationary liquid and/or solid medium that undergoes changes in its composition and volume due to diffusion

Table 29.1 Data and tools necessary for modelling high-temperature corrosion of boiler tube materials in combustion environments

Data and tools	Source	Deliverables
Material data	Material specifications Analyses	Elements for the thermodynamic system Corrosion model description
Service and process condition data	Plant data Testcorr/PREWIN data Specific analyses	Fuel and gas composition Temperature range Ash and deposit/type Exposure/service time Temperature-time excursions
Integrated thermodynamic data bank system: FactSage	Data banks for alloys, oxides and salts specially made for the project based on known data SGTE and Fact	Phase and stability diagrams for selected material and process conditions Conditions and input to CorrApp and InCorr
Kinetic data	Literature Experimental data Diffusion data	Input for CorrApp and InCorr models
Experimentation	Exposure tests TGA tests Metallography and analyses (EDS, XRD)	Model structures/ corrosion products Data for the thermodynamic system Kinetic data and fitting of results Verification of models
Corrosion model description	Model for the corrosion mechanisms*	Boundary conditions and relations to describe the mechanisms for modelling Understanding of high-temperature corrosion phenomena
Modelling Tools: – FactSage	GTT & FACT	Tailor-made stability diagrams
– ChemSheet: CorrApp	VTT & GTT & MPIE	Modelling tool/modelling cases
– InCorr: OXIDATION	UNISiegen	Modelling tool/modelling cases

* Schematic pictures of layer sequences and types, etc.

and chemical reactions. The model assumes that temperature and pressure in the medium are constant. As the diffusion model, combined with Gibbs energy minimisation, is quite slow to calculate, a simple model was developed in which the gas/salt and salt/metal layers are combined into one control volume. The model development has been presented in conference papers describing the project [12, 13].

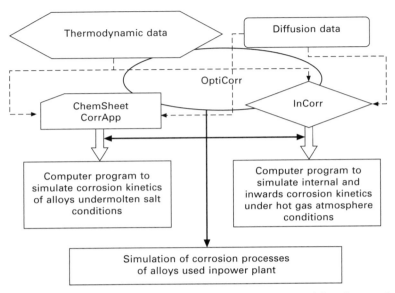

29.4 Two modelling tools developed, ChemSheet-based CorrApp and InCorr, for simulation of corrosion in power plant environments.

The first part of the modelling consists of models for the salt-induced high-temperature corrosion system. Observations confirm the model initially described by Spiegel [8, 9] (Fig. 29.5). Therefore, the corrosion sequence can be subdivided into four different parts that correspond to the kinetic steps observed in the experiment:

- dissolution of metal in the chloride melt at the melt/scale interface
- transport of dissolved chloride to the gas/melt interface
- precipitation of oxide in contact with the gas phase
- formation of a porous oxide scale.

Corresponding experimental results have been presented in [9]. A micrograph in cross-section of the Fe–salt interface exposed to 320°C is presented in Fig. 29.6.

In the salt melt model, the salt layer is formed on the metal and covered by the gas phase. The layer can be represented as a one-dimensional slab with uniform or non-uniform mesh size. The boundary conditions for the slab are the interfaces between the gas flow and the salt and the salt and the metal, (Fig. 29.7).

For the gas/salt interface it is assumed that diffusion in the salt is the limiting factor in the overall mass transport rather than the mass transfer from the gas to the interface. Then, the concentrations of gas species at the gas/salt interface can be taken to be equal to their concentrations in the gas flow.

Salt melt-induced corrosion

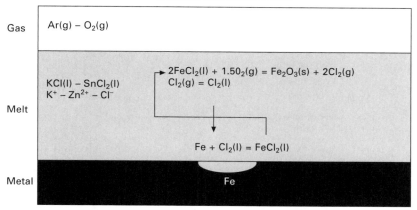

The corrosion model is based on the dissolution of the metal iron to the melt:
$$Fe(s) + Cl_2 \Leftrightarrow FeCl_2(l)$$
The chlorine source may be the equilibrium content in the melt or the following reactions:
$$2KCl(l) + 0.50_2(g) \Leftrightarrow K_2O(s) + Cl_2(g)$$
$$ZnCl_2(l) + 0.50_2(g) \Leftrightarrow ZnO(s) + Cl_2(g)$$

29.5 Model for salt melt-induced corrosion, showing iron dissolution, transport of iron chloride and oxide precipitation [9].

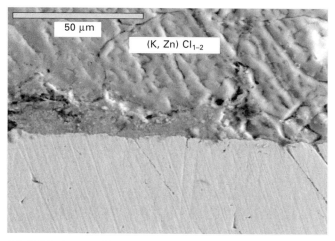

29.6 Thin layer of iron chloride (∗), formed between the salt and Fe during the incubation time of corrosion ($T = 320°C$, Ar – 4 vol.% O_2, KCl-ZnCl$_2$ deposit).

A one-dimensional (from the gas interface to the metal interface) diffusion model that takes both boundary interfaces into consideration has been produced. The model divides the salt layer into a number of control volumes and local equilibrium is calculated in each volume after the concentrations of constituents

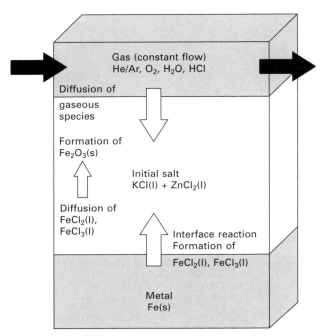

29.7 Representation of the modelled salt melt system.

in them are updated with a separate diffusion model (linear system of continuity equations).

The first data-file version is a system of seven components: Zn-K-Fe-He-O-Cl-H. It contains the gas phase, the liquid salt phase, which is composed of four constituents: KCl, $ZnCl_2$, $FeCl_2$, $FeCl_3$, and several condensed phases (oxides, chlorides, etc.) and the metal phase (Fe). As the interaction between the salt and the metal is very dependent on the interaction between the salt and the gas, it was reasonable to devise a model that could calculate the whole system. The first simulation output for the case described in Figs 29.5 and 29.6 is presented in Figs 29.8, 29.9 and 29.10. The main components in the salt are KCl (initial value 50 mol.%) and $ZnCl_2$ (initial value 50 mol.%). The simulation conditions, explained in detail in [12], are:

- temperature is 320°C
- salt amount is 15 mg/cm² and thickness of the layer 5.5×10^{-5} m.
- the atmosphere is Ar + 8 mol.% O_2, 1 bar pressure
- the $Cl_2(l)$ diffusivity is 2×10^{-9} m²/s and the metal diffusivity is 4×10^{-12} m²/s at the metal interface to allow $FeCl_{(2/3)}$ build-up in the modelling
- the mass transfer coefficient at gas/salt interface is 0.004 m/s.

Figure 29.8 shows the formation of FeO, $FeCl_2$ and $FeCl_3$ in the liquid salt after contact with the iron for 2 min. The salt composition at the metal/salt

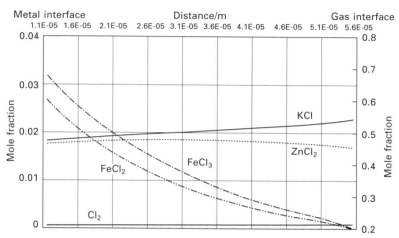

29.8 CorrApp modelling of molten salt composition after 2 min contact between Fe and KCl/ZnCl$_2$ salt at 320°C.

interface is presented in Fig. 29.9. The main compounds are the initial salts, FeCl$_2$ and FeCl$_3$, the contents of the latter species are increasing with time at the interface. The formation of oxides takes place near the gas phase on the gas/salt interface, (Fig. 29.10). The main oxide is ZnO and its content depends on the O$_2$ content in the atmosphere (8 and 16 mole.%).

The qualitative simulation results are in a good agreement with the experimental results. The quantitative results are strongly dependent on the diffusion parameters used and this will be the field for future work to improve the modelling results.

29.3.2 InCorr tool development for modelling inward/ internal oxidation

The overall simulation of high-temperature corrosion processes under near-service conditions requires both a thermodynamic model to predict phase stabilities for given conditions and a mathematical description of the process kinetics, i.e. solid state diffusion. Such a simulation has been developed by integrating the thermodynamic program library, ChemApp, into a numerical finite-difference diffusion calculation, InCorr, to treat internal oxidation and nitridation of Ni-base alloys [10]. This simulation was intended to serve as a basis for an advanced computer model for internal oxidation and sulfidation of low-alloy boiler steels.

The original computer program to calculate concentration profiles during internal-corrosion was written in FORTRAN using the standard ChemApp interface to call the Gibb's energy minimiser routine. Since these calls are

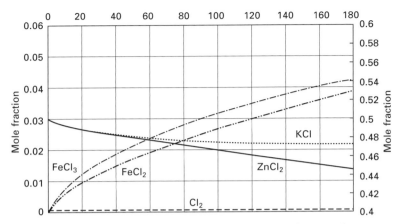

29.9 The changes in the contents of salt species at the metal interface with metal salt contact time from 0 to 180 min at 320°C, simulation using CorrApp tool.

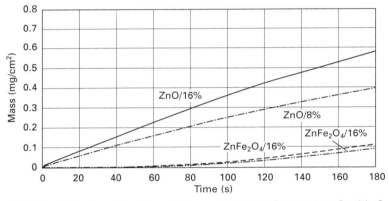

29.10 Precipitation of oxides at the gas/salt interface at 320°C with O_2 contents of 8 and 16 mol.% with time from 0 to 180 min).

required after each step in the finite-difference lattice, computation times for two-dimensional corrosion processes were unacceptably long.

Adaptation of the model to oxidation of low alloy steels

The model is capable of simulating multi-phase internal corrosion processes that are governed by solid-state diffusion in the bulk metal. Oxidation experiments in laboratory air and $He-O_2-H_2O$ mixtures revealed that internal corrosion of low-alloy steels occurs along grain boundaries as part of the inward oxide growth process. Since the low-alloy boiler steels contain small amounts of Cr, the phase composition of the inner layer exhibits a gradual transition of the $(Fe,Cr)_3O_4$ phase to islands of the spinel phase, $FeCr_2O_4$ and Cr_2O_3. To establish a suitable simulation procedure for the inward oxide growth, time-dependent boundary conditions and a subdivision of diffusion into grain boundary and bulk diffusion in two coupled finite difference lattices had to be introduced.

Kinetics calculation

Fick's second law (Eq. 29.1) was solved by means of the numerical finite-difference technique using the implicit Crank–Nicolson scheme, which is schematically represented in Fig. 29.11. It was assumed that the diffusion coefficients in the x- and y-directions, respectively, are independent of the local concentration and are only affected by temperature.

$$\frac{\partial C}{\partial t} = D_x \frac{\partial^2 C}{\partial x^2} + D_y \frac{\partial^2 C}{\partial y^2} \qquad\qquad 29.1$$

The basic idea of the finite differences method for solving partial differential equations is to replace spatial and time derivatives by suitable approximations, then to solve numerically the resulting difference equations. In other words, instead of solving for $C(x,t)$ with x and t, it is solved for $C_{i,j} \equiv C(x_i, t_j)$, where $x_i \equiv i\Delta x$, $t_j \equiv j\Delta t$. Thus, the concentrations, C_i^{j+1}, of the diffusing species for the location step, i, and the time step, $j+1$, are calculated from the neighbouring concentrations according to the implicit Crank–Nicolson solution for the diffusion differential equation (29.2):

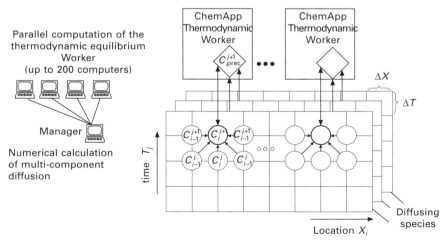

29.11 Schematic diagram of the Crank–Nicholson algorithm to solve the diffusion differential equation in combination with the ChemApp program.

$$C_i^{j+1} = C_i^j + \frac{\Delta T}{2\Delta X^2} [(C_{i+1}^{j+1} + C_{i+1}^j) - 2(C_i^{j+1} + C_i^j) + (C_{i-1}^{j+1} + C_{i-1}^j)]$$

29.2

The resulting new concentrations for the following time step, $j + 1$, are then corrected according to the thermodynamic equilibrium. The sequential calls for the ChemApp subroutines have been replaced by parallel computation, i.e. the Matlab main program (manager) operates the equilibrium calculation (thermodynamic worker) on several processors at the same time. This method allows a reduction in the computation time by a factor of more than 100.

In order to investigate the effect of grain size on the kinetics of oxidation, the finite difference technique was used to perform the diffusion calculation in such a way that it distinguishes between grain boundary diffusion and bulk diffusion (Fig. 29.12). The diffusion coefficient along grain boundaries was assumed to be 100 times higher than that in the bulk. Within the grains, the diffusion coefficient was presumed to be isotropic. Thus, it was possible to simulate intergranular oxidation as it was experimentally observed in low-alloy steels. The amount of the species depends on the initial chromium content in the steel. The diffusion coefficient for oxygen in magnetite is available from the literature only for very high temperatures.

Figure 29.13 shows a comparison of the experimental results obtained at 550°C after 72 h for a low-alloyed steel of two different grain sizes with the results of the corresponding computer simulations (only the inner oxide layer is shown). For the simulation, a mesh of 36 grains was used. There is a quantitative agreement between experiment and simulation. Increasing the alloy grain size leads to decreased inner oxide layer thickness, because the lower density of grain boundaries decreases the supply of oxygen to the oxide scale/substrate interface.

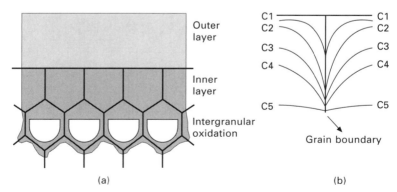

29.12 Diffusion mechanism for the simulation of intergranular corrosion; (a) model of the complete oxide scale with oxidised grain boundaries; (b) differentiation between bulk and grain boundary diffusion.

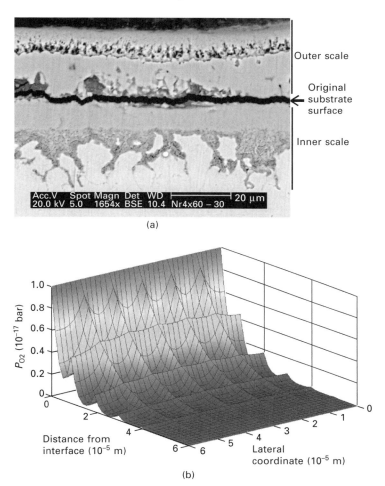

(a)

(b)

29.13 Comparison of experimental and simulated results for a low-alloyed steel oxidised at 550°C for 72 h in laboratory air.

Figure 29.14 shows simulation results for the inward oxidation of a low-alloy steel with 2.25 wt.% Cr as the alloying element at 550°C. The oxygen concentration at the outer/inner layer interface is determined by the thermodynamically stable formation of the Cr-containing spinel phase ($FeCr_2O_4$). The concentration profiles reflect the experimental results for the inward oxidation of this steel. The higher thermodynamic stability of Cr_2O_3 enables its easy formation even though the oxygen content is rather low, say with oxygen dissolved in Fe; therefore, its first appearance at the reaction front is understandable. With the depletion in Cr, Fe will take part in the reaction process, when the oxygen concentration increases to the necessary level, and combine with Cr_2O_3 to form spinel phase ($FeCr_2O_4$). As the oxygen

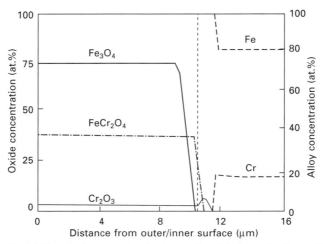

29.14 Simulation results for inward oxidation of a low-alloy steel with Cr content of 2.25wt.% at 550°C after 20 h exposure; oxide phase profiles (in at.%) and the metallic elements of a coarse-grained specimen (d = 65 μm) [14].

potential increases, part of $FeCr_2O_3$ will be further oxidised to form Fe_3O_4. The depth of the Fe_3O_4 precipitation front is responsible for the thickness of the inner scale; it was found that after 20 h exposure, the thickness of the inward oxidation layer is about 8 μm, which is in very good agreement with the experimental results [7].

29.4   Summary

The demands for more efficient and environmentally-friendly energy production will bring new challenges for boiler manufacturers and plant maintenance: bio-based fuels are aggressive and the traditional plant life assessment tools are not valid for them. Therefore, also, corrosion-based degradation mechanisms should be elucidated and modelled for the targeted service temperatures and combustion gas types.

This study presents a novel approach to modelling high-temperature oxidation and corrosion for boiler components using thermodynamic and diffusion kinetic data and libraries. The first step is to tailor stability diagrams for the given steel or deposit compositions over the ambient temperature range and for atmospheric conditions. The second development presented is a modification of ChemSheet, CorrApp tool, for calculating thermochemical processes. The modification enables use of the ChemSheet routines and interface templates for kinetic calculations combined with thermodynamic stability calculations. A salt-induced corrosion case is presented as a case that shows a promising correlation to the corresponding experimental work.

The third alternative presented is the development of the InCorr tool for internal and inward corrosion, based on finite element calculations. This method is based on thermodynamic stability calculations with solid state diffusion algorithms. The model examples describe high-temperature oxidation kinetics for a 2.25Cr steel and the results are in good agreement with the corresponding experimental studies and analyses.

29.5   Acknowledgements

The work has been supported by the EU project funding under the contract G5RD-CT-2001-00593. The authors wish to thank Dr A. Kodentsov from Technical University of Eindhoven, The Netherlands, for the fruitful comments on the use of diffusion theories and models for the starting phase of this modelling work.

29.6   References

1. L. Heikinheimo, K. Penttilä, M. Hämäläinen, U. Krupp, V. Trindade, M. Spiegel, A. Ruh and K. Hack. High temperature oxidation and corrosion modelling using thermodynamic and experimental data. *BALTICA VI Life Management and Maintenance for Power Plants*, Vol. 2. VTT Symposium 234, Espoo (2004), p. 537–551.
2. C.W. Bale *et al. CALPHAD* 26(2) (2002) pp. 189–228
3. S. Tuurna, L. Heikinheimo, M. Arponen and M. Hämäläinen. Oxidation kinetics of low alloyed ferritic steels in a moist atmosphere. *EUROCORR 2003*, 28 Sept.–Oct. 2003, Budapest, Hungary.
4. A. Ruh and M. Spiegel. Kinetic investigations on salt melt induced high-temperature corrosion of pure iron. *EUROCORR 2003*, 28 Sept.–Oct. 2003, Budapest, Hungary.
5. M. Spiegel, A. Zahs and H. J. Grabke. Fundamental aspects of chlorine induced corrosion in power plants. *Materials at High Temperatures*, 20(2) (2003) pp. 153–159.
6. S. Sroda, S. Tuurna, K. Penttilä and L. Heikinheimo: High temperature oxidation behaviour of boiler steels under simulated combustion gases. 6th International Symposium on High Temperature Corrosion and Protection of Materials, 16–21 May 2004, Les Embiez, France. Trans Tech Publications Ltd., *Materials Science Forum*, 461–464 (2004) pp. 981–988.
7. U. Krupp, V. B. Trindade, B. Z. Hanjari, S. Yang, H.-J. Christ, U. Buschmann and W. Wiechert. Experimental analysis and computer simulation on inward oxide growth during high-temperature corrosion of low-alloy boiler steels. 6th International Symposium on High Temperature Corrosion and Protection of Materials, 16–21 May 2004, Les Embiez, France. Trans Tech Publications Ltd., Materials Science Forum, 461–464 (2004) pp. 571–578.
8. M. Spiegel *Molten Salt Forum*, 7 (2003), pp 253–268.
9. A. Ruh and M. Spiegel. Kinetic investigations on salt melt induced high-temperature corrosion of pure metals. 6th International Symposium on High Temperature Corrosion and Protection of Materials, 16–21 May 2004, Les Embiez, France. Trans Tech Publications Ltd., *Materials Science Forum*, 461–464 (2004) pp. 61–68.

10. U. Krupp and H.-J. Christ. *Oxid. Met.*, 52 (1999) p. 299.
11. L. Heikinheimo, K. Hack, D. Baxter, M. Spiegel, U. Krupp, M. Hämäläinen and M. Arponen. Optimisation of in-service performance of boiler steels by modelling high temperature corrosion – EU FP5 OPTICORR project. *Materials Science Forum*, 461–464(2004) pp. 473–480.
12. K. Penttilä. Model for chloride melt induced corrosion of pure iron. Presented in EFC Workshop, Dechema 27–29 October 2004, Frankfurt, Germany.
13. K. Hack, S. Petersen, P. Koukkari and K. Penttila. CHEMSHEET – an Efficient Worksheet Tool for Thermodynamic Process Simulation. In: Y. Brechet (editor), *EUROMAT 99 – Volume 3*, Wiley-VCH Publishers, Weinheim, 2000, pp. 323–330.
14. U. Krupp, V. B. Trindade, P. Schmidt, H.-J. Christ, U. Buschmann and W. Wiechert. Computer-based Simulation of inward oxide scale growth on Cr-containing steels of high temperatures (OPTICORR), Chapter 32 this volume.

30

Influence of gas phase composition on the kinetics of chloride melt induced corrosion of pure iron (OPTICORR)

A R U H and M S P I E G E L,
Max-Planck-Institut für Eisenforschung GmbH, Germany

30.1   Introduction

The operation of waste incinerators is affected by the formation of corrosive gases and aerosols during the combustion process. This leads to severe corrosion of the metal compounds, e.g. heat exchanger tubes and water walls. As especially water walls have surface temperatures around 350°C, salt deposits are formed by condensation of aerosols within the flue gas on the metal surface. Heavy metal compounds like Pb and Zn cause the formation of eutectic salt mixtures with low melting points [1]. If salt deposits are molten, corrosion is accelerated in comparison to solid deposits. Such type of corrosion is known as 'hot corrosion'.

Corrosion phenomena beneath molten salts were initially investigated for sulphate systems [e.g. 2]. A review of fundamental work on sulphate melt-induced corrosion is given by Rapp [3]. Increased corrosion is excited by dissolution of the passivating oxide layer in the salt melt. Two major dissolution mechanisms have to be considered: basic dissolution and acidic dissolution. Basic dissolution is caused by a basic melt, meaning a high activity of O^{2-}, for example in the form of Na_2O. By reaction with the oxide, this leads to the formation of a complex oxide ion (e.g. FeO_2^{2-}) according to Eq. 30.1.

$$Fe_2O_3 + O^{2-} = 2Fe\,O_2^{-} \qquad\qquad 30.1$$

In the case of acidic dissolution, the activity of O^{2-} is low compared to the activity of SO_3 and, therefore, the metal oxide reacts becoming a metal ion (Eq. 30.2).

$$Fe_2O_3 = 2Fe^{3+} + 3O^{2-} \qquad\qquad 30.2$$

Aggravated corrosion was also observed on metals and alloys covered by molten chloride deposits. A solubility study on protective oxide films in molten chlorides at 727°C [4] has shown that its behaviour is similar to the observed dissolution of metal oxides in sulphate melt. Metal oxides can be dissolved in molten NaCl-KCl also by acidic and basic dissolution.

The effect of solid chlorides like KCl, NaCl, $MgCl_2$ and $CaCl_2$ on corrosion of alloy steels has been investigated by Reese and Grabke [5, 6]. Their experiments in $He-O_2$ and $He-O_2-SO_2$ atmospheres at 500 and 700°C have shown that the salt reacts with the oxide scale of the pre-oxidised sample forming ferrate or chromate and releasing chlorine. The chlorine diffuses through cracks and pores of the oxide scale to the metal-scale interface and reacts to solid $FeCl_2$. The vapour pressure of $FeCl_2$ reaches 10^{-4} bar at 500°C and the volatile chloride diffuses outwards through the oxide scale. At the oxide-gas atmosphere interface $FeCl_2$ reacts to Fe_2O_3, releasing further chlorine. The growth of Fe_2O_3 in cracks and pores of the oxide destroys the scale and corrosive gas can react with the unprotected metal. Thus a non-passivating scale on the metal substrate is formed and for this reason, the mechanism was termed 'active oxidation' [7]. As chlorine is not consumed in this process, it plays a catalytic role. Reese and Grabke [5] have shown by thermogravimetric studies that evaporation of $FeCl_2$ from the metal-scale interface is the rate determining step in 'active oxidation'.

Spiegel [8] reported increased mass gain in TG experiments on 2.25Cr-1Mo covered by pure $PbCl_2$ and $ZnCl_2$ at 500°C and 600°C in oxidising atmospheres in comparison with pure oxidation without deposit. At these temperatures both chlorides are molten. Using a eutectic chloride mixture like $KCl-ZnCl_2$ enhanced mass gain was also reported for lower temperatures (350–400°C). Preliminary studies on the kinetics of $KCl/ZnCl_2$ induced corrosion on pure Fe at 320°C in $Ar-O_2$ atmospheres have been carried out by Ruh and Spiegel [9]. They confirm accelerated corrosion and enhanced mass gains. The presence of three kinetic steps (incubation phase, linear stage and logarithmic/parabolic stage) has been reported in that study as well. Furthermore, it was demonstrated that the incubation time decreases with increasing oxygen concentration in the atmosphere.

Later investigations focussed on the influence of HCl and water vapour on the corrosion kinetics of Fe beneath molten $ZnCl_2/KCl$ [10]. The incubation time decreases when HCl is added to the atmosphere. For higher HCl concentrations it disappears completely, so that the corrosion starts with a linear stage. The mass gain observed after longer reaction times is clearly higher than in atmospheres without HCl. While HCl leads to increased corrosion rate water vapour did not show a steady increase of the corrosion rate. For longer reaction times the mass gain is even lower than in $Ar-O_2$ without water vapour. But the incubation time is clearly reduced.

This work is part of the EU FP5 OPTICORR project ('Optimisation of in-service performance of boiler steels by modelling high temperature corrosion'), which aims at the control and optimisation of in-service performance of boiler materials and the development of simulation tools for high temperature corrosion and oxidation of boiler steels under operating conditions. An overview of the whole project work is given by Heikinheimo et al. [11]. Parts of the thermogravimetric experiments are used for modelling of salt melt induced high temperature corrosion [12]. In particular, this work focuses on thermogravimetric experiments on chloride melt-induced corrosion of pure iron and the influence of gas phase composition on corrosion kinetics. It ties up to earlier investigations on this topic [9, 10]

30.2   Experimental methods

Thermogravimetric experiments have been carried out at 320°C on polished specimens of pure iron, covered by a 50 mol.% $ZnCl_2$–50 mol.% KCl salt deposit in Ar–O_2 and Ar–O_2–HCl atmospheres. The oxygen amount in the gas atmosphere has been varied from 4 to 16 vol.%. In the experiments with HCl, its concentration was fixed at 1000 vppm. Results from experiments in Ar–O_2, Ar–O_2–HCl and Ar–O_2–H_2O are discussed also. The corrosion products have been investigated by SEM and XRD.

30.3   Results

30.3.1 Effect of oxygen

Experiments on Fe in Ar–O_2 atmospheres generally yield TG curve as shown in Fig. 30.1 [1]. Three kinetic regimes are obvious: an incubation stage (A), a linear stage (B) and a stage that is either logarithmic or parabolic (C).

The effect of O_2 concentration in the gas atmosphere is illustrated by Fig. 30.2. The total mass gain after several hours of reaction increases slightly with increasing amount of oxygen. In Ar–16 vol.% O_2 a breakaway could be

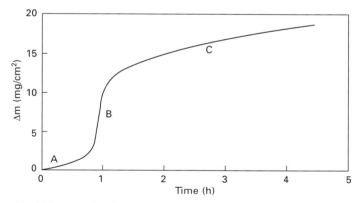

30.1 TG curve for high temperature corrosion of pure Fe covered with a 50 mol.% KCl–50 mol.% $ZnCl_2$ deposit in Ar–8 vol.% O_2 atmospheres at 320°C showing the three kinetic stages: incubation phase (A), linear stage (B) and logarithmic/parabolic stage (C). Taken from [9].

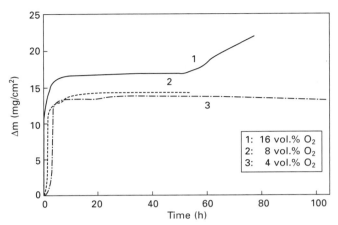

30.2 TG curves for high temperature corrosion of pure Fe covered with a 50 mol.% KCl–50 mol.% ZnCl$_2$ deposit in Ar–O$_2$ atmospheres at T = 320°C.

observed after about 53 h. The first 5 h of reaction have been extracted to illustrate the influence of O$_2$ on the incubation time (Fig. 30.3). The duration of the incubation phase strongly decreases with increasing oxygen concentrations. Thus the influence of O$_2$ on the corrosion kinetics is more pronounced at the beginning of the process and it slows down when corrosion takes place for a longer time. SEM investigations have been performed after the tests. They show scales that consist predominantly of iron oxide but also contain chloride (Fig. 30.4). The portions of chloride and oxide depend on the oxygen concentration in the atmosphere but also on the duration of the test.

30.3.2 Effect of water vapour

Thermogravimetric experiments on Fe beneath molten KCl-ZnCl$_2$ have been carried out by Ruh and Spiegel [10] in Ar–O$_2$–H$_2$O atmospheres. The results are illustrated in Fig. 30.5. After several hours of the reaction, the corrosion slows down and the mass gain is lower in comparison to tests in Ar–O$_2$ atmospheres without H$_2$O. Thus water vapour seems to reduce the corrosion rate. Figure 30.6 shows the corrosion kinetics of the first two hours. All three experiments started with an incubation phase but its duration is clearly reduced when water vapour is present in the reaction gas. A SEM micrograph of a sample after corrosion is shown in Fig. 30.7. On top of the scale, Fe-Zn oxide could be detected together with low incorporation of chloride. Iron oxide is the predominant phase in the middle zone, sometimes interspersed with low amounts of K-Zn chloride. Near the metal surface iron oxide could be observed together with iron chloride. A spinel phase (either Magnetite –

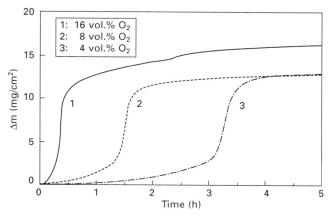

30.3 TG curves for high temperature corrosion of pure Fe covered with a 50 mol.% KCl–50 mol.% $ZnCl_2$ deposit in Ar–O_2 atmospheres at 320°C. The duration of the incubation phase decreases with increasing amount of oxygen.

30.4 Metallographic cross-section of pure iron after corrosion beneath a molten KCl-$ZnCl_2$ deposit in Ar–16 vol.% O_2 at 320°C for 77.5 h. The brighter layers consist predominantly of iron oxide. The darker areas also contain some chloride.

Fe_3O_4 or Franklinite – $ZnFe_2O_4$), ZnO and $K_2(ZnCl_4)$ have been found by XRD investigations. KCl could be detected as well. The presence of the Fe-Zn oxide on the top of the corrosion layer may well be the possible reason for the reduced mass gain after longer reaction times.

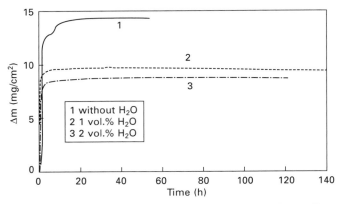

30.5 TG curves for high temperature corrosion of pure Fe covered with a 50 mol.% KCl–50 mol.% $ZnCl_2$ deposit in Ar–8% O_2–H_2O (0–2 vol.%) atmospheres at 320°C. Taken from [7].

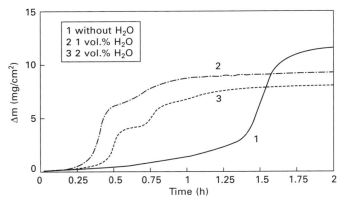

30.6 TG curves for high temperature corrosion ($T = 320$°C) of pure Fe covered with a 50 mol.% KCl–50 mol.% $ZnCl_2$ deposit in Ar–8% O_2–H_2O (0–2 vol.%) atmospheres at 320°C showing the first two hours of corrosion. Taken from [7].

30.3.3 Effect of hydrogen chloride

A first series of tests carried out in Ar–O_2–HCl atmospheres have been carried out at a fixed O_2 concentration and variable HCl concentrations [10]. These results are shown in Fig. 30.8. The addition of HCl to the reaction gas yields an increased mass gain. The most accelerated corrosion has been observed at 2000 vppm HCl. In this experiment a continuous mass loss occurs after about 55 h indicating evaporation of iron chloride, which was found as a condensate on the glass tube after the experiment. Figure 30.9 shows the mass gains occurring within the first two hours, where an incubation phase has been observed at a low HCl concentration (500 vppm) or in

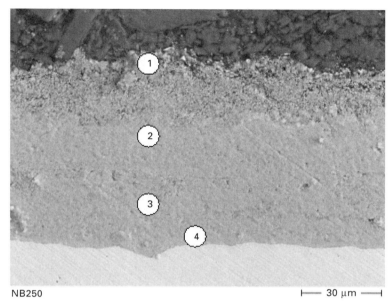

NB250   ⊢── 30 μm ──⊣

30.7 Metallographic cross-section of pure Fe covered by a molten KCl-ZnCl$_2$ salt mixture (15 mg/cm$_2$) after corrosion in Ar–8 vol.% O$_2$–1 vol.% H$_2$O showing a porous scale consisting of Fe-Zn oxide with low chloride incorporation (1), iron oxide (2), iron oxide interspersed with low amounts of K-Zn chloride (3), and iron oxide together with some iron chloride (4). Taken from [7].

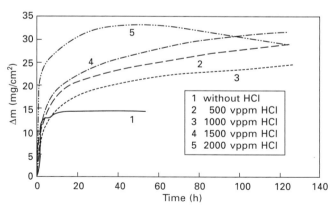

30.8 TG curves for high temperature corrosion of pure Fe covered with a 50 mol.% KCl–50 mol.% ZnCl$_2$ deposit in Ar–8% O$_2$–HCl (0–2000 vppm) atmospheres at 320°C. Taken from [7].

atmospheres without HCl. For higher concentrations of HCl the incubation phase disappears or can no longer easily be distinguished from the other kinetic stages. In these cases the linear stage is the first kinetic step when the

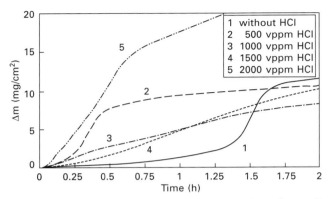

30.9 TG curves for high temperature corrosion of pure Fe covered with a 50 mol.% KCl–50 mol.% ZnCl$_2$ deposit in Ar–8% O$_2$–HCl (0–2000 vppm) atmospheres at 320°C showing the first two hours of corrosion. Taken from [7].

corrosion process starts but its slope is less steep than in the experiments with 0 and 500 vppm HCl.

In this study a series is presented at which the HCl concentration has been fixed at 1000 vppm and the amount of oxygen has been varied (Fig. 30.10). The most accelerated corrosion has been observed at 16 vol.% O$_2$. Similar to the test carried out in Ar–8 vol.% O$_2$–2000 vppm HCl, a mass loss appears after about 60 or 70 h due to the evaporation of iron chloride. Therefore the formation of volatile iron chloride is favoured in Ar–O$_2$–HCl atmospheres if either the HCl and/or O$_2$ concentration is increased. A possible reason for this may be the formation of iron chloride, subsequently evaporating. This effect has to be studied in further detail. Iron oxychloride can be excluded because it can be formed only at temperatures higher than 320°C. The stability of FeOCl in the temperature range of 325–350°C is restricted at p(Cl$_2$) > 0.0186 bar and 4.68×10^{-11} bar < p(O$_2$) < 10^{-7} bar, depending on p(Cl$_2$) at 330°C.

Figure 30.11 shows the corrosion kinetics of the same experiments within the first two hours. It clarifies that the corrosion rate increases strongly with increasing oxygen partial pressure. But after longer reaction times (Fig. 30.10) the mass gains recorded for different oxygen concentrations do not show a clear tendency. Thus the influence of oxygen on the corrosion kinetics is only distinctive at the beginning of the corrosion process.

After the tests the samples were investigated by using scanning electron microscopy (SEM). SEM micrographs of the surface of a corroded sample and a subjacent layer are shown in Fig. 30.12 and 30.13.

A SEM micrograph has also been taken from a metallographic cross-section of that sample (Fig. 30.14). It shows the corrosion scale that consists

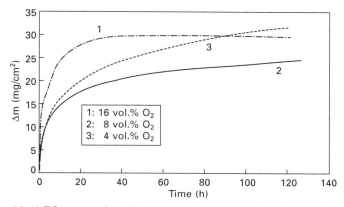

30.10 TG curves for high temperature corrosion of pure Fe covered with a 50 mol.% KCl–50 mol.% ZnCl$_2$ deposit in Ar–O$_2$ (4–16 vol.%) –1000 vppm HCl atmospheres at 320°C.

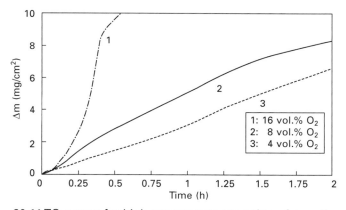

30.11 TG curves for high temperature corrosion of pure Fe covered with a 50 mol.% KCl–50 mol.% ZnCl$_2$ deposit in Ar–O$_2$ (4–16 vol.%)– 1000 vppm HCl atmospheres at 320°C showing the first two hours of corrosion.

of iron oxide and chloride. Iron oxide predominantly occurs at the upper part while iron chloride is predominantly present at the lower part of the scale.

30.4   Discussion

A model for hot corrosion beneath molten chloride is shown in Fig. 30.15 [8, 13]. In that model corrosion starts with the dissolution of Fe at the metal/salt melt interface and the formation of iron chloride. If the gas phase contains no chlorine species at the beginning, the chlorine source must be the molten chlorides. In this case chlorine can be released by the oxidation of the chloride.

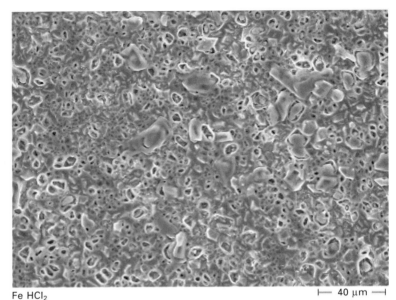

Fe HCl$_2$ ⊢— 40 µm —⊣

30.12 SEM micrograph of the surface of pure iron after corrosion beneath a molten KCl-ZnCl$_2$ deposit in Ar–8 vol.% O$_2$–1000 vppm HCl at 320°C for 127.7 h showing a molten K-Zn-Fe chloride. A small quantity of oxide is incorporated in the melt.

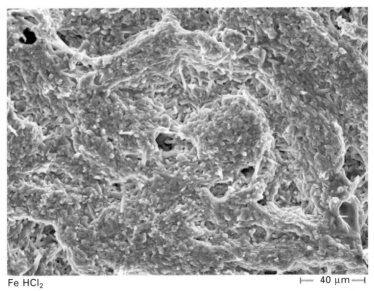

Fe HCl$_2$ ⊢— 40 µm —⊣

30.13 SEM micrograph of a dissected lower part of the scale of pure iron after corrosion beneath a molten KCl-ZnCl$_2$ deposit in Ar–8 vol.% O$_2$–1000 vppm HCl at 320°C for 127.7 h. In this zone iron chloride has formed. Qualitative EDX measurements show low concentrations of O and K.

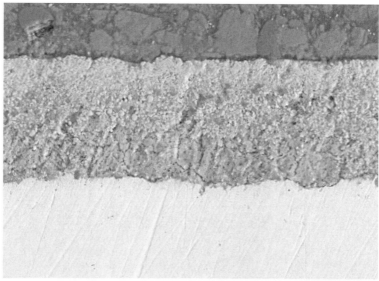

NB 144   ⊢— 40 μm —⊣

30.14 Metallographic cross-section of pure iron after corrosion beneath a molten KCl-ZnCl$_2$ deposit in Ar–8 vol.% O$_2$–1000 vppm HCl at 320°C for 127.7 h. The upper part consists of iron oxide and a low amount of chloride. The lower part predominantly consists of iron chloride.

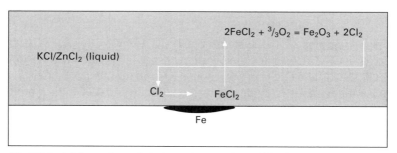

KCl/ZnCl$_2$ (liquid)

$$2FeCl_2 + {}^3/_3O_2 = Fe_2O_3 + 2Cl_2$$

Cl$_2$ → FeCl$_2$

Fe

30.15 Model for salt melt induced corrosion of Fe [8, 13], showing the reaction mechanism.

The iron chloride produced can diffuse outwards to the salt melt/gas atmosphere interface. At the outer part of the salt melt layer a higher oxygen partial pressure is present, which allows the oxidation of FeCl$_2$ to Fe$_2$O$_3$ and Cl$_2$ due to its favoured thermodynamic stability.

The recreated chlorine diffuses back to the metal/salt melt interface and allows a subsequent formation of iron chloride. Chlorine is not consumed and thus this process is catalysed by chlorine. Figures 30.16–30.18 give the

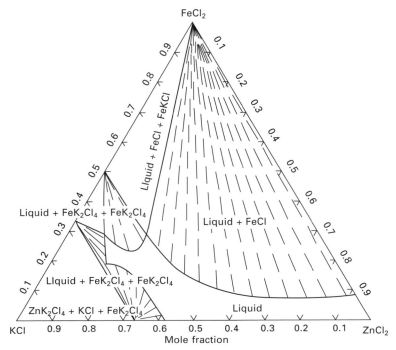

30.16 Phase diagram of the system FeCl$_2$–KCl–ZnCl$_2$ at T = 320°C. It shows that about 10 to 15 mol.% FeCl$_2$ can be dissolved in a 50 mol.% KCl/50 mol.% ZnCl$_2$ melt. Calculated with FactSage [14].

phase diagrams of the systems FeCl$_2$–KCl–ZnCl$_2$, NiCl$_2$–KCl–ZnCl$_2$ and CrCl$_3$–KCl–ZnCl$_2$ (calculated with FactSage [14]), showing the solubility of FeCl$_2$, NiCl$_2$ and CrCl$_3$ in a 50 mol.% KCl/50 mol.% ZnCl$_2$ melt. From these phase diagrams it is evident that the solubility of FeCl$_2$ in a molten KCl–ZnCl$_2$ is higher than the solubilities of NiCl$_2$, and CrCl$_3$ in molten KCl–ZnCl$_2$ with the same composition. The different solubilities explain the higher corrosion rate of Fe and the lower corrosion rates of Cr and Ni.

In oxidising atmospheres without any chlorine species in the gas atmosphere the only possible chlorine source is the chloride deposit. Cl$_2$ can be generated by oxidising one of the chloride components (Eqs 30.3 and 30.4).

$$ZnCl_2 + {}^1\!/_2O_2 = ZnO + Cl_2 \qquad\qquad 30.3$$

$$2KCl + {}^1\!/_2O_2 = K_2O + Cl_2 \qquad\qquad 30.4$$

Thermodynamic calculations show that Eq. 30.3 provides a chlorine partial pressure of about 2×10^{-4} at 320°C and 1 bar. On the other side, the oxidation of KCl to K$_2$O is only feasible from a thermodynamic point of view at very low chlorine partial pressure. But the chlorine partial pressure established by

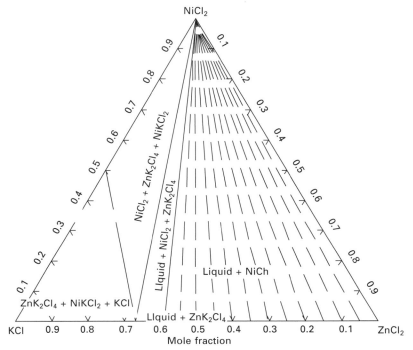

30.17 Phase diagram of the system $NiCl_2$–KCl–$ZnCl_2$ at $T = 320°C$. Almost no $NiCl_2$ can be dissolved in a 50 mol.% KCl/50 mol.% $ZnCl_2$ melt. Calculated with FactSage [14].

the oxidation of $ZnCl_2$ to ZnO is high enough to prevent KCl from oxidation, forming $ZnCl_2$ the only Cl_2 source if there is no other phase containing chlorine beside the KCl–$ZnCl_2$ deposit.

In HCl-containing atmospheres, additional Cl_2 can be provided by oxidation of HCl (Deacon reaction, Eq. 30.5).

$$2HCl + {}^1/_2O_2 = H_2O + Cl_2 \qquad\qquad 30.5$$

After Abels and Strehblow [15], the Cl_2 is the more reactive species in chlorine-induced corrosion rather than HCl. For this reason, Cl_2 is regarded as active gas component.

The chlorine partial pressures established by the oxidation of HCl (Eq. 30.5) at different HCl and O_2 concentrations is shown in Fig. 30.19. It demonstrates from a thermodynamic point of view that the amount of HCl has a greater influence on the chlorine formation than the amount of oxygen in HCl-containing atmospheres. However the TG curves do not yield comprehensible kinetic relationships. At the beginning of corrosion the corrosion rate increases strongly with increasing amount of oxygen at fixed HCl concentration (Fig. 30.11). For higher exposure times and for series

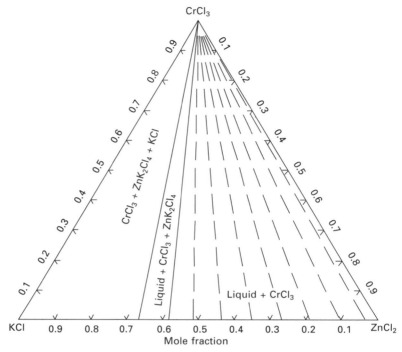

30.18 Phase diagram of the system CrCl$_3$–KCl–ZnCl$_2$ at $T = 320°$C. Almost no CrCl$_3$ can be dissolved in a 50 mol.% KCl/50 mol.% ZnCl$_2$ melt. Calculated with FactSage [14].

where the HCl concentrations have been varied at fixed O$_2$ concentrations (Figs 30.8 and 30.9), there is no plausible kinetic relationship. However, it was shown by thermodynamic calculations and experiments that corrosion is aggravated when HCl is present in the atmosphere.

The corrosion behaviour observed in H$_2$O-containing atmospheres has been discussed by Ruh and Spiegel [10]. The corrosion kinetics is somewhat retarded; compared to Ar–O$_2$ atmospheres even a shortened incubation phase is evident. But the corrosion reaction slows down quite early so that the total mass gain yielded in Ar–O$_2$–H$_2$O is lower than in Ar–O$_2$ and Ar–O$_2$–HCl atmospheres. A possible reason for this is the formation of the Fe–Zn oxide on the top of the corrosion layer, which contributes to a better corrosion resistance. In addition, another possibility should be considered. Eq. 30.5 describes the formation of water vapour and chlorine from hydrogen chloride and oxygen in the gas atmosphere by the Deacon reaction. But on the other side, water vapour in the atmosphere and chlorine generated by the oxidation of ZnCl$_2$ (Eq. 30.3) might be able to react forming HCl and O$_2$. Thus the more reactive Cl$_2$ will be consumed and the less reactive HCl will be produced. This may explain the decelerated corrosion rate after longer reaction times.

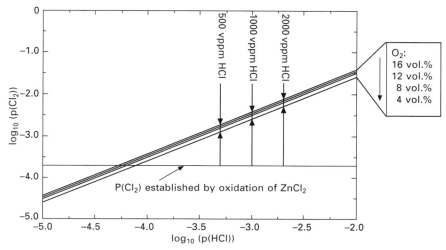

30.19 Chlorine partial pressures established by the oxidation of HCl at different $p(O_2)$ (diagonal lines) and $T = 320°C$. The arrows show the partial pressures of HCl for different experimental conditions and the horizontal line the ($p(Cl_2)$) established by the oxidation of $ZnCl_2$ in an Ar–8 vol.% O_2 atmosphere without HCl. Calculated with FactSage [14].

Thermodynamic calculations support this assumption (Fig. 30.20). The chlorine partial pressure, produced by the oxidation of $ZnCl_2$ (Eq. 30.3), decreases with increasing amount of water vapour.

30.5   Conclusions

Iron samples covered by a molten $ZnCl_2/KCl$ mixture suffer enhanced corrosion at 320°C in Ar–O_2 atmospheres. The corrosion kinetics can be described by three kinetic steps: incubation phase, linear stage and a logarithmic/parabolic stage. O_2 influences the duration of the incubation phase, which decreases with increasing amount of oxygen, but it has a relatively low influence on the total mass gain after longer reaction times.

Corrosion tests in Ar–O_2–H_2O atmospheres show reduced incubation times in comparison to tests in Ar–O_2 without H_2O, but the mass gain after a longer reaction time is comparatively low so that the total mass gain decreases with increasing amount of water vapour after some hours of corrosion. Either the formation of Fe–Zn oxide on the top of the scale or the influence of water vapour on the Deacon reaction, and thus the consumption of chlorine might be possible reasons for this behaviour.

In comparison to corrosion in Ar–O_2 and Ar–O_2–H_2O atmospheres we observed accelerated corrosion and higher mass gains in Ar–O_2–HCl

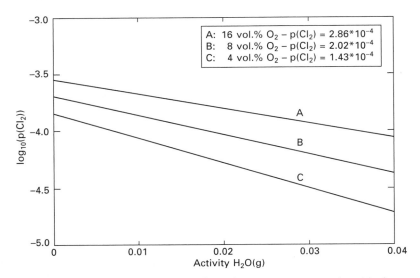

30.20 Influence of the concentration of water vapour on the chlorine partial pressure established by the oxidation of $ZnCl_2$ at different $p(O_2)$ and $T = 320°C$. The $p(Cl_2)$ established by the oxidation of $ZnCl_2$ in different $Ar-O_2$ atmospheres will be reduced with increasing H_2O concentration. Calculated with FactSage [14].

atmospheres caused by a higher Cl_2 supply due to the oxidation of HCl in the reaction gas. The higher $p(Cl_2)$ causes a higher corrosion rate, probably caused by a higher solubility of metal in the molten salt. Furthermore a mass loss could be observed after longer times of corrosion at high O_2 or HCl concentrations in the reaction gas, indicating evaporation of iron chloride.

30.6   References

1. G. Sorrell, *Materials at High Temperatures*, **1997**, *14*, 207.
2. P. Kofstad in *High Temperature Corrosion*, Elsevier Applied Science Publishers L., London, **1988**, 558.
3. R. A. Rapp, *Metallurgical and Materials Transactions A*, **2000**, *31A*, 2105.
4. T. Ishitsuka and K. Nose, *Materials and Corrosion*, **2000**, *51*, 177.
5. E. Reese and H. J. Grabke, *Materials and Corrosion*, **1992**, *43*, 547.
6. E. Reese and H. J. Grabke, *Materials and Corrosion*, **1993**, *44*, 41.
7. Y. Y. Lee and M. J. McNallan, *Metallurgical Transactions A*, **1987**, *18A*, 1099.
8. M. Spiegel, *Molten Salt Forum*, **2003**, *7*, 253.
9. A. Ruh and M. Spiegel, *EUROCORR 2003, Budapest, Hungary* (CD-ROM), **2003**.
10. A. Ruh and M. Spiegel, *EUROCORR 2004, Nice, France* (CD-ROM), **2004.**
11. L. Heikinheimo, K. Hack, D. Baxter, M. Spiegel, U. Krupp, M. Hämäläinen and M. Arponen, *Materials Science Forum*, **2004**, *461–464*, 473.
12. L. Heikinheimo, K. Penttilä, M. Hämäläinen, U. Krupp, V. Trindade, M. Spiegel, A. Ruh and K. Hack, *Baltica VI*, **2004**, 537.

13. A. Ruh and M. Spiegel, *Materials Science Forum,* **2004,** *461–464,* 61.

14. C.W. Bale, P. Chartrand, S.A. Degterov, G. Eriksson, K. Hack, R. Ben Mahfoud, J. Melançon, A.D. Pelton and S. Petersen, *Calphad,* **2002,** *26,* 2, 189.

15. J.-M. Abels and H.-H. Strehblow, *Corrosion Science,* **1997,** *39,* 115.

31

Development of toolboxes for the modelling of hot corrosion of heat exchanger components (OPTICORR)

K H A C K and T J A N T Z E N, GTT-Technologies, Germany

31.1   Introduction

In order to be able to reproduce the experimental results of the project OPTICORR, but also in order to provide a toolbox to thermochemists who wish to investigate the corrosion behaviour of heat exchanger materials used in combustion powerplants, one of the work packages of the project has been devoted to thermochemical modelling of the phase relationships and the processes leading to the formation of phases in and near the surfaces of heat exchanger materials under corrosive conditions. The general situation with respect to the interaction of materials is represented in Fig. 31.1.

On the one hand, there will be a complex gas phase comprising the major combustion products CO, CO_2 and H_2O with the respective O_2 partial pressure,

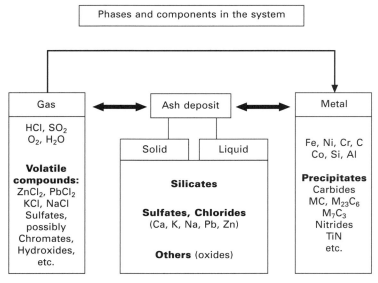

31.1 General view of the material relationships.

but also minor species such as HCl and SO_2, as well as the volatile chlorides (e.g. $ZnCl_2$, $PbCl_2$, KCl, NaCl), sulphates or hydroxides, to mention only the most important ones. On the other hand, there will be deposits of ashes such as silicates and oxides but also salts like chlorides, sulphates and others. Finally there is the metallic heat exchanger material which consists of more or less complex alloy matrices containing components such as Fe, Cr, Ni, Co, Al, Si and C and the respective phases that precipitate in smaller amounts from such alloy matrices, e.g. carbides or nitrides.

Two scenarios have been given special attention in this project: (1) the corrosion under a liquid salt layer, and (2) the internal corrosion by diffusion of oxygen into a bulk metal from a gas atmosphere.

Figure 31.2 gives a view of the situation of corrosion of the metal under a gas with a defined oxygen potential. Diffusion of oxygen into the material and diffusion of metallic components towards the surface interact with each other and lead to formation of an outside and an inside oxide layer. The inward flow of the oxygen is governed by both bulk and grain boundary diffusion in the alloy. Experimental results by Krupp *et al.* [1] show that the grain size on the metal has a strong influence on the propagation of the layer growth.

Figure 31.3 shows the sketch of the case of the corrosion under a salt layer [2]: it is assumed that the salt consisting of $ZnCl_2$ and KCl (as used in the experimental part of this project) provides a path for chlorine to attack Fe at the metal/molten salt interface. The resulting Fe-chloride is then transported to the surface where it reacts with the oxygen from the gas phase to form Fe_2O_3. The chlorine released through this reaction is then available to attack the iron in the metal/salt interface again. Experiments carried out in the

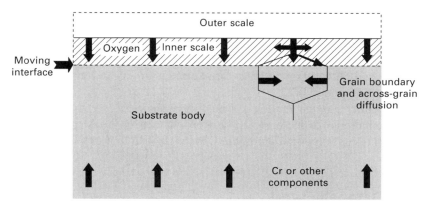

31.2 Schematic view of the gas phase corrosion process [1], including the two paths along the grain boundaries and across the grains of the metal.

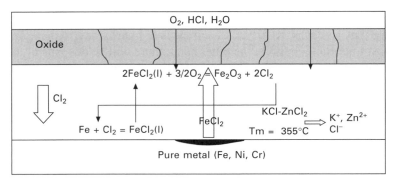

31.3 Schematic view of the salt layer corrosion process [2].

project by Ruh and Spiegel [2] confirm the layered occurrence of the salt and the oxide.

31.2   Available software tools at the onset of the project

The work carried out in the workpackage 'Toolboxes' of the OPTICORR project has been based on the availability of three software tools and two major databases. The software tools are: FactSage, the integrated thermodynamic databank system [3]; ChemApp, the programmer's library for thermochemical calculations [4]; and ChemSheet, the thermochemistry add-in [5] for Microsoft EXCEL™.

FactSage, with its modules to manipulate thermochemical databases and to calculate complex equilibrium states as well as multi-component phase diagrams, has been applied for the data compilation and assessment work described below as well as for the generation of the graphical results presented further below.

ChemApp has been employed in the framework of a parallel computing package in MatLab™ in order to handle the diffusion model that describes the inward and outward oxide layer formation under an oxygen atmosphere (see Chapter 32 and [1]).

An especially adapted version of ChemSheet has been used to execute the model calculations for the corrosion under a salt layer. In this model, too, diffusion plays an important part; here it is the diffusion of chlorine through the salt melt.

31.3   Database work

In systems presenting complex equilibria, not all the equilibria are, or even can be, determined experimentally. One of the advantages of calculated phase diagrams obtained by minimisation of the total Gibbs energy of the system is that certain equilibria are obtained as extrapolations of the analytical descriptions of the various phases into regions that are not known experimentally.

Four thermodynamic sub-databases have been created to enable the thermodynamic calculations for the OPTICORR project:

1. Database for Fe-based alloys (Al-C-Ce-Co-Cr-Cu-Fe-Mn-Mo-Nb-Ni-S-Si-V).
2. Database for Ni-based alloys (C-Cr-Fe-Mo-Ni-Si-Ti-W).
3. Salt database ($K^+,Zn^{2+},Fe^{2+},Fe^{3+},Cr^{2+},Cr^{3+},Ni^{2+}//Cl^-,(SO_4)^{2-},(CrO_4)^{2-}$)
4. Oxide, sulphate and sulphide database (MeS, MeO, $MeSO_4$, where Me = Al, C, Ce, Co, Cr, Cu, Fe, Mn, Mo, Nb, Ni, Si, Ti, V, W).

A significant number of the binary and ternary sub-systems of Fe- or Ni-based alloys have already been thermodynamically assessed. Many of these have been compiled in large databases, for example by SGTE [6]. Some new thermodynamic evaluations, for example for the Ce-Fe, Co-Cr-Fe, Co-Cr-Ni, and Al-Fe-Mn-Si systems, have been established in this work using all available experimental data. The calculated phase diagrams and the invariant reactions are in good agreement with experimental results. The Ce–Fe binary system displays two near-stoichiometric phases which form peritectically. Because of the small homogeneity ranges, they were modelled as stoichiometric phases using the formulae $CeFe_2$ and Ce_2Fe_{17} approximating the experimentally determined compositions. The experimentally determined phase diagram by Chuang et al. [7] is satisfactorily reproduced by the calculations as shown in Fig. 31.4. The calculated activity coefficients of Ce in molten Fe are presented in Fig. 31.5, where the comparison is made with the experimental data due to Teplizkij and Wladimirow [8] and Fisher and Janke [9]. The isothermal section in the Co–Cr–Fe system at 1373 K has been experimentally investigated by two different authors [10, 11]. Figure 31.6 shows the calculated isothermal section compared with these experimental data. The comparison between the calculations and experimental determined phase diagrams is good.

The database created especially for Ni-based alloys contains the thermodynamic data for all the phases significant in commercial alloys, such as carbides $M_{23}C_6$, M_3C_2, M_6C, M_7C_3, MC-eta, nitrides, silicides, Laves phases, P, R, mu and sigma phases.

The salt database comprises 21 components: $KCl–ZnCl_2–FeCl_2–FeCl_3–CrCl_2–CrCl_3–NiCl_2–K_2SO_4–ZnSO_4–FeSO_4–Fe_2(SO_4)_3–CrSO_4–Cr_2(SO_4)_3–NiSO_4–K_2CrO_4–ZnCrO_4–FeCrO_4–Fe_2(CrO_4)_3–CrCrO_4–Cr_2(CrO_4)_3–NiCrO_4$. The constituents $CrCrO_4$ and $Cr_2(CrO_4)_3$, which do not exist in nature, are required by the Gibbs energy model used. In order to make them unstable in all calculations, their Gibbs energy data have been given arbitrarily high values. This ensures that the model which needs full reciprocity between all

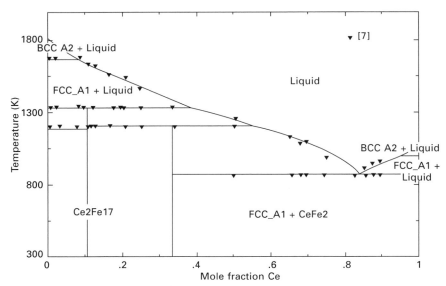

31.4 The assessed phase diagram Fe–Ce compared with experimental data [7].

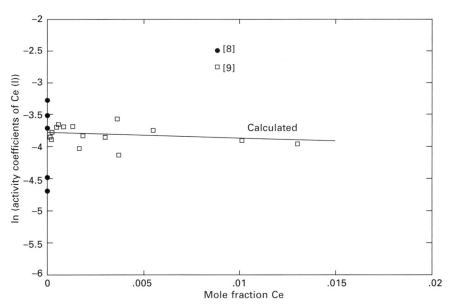

31.5 The calculated activity coefficients of Ce in molten Fe (experimental points from references [8] and [9]).

cations and anions will never give stable $CrCrO_4$ and $Cr_2(CrO_4)_3$ in the liquid phase. The thermodynamic data for the stoichiometric chlorides as well as some sulphates and chromates are taken from the FACT database

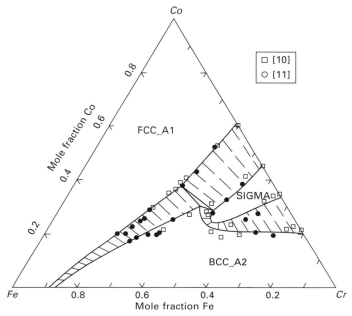

31.6 The isothermal section in the Co–Cr–Fe system at 1373 K (experimental data from references [10] and [11]).

[12] as far as available. However, there were also a number of thermodynamic datasets missing. These belong to sulphates and chromates for which work was carried out in this project. The relevant species are listed in Table 31.1. Le Van's method was used to estimate the enthalpy of formation of the compounds, Mill's method was employed to estimate the entropy of formation and the method by Kubaschewski and Ünal was used for the estimation of the heat capacity equation. The estimation methods were taken from *Materials Thermochemistry* 6th edn [13]. The binary systems $KCl–NiCl_2$, $KCl–ZnCl_2$, $KCl–K_2SO_4$, $K_2SO_4–K_2CrO_4$ were evaluated in this work using available experimental data.

The $KCl–NiCl_2$ system displays one near-stoichiometric compound $KCl.NiCl_2$ which melts peritectically at ~931 K. Figure 31.7 shows the calculated and experimentally determined phase diagrams ([14]); the agreement is good.

The system $KCl–ZnCl_2$ was investigated by different authors [15, 16] with the results being in good agreement. There are three near-stoichiometric compounds $2ZnCl_2.KCl$, $2ZnCl_2.3KCl$ and $ZnCl_2.2KCl$, two of them decompose congruently. The calculated phase diagram and the experimental data investigated by Duke and Fleming [15] are shown in Fig. 31.8, the agreement is satisfactory. The calculated activities of $ZnCl_2$ in the liquid phase for different amounts of $ZnCl_2$ are compared in Fig. 31.9 with

Table 31.1 List of relevant chromates and sulphates contained in Salt-database

Compound	Solid state	Liquid state
K_2SO_4	FACT	FACT
K_2CrO_4	FACT	FACT
$ZnSO_4$	FACT	Present work
$ZnCrO_4$	Present work	Present work
$NiSO_4$	FACT	Present work
$NiCrO_4$	Present work	Present work
$FeSO_4$	FACT	Present work
$FeCrO_4$	Present work	Present work
$Fe_2(SO_4)_3$	FACT	Present work
$Fe_2(CrO_4)_3$	Present work	Present work

FACT – the data are taken from the FACT database, Present work – the thermodynamic data of the compounds estimated using Le Van's method (enthalpy of formation), Mill's method (entropy of formation) and Kubaschewski and Ünal's method (heat capacity equation))

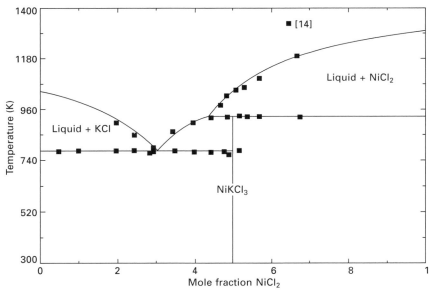

31.7 The calculated and experimentally determined [14] $KCl–NiCl_2$ phase diagrams.

experimental values carried out by Robertson and Kucharski [16]. Here too the agreement is satisfactory.

The oxide sub-database is created using the FACT [12] database and contains the thermodynamic data for all solid oxides, sulphates and sulphides of the elements contained in the databases for Fe-based and Ni-based alloys.

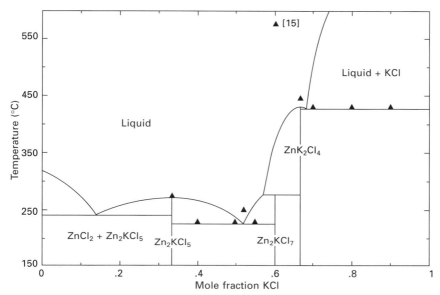

31.8 The calculated KCl–ZnCl$_2$ phase diagram compared with experimental data [15].

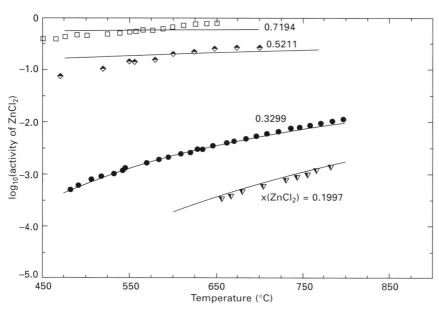

31.9 Calculated and experimental [16] activities in the KCl-ZnCl$_2$ system

31.4   Calculational results

Before complex thermodynamic models including explicit kinetics (here the diffusional transport of reactive elementary or molecular species) are employed, it is most useful to obtain, with the aid of classical thermochemical calculations, a picture of the momentary situation, i.e. of the frozen-in state at a certain moment in time. For that purpose the databank system FactSage provides two modules which are particularly suited: Equilib and Phase Diagram.

The Equilib module is meant for the calculation of so-called complex equilibria, i.e. equilibrium states of multi-component multi-phase systems comprising a gas phase, many non-ideal condensed solutions as well as many stoichiometric condensed phases. Such states prevail in the corrosion systems investigated here. In the calculation all parameters, e.g. composition, total pressure and temperature, except one are kept constant. Under variation of the one independent parameter the behaviour of the different phases, i.e. phase amounts and phase internal composition as well as species activities, can be investigated.

The Phase Diagram module permits the generation of phase diagrams of three basic types: (1) with two potential axes (e.g. T vs $P(O_2)$ for constant alloy composition, or $P(S_2)$ vs $P(O_2)$ for constant temperature and composition of the alloy components), (2) with one potential axis and one ratio of extensive properties (e.g. T vs X (mole fraction), or $P(O_2)$ vs X with all other variables constant), and (3) with two ratios of extensive properties on the axes (e.g. X_i (mole fraction of component i) vs X_j (mole fraction of component j) with all other variables constant).

As a short overview of the multitude of possible calculations, below are given several of the calculational results using the newly compiled and assessed databases of this project in the Equilib and Phase Diagram modules, respectively. Note that the results of the kinetic simulations, i.e. the time dependent gas and salt corrosion respectively, are given in Chapters 32 and 29.

31.4.1 One-dimensional mappings for gases and alloys

Two typical results are given here: one on the analysis of one of the corrosive atmospheres used in the experimental part of the project (15 vol.% H_2O, 8% O_2, 0.02% SO_2, rest argon), the other on the phase content of an X20 steel (10.3 wt.% Cr, 0.72% Ni , 0.87% Mo, 0.26% V, 0.18% C, 0.23% Si, 0.62% Mn, 0.003% S, rest Fe) which has also been investigated under corrosive conditions in the project.

Figure 31.10 shows the major gas components as entered into the corrosion experiment. The curves indicate that the gas has the same composition with

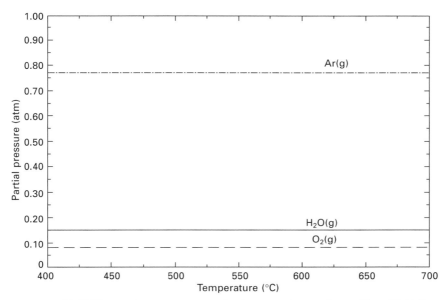

31.10 The major components in the experimental gas 15 vol% H_2O, 8% O_2 and 0.02%. SO_2, rest argon, temperature range 400 to 700°C and a total pressure 1 atm.

respect to Ar, H_2O and O_2 for all temperatures, i.e. the partial pressures of these are independent of temperature and agree with the nominal composition. However, in Fig. 31.11 it can be seen that SO_2 is not the highest among the minority components. Instead SO_3 has an almost constant partial pressure of about 1×10^{-4} atm and the pressure of SO_2 varies from 1×10^{-6}–1×10^{-4} atm over the temperature range from 400 to 700°C. In other words, the gas is not set for a fixed partial pressure of SO_2 but of SO_3.

Figure 31.12 shows the phase distribution in the X20 steel. At temperatures above approximately 460°C the alloy consists of a BCC_A2-matrix from which about 3.5 wt.% of $M_{23}C_6$ will precipitate together with a very small amount of MnS (because of the high affinity of Mn to S and according to the very small amount of S in the steel). At lower temperatures the present database predicts under equilibrium conditions the occurrence of a small amount of SIGMA phase stabilised not only by Cr but also by Mo and Ni (Fig. 31.13). This may in reality not be found because of kinetic inhibitions. Figure 31.14 shows the phase internal composition of the $M_{23}C_6$ precipitate, which contains mainly Cr as the metallic component but also Mo and Fe. The Fe content drops with decreasing temperature while the Cr content increases and Mo remains constant.

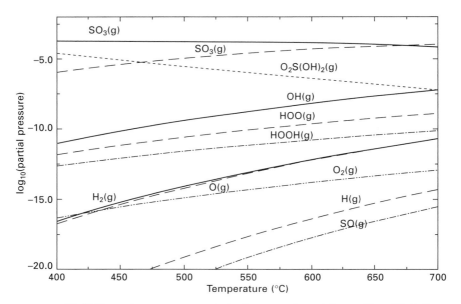

31.11 The minor components in the experimental gas 15 vol% H_2O, 8% O_2 and 0.02% SO_2, rest Argon, temperature range 400 to 700°C and a total pressure 1 atm

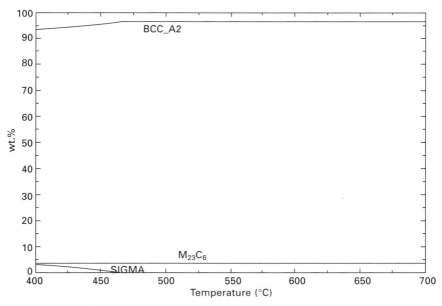

31.12 The matrix phase and the precipitate phases of the X20 steel in the range 400 to 700°C.

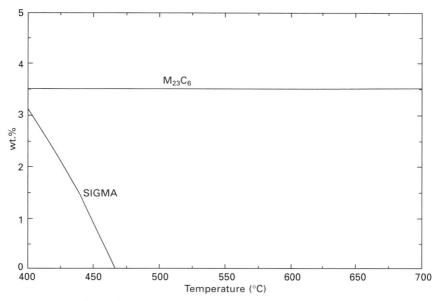

31.13 The minority phases of the X20 steel in the range 400 to 700°C.

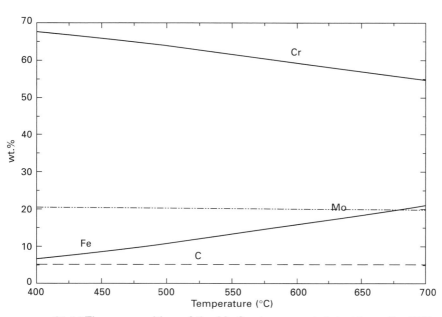

31.14 The composition of the $M_{23}C_6$ phase precipitated from the X20 steel.

31.4.2 Two-dimensional mappings (phase diagrams) for salts and alloys under corrosive atmospheres

As examples for phase diagrams obtained from the new databases, Type 3 diagrams (X_i vs X_j) for the ternary chloride salt systems with $FeCl_2$, $CrCl_2$, $NiCl_2$ and $KCl–ZnCl_2$ respectively are given. Furthermore, there are Type 1 diagrams (T vs log P_i and log P_i vs log P_j) for Fe–Cr and Ni–Mo alloys using combinations of oxygen, chlorine and SO_2 to define the gas phase.

Figures 31.15–31.17 show a series of isothermal ($T = 320°C$) salt systems which show for $FeCl_2$, $CrCl_2$ and $NiCl_2$, respectively, the trends of their solubility in a $KCl–ZnCl_2$ melt. It is obvious from the graphs that for $FeCl_2$ the solubility is fairly high (reaching about 12 at.% for a 50/50 $KCl–ZnCl_2$ melt), while $CrCl_2$ only shows a medium solubility of about 3 at.% and the $NiCl_2$ solubility is less than 1%. This trend is in good qualitative agreement with the salt layer corrosion experiments (see Chapter 30) which showed that samples of pure iron are far more attacked under a $KCl–ZnCl_2$ salt layer than those of pure Cr. Samples of pure Ni showed virtually no corrosion during the exposure times of the present experiments.

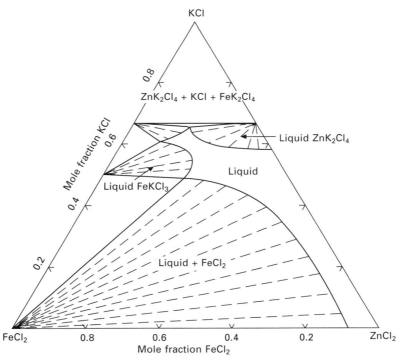

31.15 The isothermal section of the system $FeCl_2–KCl–ZnCl_2$ for 320°C.

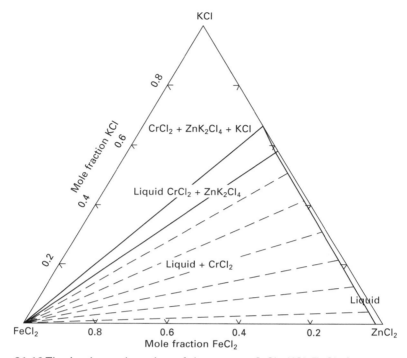

31.16 The isothermal section of the system $CrCl_2$–KCl–$ZnCl_2$ for 320°C.

Figure 31.18 shows a phase diagram in which for variable temperature and variable oxygen potential ($\log P(O_2)$) the phase fields are shown for an Fe–Cr alloy of 20 wt.% Cr. Two lines (lines 1 and 2) are noteworthy in this diagram. Line 1 depicts the outer line of stability of the pure metallic state. One can see that the ferritic state of the alloy changes to austenite because of the loss of Cr by formation of Me_2O_3 solid solution which is under these conditions almost pure Cr_2O_3. Line 2 depicts the outer line of existence of metal. Beyond that line, all metal will be consumed by the formation of oxide solid solutions. At lower temperatures and very low oxygen potentials (partial pressures) the primary phase will be Fe-Spinel, i.e. the solid solution between $(FeO)(Fe_2O_3)$ and $(FeO)(Cr_2O_3)$, replacing trivalent Fe by Cr. At higher temperatures and intermediate oxygen potentials there is additional formation of Wustite solid solution. It should be noted that for high oxygen potential the solid oxide is a solid solution of Fe_2O_3 and Cr_2O_3 with the initial 80/20 composition of the two metals.

Figure 31.19 shows for a given temperature (600°C) and a fixed alloy composition (20 wt.% Cr) the behaviour of the Fe–Cr alloy under variable $P(O_2)$ and $P(Cl_2)$. Interestingly, for low oxygen potentials the first phase to

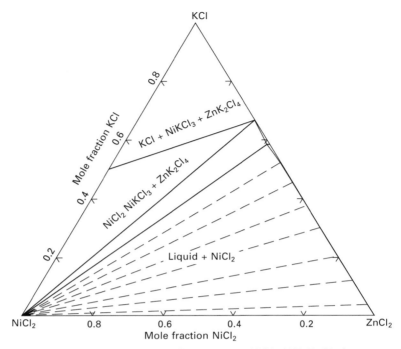

31.17 The isothermal section of the system $NiCl_2–KCl–ZnCl_2$ for 320°C.

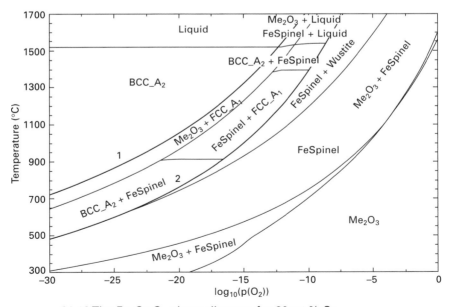

31.18 The Fe–Cr–O_2 phase diagram for 20 wt.% Cr.

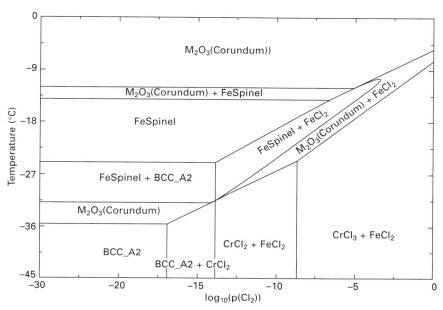

31.19 The Fe–Cr–O$_2$–Cl$_2$ phase diagram for T = 600°C and 20 wt.% Cr.

form from the alloy is CrCl$_2$, while for low chlorine potential Me$_2$O$_3$ (=Cr$_2$O$_3$) solid solution is formed. For high chlorine potentials the chlorides are fairly stable in comparison to the oxides and it is the Me$_2$O$_3$ solid solution which forms next.

Figure 31.20 shows for a given temperature (600°C) and a fixed alloy composition (10 wt.% Mo) the behaviour of the Ni–Mo alloys under variable P(O$_2$) and P(SO$_2$). In comparison to the Fe–Cr alloy under O$_2$/Cl$_2$ there is a remarkable difference here because the phase field for the pure FCC_A1 metal matrix is rather small. For very low oxygen potentials and low SO$_2$ potentials there is already formation of sulphide phases starting with MoS$_2$. When the metal matrix is completely consumed, Ni$_3$S$_2$ will co-exist with MoS$_2$. Increasing the SO$_2$ potential will finally lead to NiS$_2$ and MoS$_3$ co-existence. When increasing the oxygen potential for low SO$_2$ potentials again the Mo component in the alloy will be the first to react. MoO$_2$ is precipitated from the FCC_A1 matrix. Further increase of the oxygen potential will finally lead to a co-existence of NiO and MoO$_3$. If both O$_2$ and SO$_2$ have a high potential the formation of NiSO$_4$ together with MoO$_3$ occurs.

The above diagrams represent only a small number of several series of figures which have all be compiled into a 'GuideBook' which represents one of the products of the OPTICORR project (ISBN 951-38-6739-0).

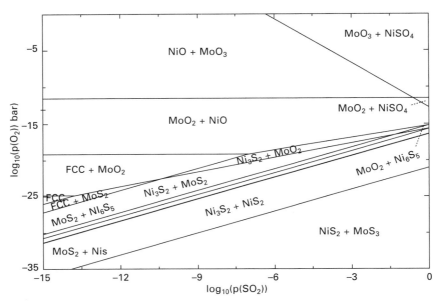

31.20 The Ni–Mo–O$_2$–SO$_2$ phase diagram for $T = 600°C$ and 10 wt.% Mo.

31.5   Summary

Special databases have been assembled dedicated to the field of corrosion of heat exchanger materials (steels and nickel-based alloys) under liquid salt layers and by direct gas diffusion through the bulk and along grain boundaries in the metal. These databases expand the scope of the standard databases for alloys (SGTE) and also for salts and oxides (FACT). The new databases have been successfully applied in the generation of classical thermodynamic one- and two-dimensional mappings but also in the kinetic models developed in the OPTICORR project.

The results are very well suited to understand the processes of corrosion leading to deterioration of heat exchanger materials under experimental conditions that simulate the real situation in power plants. However, for the description of the full situation in a power plant, the databases need further extension, especially with respect to the high complexity of real world salt deposits and their interaction with silicate deposits which will also form.

31.6   Acknowledgement

The authors wish to thank the Commission of the European Community for the funding of the project under contract No. G5RD-CT-2001-00593,

'Optimisation of In-Service Performance of Boiler Steels by Modelling High Temperature Corrosion'.

31.7   References

1. U. Krupp, V.B. Trindade, B.Z. Hanjari, H.-J. Christ, U. Buschmann, W. Wiechert, *Materials Science Forum*, 461–464 (2004), pp 571–578.
2. A. Ruh and M. Spiegel, *Materials Science Forum*, 461–464 (2004) pp 61–68.
3. C.W. Bale, P. Chartrand, S.A. Degterov, G. Eriksson, K. Hack, R. Ben Mahfoud, J. Melancon, A.D. Pelton and S. Petersen, *CALPHAD*, 26(2) (2002), pp 189–228.
4. G. Eriksson, K. Hack and S. Petersen Werkstoffwoche '96, Symposium 8, Simulation, Modellierung, Informationssysteme, J. Hirsch (ed.), DGM Informationsgesellschaft Verlag, 1997.
5. P. Koukkari, K. Pentillä, K. Hack and S. Petersen, Microstructure, Mechanical Properties and Processes Computer Simulation and Modelling, *EUROMAT 99 – Volume 3* (2000) 323, Wiley-VCH, Weinheim.
6. SGTE Pure Substance Database, Edition 2002 and Solution Database, Edition 2004.
7. Y.C. Chuang, C.H. Wu and Z.B. Shao, *J. Less-Commn Met.*, 136 (1987), pp. 147–153.
8. E.W. Teplizkij and L.P. Wladimirow *Zhurnal fis. chim.*, 46 (1972), pp. 762–763.
9. W.A. Fisher and D. Janke, *Arch. Eisenhüttenwesen*, 49 (1978), pp. 425–430.
10. J. Zhangpeng and B. Jansson, TRITA-MAC-0189, Royal Institute of Technology, Stockholm, May 1981.
11. M. Dombre, O.S. Campos, N. Valignat, C. Allibert, C. Bernard and *J. Driole, J. Less-Comon Met.*, 66 (1979), 1.
12. FACT Database, Edition 2002 (in [3]).
13. O. Kubaschweski, C.B. Alcock and P.J. Spencer, *Materials Thermo-Chemistry*, 6th edn, Pergamon Press, 1993.
14. A. Tanaka and H.G. Katayama, Private Communication, 1996.
15. F.R. Duke and R.A.J. Fleming, *J. Electrochem. Soc.*, 104 (1957), pp 774–785.
16. R.J. Robertson and A.S. Kucharski, *Can. J. Chem*, 51 (1973), pp 3114–3122.

Computer-based simulation of inward oxide
scale growth on Cr-containing steels at
high temperatures (OPTICORR)

U K R U P P, University of Applied Sciences Osnabrück,
V B T R I N D A D E, P S C H M I D T, S Y A N G,
H - J C H R I S T, U B U S C H M A N N and
W W I E C H E R T, Universität Siegen, Germany

32.1   Introduction

The distribution and structure of grain boundaries play an important role for
the kinetics of many high-temperature degradation processes since the transport
of matter along interfaces is by order of magnitudes faster than throughout
the bulk. Therefore, reducing the grain size, i.e., increasing the fraction of
fast diffusion paths, may have a detrimental effect, as is known for the creep
behavior of metals and alloys [1]. On the other hand, the high-temperature
oxidation resistance of CrNi 18 8-type stainless steels, which are widely
used for superheater tubes in power plants, can benefit from smaller grain
sizes. As reported by Teranishi *et al.* [2] and in Chapter 5 of this volume [3],
the formation of a protective, Cr-rich oxide scale ($FeCr_2O_4$ and/or Cr_2O_3) is
promoted by the fast outward flux of Cr along the substrate grain boundaries.
A similar effect can be used by providing nanocrystalline surface layers on
Ni-based superalloys. Wang and Young [4] have shown that an increase in
the fraction of grain boundaries can decrease the critical Al concentration
required for the establishment of a superficial Al_2O_3 scale on a material that
usually forms a Cr_2O_3 scale.

As shown in Chapter 5, it is obvious that in the case of low-Cr steels,
typically used for cooling applications in power generation up to temperatures
of approximately 550°C, the beneficial effect of grain refinement disappears.
Here, the grain boundaries seem to act as fast-diffusion paths for the oxygen
transport into the substrate. From inert-gold marker experiments (see [3]) we
know that oxide scale growth occurs by both outward Fe diffusion leading to
the formation of hematite (Fe_2O_3, outermost) and magnetite (Fe_3O_4) and
inward O transport leading to $(Fe,Cr)_3O_4$ formation. As a consequence of the
Cr content in the substrate, a gradient in the Cr concentrations establishes
reaching from the outer/inner scale interface ($c_{Cr} = 0$) to the inner-scale/
substrate interface, where the Cr concentration corresponds to the sole formation
of the spinel phase $FeCr_2O_4$. The inward oxide growth itself is governed by
an intercrystalline oxidation mechanism as described in Chapter 5: oxygen

atoms that have reached the scale/substrate interface by short-circuit diffusion through cracks, pores (see [5, 6]) or by O anion transport (see [7]) penetrate along the substrate grain boundaries leading to the formation of Cr_2O_3 and, consequently, $FeCr_2O_4$. Progress of the scale/substrate interface occurs as soon as the bulk of the grains are oxidized also. It was shown that the thickness of the inner scale decreases when the grain size increases (Fig. 32.1(a)–(c)) and, accordingly, the overall oxidation kinetics becomes slower (Fig. 32.1(d)).

On the basis of the transport mechanisms mentioned above, the objective of this study was the fundamental modification of a computer model originally

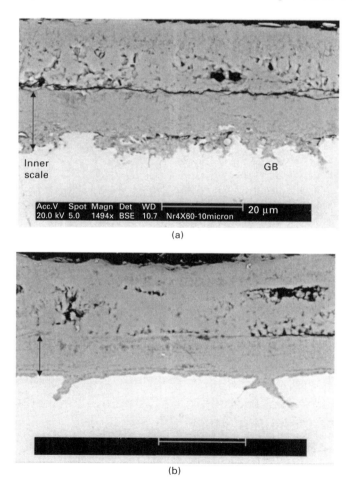

(a)

(b)

32.1 Cross-section of the low-alloy steel B (X60, c_{Cr} = 1.44 wt.%) with different grain sizes of (a) d = 10 μm, (b) d = 30 μm and (c) d = 100 μm after 72 h exposure at T = 550°C to air, and (d) corresponding oxidation kinetics as parabolic rate constants k_p vs grain size.

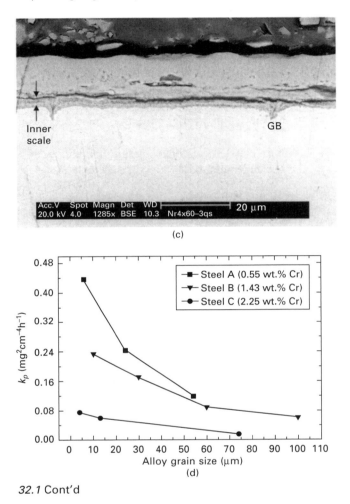

32.1 Cont'd

developed for internal corrosion processes [8] and its adaptation to the oxidation of Cr-containing steels.

32.2   Numerical modeling of diffusion-controlled corrosion processes

Generally, the generic driving force of high temperature corrosion processes can be separated into (i) transport mechanisms, i.e., solid-state diffusion in most cases, and (ii) thermodynamics of chemical reactions. The commonly used, phenomenological way to treat diffusion processes is the application of a second-order partial differential equation (Fick's second law) formulating a relationship between the derivative of the concentration of a species c after the time t and its gradient by means of the location- and temperature-dependent diffusion coefficient D, which represents the jump frequency of the species within a substrate.

$$\frac{\partial c}{\partial t} = \nabla(D\nabla c).$$ 32.1

This equation can be rewritten in a simplified form for two-dimensional diffusion problems by neglecting any cross terms in the following form:

$$\frac{\partial c}{\partial t} = D_x\frac{\partial^2 c}{\partial x^2} + D_y\frac{\partial^2 c}{\partial y^2},$$ 32.2

where D_x and D_y are the diffusion coefficients in the x and y directions, respectively. As a first step to transform the differential equation (32.2) into a difference equation, the first derivative of the concentration c after the time t at the location x and y can be approximated as the difference quotient of the concentrations at t and $t + \Delta t$. The second derivatives of the right-hand side of Eq. 32.3 are replaced by the mean values of the second derivatives at the time t and the time $t + \Delta t$.

$$\left.\frac{\partial c}{\partial t}\right|_{x,y} = \frac{c(x, y, t + \Delta t) - c(x, y, t)}{\Delta t}$$

$$= \frac{1}{2}\left(D_x\left.\frac{\partial^2 c}{\partial x^2}\right|_{y,t} + D_y\left.\frac{\partial^2 c}{\partial y^2}\right|_{x,t} + D_x\left.\frac{\partial^2 c}{\partial x^2}\right|_{y,t+dt} + D_y\left.\frac{\partial^2 c}{\partial y^2}\right|_{x,t+dt}\right)$$ 32.3

By substituting the second derivatives in Eq. 32.3 by the corresponding difference quotients of the concentrations at the locations $x - \Delta x$, x, and $x + \Delta x$, as well as at $y - \Delta y$, y, and $y + \Delta y$, respectively, at the times t and $t + \Delta t$, one obtains the implicit Crank–Nicolson scheme [9] of the finite-difference approach for two-dimensional diffusion.

$$\frac{c(x, y, t + \Delta t) - c(x, t, t)}{\Delta t}$$

$$= \frac{D_x(x, y)}{2} \cdot \frac{c(x - \Delta x, y, t) - 2c(x, y, t) + c(x + \Delta x, y, t)}{\Delta x_l(x, y) \cdot \Delta x_r(x, y)}$$

$$+ \frac{D_x(x, y)}{2}$$

$$\times \frac{c(x - \Delta x, y, t + \Delta t) - 2c(x, y, t + \Delta t) + c(x + \Delta x, y, t + \Delta t)}{\Delta x_l(x, y) \cdot \Delta x_r(x, y)}$$

$$+ \frac{D_y(x, y)}{2} \cdot \frac{c(x, y - \Delta y, t) - 2c(x, y, t) + c(x, y + \Delta y, t)}{\Delta y_l(x, y) \cdot \Delta y_r(x, y)}$$

$$+ \frac{D_y(x, y)}{2}$$

$$\times \frac{c(x, y - \Delta y, t + \Delta t) - 2c(x, y, t + \Delta t) + c(x, y + \Delta y, t + \Delta t)}{\Delta y_l(x, y) \cdot \Delta y_r(x, y)}$$ 32.4

The concept of the finite-difference equation (32.4) with locations of variable step widths Δ_{xl} and Δ_{xr} at the left- and right-hand side of x and Δ_{yl} and Δ_{yr} at the left- and right-hand side of y is illustrated schematically in Fig. 32.2.

If applied to an area of the size $n \cdot \Delta x \cdot m \cdot \Delta y$, Eq 32.4 can be rewritten as a matrix equation, valid for the concentrations $c(x, y)$ at each location, which can be expressed as concentration vectors $\vec{c}(t)$ and $\vec{c}(t + \Delta t)$ for successive time steps t and $t + \Delta t$. Since the right-hand side of Eq. 32.4 consists of elements of the concentration vectors at different locations according to Fig. 32.2, it can be expressed by using the complete concentration vectors $\vec{c}(t)$ and $\vec{c}(t + \Delta t)$ multiplied by the corresponding matrices $\mathbf{M_x}$ and $\mathbf{M_y}$. The product of the matrix $\mathbf{B_x}$ and $\mathbf{B_y}$ with the boundary concentration vectors $\vec{c}_b(t)$ and $\vec{c}_b(t + \Delta t)$ is required to establish the boundary conditions in Eq. 32.4, e.g., at locations ahead of the boundary x_1 no valid concentration value for $x_1 - \Delta x$ within the finite-difference mesh does exist, therefore, it has to be defined by $c_b(x_1 - \Delta x, y, t)$. Finally, the location-dependent diffusion coefficients $D(x, y)$ and the flexible step widths Δx_l and Δx_r, and Δy_l and Δy_r are expressed by the matrices $\mathbf{R_x}$ and $\mathbf{R_y}$, respectively:

$$\mathbf{R_x} = \text{diag}\left[\frac{D_x(x, y)}{2\Delta x_l \Delta x_r}\right] \quad \text{and} \quad \mathbf{R_y} = \text{diag}\left[\frac{D_y(x, y)}{2\Delta y_l \Delta y_r}\right] \qquad 32.5$$

Then, the matrix equation representing Eq. 32.4 for each location step of the finite-difference mesh can be written as follows:

$$\frac{1}{\Delta t}[\vec{c}(t + \Delta t) - \vec{c}(t)]$$

$$= \mathbf{R_x}[\mathbf{M_x}\vec{c}(t) + \mathbf{B_x}\vec{c}_b(t)] + \mathbf{R_x}[\mathbf{M_x}\vec{c}(t + \Delta t) + \mathbf{B_x}\vec{c}_b(t + \Delta t)]$$

$$+ \mathbf{R_y}[\mathbf{M_y}\vec{c}(t) + \mathbf{B_y}\vec{c}_b(t)] + \mathbf{R_y}[\mathbf{M_y}\vec{c}(t + \Delta t) + \mathbf{B_y}\vec{c}_b(t + \Delta t)].$$

$$32.6$$

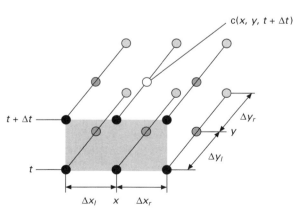

32.2 Schematic representation of the two-dimensional finite-difference mesh applied within the implicit Crank–Nicolson approach.

When the boundary concentrations are assumed as (i) to be homogeneous and (ii) to experience only small changes, i.e., $\vec{c}_b(t) \approx \vec{c}_b(t + \Delta t)$, and the matrices in Eq. 32.6 are multiplied according to $\mathbf{M} = \mathbf{R_x M_x} + \mathbf{R_y M_y}$ and $\mathbf{B} = 2(\mathbf{R_x B_x} + \mathbf{R_y B_y})$, we obtain the following governing equation for the concentration vector $\vec{c}(t + \Delta t)$ as a function of the concentration vector at the preceding time step $\vec{c}_b(t)$ and $\vec{c}(t)$:

$$\left[\frac{1}{\Delta t}\mathbf{E} - \mathbf{M}\right]\vec{c}(t + \Delta t) = \left[\frac{1}{\Delta t}\mathbf{E} - \mathbf{M}\right]\vec{c}(t) + \mathbf{B}\vec{c}_b(t), \qquad 32.7$$

with the unit matrix $\mathbf{E}$.

Implemented in the commercial simulation design environment MatLab, Eq. 32.7 is solved for all species participating in the corrosion reaction and stepwise for the complete reaction time $p \cdot \Delta t$ according to the schematic representation in Fig. 32.3 (simplified for one-dimensional diffusion).

To accommodate the possible chemical reactions of the ongoing corrosion process, the calculated concentrations at $\vec{c}(t + \Delta t)$ ($c_{i,j}^{k+1}$ in Fig. 32.3) must be corrected according to the local thermodynamic equilibrium. For this purpose, the concentrations $\vec{c}(t + \Delta t)$ are transferred into a thermodynamic subroutine ThermoScript [10], which contains the commercial program ChemApp [11]. ChemApp is based on a numerical Gibbs' energy minimization routine in combination with tailor-made databases [12]. In order to avoid excessive calculation times, the parallel-computing system PVM (parallel virtual machine) is used, i.e., ThermoScript distributes the individual

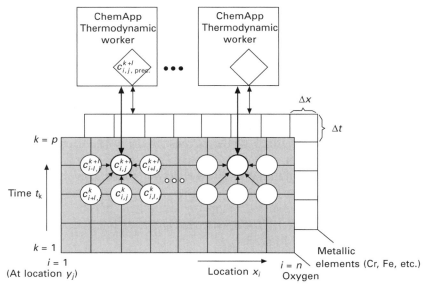

32.3 Schematic representation of the implicit finite-difference technique in combination with the thermodynamic program ChemApp.

equilibrium calculations to thermodynamic workers according to the schematic representation in Fig. 32.3. Therefore, by using, e.g., 200 parallel-working processing units (CPUs) the calculation time can be reduced by a factor of up to 200 as compared to a conventional PC. The complete sets of concentrations of all the participating species in equilibrium form the new starting concentration vector $\vec{c}(t)$ for the application of Eq. 32.7 at the following time step $k + 2$.

To apply the finite-difference approach mentioned above to oxidation processes of Cr-containing steels, it has to be taken into account that, generally, the oxide scale consists of three or more separate layers, e.g., in the case of low-Cr steels an outer magnetite scale (below an outermost hematite layer) on top of an inward-growing magnetite/spinel-phase scale and an intercrystalline oxidation zone below the scale/substrate interface.

Since it was shown in Chapter 5 that oxidation processes of Cr-containing steels are mainly governed by grain-boundary transport of the reacting species, i.e, Cr and O grain-boundary diffusion, a two-dimensional finite-difference model has been established that distinguishes between fast diffusion along substrate grain boundaries and slow transport through the bulk. Due to the lack of data available for interface diffusivities, on the base of an estimate value for the grain boundary width of $\delta = 0.5$ nm [13], the grain-boundary diffusion coefficient was assumed to be 100 times higher than the bulk diffusion coefficient, which has a value of $D_b = 5.39 \times 10^{-13}$ m^2/s for oxygen in iron at $T = 550°C$ [14].

According to the schematic representation for the intercrystalline oxidation process in low-Cr steels in Fig. 32.4, oxygen firstly penetrates along the substrate grain boundaries, and hence, the diffusivity elements D_y of the matrix $\mathbf{R_y}$ along the grain boundaries are set to the grain boundary diffusion coefficient D_{GB}. At the same time, oxygen bulk diffusion takes place driven by the two-dimensional concentration gradient. Hence, the respective elements D_x and D_y of the matrices $\mathbf{R_x}$ and $\mathbf{R_y}$ are set to the bulk diffusion coefficient D_b. The first oxidation product that becomes thermodynamically stable during this process is Cr_2O_3. The corresponding depletion in Cr in combination with the increase of the oxygen activity leads to the subsequent formation of the Fe–Cr spinel $FeCr_2O_4$. Further Cr depletion manifests itself in a Cr gradient in the $(Fe,Cr)_3O_4$ phase over the inward-growing oxide scale, finally resulting in almost pure magnetite (Fe_3O_4). As soon as the metallic substrate is completely consumed (except the residual Fe in equilibrium with Fe_3O_4), the inner-oxide/substrate interface moves one location step inward (see Fig. 32.4).

Diffusion through the inner $Fe_3O_4/FeCr_2O_4$ scale, the mechanism of which is not fully understood since it depends strongly on the Cr concentration [15] and is affected by the pronounced porosity, is treated in this study by an effective diffusion coefficient D_{eff}. The observed value of D_{eff} was estimated

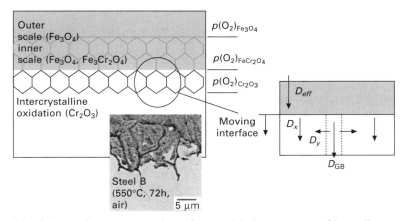

32.4 Schematic representation of the oxidation process of low-alloy Cr-containing steels and the two-dimensional model using a moving-interface approach and distinguishing between grain boundary and bulk diffusion.

by means of the experimentally determined k_p value for the inner-scale growth kinetics in combination with Wagner's theory of oxidation [16]:

$$k_p = \int_{p(O_2)Cr_2O_3}^{p(O_2)Fe_3O_4} D_O \, d \ln p(O_2) \qquad 32.8$$

where $p(O_2)Fe_3O_4$ is the oxygen partial pressure at the outer/inner scale interface (see Fig. 32.4) and $p(O_2)Cr_2O_3$ the estimated, respective pressure at the scale/substrate interface.

Figure 32.4 summarizes the concept of the simulation procedure: (i) the origin of the grid is starting at the outer/inner scale interface, where the spinel phase completely disappears; (ii) inward growth of the inner scale is defined by the condition that the metallic substrate is completely consumed, and finally, (iii) the depth of the intercrystalline oxidation attack depends on the Cr_2O_3 equilibrium.

The concept of the simulation program that has been developed within the present study is shown in Fig. 32.5, highlighting the main elements of the model.

32.3   Simulation results and discussion

The finite-difference model described above was applied to the oxidation processes of low- and high-Cr steels as discussed in Chapter 5 using diffusion data given as:

- $D_{Cr \ in \ \alpha Fe} = 3.59 \ m^2/s \cdot 10^{-6} \exp(-179 \ kJ/RT)$ ref. [17]
- $D_{O \ in \ \alpha Fe} = 3.78 \ m^2/s \cdot 10^{-7} \exp(-92.1 \ kJ/RT)$ ref. [14]
- $D_{Fe \ in \ \alpha Fe} = 2.8 \ m^2/s \cdot 10^{-4} \exp(-251 \ kJ/RT)$ ref. [18]

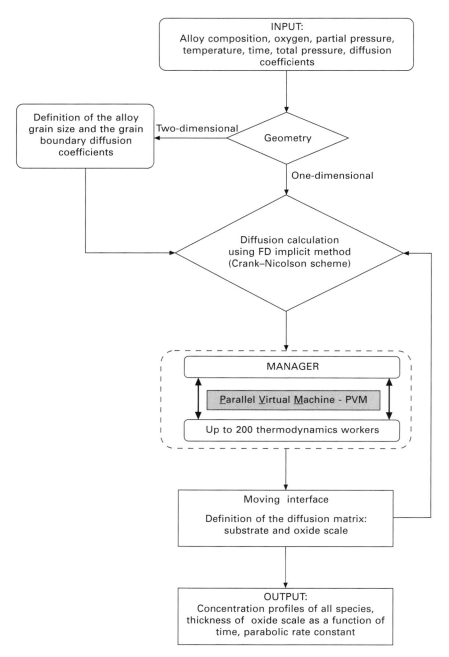

32.5 Main elements of the computer model to simulate the inner oxide scale growth on Cr-containing steels.

Figure 32.6 shows the calculated two-dimensional concentration profiles for magnetite (Fe_3O_4), the Fe–Cr spinel phase ($FeCr_2O_4$) and chromia (Cr_2O_3) for steel B containing 1.44 wt.% Cr and having a grain size of $d = 30$ µm ([3]). It should be mentioned here that the thermochemical database of the system treats Fe_3O_4 and $FeCr_2O_4$ as separate but coexisting phases, i.e., an increase in the Cr concentration leads to a relative increase in the fraction of the spinel phase $FeCr_2O_4$.

The results support the experimental observation that the first oxide phase being formed is Cr_2O_3 along the substrate grain boundaries, followed by $FeCr_2O_4$ and finally, Fe_3O_4. Since oxygen penetration into the alloy occurs by both grain boundary as well as bulk diffusion, the oxidation process proceeds from the grain boundaries into the bulk until the complete metallic phase is consumed. This situation defines the progress of the inner oxide scale (see arrow in Fig. 32.6(a)).

The simulated growth kinetics of the inner oxide scale as a function of the exposure time is shown in Fig. 32.7 for three different grain sizes. In the case of the material with a grain size of $d = 10$ µm, it could be demonstrated that the simulated data are in excellent agreement with the experimentally determined values of the inner-scale thickness for various exposure times.

While for low-alloy steels with initial Cr concentrations between 0.5 and 2.3 wt.% an increase in the grain size results in a decrease in the oxidation kinetics, high-Cr austenitic steels exhibit a contrary effect: a smaller grain size results in an increase of the outward Cr transport and hence, to a transition in the oxidation mechanism from non-protective magnetite/spinel phase $(Fe,Cr)_3O_4$ to protective chromia (Cr_2O_3) formation. This is supported by first simulation results using the two-dimensional finite-difference approach as described in Section 32.2 in combination with diffusion data taken from [17]. Figure 32.8(a) shows qualitatively the Cr enrichment and gradient along a substrate grain boundary for high-temperature exposure of the austenitic stainless steel TP347 (cf. [3]). Indeed, the simulation revealed for an increase in the grain size, while keeping all the other parameters the same, that instead of pure Cr_2O_3 a mixture of oxide phases is formed as shown in Fig. 32.8(b).

The examples presented above demonstrate the applicability of the combination of the finite-difference technique and computational thermodynamics to complex corrosion processes which depend substantially on the material's microstructure. Of course, much more experimental work is required to describe in detail the variety of possible transport processes, e.g., taking place within the porous inner oxides scale and along the substrate grain boundaries. Strictly speaking, the structure of the interfaces should change as soon as the grain boundaries are covered by an oxide phase which is in contact with the matrix on both sides. Also, the actual value of the grain boundary thickness and its implementation in the finite-difference approach

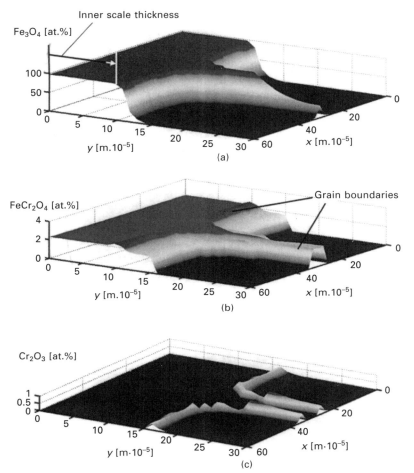

32.6 Simulated lateral concentration profiles of the oxide phases (a) Fe_3O_4, (b) $FeCr_2O_4$, and (c) Cr_2O_3 formed during exposure of the low-alloy steel B with a grain size of $d = 30$ μm (X60, $c_{Cr} = 1.44$ wt.%) at $T = 550°C$ to air ($y = 0$ corresponds to the original inner-scale/metal interface at $t = 0$ s).

remains an open question. In the present study, the grain boundary thickness is assumed to be $\delta = 0.5$ nm according to [13] and the segregation factor to be $s = 1$.

32.4   Summary

The oxidation behavior of steels depends strongly on the Cr concentration. While the magnetite scale formed on low-alloy steels provides only moderate

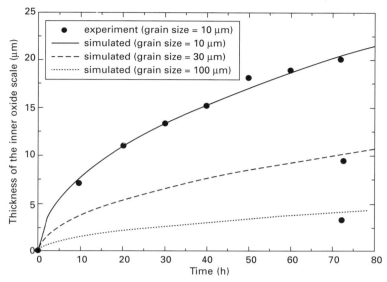

32.7 Comparison of the simulated inner-oxide growth kinetics for the low-alloy steel B (X60, c_{Cr} = 1.44 wt.%) with three different grain sizes and with the experimentally measured inner-oxide thickness for specimens having a grain size of d = 10 μm.

protection, steels with a Cr concentration above 20 wt.% are capable of forming a slow-growing and protective Cr_2O_3 scale. In [3] it was shown that this capability is promoted by fine-grained microstructures and in the present study a computer model is introduced and described which allows the simulation and prediction of the high-temperature oxidation process of Cr-containing steels. The model combines a numerical two-dimensional finite-difference diffusion calculation with computational thermodynamics making use of a parallel computing technique.

The application of the model to Cr-containing steels in combination with a tailored thermodynamic database yielded results that are in excellent agreement with the experimental observation, that the oxidation process is substantially governed by oxygen and chromium grain boundary diffusion. In particular, the model shows that the case of low-alloy steels with Cr concentrations between 0.5 and 2.3 wt.%, the oxidation rate increases with decreasing grain size due to a more pronounced intercrystalline oxidation attack. In contrast, for high-Cr steels (Cr concentrations above approximately 18 wt.%) the oxidation rate decreases strongly with decreasing grain size. This can be attributed to an increase in the outward Cr flux establishing a protective chromia scale.

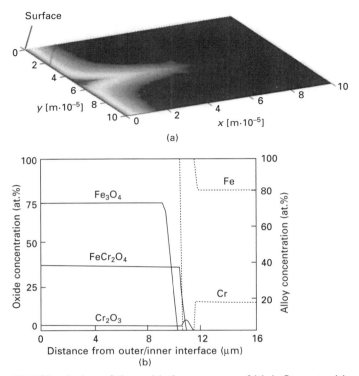

32.8 Simulation of the oxidation process of high-Cr austenitic steel TP347: (a) qualitative Cr enrichment along substrate grain boundary on a fine-grained material (d = 11 μm) forming a superficial Cr_2O_3 scale and (b) one-dimensional concentration profiles of the oxidation products and the metallic elements of a coarse-grained specimen (d = 65 μm).

32.5   Acknowledgements

This research has been supported by the EU project OPTICORR, Deutsche Forschungsgemeinschaft (DFG), and by the Brazilian research foundation (CAPES) through a fellowship to one of the authors (V.B. Trindade). This support is gratefully acknowledged.

32.6   References

1. A.P. Sutton, R.W. Balluffi, *Interfaces in Crystalline Materials*, Oxford University Press, Oxford, 1995.
2. H. Teranishi, Y. Sawaragi and M. Kubota, *The Sumitomo Research*, **1989**, *38*, 63.
3. V.B. Trindade, U. Krupp, Ph. E.-G. Wagenhuber, S. Yang and H.-J. Christ, **2008**, this volume (Chapter 5).
4. F. Wang and D.J. Young, *Oxidation of Metals*, **1997**, *48*, 497.

5. R.Y. Chen and W.Y.D. Yuen, *Oxidation of Metals*, **2003**, *59*, 433.
6. M. Schütze, *Journal of Corrosion Science and Engineering*, **2003**, *6*.
7. T. Maruyama, N. Fukagai, M. Ueda and K. Kawamura, *Materials Science Forum* **2004**, *461–464*, 807.
8. U. Krupp and H.-J. Christ, *Oxidation of Metals* **1999**, *52*, 299.
9. J. Crank, *The Mathematics of Diffusion*, 2nd edn, Clarendon Press, Oxford, **1986**.
10. U. Buschmann, unpublished research, University of Siegen, **2004**.
11. G. Erickson and K. Hack, *Metallurgical Transactions*, **1990**, *21B*, 1013.
12. K. Hack and T. Jantzen, **2008**, this volume (Chapter 31).
13. J.C. Fisher, *Journal of Applied Physics*, **1951**, *22*, 74.
14. T. Heumann, *Diffusion in Metallen*, Springer-Verlag, Berlin, Germany, **1992**.
15. R. Dieckmann, *Journal of Physics and Chemistry of Solids*, **1998**, 59, 507.
16. C. Wagner, *Zeitschrift für Physikalische Chemie*, **1933**, *21*, 25.
17. I. Kaur, W. Gust, L. Kozma, *Handbook of Grain and Interphase Boundary Diffusion Data*, Ziegler Press, Stuttgart, Germany, **1989**.
18. E.A. Brandes, G.B. Brooks (eds), *Smithells Materials Reference Book*, 7th edn, Butterworth-Heinemann, Oxford, **1992**.

High temperature oxidation of γ-NiCrAl modelling and experiments

T J N I I D A M, N M V A N D E R P E R S and
W G S L O O F, Delft University of Technology,
The Netherlands

33.1   Introduction

The protection offered by MCrAlY (M = Ni,Co) alloys against high temperature oxidation relies on the ability of the alloy to develop and maintain a continuous, dense and slow growing α-Al$_2$O$_3$ scale on its surface [1]. Generally, the formation of such a continuous alumina layer is preceded by an initial period of fast oxidation, associated with the simultaneous formation of α-Al$_2$O$_3$ and the fast growing, non-protective oxide phases Cr$_2$O$_3$, NiO, NiCr$_2$O$_4$ and/or NiAl$_2$O$_4$ [1]. The formation of these oxides usually results in detrimental effects. For instance, the development of NiAl$_2$O$_4$ spinel on MCrAlY bond coatings at the interface with a ceramic topcoat in thermal barrier coating (TBC) systems promotes the spalling of the ceramic topcoat upon thermal cycling [2]. Therefore, ideally the formation of an oxide scale constituted solely of α-Al$_2$O$_3$ is desired.

Recently a coupled thermodynamic-kinetic (CTK) oxidation model was developed for the thermal oxidation of ternary alloys [3]. Calculations with this model revealed a close relation between the formation of undesired, non-protective oxides and the compositional changes within the underlying alloy [3]. It was also shown by both model calculations and experiments that the *total* amount of non-protective oxides (i.e. all oxide phases excluding α-Al$_2$O$_3$) is suppressed by maintaining a sufficiently high Al concentration in the alloy at the oxide/metal (O/M) interface during the early stages of oxidation as realised by, e.g., a low partial oxygen pressure (pO_2) upon oxidation or a small alloy grain size [4].

In this study it is shown that the amount of each *individual* oxide phase developed is also determined by the composition in the alloy at the O/M interface. The formation of α-Al$_2$O$_3$, Cr$_2$O$_3$, NiO, NiCr$_2$O$_4$ and NiAl$_2$O$_4$ was determined as a function of oxidation time for the thermal oxidation of a γ-Ni-27Cr-9Al (at.%) alloy at 1353 K and 1443 K and a partial oxygen pressure of 20 kPa using *in-situ* high temperature X-ray diffractometry (XRD) [5–7]. The intensities recorded in the XRD experiment can be used to assess the oxidation kinetics quantitatively because virtually no spallation of the oxide scale occurs at the oxidation temperature [5]. The results obtained by XRD are compared with microstructural observations from scanning electron microscope (SEM) backscattered electron (BSE) images, and model calculations using the CTK oxidation model [3, 4].

33.2   Experimental methods

A single-phase γ-Ni-27Cr-9Al (at.%) alloy was cast in the form of a cylindrical rod and subsequently annealed in a sealed quartz tube filled with Ar at 1373 K for 400 h in order to homogenise and recrystallise the alloy. The mean linear grain size of the alloy after annealing is about 750 μm. Disc-shaped specimens (diameter 10 mm and thickness 2 mm) were cut from the rod using spark-erosion. Prior to high temperature XRD, the specimen surface was prepared by successively grinding and polishing. Polishing was performed with paste of 0.25 μm diamond grains as a final step. After each preparation step, the specimens were thoroughly cleaned ultrasonically with isopropanol and dried by blowing with pure compressed nitrogen gas.

High-temperature XRD was employed to determine the evolution of the crystalline phases present in the oxide scale as function of oxidation time. To this end, a γ-Ni-27Cr-9Al (at.%) alloy was heated in a high-temperature chamber (Bruker AXS MRI) within 3 min to 1353 K or 1443 K in a continuous gas flow of about 50 ml/min of He with 20 vol.% O_2. The temperature was controlled with a Pt-10%Rh strip within about 2 K and simultaneously monitored with a chrome/alumel thermocouple spot welded to the specimen. After reaching the desired oxidation temperature, diffractograms were recorded *in-situ* from the sample surface for selected time intervals up to 32 h of oxidation with a Bruker AXS D5005 θ-θ diffractometer using monochromatic Cu K_α radiation. These diffractograms were acquired for the Bragg-Brentano geometry in the 2θ range of 22° to 54° with a step size of 0.1° and counting times per step of 2 s, 6 s, 11 s and 22 s for oxidation times between 0–1 h, 1–4 h, 4–16 h and 16–32 h, respectively. The use of low counting times at short oxidation times was necessary to follow the oxide phase evolution during the early oxidation stages. After oxidation, the alloy was allowed to cool to room temperature in the high temperature chamber (after 1 min of cooling the specimen temperature was dropped below 700 K).

After high-temperature XRD, an oxide/alloy cross-section was prepared from the oxidised alloy as follows. First a gold layer was deposited on the oxide scale by sputtering. Then, a nickel layer was plated on top of the gold layer using a Watts bath. Next, the specimen was cut perpendicular to the specimen surface with a diamond saw. Finally, the oxide/alloy cross-section was ground and polished as described above.

The obtained oxide/alloy cross-section was investigated with a JEOL

JSM 6500F SEM, equipped with an Autrata [8] detector for observation of BSE images and a Noran Pioneer 30 mm^2 Si(Li) detector for energy-dispersive X-ray spectroscopy (EDXS). The X-ray spectra acquisition and processing was performed with a ThermoNoran Vantage system (version 2.3).

33.3   Data evaluation

33.3.1   High-temperature XRD

From the recorded diffractograms the amount of each oxide phase developed was determined as follows.

First, the phases present in the oxide scale were identified using the ICDD database (ICDD – International Centre for Diffraction Data, Newton Square, PA). For oxidation at 1353 K and 1443 K the following oxide phases were observed in the oxide scale during oxidation (Fig. 33.1): randomly oriented α-Al_2O_3 [9], (110) textured Cr_2O_3 [10], (012) textured NiO [11], (220) and (400) textured $NiCr_2O_4$ at 1353 K and 1443 K, respectively [12] and randomly oriented $NiAl_2O_4$ [13]. No reflections originating from the course-grained alloy were observed. The amount of metasable aluminas within the oxide scale (which are expected to be present for short oxidation times [14–16]) was too low to be detected with XRD. The texture in the oxide scale is maintained for all oxidation times, as determined for a given oxide phase by the intensity ratios of the strongest reflection and the other reflections in the diffractograms (i.e. these intensity ratios remain virtually constant, see Fig. 33.1). This implies that a change in the peak intensity of a given oxide phase reflects a change in the amount of the oxide phase developed.

Next, the integrated peak intensities and the peak positions, corresponding to the Cu $k_{\alpha1}$ radiation (wavelength Cu $k_{\alpha1}$ = 1.540562 Å), were resolved for all visible reflections with the profile-fitting program PROFIT (Philips Analytical BV, Almelo, The Netherlands). A pseudo Voigt profile shape, and adopting a linear background, was used to fit simultaneously the $k_{\alpha1}$ and $k_{\alpha2}$ components of all the diffraction lines (a value of 0.5 was used for the ratio of their intensities).

Then the amounts of α-Al_2O_3, Cr_2O_3, NiO, $NiCr_2O_4$ and $NiAl_2O_4$ formed were calculated as a function of oxidation time. To this end, a model for the absorption of the X-rays by the oxide scale was adopted. As illustrated in Fig. 33.2, for all oxidation times the oxide scale is assumed to consist of n (= 5) different layers of uniform thickness, with each oxide layer i constituted solely of a single oxide phase. However, the actual growth and disappearance of oxide phases as a function of oxidation time differs from this model when an oxide phase occurs as isolated crystallites, e.g. α-Al_2O_3 for short oxidation times and NiO for long oxidation times (see Ref. [4] and Section 33.4).

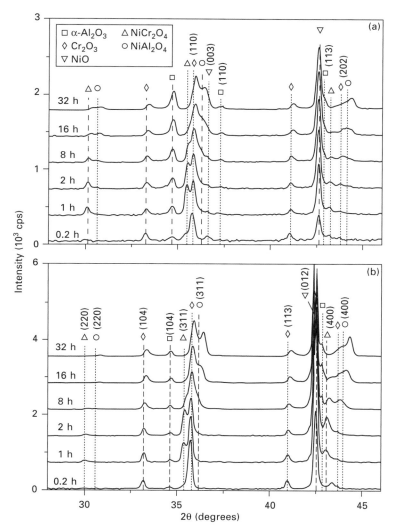

33.1 Diffractograms recorded during oxidation of a γ-Ni-27Cr-9Al alloy at pO_2 = 20 kPa and at (a) 1353 K and (b) 1443 K. Phase identification was performed using the ICDD database, while adopting a constant multiplication factor for the lattice parameters of each oxide phase (> 1 due to thermal expansion of the specimen). The dashed lines indicate the reflections that were used to calculate the amount of each oxide phase developed (see Section 33.3.1 for details).

Considering Fig. 33.2, for the oxide layers 2 to *n*, the thickness d_i^t of oxide layer *i* at time *t* can be written as:

$$d_i^t = -\frac{\sin \theta_i^t}{2\bar{\mu}_i} \ln \left(1 - \frac{2\bar{\mu}_i}{f_{i,c} \sin \theta_i^t} I_{i,c}^t \right),$$ 33.1

where θ_i^t is the Bragg angle at which the X-rays are scattered from layer i at time t, $\overline{\mu}_i$ is the linear absorption coefficient of layer i, $f_{i,c}$ is the scaling constant for layer i (see below) and $I_{i,c}^t$ is the intensity of the X-rays scattered from oxide layer i at time t in the *absence* of oxide layers 1 to $i-1$. $I_{i,c}^t$ is related to I_i^t, the intensity of X-rays scattered from oxide layer i at time t in the *presence* of oxide layers 1 to $i-1$, according to:

$$I_{i,c}^t = I_i^t / A_i^t,$$

33.2

where A_i^t is the correction factor for the absorption of the X-rays in the layers 1 to i-1, given by [17]:

$$A_i^t = \exp\left(\frac{-2}{\sin \theta_i^t} \sum_{j=1}^{i-1} \mu_j d_j^t\right).$$

33.3

Since there are no oxides formed on top of NiO, for oxide layer 1, $A_i^t = 1$ and thus $I_{1,c}^t = I_1^t$. Then, for known values of I_i^t, θ_i^t, $\overline{\mu}_i$ and $f_{i,c}$ the amount of each oxide phase developed can be calculated using the following calculation sequence: (i) determination of the amount of NiO (i.e. d_1^t) using Eq. (33.1) with $I_{1,c}^t = I_1^t$, (ii) calculation of the intensity of NiAl$_2$O$_4$ in the absence of the NiO layer (i.e. $I_{2,c}^t$) using Eqs 33.2 and 33.3, (iii) computation of the amount of NiAl$_2$O$_4$ using Eq. (33.1) (i.e. d_2^t). Finally, the thickness of layers 3, 4 and 5 are calculated analogously.

For oxide layers 1, 2, 4 and 5, the values for I_i^t and θ_i^t were obtained from the integrated peak intensities and peak positions of the NiO (012), NiAl$_2$O$_4$ (311), Cr$_2$O$_3$ (104), and α-Al$_2$O$_3$ (104) reflections, respectively (as indicated by the dashed lines in Fig. 33.1). For oxide layer 3, the values for I_i^t and θ_i^t were obtained from the NiCr$_2$O$_4$ (220) reflection at 1353 K and from the NiCr$_2$O$_4$ (400) reflection at 1443 K, due to differences in texture within the NiCr$_2$O$_4$ phase (cf. Fig. 33.1(a) and 33.1(b)). The time t corresponds with the average time for which the recording of a diffraction pattern started and ended.

The value for $\overline{\mu}_i$ was obtained using the following equation [17]:

$$\overline{\mu}_i = \rho_i \sum_{k=1}^{m} (\mu/\rho)_k w_{ik},$$

33.4

where $(\mu/\rho)_k$ is the mass absorption coefficient of element k for Cu k_α radiation, ρ_i is the density of layer i, w_{ik} is the weight fraction of element k in oxide layer i, and m is the total number of elements in oxide layer i. The values for $(\mu/\rho)_k$ and ρ_i, used to calculate $\overline{\mu}_i$ for the oxide phases considered, are listed in Tables 33.1 and 33.2, respectively.

The scaling parameters $f_{i,c}$ were solved from Eq. (33.1) using the values for $I_{i,c}^t$ and d_i^t after 32 h of oxidation ($f_{i,c}$ is taken as constant during oxidation, i.e. it is assumed that the texture in the oxide scale does not change during oxidation, see above). The thickness d_i^{32} of oxide layer i after 32 h of

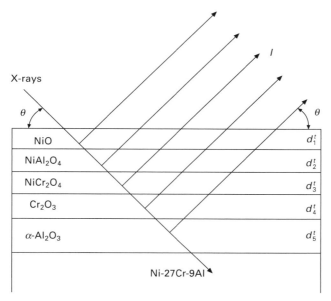

33.2 Schematic model adopted for the absorption of X-rays during oxidation of a γ-Ni-27Cr-9Al alloy at $pO2 = 20$ kPa and a temperature of 1353 K and 1443 K. This model is used to calculate the amount of each oxide formed as function of oxidation time (see Section 33.3 for details). Note that d_2^t and d_3^t are zero for oxidation times shorter than about 0.5 h, and that d_3^t remains zero for oxidation times shorter than about 3 h.

Table 33.1 Values for the mass absorption coefficient $(\mu/\rho)_k$ of element k for Cu k_α radiation [cm²/g]. Data were taken from Ref. [18]

k	Al	Cr	Ni	O
$(\mu/\rho)_k$	49.6	247	48.8	11.5

Table 33.2 Values for the density ρ_i of oxide layer i (g/cm³). Data were taken from Ref. [19]

i	1 = NiO	2 = NiAl₂O₄	3 = NiCr₂O₄	4 = Cr₂O₃	5 = α-Al₂O₃
ρ_i	6.85	4.52	5.27	4.27	3.98

oxidation was determined accurately from BSE images recorded from an oxide/alloy cross-section prepared after the XRD experiment. As illustrated in Fig. 33.3(a) and (b), each BSE image revealed a high contrast between the different oxide phases that were present within the oxide scale. Therefore, the oxide scale could be easily divided into separate parts by conversion of the BSE images into greyscale images using standard image analysis software (cf. Fig. 33.3(a) and 33.3(c)). EDXS analysis revealed that the black, white, light grey and dark grey areas corresponded with the oxide phases α-Al_2O_3, Cr_2O_3, $NiAl_2O_4$ and NiO, respectively (Fig. 33.3(c)). Finally the amount of each oxide phase developed after 32 h of oxidation was established with the image analysis software using the procedure described in Ref. [4].

$NiCr_2O_4$ could not be detected clearly with EDXS after 32 h of oxidation at both 1353 K and 1443 K (then $d_3^{32} \sim 0$). Therefore, a separate high temperature XRD experiment was executed for 8 h at 1353 K. From this experiment, the amount of $NiCr_2O_4$ was determined in a similar manner as discussed above. The scaling factor $f_{3,c}$ obtained for $NiCr_2O_4$ at 1353 K was also used for oxidation at 1443 K.

33.3.2 Modelling

The experimentally determined values for the amount of each oxide phase formed as function of oxidation time were compared with their corresponding calculated values, as obtained from the CTK oxidation model. This oxidation model is a numerical model, based on the finite-difference approach, which describes the thermal oxidation of a single-phase ternary alloy [3]. The model assumes local equilibrium at the O/M interface during oxidation. From a comparison of the equilibrium partial oxygen pressures of the concerned oxide phases at the O/M interface, it follows which *pure* oxide phases are formed during a given time step in the calculations. The concentrations of the alloy constituents in the alloy at the O/M interface are calculated by applying a mass balance for all alloy constituents at the O/M interface, which states that the total amount of the alloy constituent that is released by solid-state volume diffusion from the bulk of the alloy towards the O/M interface equals the total amount of the alloy constituent that is received by the developing oxide scale [3]. It is noted that the CTK oxidation model does not address the morphology and distribution of the evolved oxide phases in the developing oxide scale. Further, the formation of mixed oxide phases (e.g. $NiCr_2O_4$ and $NiAl_2O_4$) due to the reaction between pure oxide phases within the oxide scale was not taken into account [3].

The model was applied for the oxidation of a γ-Ni-27Cr-9Al alloy at 1353 K and a pO_2 of 20 kPa to calculate the amounts of α-Al_2O_3, Cr_2O_3 and NiO formed as well as the course of the Al concentration in the alloy at the O/M interface as a function of oxidation time. The total oxide scale thickness d_{ox}^t, as obtained from the XRD experiment (see Fig. 33.4), was used as input in the model calculations.

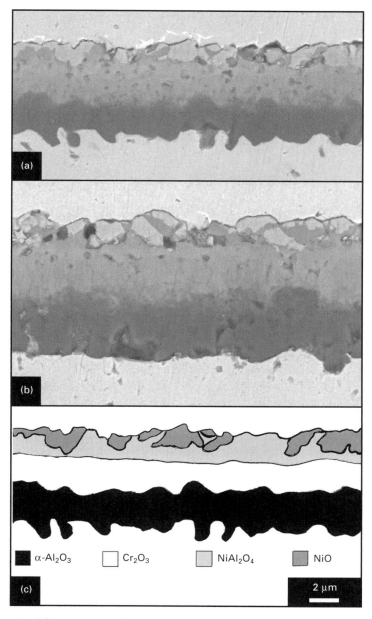

33.3 BSE images of the oxide scale developed after 32 h of oxidation of a γ-Ni-27Cr-9Al alloy at pO_2 = 20 kPa and (a) 1353 K and (b) 1443 K. Corresponding greyscale image (c) of the oxide scale at 1353 K. The black, white, light grey and dark grey areas correspond with the oxide phases α-Al$_2$O$_3$, Cr$_2$O$_3$, NiAl$_2$O$_4$ and NiO, respectively.

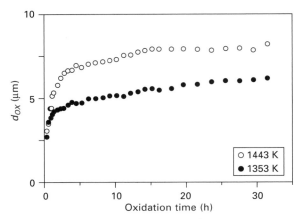

33.4 Total oxide scale thickness d_{ox}^t as a function of oxidation time for the oxidation of a γ-Ni-27Cr-9Al alloy at $pO2 = 20$ kPa and a temperature of 1353 K and 1443 K. For each oxidation time, the value for d_{ox}^t was obtained from the XRD experiments as the sum of the amounts of α-Al$_2$O$_3$, Cr$_2$O$_3$, NiO, NiCr$_2$O$_4$ and NiAl$_2$O$_4$ formed (see Section 33.3.1 for details).

33.4   Results and discussion

The XRD results for the thermal oxidation of a γ-Ni-27Cr-9Al alloy at 1353 K and 1443 K are shown in Figs 33.4, 33.5 and 33.6, respectively. The amounts of α-Al$_2$O$_3$, Cr$_2$O$_3$, NiO, NiCr$_2$O$_4$ and NiAl$_2$O$_4$, as well as the *total* amount of non-protective oxides (i.e. all oxide phases excluding α-Al$_2$O$_3$), formed as function of oxidation time are displayed in Fig. 33.5. These experimentally determined values are compared with their corresponding calculated values from the CTK oxidation model in Fig. 33.6.

Based on the XRD results, as well as on microstructural observations from BSE images, three different regimes were identified upon oxidation of the alloy at 1353 K and 1443 K (corresponding with the growth stages II, III and IV in Fig. 33.7 [4]):

1. Internal oxidation of Al and simultaneous overgrowth of Cr$_2$O$_3$ and NiO.
2. Successive formation of continuous Cr$_2$O$_3$ and α-Al$_2$O$_3$ layers.
3. Exclusive, parabolic growth of α-Al$_2$O$_3$.

In the following sections, the XRD results will be discussed for each of these three regimes.

33.4.1 Internal oxidation of Al and simultaneous overgrowth of Cr$_2$O$_3$ and NiO

For very short oxidation times (i.e. 0.2 h), the pure oxides α-Al$_2$O$_3$, Cr$_2$O$_3$ and NiO were the only phases that were detected with XRD (Figs 33.1 and 33.5).

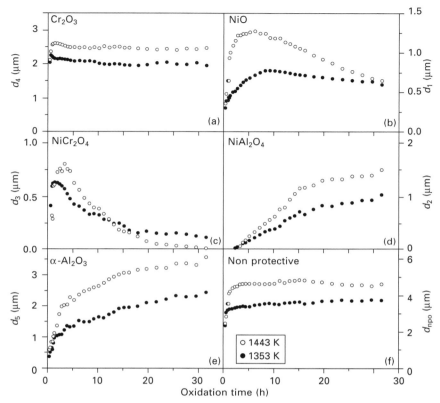

33.5 Experimentally determined amount of (a) Cr_2O_3, (b) NiO, (c) $NiCr_2O_4$, (d) $NiAl_2O_4$, (e) $\alpha\text{-}Al_2O_3$ and (f) the total amount of non-protective oxides d_{npo} developed upon oxidation of a γ-Ni-27Cr-9Al alloy at pO_2 = 20 kPa and a temperature of 1353 K and 1443 K. Data were obtained from the diffractograms using the procedure described in Section 33.3.1.

These XRD results are in good agreement with the microstructural observations from BSE images, which revealed the formation of almost pure NiO and Cr_2O_3 layers on top isolated elongated internal $\alpha\text{-}Al_2O_3$ precipitates for short oxidation times (Fig. 33.8 and Ref. [4]). Between the Cr_2O_3 and the isolated $\alpha\text{-}Al_2O_3$ precipitates a metallic layer of almost pure Ni (i.e. heavily depleted in Al and Cr [4]) is present for very short oxidation times (i.e. 0.2 h, see Fig. 33.7), since the $\alpha\text{-}Al_2O_3$ is oxidised at a larger depth within the alloy than the Cr_2O_3, (i.e. the activity of oxygen required to oxidise the Al within the alloy is smaller than its activity to oxidise the Cr within the alloy [20]).

No spinel oxides were detected within the oxide scale after very short oxidation times (Figs 33.1 and 33.5), in contrast with the high temperature

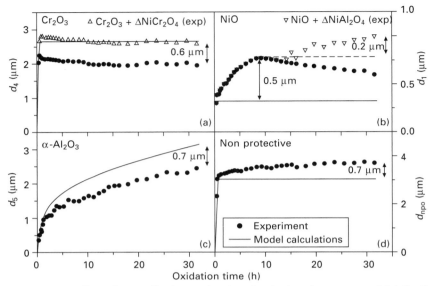

33.6 Experimentally determined and calculated amounts of (a) Cr_2O_3, (b) NiO, (c) α-Al_2O_3 and (d) the total amount of non-protective oxides d_{npo} developed upon oxidation of a γ-Ni-27Cr-9Al alloy at $pO_2 = 20$ kPa and a temperature of 1353 K. Experimental data are identical to Fig. 33.4. $\Delta NiCr_2O_4$ corresponds with the maximum amount of $NiCr_2O_4$ developed. $\Delta NiAl_2O_4$ is equal to d_2^t - $\Delta NiCr_2O_4$ for $t = 8$ h, i.e. the amount of $NiAl_2O_4$ that develops at the expense of NiO.

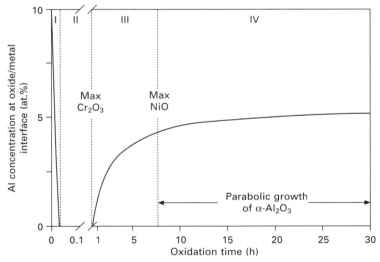

33.7 Calculated Al concentration in the alloy at the O/M interface as a function of oxidation time for the oxidation of a γ-Ni-27Cr-9Al alloy at $pO_2 = 20$ kPa and a temperature of 1353 K. The roman numerals denote the four successive growth stages that can be identified upon oxidation of the alloy (see Ref. [4] for details).

oxidation of a NiCrAl alloy with a higher Al content of 12 at.%, for which spinel oxides developed already after 1 min of oxidation [14, 20]. Clearly, $NiCr_2O_4$ spinel starts to develop after only about 0.2 h of oxidation (Fig 33.1 and 33.5), i.e. right after the formation of the NiO and Cr_2O_3 overlayers. BSE images revealed that the $NiCr_2O_4$ is formed as a continuous layer between the Cr_2O_3 and the NiO [4]. Also, between 1 h and 2 h of oxidation, the formation of $NiCr_2O_4$ is associated with a decrease in the amount of Cr_2O_3 formed (Fig. 33.5(a) 5(c)). All this indicates that the formation of $NiCr_2O_4$ is due to a solid-state reaction between the Cr_2O_3 and the NiO, in agreement with thermodynamic predictions [21].

The values (as determined with XRD) for the amounts of α-Al_2O_3, Cr_2O_3 and NiO (Fig. 33.5) developed, as well as for the total oxide scale thickness (Fig. 33.4), increase very rapidly during the early stages of oxidation in accordance with the large oxide scale thickness observed after 0.2 h of oxidation (Fig. 33.8). This rapid increase in the oxide scale thickness is associated with a severe Al and Cr depletion in the alloy near the O/M interface [4]. As shown in Fig. 33.7, immediately after the onset of oxidation, the calculated Al concentration in the alloy at the O/M interface drops rapidly towards a value close to zero (in our opinion due to a very short period of exclusive α-Al_2O_3 formation at the onset of oxidation, stage I in Fig. 33.7, see Ref. [4] for details). Such a low value for the Al concentration in the alloy at the O/M interface was also observed experimentally after 10 min of oxidation [4].

The calculated amounts of α-Al_2O_3 and NiO formed are in good agreement with their corresponding experimental values for this oxidation stage (i.e.

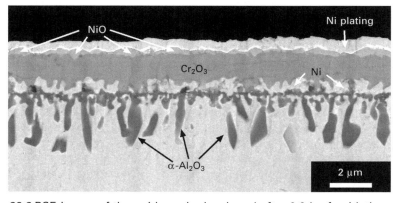

33.8 BSE image of the oxide scale developed after 0.2 h of oxidation of a γ-Ni-27Cr-9Al alloy at pO_2 = 20 kPa and a temperature of 1373 K in a tube furnace (after Ref. [4]). For short oxidation times, the oxide scale consists of pure NiO and Cr_2O_3 layers on top of an internal oxidation zone of isolated α-Al_2O_3 precipitates.

stage II in Fig. 33.7), except for the amount of Cr_2O_3, which is overestimated by the model (Fig. 33.6). This discrepancy is largely attributed to the consumption of Cr_2O_3 by the developing $NiCr_2O_4$ spinel (the reaction between Cr_2O_3 and NiO was not considered in the CTK oxidation model, see Section 33.3.2). Adding the maximum amount of $NiCr_2O_4$ ($\Delta NiCr_2O_4 \sim 0.6$ μm, see Fig. 33.6) to the amount of Cr_2O_3 formed results in a much better agreement between model predictions and experiment (Fig. 33.6). It is noted that the actual amount of Cr_2O_3 consumed by $NiCr_2O_4$ is expected to be slightly lower than 0.6 μm (i.e. about 0.4 μm, since the molecular volume of Cr_2O_3 is about twice as large as the molecular volume of NiO [19]).

33.4.2 Successive formation of continuous Cr_2O_3 and α-Al_2O_3 layers

The internal oxidation of Al and the formation of Cr_2O_3 stop as soon as a continuous Cr_2O_3 layer has been developed on top of the isolated α-Al_2O_3 precipitates (after about 1 h of oxidation, as observed with SEM [4]). Then, at the O/M interface, only lateral growth of α-Al_2O_3 precipitates occurs until they coalesce into a continuous α-Al_2O_3 layer [1, 4, 20]. After the formation of a continuous Cr_2O_3 layer, which corresponds with the time at which the amount of Cr_2O_3 in the oxide scale attains its maximum value, the Al concentration in the alloy at the O/M interface increases rapidly (Figs 33.5, 33.7 and Ref. [4]).

The time required for the formation of a continuous Cr_2O_3 layer increases with increasing oxidation temperature (Fig. 33.5). Moreover, the initial amount of α-Al_2O_3, Cr_2O_3 and NiO formed is much larger at the high oxidation temperature (Fig. 33.5). This indicates that for oxidation at higher temperature, (i) the internal oxidation of Al is more extensive and takes place at a larger depth in the alloy, and (ii) the growth of the Cr_2O_3 and NiO overlayers is more rapid (cf. Fig. 33.3(a) and (b)).

The lateral growth of α-Al_2O_3 at the O/M interface is accompanied by the continuous thickening of the NiO layer at the oxide surface (Fig. 33.5(b)). This was confirmed by the microstructural observations. The BSE images show that larger NiO crystallites are present at the oxide surface after 32 h than after 0.2 h of oxidation (cf. Figs 33.3 and 33.8). Thus, as long as a continuous α-Al_2O_3 layer has not been developed, Ni is able to diffuse from the alloy through the developing Cr_2O_3 layer to the oxide surface, where it reacts with oxygen to form NiO.

Since the CTK oxidation model only predicts the formation of oxides at the O/M interface, reactions within the oxide scale or at the surface are not accounted for (see Ref. [3] and Section 33.3.2). The model predicts the exclusive formation of α-Al_2O_3 after the formation of a continuous Cr_2O_3 layer (i.e. during stages III and IV in Fig. 33.7). As a result, the amount of

α-Al$_2$O$_3$ formed is overestimated and the amount of NiO formed (and thus also the total amount of non-protective oxide phases) is underestimated for this oxidation stage (Fig. 33.6(b)–(d)).

Besides the continuous thickening of the NiO layer, a continuous NiAl$_2$O$_4$ spinel layer developed in between the Cr$_2$O$_3$ and the NiO during this oxidation stage (Fig. 33.3 and 33.5). Apparently, Al diffuses also from the alloy though the Cr$_2$O$_3$ layer during this oxidation stage (stage III in Fig. 33.7). The onset of NiAl$_2$O$_4$ formation corresponds with the time at which the amount of NiCr$_2$O$_4$ reaches its maximum value (Fig. 33.5). Moreover, the total amount of NiCr$_2$O$_4$ *plus* NiAl$_2$O$_4$ formed remains relatively constant at a value of 0.6 µm (equal to the maximum amount of NiCr$_2$O$_4$ formed) as long as a continuous α-Al$_2$O$_3$ layer has not been developed (Fig. 33.5). This suggests that NiCr$_2$O$_4$ transforms into NiAl$_2$O$_4$. Additional evidence for this conclusion was provided by EDXS analysis of the spinel layer. NiCr$_2$O$_4$ was never constituted of only Ni, Cr and O, but always contained some Al (and vice versa NiAl$_2$O$_4$ always contained some Cr). Furthermore, large shifts in peak position (towards higher 2θ) were observed in the diffractograms for the reflections of NiCr$_2$O$_4$ and NiAl$_2$O$_4$ (cf. Fig. 33.1 with the results presented in Ref. [15]). This indicates that a gradual shift in the composition of both spinels (towards higher Al and lower Cr content) happens upon increasing oxidation time. Indeed, as measured with EDXS, the outer oxide layer becomes more enriched in Al with increasing oxidation time.

Since the amount of Cr$_2$O$_3$ remains constant while the NiCr$_2$O$_4$ transforms into NiAl$_2$O$_4$ (cf. Figs 33.5(a), (c) and (d)), the Cr that is released by this transition is thought to diffuse from the oxide scale to the O/M interface. An enrichment of Cr was observed in the alloy at the O/M interface after about 4 h of oxidation (cf. Refs. [3, 4]). This enrichment is mainly due to the recession of the O/M interface into the alloy [3], but the diffusion of Cr from the oxide scale may have also contributed. Alternatively, as suggested in Ref. [15], the Cr may diffuse to the oxide surface where it subsequently evaporates as Cr$_2$O$_3$.

33.4.3 Exclusive, parabolic growth of α-Al$_2$O$_3$

As soon as a continuous α-Al$_2$O$_3$ layer has been developed on the alloy surface, a transition from very fast to slow parabolic growth kinetics occurs, because the growth kinetics of the oxide scale become limited by the diffusion of Al and/or O through the developing α-Al$_2$O$_3$ layer (see Fig. 33.4 and Refs. [1, 4, 15, 20]). At this point the Al concentration in the alloy at the O/M interface gradually attains a constant value as imposed by the parabolic growth kinetics (cf. stage IV in Fig. 33.7 and Refs. [3, 4]).

The onset of this final oxidation stage corresponds with the time at which the amount of NiO in the oxide scale attains its maximum value (Fig. 33.5(b) and 33.7). Then the quantity of NiO in the oxide scale gradually decreases upon prolonged oxidation as a result of fragmentation of the continuous NiO layer into isolated NiO crystallites, in agreement with microstructural observations (cf. Figs 33.3 and 33.8).

For the higher oxidation temperature, the maximum amount of NiO formed is larger, but the time required to reach this maximum amount is shorter (i.e. 0.8 μm after 8 h and 1.2 μm after 5 h of oxidation at 1353 K and 1443 K, respectively, see Fig. 33.5(b)). It is therefore concluded that the lateral growth of α-Al_2O_3 precipitates at the O/M interface and the overgrowth of NiO crystals proceeds faster at 1443 K than at 1353 K.

The decrease in the amount of NiO formed is associated with the continued growth of the $NiAl_2O_4$ spinel phase, as illustrated in Fig. 33.6(b). The transformation of NiO into $NiAl_2O_4$ is more pronounced at higher oxidation temperatures (cf. Figs 33.5(b) and (d)). The total amount of $NiAl_2O_4$ that forms is small compared to the amount of α-Al_2O_3 that develops during this growth stage (cf. Fig. 33.5(d) and (e)). Therefore, after the development of a continuous α-Al_2O_3 layer, the total amount of non-protective oxides (i.e. all oxides except α-Al_2O_3) in the oxide scale maintain an approximately constant value (Fig. 33.5(f)). Then, the increase in the total oxide scale thickness is almost solely determined by the increase in the amount of α-Al_2O_3 developed at the O/M interface, as determined by the value for the parabolic growth constant (Fig. 33.4). Apparently, the continuous α-Al_2O_3 layer significantly reduces the outward diffusion of Ni and Al though the developing oxide scale.

33.5   Conclusions

The amounts of α-Al_2O_3, Cr_2O_3, NiO, $NiCr_2O_4$ and $NiAl_2O_4$ developed, as obtained from high temperature X-ray diffractometry (XRD) experiments upon thermal oxidation of a α-Ni-27Cr-9Al (at.%) alloy at 1353 K and 1443 K, are in good agreement with microstructural observations from scanning electron microscope backscattered electron images, and model calculations using a coupled thermodynamic-kinetic oxidation model.

For short oxidation times, the oxide scale consists of an outer layer of NiO on top of an intermediate layer of Cr_2O_3 and an inner zone of isolated α-Al_2O_3 precipitates in the alloy. The amounts of Cr_2O_3 and NiO in the oxide scale attain their maximum values when successively continuous Cr_2O_3 and α-Al_2O_3 layers are formed. The development of a continuous α-Al_2O_3 layer results in a transition from very fast to slow parabolic growth kinetics. Then the total amount of non-protective oxides (i.e. all oxide phases excluding α-Al_2O_3) in the oxide scale maintain an approximately constant value.

The formation of $NiCr_2O_4$ spinel as a continuous layer in between NiO and Cr_2O_3 is a result of the solid-state reaction between these pure oxide phases. After the amount of $NiCr_2O_4$ reaches its maximum value, a transition from $NiCr_2O_4$ into $NiAl_2O_4$ spinel takes place in between the Cr_2O_3 and NiO. During the slow parabolic growth, $NiAl_2O_4$ also develops at the oxide surface as a result of the fragmentation of the continuous NiO layer into isolated crystallites.

Enhancing the oxidation temperature increases the rate of formation of α-Al_2O_3, Cr_2O_3, NiO, $NiCr_2O_4$ and $NiAl_2O_4$ during the early stages of oxidation, but decreases the time required for the formation of a continuous α-Al_2O_3 layer.

33.6   Acknowledgement

Financial support of the Technology Foundation STW is gratefully acknowledged.

33.7   References

1. P. Kofstad, *High Temperature Corrosion*, Elsevier Applied Sciences, London, **1988**.
2. M.J. Stiger, N.M. Yanar, M.G. Topping, F.S. Pettit, G.H. Meier, *Zeitschrift für Metallkunde*, **1999**, *90*, 1096.
3. T.J. Nijdam, L.P.H. Jeurgens and W.G. Sloof, *Acta Materialia*, **2003**, *51*, 5297.
4. T.J. Nijdam, L.P.H. Jeurgens and W.G. Sloof, *Acta Materialia*, **2005**, *53*, 1643.
5. V. Kolarik, W. Engel and N. Eisenreich, *Materials Science Forum*, **1993**, *133–136*, 563.
6. N. Czech, V. Kolarik, W.J. Quadakkers and W. Stamm, *Surface Engineering*, **1997**, *13*, 384.
7. E. Caudron and H. Buscail, *Applied Surface Science*, **2000**, *158*, 310.
8. R. Autrata, P. Schauer, J. Kvapil and J. Kvapil, *Journal of Physics E: Scientific Instruments*, **1978**, *11*, 707.
9. *Powder Diffraction file, 46-1212*, from International Centre for Diffraction Data (ICDD).
10. *Powder Diffraction file, 38-1479*, from International Centre for Diffraction Data (ICDD).
11. *Powder Diffraction file, 22-1189*, from International Centre for Diffraction Data (ICDD).
12. *Powder Diffraction file, 23-1271*, from International Centre for Diffraction Data (ICDD).
13. *Powder Diffraction file, 10-0339*, from International Centre for Diffraction Data (ICDD).
14. B.H. Kear, F.S. Pettit, D.E. Fornwalt and L.P. Lemaire, *Oxidation of Metals*, **1971**, *3*, 557.
15. I.A. Kvernes and P. Kofstad, *Metallurgical Transactions*, **1972**, *3*, 1511.
16. J.L. Smialek and R. Gibala, *Metallurgical Transactions*, **1983**, *14A*, 2143.
17. P. Scardi, L. Lutterotti and A. Tomasi, *Thin Solid Films*, **1993**, *236*, 130.
18. *International Tables for Crystallography C: Mathematical, Physical and Chemical*

Tables, 2nd edn, A.J.C. Wilson, E. Prince (eds), Kluwer Academic Publishers, Dordrecht, **1999**, 230.

19. O. Kubaschewski and B.E. Hopkins, *Oxidation of Metals and Alloys*, 2nd edn, Butterworths, London, **1962**, 9.
20. C.S. Giggins and F.S. Pettit, *Journal of the Electrochemical Society*, **1971**, *118*, 1782.
21. N. Birks and H. Rickert, *Journal of the Institute of Metals*, **1962**, *91*, 308.

Index

1.4910 austenitic steel 193–200
180°-return bend fittings 470, 472

abradeable gas turbine seal materials
    36–64
  analysis 57–62
  cross-sectional examinations 46–7,
    51–4, 56–7, 58, 59
  experimental methods 39–47
  laboratory furnace screening tests
    36–8, 39–43, 47–54
  post exposure examinations 46–7
  results 47–62
  thermal cycling tests 38, 43–6, 54–7,
    58, 59
  visual examinations 46, 48–51, 56, 57
  weight change data 47–8, 49
acidic dissolution 533
active ensembles of nickel atoms 31–3
active oxidation 534
adhesive mode of spallation 115, 117,
    118–21, 173
ADSEALS project 36
aerospace industry 249
air ventilation, duration of 418, 420–3,
    425
AISI304 297, 298
AISI441 288, 289, 291
  parameter variation 335, 337, 338, 353
    analysis of net weight change
      curves 348, 350
    ANOVA 357–9, 360–1
    oxide scale morphology and
      growth rate 340, 342, 343, 344,
      359–60

alkali chlorides 442
  in-situ alkali chloride monitor (IACM)
    428, 430–2
  see also under individual names
alkali corrosion 501–13
alkali hydroxides 502, 504
  see also free alkali index
alkali metals 442, 502, 503
alkali sulphates 442, 504
Alloy 602 CA 418, 420, 421, 422–3,
    424–6
Alloy 603 418, 420, 422–3, 424–6
Alloy 617 475, 476, 481, 485, 487
  steam oxidation tests 193–200
Alloy 693 418, 420, 422–3, 424–6
Alloy 800 288, 290, 291
  creep tests on oxidised Alloy 800
    204–7, 208
  statistical experiment design 317–20,
    328–9
  steam oxidation tests 193–200
Alloy 800H 289, 291
  parameter variation 335, 337, 338,
    339
    analysis of net weight change
      curves 348, 351
    ANOVA 353–7, 360–1
    oxide scale morphology and
      growth rate 340–2, 344, 345,
      346, 359–60
alloy design strategies 3–18
  functional surface formation by gas
    reactions 13–15
  self-graded metallic precursor coating
    concepts 4–8

surface modification approaches 8–13
    continuous protective nitride and
        carbide surface layer 8–11
    ternary nitride and carbide
        protective surface layer 11–13
alloy grain boundaries *see* grain
    boundaries
α-alumina scale 60
    FeCrAlRE alloys 129–30, 161
        pre-forming by gas annealing 134,
            143–6, 157
    minimum necessary chromium and
        aluminium contents for
        formation in Fe-20Cr-5Al
        alloys 69–70
    thermal oxidation of ternary alloys
        582, 584–97
        exclusive, parabolic growth 590,
            595–6
        successive formation of continuous
            chromia and α-alumina layers
            590, 594–5
    see also alumina scales
Aluchrom P 131, 132
    model catalyst supports 149–55
Aluchrom Y 401, 403
Aluchrom YHf 131, 132, 155–7
    gas annealing 145, 146
    influence of aluminising 140–3, 144,
        145
    model catalyst supports 149–55
    oxidation behaviour
        compared with Aluchrom YHfAl
            136–40
        during resistance and furnace
            heating 401, 402–13
    seals 43, 44, 56, 58, 59, 60–1
Aluchrom YHfAl 131, 132, 155–6, 170
    gas annealing 145, 146
    model catalyst supports 149–55
    oxidation behaviour compared with
        Aluchrom YHf 136–40
alumina scales 4–8, 93
    α-alumina scale *see* α-alumina scale
    CMSX-4 494–5, 497, 498
        spallation 496–7, 498
    nickel-base alloys 91
    recirculation burners 420, 421, 423,
        424-5, 426

TBCs on γ-TiAl 252–3, 259–60, 262
aluminides 255–62
aluminising
    aluminised 12% chromium ferritic
        steels 211, 221–30, 232–3, 234
    FeCrAlRE alloys 131, 132–4, 135,
        140–3, 144, 145, 155–7, 158–9
        combined effect of gas annealing
            and 146–7, 157
        model catalyst supports 149–55,
            158
aluminium
    critical
        content 65–6, 67–70, 71, 77,
            130, 161
    diffusion coatings on ferritic-
        martensitic steels
        co-diffusion Al-B coating 177,
            183–4, 186, 189–90
        co-diffusion Al-Si coating 181–2,
            185, 189
        role of aluminium diffusion rate
            190
        single element coating 179–80,
            181, 182, 183, 188–9
        two-step chromium and aluminium
            coating 184–8, 190–1
    FeCrAl alloys *see* iron-chromium-
        aluminium (FeCrAl) alloys
    FeCrAlRE alloys *see* iron-chromium-
        aluminium-reactive element
        (FeCrAlRE) alloys
    internal oxidation in ternary alloys
        and simultaneous overgrowth of
        chromia and nickel oxide 590–4
    reservoir in FeCrAlRE alloys 159
    influence on oxidation behaviour
        140–3, 144, 145, 155–7
    surface modification to increase
        130–1, 132–4, 135
    slurry coatings 176
aluminium chloride 179, 188
aluminium nitride 11, 253
ammonium chloride 179, 188
ammonium sulphate (ChlorOut) 428–9,
    430, 431, 432–8, 439, 442–3
analysis of variance (ANOVA) 319, 324
    parameter variation 349–59, 360–1
    rapid thermal cycle tests 379–81, 382
atmosphere *see* gas composition

austenite 28
austenitic steels 176
    NiCr steels 289–90, 299, 307
    role of alloy grain boundaries in
        oxidation 81–2, 86–8, 89, 90
    steam oxidation tests 193–200
automotive industry 249

balance, in experimental design 329–30
basic dissolution 533
bending 48–51
biomass combustion 501–13
    biomass vs fossil fuels 502–10
        potassium hydroxide diffusion
            through the boundary layer vs
            reaction with carbon dioxide
            504–10
        thermodynamic equilibrium
            calculations 502–4
    see also wood-fired boilers
bipolar plates 8–10
Blyholder mechanism for absorption of
        carbon monoxide 104–7
boron-aluminium co-diffusion coatings
        177, 183–4, 186, 189–90
boundary layer 504–10
brazing 61
breakaway oxidation 39, 60, 61
    critical aluminium content in
        Fe-20Cr-5Al alloys 65–79
    time to breakaway 403–5
bright carbon deposits 21, 23, 24, 26
brittle behaviour 234
broccoli effect 118, 121
BSE images 587–8, 589, 591, 593
bulk diffusion 528, 574–5
burner rig 416–17
    see also recirculation burners

calcium carbonate 504
calibration periods for thermocouples
        301–2
carbide surface layer
    chromium carbide on FeCrAl alloys
        118–21, 127
    continuous 8–11
    ternary 11–13
carbon
    bright carbon deposits 21, 23, 24, 26

filaments 21–8, 29, 31, 33–4
formation (coking) and metal dusting
        93–109
    nickel-copper binary alloys 94–5,
        101–3, 107, 108
    transition metals 94–5, 97–100,
        104, 105–7, 108
    impurity effect on FeCrAl alloys 115,
        116, 117, 118–21, 127
    nanotubes 28, 29, 33–4
carbon activity 478–9
carbon dioxide 504–10
carbon monoxide 108
    dissociative and non-dissociative
        adsorption at metal surfaces
        104–7
    initial stage of metal dusting 103–4
    metal dusting and coking behaviour
        94, 95–6
carburisation
    ethylene pyrolysis 445–73
        carburisation resistance 467–8
        and erosion wear 452, 459–66,
            468–9
    GCR helium and NiCrMo alloys 480
    nickel-copper alloys 19–35
catalysis 31–3
catalyst supports, model 149–55, 157,
        158
cementite 19, 93
centre point, orthogonal contrasts with
        332–3
cerium 162, 553, 554
ChemApp 519, 525, 530, 552, 573
chemical vapour deposition (CVD)
    coatings on 9–12% chromium
        ferritic-martensitic steels
        176–92
    silicon surface treatment of inner
        surface of steel pipes 236–48
ChemSheet 519, 520–5, 526, 530, 552
chloride melt induced corrosion see salt
        melt corrosion
ChlorOut 428–9, 430, 431, 432–8, 439,
        442–3
chromia scales 93
    CMSX-4 495, 497, 498
    formed in recirculation burners 420,
        421, 423, 424–6

InCorr modelling tool 529–30
mixed chromium-reactive element
    oxide YCrO₃ 273, 274
role of alloy grain boundaries 80–1
    high-Cr steel 86–8, 89, 90
    minimum concentration of
        chromium necessary to form a
        chromia scale 81, 87–8
    nickel-base alloys 90, 91
simulation of inward oxide scale
    growth 574, 575, 577, 578–9,
    580
subscale on FeCrAl/FeCrAlRE alloys
    130, 405
thermal oxidation of ternary alloys
    584–90, 591, 592, 596–7
    internal oxidation of aluminium
        and simultaneous overgrowth of
        chromia and nickel oxide 590–4
    successive formation of continuous
        chromia and α-alumina layers
        590, 594–5
chromium
    binding with yttrium 272–3
    FeCrAl alloys *see* iron-chromium-
        aluminium (FeCrAl) alloys
    FeCrAlRE alloys *see* iron-chromium-
        aluminium-reactive element
        (FeCrAlRE) alloys
    metal dusting and coking behaviour
        94, 97–8
    minimum concentration to form a
        chromia scale 81, 87–8
    two-step chromium and aluminium
        coating on P91 184–8, 190–1
chromium carbide 118–21, 127
chromium chloride–potassium chloride–
    zinc chloride system 544, 546,
    562, 563
chromium-containing steels
    9–12% Cr 287, 299, 307
    16–18% Cr 289, 299, 307
    erosion-corrosion resistance of 12%
        chromium ferritic steels 210–35
    minimum content in iron-based alloys
        to form α-alumina scale 69–70
    novel diffusion coatings for 9–12% Cr
        ferritic-martensitic steels
        176–92

role of alloy grain boundaries 80–92
silicon surface treatment of
    martensitic 9–11% Cr steels
    236–48
simulation of inward oxide scale
    growth 568–81
see also high-Cr steels; low-Cr steels
chromium-nickel-iron steels 446, 447
    carburisation 462, 463, 467
    erosion wear and carburisation 468–9,
        472
    erosion wear behaviour 452–3, 462–6
chromium nitride 11
chromium platinum nitride 12
cleaning of specimens 304–5
CM247 288, 290, 291
    parameter variation 335, 337, 338,
        339, 359–60
    analysis of net weight change data
        352
    oxide scale morphology and
        growth rate 342–3, 347, 348
CMSX-4 491–500
CNOSH (no self healing) 66
coatings 3
    FeCrAl alloys with minor additions
        and impurities 124–6
    FeCrAlRE alloys 162, 170–1, 172,
        174
    novel diffusion coatings for 9–12% Cr
        ferritic-martensitic steels *see*
        diffusion coatings
    Q1100 *see* Q1100 coating
    self-graded metallic precursor coating
        concepts 4–8
    thermal barrier coatings *see* thermal
        barrier coatings (TBCs)
coaxial thermocouples 387–90
cobalt
    Co-Cr-Fe system 553, 555
    CoMo intermetallic alloys 14–15
    metal dusting and coking behaviour
        94, 97–8
cobalt oxide 494, 498
code of practice
    thermocyclic oxidation testing 293–5
    draft 293
    experimental validation 293–4,
        295

formulation and fine-tuning 294–5
TMF testing 398
co-diffusion coatings 179, 180, 189–90
    Al-B on P91 177, 183–4, 186,
        189–90
    Al-Si on P91 181–2, 185, 189
coefficient of thermal expansion (CTE)
    447–8, 456–7
cohesive mode of spallation 115, 117,
    121–4, 125, 126, 173
coking behaviour 93–109
column collapsing 315
CoMo intermetallic alloys 14–15
complex equilibria 558–61
constant power tests 365, 370, 371
constant voltage tests 365
continuous isothermal exposure testing
    281
continuous nitride and carbide layers
    8–11
contrast values 320, 322
cooling practice 299, 302–3, 309, 310
copper
    alloying with to reduce metal dusting
        see nickel-copper alloys
    metal dusting and coking behaviour
        94, 97–8, 100
        adsorption of carbon monoxide on
            surface 105, 106
        surfactant-mediated suppression
            106–7, 108
corkscrew plastic deformation 411–12
corners, breakaway oxidation at 70–6
CorrApp model 519, 520–5, 526, 530
corrosion-dominated wear 469
COTEST 279–96
    code of practice 293–5
        draft 293
        experimental validation 293–4,
            295
        final version 294–5
    concept 284
    contractors and subcontractors 283
    development of a set of test
        procedures 286–7, 288–9
    evaluation of existing cyclic oxidation
        data 284
    evaluation of presently used test
        procedures 284

experimental investigation of selected
    materials and evaluation of
    oxidation behaviour 290–3
parameter variation see parameter
    variation
rapid cyclic oxidation tests see rapid
    cyclic oxidation tests
reference materials 287–90, 291
statistical experiment design see
    statistical experiment design
coupled thermodynamic-kinetic (CTK)
    oxidation model 582, 588,
    594–5
cracking
    abradeable gas turbine seal materials
        56–7, 59
    Q1100 coating 456–7
    scale through cracking 122–3
    steam oxidation and 201
        creep tests on oxidised Alloy 800
            206, 208
Crank-Nicolson approach, implicit 527,
    571–4
creep
    CMSX-4 495–6, 499
    FeCrAl 409–10
    power station materials and effects of
        steam oxidation on creep
        strength 193, 203–7
        creep tests on oxidised Alloy 800
            and P92 204–7, 208
        effect of oxidation on creep
            behaviour 203–4, 205
    strength 475, 476
critical aluminium content 65–6, 67–70,
    71, 77, 130, 161
cross-sectional examinations 46–7, 51–4,
    56–7, 58, 59
cycle duration 305, 306, 310
cyclic oxidation testing 281, 282
    COTEST see COTEST
    parameter variation see parameter
        variation
    statistical experiment design see
        statistical experiment design
    survey of current practice 297–311
        cyclic oxidation data 305–10
        differences in testing practice
            299–305

methodology 297–9

d-state band 106–7, 108
databases 552–7, 566
Deacon reaction 545, 546
decarburisation 480
decoking 445
deposits, in wood-fired boilers 428,
    432–5
diffusion
    Fick's laws 73, 527, 570–1
    modelling high-temperature corrosion
        519–30
    potassium hydroxide through the
        boundary layer 504–10
    simulation of inward oxide scale
        growth 568–81
        numerical modelling of diffusion-
            controlled corrosion processes
            570–5, 576
        simulation results 575–8, 579, 580
diffusion coatings 176–92
    co-diffusion coatings 181–4, 185, 186,
        189–90
    combination of coating process with
        heat treatment of the steels
        184–8, 190–1
    experimental methods 177–9
    single element coatings 179–81, 182,
        183, 184, 188–9
    two-step coating 184–8, 190–1
diffusion coefficients 68, 575
    aluminium diffusion rate and diffusion
        coatings 190
    effective 574–5
    potassium hydroxide 505
discontinuous isothermal exposure testing
    281
Discrete Variational Xα method (DV-Xα)
    96, 105, 106
dissociative adsorption of carbon
    monoxide 104–7
ductility 233, 234, 454–6
duration
    cycle 305, 306, 310
    test 291, 292
dusting, metal see metal dusting

effective diffusion coefficient 574–5

elbow fittings 470, 471
electrodeposition 265
    determination of deposition conditions
        266–9, 270
    electron-beam physical vapour deposition
        (EB-PVD) 250, 251
    electronic density of states (DOS) 105,
        106
    electronic structures 96
        dissociative and non-dissociative
            adsorption of carbon monoxide
            at metal surface 104–7
electroplating 53–4
environmental humidity
    parameter variation 336, 349–59, 360,
        361
    statistical experiment design 315,
        317–19, 320–5
    survey of cyclic oxidation testing 307
Equilib module 558–61
erosion-corrosion dominated wear 469
erosion-corrosion map 469
erosion-corrosion resistance 210–35
    aluminised ferritic steels 211, 221–30,
        232–3, 234
    erosion-corrosion tests 212, 213, 214
    experimental methods 211–14
    uncoated ferritic steels 214–21,
        230–1, 233–4
erosion-dominated wear 469
erosion-enhanced oxidation 233, 234
erosion wear 445–73
    behaviour 452–3
    carburisation and 452, 459–66, 468–9
    high temperature materials
        degradation facility 449–52
errors, temperature 391–2
Esshete 1250 steel 433, 437, 438, 439,
    443
ethylene pyrolysis 445–73
    carburisation resistance 467–8
    coating characterisation 453–7
    coating microstructure 447, 466–7
    erosion wear 449–53
        influence of carburisation on 452,
            459–66, 468–72
    oxidation 448, 457–9, 460, 467, 469
    performance characterisation 447–9
eutectic mixtures 501, 511, 533

experimental variables 308–10
external oxidation–internal oxidation
        transition 67–8

F-values 380, 381
FactSage 518–19, 552, 558
fast initial oxide growth 171–2, 173
Fermi level 105, 106
    shifting in the d-state band 106–7, 108
ferritic-martensitic steels
    novel diffusion coatings for 176–92
    steam oxidation tests 193–200
ferritic steels
    9–12% Cr 287, 299, 307
    16–18% Cr 289, 299, 307
    erosion-corrosion resistance of 12%
        Cr steel 210–35
    role of alloy grain boundaries in
        oxidation 81–6, 90
fibre mats
    air oxidation of FeCrAlRE alloys
        147–9, 157–8
    seals 38, 44, 56, 60, 62, 63
Fick's laws of diffusion 73, 527, 570–1
finite difference technique 527–8
    inward oxide scale growth 571–5, 576
        simulation results 575–8, 579, 580
Fisher F statistics 380, 381
flame tube see recirculation burners
flue gases 428–9, 430–2, 432–5
fluidised bed combustion 210
    erosion-corrosion resistance of 12%
        chromium ferritic steels 210–35
    reducing superheater corrosion in
        wood-based boilers 428–44
fluidised bed CVD (FBCVD) 176, 177,
        178–9, 191
focused light heating 365–7
fossil fuels
    biomass vs 502–10
    industrial gas oil (IGO) 415, 416, 423,
        424, 425–6
fracture toughness 447, 448, 456, 466
free alkali index 502, 503
    thermodynamic equilibrium 504,
        506–9
Full-Potential Linear Muffin-Tin Orbital
        (FP-LMTO) method 96, 106
furnace heating 400–14

gamma titanium aluminides (γ-TiAl)
        249–63
gas annealing treatment 131–2, 134,
        158–9
    combined effect of aluminising and
        146–7, 157
    fibre mats 147–9, 157–8
    model catalyst supports 149–55, 158
    preforming an α-alumina scale 134,
        143–6, 157
gas blast cooling 301–2, 309, 310
gas composition
    effect on lifetime of surface-treated
        FeCrAlRE alloys 161–75
    influence on erosion wear behaviour
        462–6
    influence on kinetics of chloride melt
        induced corrosion 533–49
    laboratory furnace screening tests of
        seal materials 41–2
    recirculation burners 418, 419, 420
gas-cooled reactor (GCR) helium 474–90
    characteristics of 475–7
    comparing Haynes 230 to reference
        NiCrMo alloys 485–8
    conditions of tests in helium 482
    corrosion at 750°C 484–5
    corrosion at 950°C 485, 486
    experimental methods 480–3
    NiCrMo alloy corrosion 475–80
        corrosion damage 479–80
        critical conditions for surface
            reactions 479
        gas/metal surface reactions 477–8
        kinetics of surface reactions 478–9
        oxidation of NiCrMo alloys 480
    oxidation kinetics 483–4
gas flow 299, 303
gas phase corrosion 551
    development of toolboxes for
        modelling 558–61
gas reactions 3
    functional surface formation by
        13–15
gas turbine abradeable seal materials see
        abradeable gas turbine seal
        materials
geometry, specimen see specimen
        geometry

glow discharge optical emission
    spectroscopy (GDOES) 432
gold marker experiments 82–4, 85
Graeco Latin square experimental design
    315, 316, 371–2
grain boundaries 568
    role of alloy grain boundaries in
        oxidation mechanisms 80–92
    high-Cr austenitic steel 81–2,
        86–8, 89, 90
    low-Cr ferritic steels 81–6, 90
    materials and experimental
        methods 81–2, 83
    nickel-base alloys 81–2, 88–90, 91
grain boundary diffusion 528, 574–5
grain boundary grooving 25–7
grain size
    InCorr modelling tool 528–9
    simulation of inward oxide scale
        growth 568–9, 577, 579, 580
graphite
    nucleation 31–3
    particle clusters 21–8, 29, 30–1, 33,
        34
greyscale images 588, 589

haematite 80
hafnium
    FeCrAlRE alloys 162, 168, 169
    Y+Zr+Hf coating 170–1, 172
    minor element effect on FeCrAl
        alloys 115, 116, 117, 123–4,
        125, 126, 127
    coatings 124–6
half normal plots 319–28, 329–32
    balance 329–30
    full set of effects and interactions
        330
    interactions between two- and three-
        level factors 331–2
    linear effects 320–1, 322–7, 330–2
    method 322–4
    orthogonal contrasts 320–1, 322,
        330–3
    prediction of oxidation trends 324–7
    quadratic effects and interactions
        320–1, 322–7, 330–1
    synergies 321, 330–2
hardness 453–6

silicon surface treatment of inner
    surface of steel pipes 242–3,
    244
Hastelloy X 475, 476, 480, 481, 485,
    487, 488
Haynes 214 honeycombs 39, 40, 43, 44,
    57, 60, 61–3
    laboratory furnace screening tests
        47–54
    ribbon furnace tests 56, 58
Haynes 230 475, 476, 480–2, 489
    comparison with reference NiCrMo
        alloys 485–8
    corrosion at 750°C 484–5
    corrosion at 950°C 485, 486
    oxidation kinetics 483–4
HCM12A
    erosion-corrosion resistance 210–35
        aluminised 221–30, 232–3, 234
        uncoated 214–21, 230–1, 233–4
    steam oxidation tests 193–200, 201–3
heat exchanger materials 210–11
    toolboxes for modelling hot corrosion
        of 550–67
heat treatment
    and coating combined 184–8, 190–1
    NiCr alloys 265, 269, 271
heating practice 299, 302–3
helium see gas-cooled reactor (GCR)
    helium
high-Cr steels
    role of alloy grain boundaries in
        oxidation 81–2, 86–8, 89, 90
    simulation of internal oxide scale
        growth 575–8, 579, 580
high temperature materials degradation
    facility 449–52
high-temperature X-ray diffraction
    (XRD) 583, 584–8, 589, 590,
    591, 592
hollow cylindrical specimens 395, 396
hollow sphere seals 38, 44, 56, 60, 62,
    63
homogeneous nucleation 510
honeycomb materials and seals 36–64
hot dwell time see upper dwell time
hot-tube plastic deformation 411–12
'hourglass' waviness 405, 406–7, 408,
    409, 410–11, 413

hydrogen
    bubbling 268
    silicon surface treatment of inner
        surface of steel pipes 237–8,
        245
hydrogen chloride 534, 538–41, 542,
        543, 545–6, 547–8

impact angle 212, 217–18, 225, 226,
        230–1, 232–4
implicit Crank-Nicolson approach 527,
        571–4
impurities 113–28
in-situ alkali chloride monitor (IACM)
        428, 430–2
Incoloy 800H 297, 298, 469
InCorr tool 519, 522, 525–30, 531
    adaptation of model to oxidation of
        low alloy steels 526–7
    kinetics calculation 527–30
induction heating 365
industrial gas oil (IGO) 415, 416
    sulphur content 423, 424, 425–6
    see also recirculation burners
inner surfaces of steel pipes 236–48
intercrystalline oxidation zone 574–5
interfacial barrier 459–62, 463, 472
intergranular oxidation 81, 90, 91, 528
intermetallic alloys 13, 14–15
    Al_5Fe_2 232, 233
internal graphite precipitation 30–1
internal oxidation 67, 81
    aluminium in recirculation burners
        420, 425
    aluminium in ternary alloys 590–4
    transition between external oxidation
        and 67–8
intrinsic chemical failure 66
inward oxide scale growth 568–81
    numerical modelling of diffusion-
        controlled corrosion processes
        570–5, 576
    simulation results 575–8, 579, 580
iron
    chloride melt induced corrosion
        533–49
    metal dusting and coking behaviour
        94, 97–9

adsorption of carbon monoxide on
        surface 105, 106
adsorption of sulphur on surface
        104
iron aluminides 179–80, 188
iron-based alloys
    database 553
    metal dusting 19
iron-cerium system 553, 554
iron chloride 551–2
    CorrApp model 523, 524–5, 526
    gas phase composition and chloride
        melt induced corrosion 536,
        537, 539, 540–1, 541–4
    wood-fired boilers 435, 437, 440–2
iron chloride–potassium chloride–zinc
        chloride system 544, 562
iron-chromium-aluminium (FeCrAl)
        alloys
    critical Al content 65–6, 67–70, 71,
        77, 130, 161
    effects of minor additions and
        impurities on oxidation
        behaviour 113–28
        adhesive mode of spallation 115,
            117, 118–21
        cohesive mode of spallation 115,
            117, 121–4, 125, 126
        experimental methods 114–15
        implications for alloy/coating
            development 124–6
        results 115–17
    Fe-20Cr-5Al alloys 65–79, 290
        critical aluminium content 65–6,
            67–70, 71, 77
        three-dimensional aluminium
            depletion profile 66–7, 70–6,
            77
    oxidation behaviour during resistance
        and furnace heating 400–14
        analysis 407–13
        experimental methods 401–2
        results 402–7
iron-chromium-aluminium-reactive
        element (FeCrAlRE) alloys
    cyclic oxidation testing 299, 307
        cycle duration 310
        specimen thickness 308, 309

effect of gas composition and
   contaminants on lifetime
   161–75
  materials 162, 163
  mixed gas corrosion 162, 164–70,
   171–3
  sample preparation and
   characterisation 164
  surface-treated alloys 162, 170–1,
   172, 174
  testing atmospheres 162–4
lifetime extension 129–60
  air oxidation of foil and sheet
   coupons 136–47
  aluminising 131, 132–4, 135,
   140–3, 144, 145, 155–7, 158–9
  combined effect of aluminising and
   gas annealing 146–7, 157
  comparison of oxidation behaviour
   of Aluchrom YHf and
   Aluchrom YHfAl 136–40
  experimental methods 132–6
  fibre mats 147–9, 157–8
  model catalyst supports 149–55,
   157, 158
  oxidation testing 135–6
  preforming of α-alumina scale
   with gas annealing 134, 143–6,
   157
  surface modification to increase Al
   reservoir 130–1, 132–4, 135
iron-chromium-carbon-oxygen phase
  stability diagram 95–6
iron-chromium model alloys 200–3
iron-chromium-oxygen-chlorine system
  563–5
iron-chromium-oxygen system 518, 519,
  563, 564
iron oxide 536, 537, 539, 540–1, 542,
  543
iron-silicon phase diagram 240, 241
iron-sodium-potassium-chlorine phase
  diagram 518–19, 520
iron sulphide 437, 442, 443
ISO 'High Temperature Corrosion
  Testing' work group 280–2

joule heating 364–5, 366

oxidation behaviour of FeCrAl alloys
  400–14
rapid cyclic oxidation tests 367–82
  design of test facility 367–71
  statistical analysis 375–81
  statistical design of test matrix
   371–5

Kanthal A1 288–9, 290, 291
  rapid cyclic oxidation tests 369, 370,
   371
  statistical analysis 375–81
  test matrix 371–5
Kanthal AF 75, 131, 132
  air oxidation behaviour 136–9
  influence of aluminising 140–3,
   144, 145, 156–7
Kyoto Protocol 210, 517

laboratory furnace screening tests 36–8,
  39–43, 47–54
  cross-sectional examinations 51–4
  materials and sample preparation 39,
   40
  procedure 42–3
  test facilities 41
  test gas composition 41–2
  test programme 42
  visual examinations 48–51
  weight change data 47–8, 49
lanthanum 162
Latin square approach 315, 316, 371–2
layout of test facilities 298, 299, 300
lead chloride 534
LEAFA project 65, 113
lifetime
  effect of gas composition and
   contaminants on lifetime of
   surface-treated FeCrAlRE
   alloys 161–75
  extension for FeCrAlRE alloys in air
   129–60
  resistance and furnace heating of
   FeCrAl alloys 403–6, 413
linear effects 320–1, 322–7, 330–2
long dwell time tests 286, 288, 312, 335,
  363
  specimen arrangement 335–6

survey of cyclic oxidation testing
    practice 299–307
low-Cr steels
    9–12% Cr 287, 299, 307
    adaptation of InCorr model to
        oxidation of 526–30, 531
    erosion-corrosion resistance of 12%
        Cr steels 210–35
    role of alloy grain boundaries in
        oxidation 81–6, 90
    simulation of inward oxide scale
        growth 574, 575–8, 579
lower dwell time
    parameter variation 336, 349–59, 360,
        361
    statistical experiment design 315,
        317–19, 320–5

$M_{23}C_6$ precipitate 559, 561
magnetite 80
    simulation of inward oxide scale
        growth 574–5, 577, 578
main effects plots 317–19
manganese 420, 421, 425
    metal dusting and coking behaviour
        94, 97–8
mass change see weight/mass change
material type
    reference materials 287–90, 291
    survey of cyclic oxidation testing
        299, 307
MAX phases 11–13
MCrAlY electroplating 53–4
measurement techniques, survey of
        299, 305–7
mechanically induced chemical failure
        66
metal dusting
    and coking behaviour 93–109
        dissociative and non-dissociative
            adsorption of carbon monoxide
            at metal surfaces 104–7
        experimental methods 94–6
        initial stage of metal dusting 102–4
        NiCu binary alloys 94–5, 101–3,
            107, 108
        theoretical analysis 96–7
        transition metals 94–5, 97–100,
            104, 108

nickel-copper alloys and resistance to
        19–35
    experimental methods 20
    mechanism 29–34
    results 20–9, 30
metallic precursor coatings 4–8
micro-indentation survey 447, 448, 453–6
mixed gas corrosion 162, 164–70, 171–3
model catalyst supports 149–55, 157, 158
modelling
    high-temperature corrosion 517–32
        ChemSheet-based modelling tool
            519, 520–5, 526, 530
        InCorr tool development 519, 522,
            525–30, 531
        tailored stability diagrams 518–19,
            520, 530
        thermodynamic data collection
            518–19
    simulation of inward oxide scale
        growth 568–81
    thermal oxidation of ternary alloys
        582–98
    toolboxes for modelling hot corrosion
        of heat exchanger materials
        550–67
multi-phase alloys 13, 14
multiple regression 325–7
Munksund boiler 429–30, 431, 432–3,
        434, 435, 436, 437, 439, 443

NASA 'winCOSP' programme 313
natural cooling 301–2, 309, 310
network spallation 171–2, 173
nickel
    active ensembles of nickel atoms 31–2
    metal dusting and coking behaviour
        94, 97–100
nickel-aluminium powder-fill 53
nickel-base alloys 176, 290, 299, 307
    CMSX-4 491–500
    database 553
    metal dusting 19
    role of alloy grain boundaries in
        oxidation 81–2, 88–90, 91
    steam oxidation tests 193–200
nickel chloride-potassium chloride-zinc
        chloride system 544, 545, 562,
        564

nickel-chromium alloys 11
  electrodeposition of yttrium-
    containing thin films 265,
    266–9, 270
  high temperature behaviour of
    uncoated or yttria-coated
    269–74
  thermal treatments 265, 269
nickel-chromium-aluminium alloys
    582–98
nickel-chromium-carbon-oxygen phase
    stability diagram 95–6
nickel-chromium-molybdenum alloys
    475–80, 489
  comparison with Haynes 230 485–8
  corrosion in GCR helium 475–80
    corrosion damage 479–80
    gas/metal surface reactions 477–9
    oxidation 480
nickel-chromium-tungsten alloys 475,
    489
  see also Haynes 230
nickel-copper alloys
  metal dusting and coking behaviour
    94–5, 101–3, 108
    surface segregation 107
  resistance to metal dusting 19–35
    experimental methods 20
    mechanism 29–34
    results 20–9, 30
nickel-molybdenum-oxygen-sulphur
    dioxide system 565, 566
nickel oxide 494, 495, 498
  thermal oxidation of ternary alloys
    584–90, 591, 592, 595–7
    internal oxidation of aluminium
    and simultaneous overgrowth
    of chromia and nickel oxide
    590–4
Nimonic 86 475
niobium 162, 249
  TBCs on γ-TiAl 252–3
nitride surface layer
  continuous 8–11
  TBCs on γ-TiAl 252–4, 255, 256, 257
  ternary 11–13
non-dissociative adsorption of carbon
    monoxide 104–7
number of cycles 305, 306

rapid cyclic oxidation tests 371–8
Nusselt number 505
Nyköping boiler 429, 430, 433, 435–8,
    443

oil combustion see industrial gas oil
    (IGO); recirculation burners
OPTICORR 517–32
  CorrApp 519, 520–5, 526, 530
  gas phase composition and chloride
    melt induced corrosion of iron
    533–49
  InCorr tool development 519, 522,
    525–30, 531
  tailored stability diagrams 518–19,
    520, 530
  thermodynamic data collection
    518–19
  toolboxes for modelling of hot
    corrosion 550–67
orthogonal contrasts 320–1, 322, 330–3
oxidation
  effects of minor additions and
    impurities on oxidation
    behaviour of FeCrAl alloys
    113–28
  ethylene pyrolysis 448, 457–9, 460,
    467
    erosion-corrosion map 469
  evaluation of oxidation behaviour
    291–3
  GCR helium and NiCrMo alloys 480
    kinetics 483–4
  modelling oxidation of ternary alloys
    582–98
  oxidation behaviour of FeCrAl alloys
    during resistance and furnace
    heating 400–14
  pre-oxidation treatment of seal
    materials 51–3, 61
  role of alloy grain boundaries 80–92
  steam oxidation see steam oxidation
oxidation-affected erosion 231, 232,
    233–4
oxidation-dominated wear 469
oxidation rate constant 345–8, 350, 351
  ANOVA 352, 353–4, 355, 357, 358
oxidation testing
  COTEST see COTEST

FeCrAlRE alloys in air 135–6
silicon surface treatment of inner
    surface of steel pipes 243–7
TBCs on γ-TiAl 250–1, 252–8
uncoated and yttria-coated NiCr alloys
    265, 269–74
oxide coatings 4–8
oxide scales
    12% chromium ferritic steels
        aluminised 228–30, 234
        uncoated 219–21, 230–1, 234
    geometry effects on integrity of oxide
        scales of CMSX-4 491–500
    inner and outer 551
    morphologies and growth rates and
        parameter variation 339–45,
        346, 347, 348, 359–60
    prevention of metal dusting 93–4
    rapid cyclic oxidation tests 374–5
    recirculation burners 420–6
    role of alloy grain boundaries 80–92
    simulation of inward oxide scale
        growth 568–81
    thermal oxidation of ternary alloys
        584–97
    see also alumina scales; chromia
        scales; silica scales
oxide, sulphate and sulphide database
    553, 557
oxygen
    chloride melt induced corrosion
        535–6, 537
    partial pressure 478–9
    solubility 68–9
    transport paths 122–3

P91 177–8, 287, 288, 291, 297, 298
    coatings on 179–88
    parameter variation 335, 337, 338,
        359–60
        analysis of net weight change data
            352
        oxide scale morphology and
            growth rate 339–40, 341, 342
    steam oxidation tests 193–200, 201–3
P92
    creep tests on oxidised P92 204–6
    oxidation and creep behaviour 203–4,
        205

pack cementation coating process 176,
    177, 178, 191, 211
parabolic rate constant 82, 84
parameter variation 334–62
    experimental methods 335–6
    results 336–59
        analysis of net weight change
            curves 345–9, 350, 351, 352
        oxide scale morphologies and
            growth rates 339–45, 346, 347,
            348, 359–60
        statistical evaluation using ANOVA
            349–59, 360–1
parameters
    influential 315
    survey of cyclic oxidation testing data
        305–7
petrochemical furnaces see ethylene
    pyrolysis
phase angle deviations 392–4
Phase Diagram module 558, 562–6
pinhole defects 9
pipes, inner surface treated 236–48
plastic deformation 232
    FeCrAl alloys during resistance and
        furnace heating 405–7, 408,
        409, 410–12
platinum 94, 97–8, 100
PM2000
    honeycomb seals 39, 40, 47–53, 57,
        58, 60–1
    three-dimensional aluminium
        diffusion profile 71–6
PM2Hf seals 43, 44, 56, 58, 60–1
pores 201
potassium 442
potassium carbonate 510–11
potassium chloride 551–2
    biomass combustion 501–2, 510–12
    CorrApp model 524, 525, 526
    wood-fired boilers 428, 429, 430–5,
        442
potassium chloride-zinc chloride system
    555–6, 557
    with chromium chloride 544, 546,
        562, 563
    with iron chloride 544, 562
    with nickel chloride 544, 545, 562,
        564

potassium hydroxide 442, 510–11, 512
  diffusion through the boundary layer
    vs reaction with carbon dioxide
    504–10
potassium sulphate 429, 433, 501–2,
    510–12
power plants 80, 210–11
  steam oxidation and creep strength of
    power station materials
    193–209
prediction of oxidation trends 324–7
pre-oxidation 51–3, 61
protective oxide growth time 348–9, 350,
    351, 352
  AISI441 357, 358, 359
  Alloy 800H 353, 354
proton exchange membrane fuel cells
    (PEMFCs) 8–10
pseudo-dynamic method for Young's
    modulus 396
pyrometry 391

Q1100 coating 446–9, 472
  carburisation resistance 448–9,
    467–8
  commercial opportunities and future
    research 470–2
  erosion wear behaviour 452–3
  influence of carburisation on erosion
    wear 459–66, 468–9
  influence of oxidising atmosphere on
    erosion behaviour 465, 466
  microstructure 447, 466
  oxidation behaviour 448, 457–9, 460,
    467
  performance characterisation 447–9,
    466–7
  structure and properties 453–7
quadratic effects 320–1, 322–7, 330–1

random orders 316
random spallation 171–2, 173
rapid cyclic oxidation tests 363–83
  design of a joule heated test facility
    367–71
  possible test procedures 364–7
  statistical analysis of tests on Kanthal
    A1 wire/foil samples 375–81

ANOVA 379–81, 382
  statistical design of a test matrix
    371–5
ratcheting mechanism 410–11, 413
reactive elements
  effects on oxidation behaviour of
    FeCrAl alloys 113–28
  see also iron-chromium-aluminium-
    reactive element (FeCrAlRE)
    alloys
recirculation burners 415–27
  alloys and exposure tests 418
  burner rig 416–17
  characterisation of oxide scale 420,
    421, 425
  effect of duration of air ventilation
    after burner shutdown 418,
    420–3, 425
  effect of sulphur content in fuel 423,
    424, 425–6
  effect of temperature 424, 426
  measurement of temperature and
    gaseous environment 417–18,
    419–20
  metallographic analysis 418–19
reference materials 287–90, 291
regression 319, 324
  multiple 325–7
resistance heating see joule heating
ribbon furnace thermal cycling tests 38,
    43–6, 54–7
  cross-sectional examinations 56–7, 58,
    59
  materials 43–4
  procedure 46
  temperature logging 54, 55
  test facilities 44, 45
  test programme 46
  visual observations 56, 57
ribbon type thermocouples 387–90
rigid band model 105–7
rupture life 203–4

salt database 553–6, 557
salt melt corrosion 551–2
  CorrApp model 522–5, 526
  influence of gas phase composition
    533–49

effect of hydrogen chloride 534,
    538–41, 542, 543, 545–6,
    547–8
effect of oxygen 535–6, 537
effect of water vapour 534, 536–8,
    539, 546–7, 548
experimental methods 535
model 541–4
toolboxes for modelling 562–6
seal materials *see* abradeable gas turbine
    seal materials
selective oxidation 496
self-graded metallic precursor coating
    concepts 4–8
short dwell time tests 286, 288, 312, 363
furnace design 335
survey of cyclic oxidation testing
    practice 299–307
silane 236–7, 245
silica scales 4–5, 93
nickel-base alloys 90, 91
recirculation burners 420, 423, 425,
    426
silicon 115, 116, 117
coatings
    co-diffusion Al-Si coating 181–2,
        185, 189
    single element coating 180–1, 184,
        189
surface treatment via CVD of inner
    surface of steel pipes 236–48
    experimental methods 237–9
    results 240–7
silver-silicon thin film 14
silicon chloride 237–9, 245–7
silicon nitride 7–8
silver
metal dusting and coking behaviour
    94, 97–8, 100
silver–silicon thin film 14
simulated exhaust gas 164–6
single-colour pyrometry 391
single element diffusion coatings 188–9
aluminium on P91 179–80, 181, 182,
    183, 188–9
silicon on P91 180–1, 184, 189
slow initial oxide growth 171–2, 173
SMILER project 113–14, 129
sodium chloride 430–2

solid cylindrical specimens 395
solid flat specimens 394–5
solubility
carbon in nickel 31
oxygen in iron 68–9
spallation 66
12% chromium ferritic steels 220–1
adhesive mode 115, 117, 118–21, 173
CMSX-4 496–7, 498
cohesive mode 115, 117, 121–4, 125,
    126, 173
FeCrAlRE alloys 166–7, 168, 169,
    171–2, 173
steam oxidation 198–200, 201, 204
time to spall *see* time to spall
specimen geometry
and oxide scale integrity for CMSX-4
    491–500
and preparation and handling in cyclic
    oxidation testing 299, 303–5
rapid cyclic oxidation tests 371–81,
    382
temperature gradient and in TMF
    testing 394–5, 396
specimen thickness *see* thickness,
    specimen
speed, in erosion-corrosion tests 212,
    214–17, 221–5, 230–1, 232,
    233–4
spinels 195, 196, 536–7
$FeCr_2O_4$ 80, 529–30
    simulation of inward oxide scale
        growth 574–5, 577, 578
    modelling oxidation of ternary alloys
    $NiAl_2O_4$ 584–90, 591, 592, 595,
        596–7
    $NiCr_2O_4$ 584–90, 591, 592, 593,
        594, 595, 596–7
wood-fired boilers 437, 438
spot-welded thermocouples 387–91
spray coatings 176
stability diagrams, tailored 518–19, 520,
    530
standardisation of test methods 279–80
development of a set of test
    procedures suitable for
    standardisation 286–7, 288–9
see also COTEST
static method for Young's modulus 396

statistical experiment design 312–33
    alternative design 332–3
    analysis of results for Alloy 800
        317–20
    data for Alloy 800 328–9
    design of experiments, results and
        operational observations
        315–16
    experimental method and response
        summaries 313–14
    half normal plotting 319–28, 329–32
    influential parameters 315
    orthogonal contrasts 320–1, 322,
        330–3
    prediction of oxidation trends 324–7
    synergies 321, 330–2
    test matrix for rapid cyclic oxidation
        testing 371–5
steam oxidation 193–209
    creep tests on oxidised Alloy 800 and
        P92 204–7, 208
    effect on creep behaviour 203–4, 205
    model alloys 200–3
    tests 193–200
stress
    in-plane stress evolution in FeCrAl
        409–10, 412–13
    thermal mismatch stresses 495–6, 498
stress rupture life 204, 205
sulphur 443
    adsorption on iron surfaces and metal
        dusting 104
    content of IGO fuel 423, 424, 425–6
sulphur dioxide 442–3
superheater corrosion, reducing 428–44
support, specimen 305
surface area of specimen 304
surface finish, specimen 304
surface modification 3–4, 8–13
    continuous protective nitride and
        carbide layer formation 8–11
    to increase aluminium reservoir in
        FeCrAlRE alloys 130–1, 132–4,
        135
    silicon surface treatment of inner
        surface of steel pipes 236–48
    ternary nitride and carbide protective
        layer formation 11–13

surface segregation 107
surfactant-mediated suppression 106–7
synergy
    between carburisation and erosion
        wear 462, 464, 468–9, 472
    statistical experiment design 321,
        330–2

T92 433, 435–7
Taguchi experimental design 315, 316,
    372
tailored stability diagrams 518–19, 520,
    530
tantalum 162
temperature
    control in cyclic oxidation testing 297,
        300–2
    dependence of weight gain on for
        ferritic-martensitic steels 198–9,
        201–3
    effect on oxide scales in recirculation
        burners 424, 426
    gradient and oxidation behaviour of
        FeCrAl alloys during resistance
        and furnace heating 411–13
    logging in ribbon furnace testing 54,
        55
    measurement in flame tube 417, 419
    parameter variation 336, 349–61
    rapid cyclic oxidation tests 371–81,
        382
    TMF testing
        dynamic temperature measurement
            and control 386–91
        influence of temperature gradients
            in different specimen
            geometries 394–5, 396
        influence of temperature-strain
            phase deviations 392–4
        influence of temperature tolerances
            391–2
temperature-based strain compensation
    393
temperature/power density calibration
    curves 369
ternary alloys 582–98
    exclusive, parabolic growth of
        α-alumina 590, 595–6

experimental methods 583–4
high-temperature XRD 583, 584–8,
    589, 590, 591, 592
internal oxidation of aluminium and
    simultaneous overgrowth of
    chromia and nickel oxide
    590–4
modelling 588–90
successive formation of continuous
    chromia and α-alumina layers
    590, 594–5
see also iron-chromium-aluminium
    (FeCrAl) alloys; iron-
    chromium-aluminium-reactive
    element (FeCrAlRE) alloys
ternary nitride and carbide surface layers
    11–13
test duration 291, 292
test matrices 286–7, 288–9
    rapid cyclic oxidation testing 371–5
test parameters see parameter variation;
    parameters
test procedures
    COTEST see COTEST
    development of a set suitable for
        standardisation 286–7, 288–9
    evaluation of presently used
        procedures 285
test start procedure (TMF testing) 395–7
TESTCORR 280
thermal barrier coatings (TBCs) 4
    on γ-TiAl 249–63
    metal grid seals 38, 44, 56–7, 59, 60,
        62, 63
thermal cycling ribbon furnace tests 38,
    43–6, 54–7, 58, 59
thermal mismatch stresses 495–6, 498
thermal shock resistance tests 447–8,
    456–7
Thermo Calc modelling 438–42
thermocouples 300–2
    TMF testing 386–91
thermocycle, defining 290–1, 292
thermocyclic oxidation testing see
    COTEST; cyclic oxidation
    testing
thermodynamic equilibrium 502–4,
    506–9

thermogravimetric testing 281
thermo-mechanical fatigue (TMF) testing
    384–99
    best practice procedures for starting a
        TMF test 395–7
    pre-normative research work 386–97
        dynamic temperature measurement
            and control 386–91
        influence of temperature gradients
            in different specimen
            geometries 394–5, 396
        influence of temperature-strain
            phase deviations 392–4
        influence of temperature tolerances
            391–2
        reference TMF cycle 386, 387
        test material 386
ThermoScript 573–4
thickness, specimen 308, 309, 405, 406
    thickness changes 212
        aluminised ferritic steels 221–5,
            226, 232
        uncoated ferritic steels 214–18,
            230–1
three-dimensional aluminium depletion
    profile 66–7, 70–6, 77
time-based strain compensation 393
time-of-flight secondary ion mass
    spectrometry (TOF-SIMS) 432
time to spall 313–14, 317, 322–4, 328
    model 325–7
titanium 420, 421, 425
    FeCrAlRE alloys 162, 168
    minor element effect on FeCrAl alloys
        115, 116, 117
titanium aluminides 249–63
titanium aluminium carbides 11–12, 13
titanium aluminium chromide 7–8
titanium aluminium nitride 253
titanium nitride 253, 494–5, 497, 498
TMF-STANDARD project 384–99
tolerances, temperature 391–2
toolboxes for modelling hot corrosion
    550–67
    available software tools 552
    calculational results 558–66
        one-dimensional mappings for
            gases and alloys 558–61

two-dimensional mappings for
    salts and alloys 562–6
databases 552–7, 566
TP347 H steel 433, 437
transition metals
    electronic structure and adsorption of
        carbon monoxide at surface
        105–7
    metal dusting and coking behaviour
        94–5, 97–100, 104, 108
transition zone 252–3, 254
two-colour pyrometry 391
two-step coating 184–8, 190–1

ultra-short dwell time testing 286–7,
    288–9, 312, 363–4
    survey of cyclic oxidation testing
        practice 299–307
    see also rapid cyclic oxidation tests
upper dwell temperature 315, 317–19,
    320–5
upper dwell time (hot dwell time)
    parameter variation 336, 349–59,
        360–1
    rapid cyclic oxidation tests 371–81,
        382
    statistical experiment design 315,
        317–19, 320–5

validation
    code of practice for thermocyclic
        oxidation testing 293–4, 295
    validation testing of TMF testing 398
vanadium 162
variability of test results 308
very high temperature reactor (VHTR)
    474
visual examinations 46, 48–51, 56, 57
voids 201

water vapour
    chloride melt induced corrosion 534,
        536–8, 539, 546–7, 548
    FeCrAlRE alloys 161, 163–6, 172–3,
        174
waviness 405, 406–7, 408, 409, 410–11,
    413
wedge-shaped specimens 491–500
weight/mass change 306–7

abradeable gas turbine seal materials
    47–8, 49
cycling parameter variation 339–45,
    346, 347, 348
    analysis of net weight change
        curves 345–9, 350, 351, 352
    ANOVA 354–7
    effects of reactive elements and
        impurities on FeCrAl alloys
        115, 116
    FeCrAlRE alloys 164–6, 171, 172
    nickel-copper alloys 20, 21
    oxidation of uncoated and yttria-
        coated NiCr 273
    resistance and furnace heating of
        FeCrAl alloys 402–3
    role of alloy grain boundaries in
        oxidation mechanisms 86, 87,
        88–9, 90
    specimen geometry and CMSX-4 493
    statistical experiment design 313–14,
        317–19, 322–4, 329
    steam oxidation tests 195, 198–9,
        200–1
    TBCs on γ-TiAl 252, 255–7, 258
'winCOSP' programme 313
wood-fired boilers 428–44
    corrosion results 435–8, 439
    deposit and flue gas chemistry 432–5
    plants used for testing 429–32
    Thermo Calc modelling 438–42
wustite 80

X20 steel 433, 437, 438, 439, 559, 560,
    561
X-ray diffraction, high-temperature 583,
    584–8, 589, 590, 591, 592

Young's modulus 395–6
yttria
    coatings on NiCr alloys 264, 269–74
    mixed chromium-reactive element
        oxide YCrO₃ 273, 274
yttrium 162, 168, 169
    Y+Zr+Hf coating 170–1, 172
yttrium-containing films 264–9
    determination of deposition conditions
        266–9, 270
    electrodeposition 265

zinc chloride 524, 525, 526, 534, 551–2
zinc oxide 525, 526
zirconium
    FeCrAlRE alloys 162, 168–70, 172–3,
        174

Y+Zr+Hf coating 170–1, 172
minor element effects on FeCrAl
    alloys 115, 116, 117, 121–3,
    124, 127
    coatings 124–6